THE TECHNOLOGY OF
FOOD PRESERVATION
FOURTH EDITION

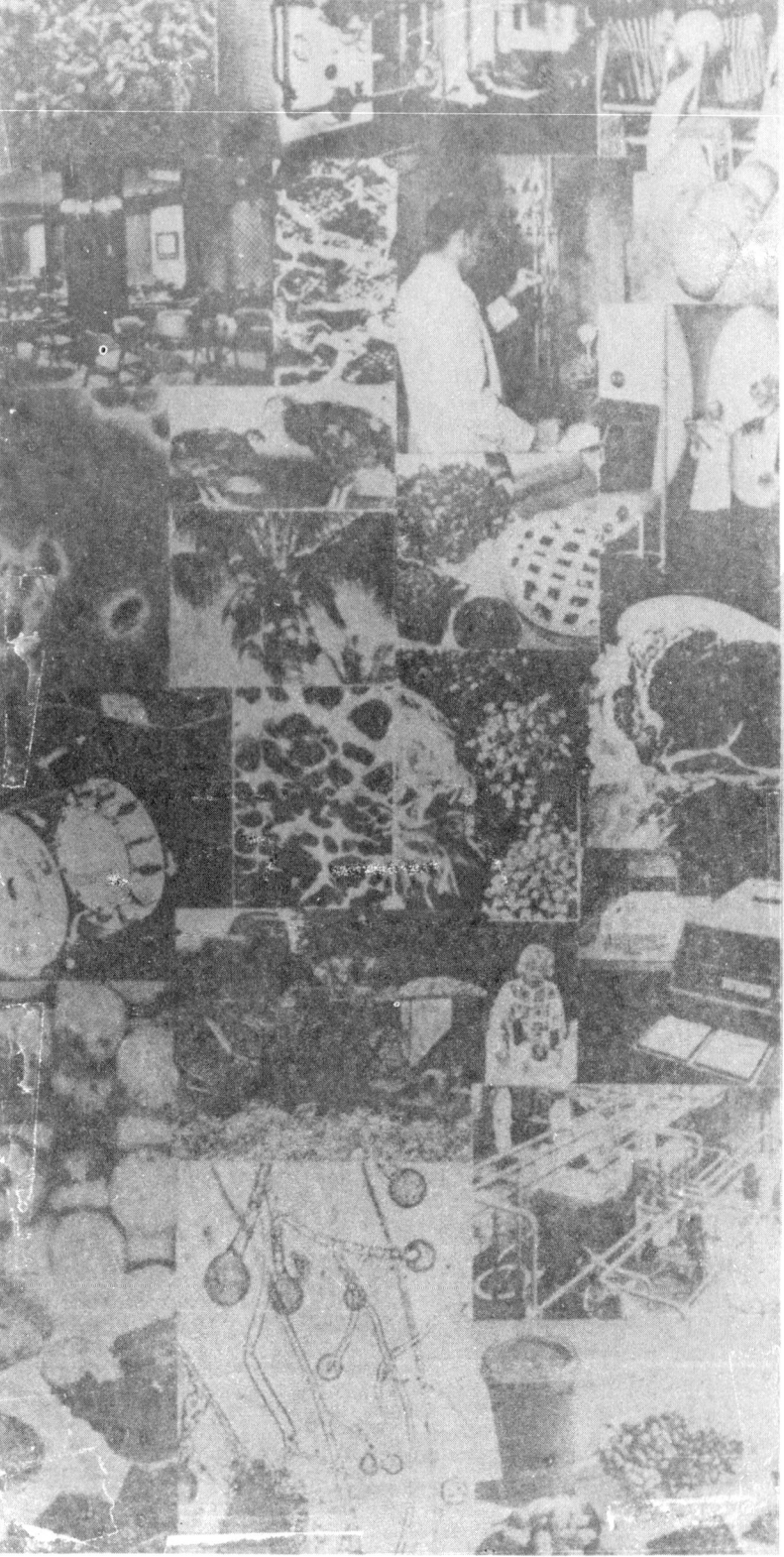

THE TECHNOLOGY OF FOOD PRESERVATION

Fourth Edition

Norman W. Desrosier
President, Overseas Book Co-Publishers, Ltd.

James N. Desrosier
The New Product Company

CBS

CBS Publishers & Distributors Pvt. Ltd.

New Delhi • Bengaluru • Chennai • Kochi • Kolkata • Mumbai
Hyderabad • Uttarakhand • Nagpur • Patna • Pune • Jharkhand

ISBN: 81-239-1128-9

First Indian Edition: 1987
Reprint: 1998, 2002, 2004

Published by **Satish Kumar Jain** and produced by **Varun Jain** for
CBS Publishers & Distributors Pvt. Ltd.,
4819/XI Prahlad Street, 24 Ansari Road, Daryaganj, New Delhi - 110002
delhi@cbspd.com, cbspubs@airtelmail.in • www.cbspd.com
Ph.: 23289259, 23266861, 23266867 • Fax: 011-23243014

Corporate Office: 204 FIE, Industrial Area, Patparganj, Delhi - 110 092
Ph: 49344934 • Fax: 011-49344935
E-mail: publishing@cbspd.com • publicity@cbspd.com

Branches:
• *Bengaluru:* 2975, 17th Cross, K.R. Road, Bansankari 2nd Stage,
  Bengaluru - 70 • Ph: +91-80-26771678/79 • Fax: +91-80-26771680
  E-mail: cbsbng@gmail.com, bangalore@cbspd.com
• *Chennai:* No. 7, Subbaraya Street, Shenoy Nagar, Chennai - 600030
  Ph: +91-44-26681266, 26680620 • Fax: +91-44-42032115
  E-mail: chennai@cbspd.com
• *Kochi:* Ashana House, 39/1904, A.M. Thomas Road, Valanjambalam,
  Ernakulum, Kochi • Ph: +91-484-4059061-65
  Fax: +91-484-4059065 • E-mail: cochin@cbspd.com
• *Kolkata:* 6-B, Ground Floor, Rameshwar Shaw Road, Kolkata - 700014
  Ph: +91-33-22891126/7/8 • E-mail: kolkata@cbspd.com
• *Mumbai:* 83-C, Dr. E. Moses Road, Worli, Mumbai - 400018
  Ph: +91-9833017933, 022-24902340/41 • E-mail: mumbai@cbspd.com

Representatives:

• Hyderabad: 0-9885175004	• Nagpur: 0-9021734563
• Patna: 0-9334159340	• Pune: 0-9623451994
• Jharkhand: 0-9811541605	• Uttarakhand: 0-9716462459

Printed at:
Neekunj Print Process, Delhi

Preface To
The Fourth Edition

There were few texts available in food technology when this book was first written. Now, there are a variety of excellent books in various areas of the field.

Since this book continues to serve a useful purpose, and a revision was needed, we undertook it with the concept in mind of holding the essential nature of the book while at the same time tightening our focus on the technology of food preservation itself. We felt this was justifiable due to the enormous growth of knowledge in the field and our interest in holding this text to a reasonable size.

Therefore, we added chapters dealing with the nature of hazards involved in foods, the storage of all kinds of fresh foods, expanded the coverage to include cereals and beans, and brought together the newly evolving area of gas atmosphere storage of fresh foods of both plant and animal origin.

We added a new chapter on quality assurance and good manufacturing practices. And, we reorganized the presentation on food additives, giving attention to means of understanding government regulations for using chemical additives in food preservation, what to use, when and where.

Since the last edition, there has been a mushrooming development in the field of semi-moist or intermediate moisture foods. This field has grown into an area in its own right.

As is necessary in such an effort, we have drawn heavily from certain sources and these are noted in the text. However, a special note of thanks is due to:

S.R. Adams, J.A. Anderson, R. Angelotti, W.S. Arbuckle, D.H.

Ashton, J.B. Biale, E.J. Bond, D.P. Bone, G.N. Bookwalter, M.L. Brockman, C.K. Brown, L.B. Bullerman, S.P. Burg, E.A. Burg, F.L. Byran, C.H. Byrne, R.B. Davis, C.B. Denny, J.W. Eckert, J.L. Etchells, G.M. Evancho, L.J. Filer, H.P. Fleming, R.E. Gunnerson, N.F. Haard, R.L. Hall, C.W. Hall, T.E. Hartung, Y.S. Henig, D.L. Huffman, K. A. Ito, A.D. Johnson, A. Kramer, W.E. Kramlich, F. Dwai, W.F. Kwolek, T.P. Labuza, R.A. Lampi, H. Leung, W.J. Lipton, N. Luh, D.B. Lund, R.F. McFeeter, R.D. Middlekauf, H. Mitsuda, H.A. Morris, W.E. Muir, F.J. Olivigni, M.P.S.A. Palumbo, E.B. Pantastico, H.M. Pearce, M. Pearson, W.E. Perkins, A.C. Peterson, M.S. Peterson, C.T. Phan, J. Pominski, N.N. Potter, K.S. Putohit, T.V. Raurakrishnan, K.D. Ross, A.L. Ryall, D.K. Salunkhe, P.M. Scott, M. Seeger, R.N. Sinha, D.S. Skene, E.A. Sloan, C.A. Smith, Jr., J.D. Smith, J.J. Spadaro, C.R. Stumbo, F.W. Tauber, R.R. Tompkin, D.K. Tressler, H.L.E. Vix, E.W. Williams, M.T. Wu, A. Yamamota and M. Yao.

We also wish to make special note of the assistance given us by Mr. John B. Klis, Editor and Director of Publications, Institute of Food Technologists, and to the Institute itself, for granting us permission to abstract and modify a number of articles in the journal *Food Technology* and in the *Journal of Food Science* for inclusion in this book. Special note is made of each work as it appears in this text.

Also, Dr. J.G. Woodroof, Dr. J.A. Jaynes and Dr. W.S. Arbuckle granted permission to incorporate sections of their works on peanut butter, sweetened condensed milk and ice cream technologies, respectively, and to Dr. W.F. Kwolek, Dr. G.N. Bookwalter and Dr. T.P. Labuza for their contributions to the sections on predicting the storage life of foods, all of which greatly expanded the scope of this text.

To each, we extend our appreciation for their courtesy in allowing us to quote from their works.

We have been helped in this revision by suggestions from a wide spectrum of users, and for this assistance and guidance, we are most grateful. In particular, we wish to thank the members of the Editorial Advisory Board in Food Science and Technology for their various critiques.

We have used the International System of Notations for Weights and Measures (Metric System) where ever possible. Convenient conversion tables are given in the appendix for ready interpretation into British Units and vice versa.

Since this book has found wide use in a large number of educational institutions in many countries, we hope that the revision will also make it a more complete and convenient unit of study.

The senior author would like to note the distinct pleasure in having his son James join in this writing, for without his assistance, it would not have been possible.

A special note of thanks is due Arlene L. Hoeppner, Editor, AVI Publishing Co., for her constructive and untiring efforts in bringing this new edition into being.

We also wish to acknowledge the help and encouragement given us by Ann C. Desrosier, wife and mother.

NORMAN W. DESROSIER
JAMES N. DESROSIER

Westport, Conn.
January 3, 1977

Preface To
The Third Edition

As this book enters its second decade of use, it seems appropriate not only to update the material presented and expand it to keep abreast of an unfolding food technology, but to bring together certain matters related to the application of this technology which might be useful to the reader.

In order to make the book more complete, new chapters have been added on the subjects of semi-moist foods, the principles of baking and the preservation of bakery products, and the stability of preserved foods over long periods of time under various environmental conditions. In the fourth chapter added, I have tried to present an insight into the application of the technology of food preservation to new product development. Essentially, the system presented is that being used successfully by the food industry in one form or another to yield new products. An important driving force for new technologies in food preservation is found in new product development, and I believe all will agree that most significant new developments in the food industry are technologically based.

An undertaking of this kind must necessarily draw upon the technical literature. I have drawn heavily from certain sources and these are noted in the text. However, a special note of obligation is due to Mr. H.M. Burgess, Mr. S.R. Cecil, Miss C.J. Doherty, Mr. H. Flick, Mr. J. Halik, Mr. F.J. Hallinan, Mr. F. Hollis, Dr. E.S. Josephson, Mr. M. Kaplow, Mr. R. Klose, Dr. G.D. LaBaw, Dr. S.A. Matz, Mr. J.J. O'Neil, Dr. N.W. Potter, Mr. E.W. Pyler, Dr. L.S. Spiegel, Dr. D.K. Tressler, and Dr. J.G. Woodroof. I am most obliged to all involved for allowing me to draw so generously from their published works, and to Dr. I.I. Rusoff and Dr. O.G. Jensen for assistance in this revision.

I wish to make a special note of tribute to the late Dr. L.B. Parsons, former Vice President-Research for Lever Bros., who had been my tutor for several years on industrial research and new product development, following his retirement. The system of new product development presented in this book is widely used now and he was an important contributor to it.

It is also a personal pleasure to acknowledge the professional opportunities given me by Mr. John J. Toomey, Vice President, Operations, National Biscuit Company (now retired), and to Mr. Alger B. Chapman, former Chairman, Beech Nut Lifesavers, Inc. Their kindness and encouragement have been most appreciated.

It is also a special pleasure to acknowledge the assistance given me in this edition by John, James, Brian, Nancy and Ann Desrosier.

Since this book has found some use in a good number of educational institutions in several countries, it is hoped that the additional chapters will also make a more complete and convenient unit of study.

NORMAN W. DESROSIER

Pelham Manor, New York
September 15, 1969

Preface To
The Second Edition

The people on earth now need about 3.6 trillion pounds of food per year, some 200 billion pounds more than was needed at the time this book was published in 1959. Since that time the world has added a human population equal to that of the United States and equal to the world population 2,000 years ago. Conservative estimates indicate that more than half the people of the world are now malnourished; certainly food technologists have before them a challenge of immense proportions. Time will reveal if we are equal to the challenge.

Over the past four years I have been testing the strengths and weaknesses of this book with the hope of improving the presentation. As a result, there have been substantial changes in a number of chapters, and graphs have for the most part been replaced with tables.

If I have been successful in this revision it has been possible in large part because of the suggestions given me by the following, whose generosity it is my pleasure to acknowledge:

Prof. W.B. Esselen, University of Massachusetts; Prof. S.A. Goldblith, Massachusetts Institute of Technology; Prof. H.O. Hultin, University of Massachusetts; Dr. Roy E. Morse, Lipton Foods, Inc.; Prof. R.C. Nicholas, Michigan State University; Prof. J.J. Powers, University of Georgia and Prof. B.E. Schweigert, Michigan State University.

I would also like to acknowledge the assistance given me by Mr. S.N. Deshpande, Mr. K.W. Lucas, Mr. M.J. Myers, Mr. P.E. Nelson and Mr. R.W. Trudeau of Purdue University, who have been most helpful in preparing this revision.

To the above and to all those who have contributed to the improvement of this book I am most deeply obliged.

NORMAN W. DESROSIER

West Lafayette, Indiana
January 15, 1963

Preface To
The First Edition

The people on earth require about 3.4 trillion pounds of food, one mouthful at a time, during the period from one food harvest to the next. Packaged into one-pound units, this one-year's food supply for these 2.8 billion people would have a path around the earth, off into space to the moon, then around the moon in a continuous band. While this much food is available on earth in one place or another, a goodly portion never benefits mankind and most human beings are malnourished.

All of man's foods are perishable commodities; they begin to deteriorate shortly after harvest, gather or slaughter. Some deterioration is accompanied by the production of poisonous agents; other deteriorations inflict losses in the nutritive value of foods. Man has learned to control some of these natural destructive forces and withhold the fruits of nature as his own food supply. Toward this end, mankind has accumulated a technology for preserving foods. The successful application of present information could go far in reducing the number of hungry mouths in the world.

The purpose of this text is to present the elements of the technology of food preservation. It is founded in the physical and biological sciences. For this reason, *Food Technology* offers opportunities for the integration of knowledge for the betterment of mankind, and offers great challenges to researchers, teachers, and writers of texts.

I am grateful to Dr. Carl R. Fellers, Dr. William B. Esselen, and Dr. Arthur S. Levine, of the University of Massachusetts, for they introduced me to this subject. I also wish to acknowledge the opportunity given me by Dr. N.K. Ellis and Purdue University in pursuit of such interests.

I have drawn freely from the published and unpublished material developed by my former graduate students at Purdue University, including Dr. G.R. Ammerman, Dr. F.W. Billerbeck, Dr. E.E. Burns, Mr. J.F. Farley, Dr. M.L. Fields, Dr. F. Heiligman, Dr. K.R. Johnson, Dr. G.D. LaBaw, Dr. F.J. McArdle, and Mr. W.L. Porter. In Chapter 4, I have relied heavily on the publications of the U.S. Department of Agriculture by Mr. W.T. Pentzer and Dr. G.B. Ramsey and their coworkers.

Dr. Martin S. Peterson, Editor of *Food Technology and Food Research*, and Chief, Technical Services Office, QM. Food and Container Institute for the Armed Forces, Chicago, Ill., has been most generous in his suggestions on the manuscript and aiding me in securing the many photographs from the Institute. The following companies were most kind in making available several photographs for this text: American Can Co., Food Machinery and Chemical Corp., The Foxboro Company, The Putnam Publishing Co., The Vilter Manufacturing Company, W. F. and John Barnes Co., and Libby McNeil & Libby. I have acknowledged the sources of all photographs as they appear in the text.

A suggested reading list is presented at the end of each chapter which greatly expands each of the major subject areas.

The horizon of information in the technology of food preservation is to be found in the current literature. At the end of this manuscript a list of journals is presented which covers the topical areas embraced in this text, and which will be found invaluable to a reader.

It is a pleasure to acknowledge also the aid given me by Dr. Donald K. Tressler, for without his patient encouragement this manuscript would probably never have been completed.

I am indebted to all researchers and teachers in the many fields of knowledge touched in this text. It represents a portion of the cumulative knowledge of generations of contributors who developed our present understanding of man's foods. I have attempted to bring this information together, with as much visual material as practical, within these two covers. It is obvious that all the information available on the subjects covered cannot be presented in a text of this size. This presentation is my compromise.

It is a pleasure to acknowledge the assistance given me by my wife and family in this undertaking.

West Lafayette, Indiana   NORMAN W. DESROSIER
March 31, 1959

Contents

Chapter 1

Food and Its Preservation

SOURCE OF FOOD PROBLEMS

Our nutrient needs in the whole have been identified and are generally known.[1] These needs are drawn from the plant and animal kingdoms in the form of food. Harvests of food from these kingdoms occur in rhythmic waves, related to the movement of the earth around the sun. Our hunger and our food harvests are not usually in harmony, throughout the year, in any one location on earth.

A further complication exists in that plant and animal "crops" begin to deteriorate shortly after harvest, gathering or slaughter. Some deteriorations are accompanied by the production of poisonous agents, while others inflict losses in the essential nutritive values of the foods. We must learn to control these forces in order to retain selected products of nature as our food supply, to be consumed at a time and place of our choosing, such that we may enjoy the rewards of proper nutrition in terms of bouyant health. This is the subject of this book.

Impact of Science and Technology

The successful application of modern food technology permits the conservation of desirable qualities in stabilized food supplies. Such stabilized foods permit their widespread distribution to meet the needs of people wherever they may be—on earth, on the moon—and as needed (Fig. 1.1).

Factors influencing the storage stability of foods include the type and quality of raw materials used, the method and effectiveness of processing, the type and manner of packaging, the mechanical abuse the packaged products receive in storage and distribution, as well as

---

[1] See Appendix Table A.1.

Courtesy of NASA

FIG. 1.1. POT ROAST AND GRAVY MEAL FOR ASTRONAUT

the influence exerted by the temperature and humidity of storage.

Each food system or type has, under the best of conditions, a potential storage life. This potential may be consumed quickly by mechanical abuse, inadequate packaging, and adverse storage conditions. The potential may be conserved by judicious selection and application of conditions of processing, packaging and storage, which will indeed protect and prolong the retention of desirable qualities in both the product and its package.

NATURE OF PLANTS AND ANIMALS

Since we seek to maximize human gains over natural destructive forces, we could profit from a brief examination of men who preceded us on earth. We see that the greatest developments of the human race, considering achievements, have been located in restricted areas. These form a belt bordered on one side by the frozen tundra and the other by the disease-infested tropics. The two hemispheres offered equally good opportunities for advancement where agriculture was rewarding. Those regions on earth where civilizations occurred at an early date were also areas where food production and food storage from one harvest to next were possible because the climate was favorable for such activity. The preservation and storage of foods was an important factor involved in the civilization of man and improvements in the technology of food preservation played no small part in the spread of civilization.

We, too, must make our contribution toward an adequate food supply base for all; we, too, must continue to inquire into natural phenomena. The desired results will be obtained by the application of the new information, but history reveals that the adoption of changes will not be rapid.

For example, knowledge of food production extends for some 8000 years, from the Neolithic Revolution in the Near East to the present. Food chemistry, on the other hand, is slightly over a century old. Only during the last half-century have our major nutrient needs been identified. What then became obvious was that for most of our history our ideas of good food were not necessarily related to our nutrient needs from a biological or chemical sense. With this in mind, what is the general chemical composition of the major plant and animal products that we eat and how effective is the present food supply system from a nutrient standpoint? First we might explore the chemical composition of food, then assay the efficiencies of our inherited methods of conserving these nutrients.

Foods Alike in the Elements They Contain

The general life chemistries of plants and animals have much in common. Hence, it is to be expected, and indeed found, that living entities require much the same chemical elements. These include hydrogen, carbon, oxygen, nitrogen, calcium, phosphorus, sulfur, sodium, potassium, iron and a list of minor elements.

Foods Differ in Molecular Constituents

An important difference is found in the molecular form in which these specific elements occur. The molecular forms vary among various plants and animals.

Information concerning the composition of plant and animal tissues in terms of molecules is recognized to be more useful than their composition solely in terms of the chemical elements they contain.

Because animals are incapable of the elementary syntheses performed by plants, the animals require specific, rather complex molecules as nutrients. This is reflected in the chemical composition of their tissues.

Components of Universal Nutrient Called Food

The idea that foods vary in terms of molecular structure developed in 1834 when the universal nutrient for man called "food" was found to contain three major molecular groupings or components: the carbohydrates, proteins, and fats. Since then and up to the recent discovery of vitamin B-12, there have been more than 50 essential molecules or nutrients identified in foods. These chemical compounds, which include the vitamins and minerals, comprise the material present in living substances of plants and animals which man requires in his food. These materials are: arginine, histidine, isoleucine, leucine, lysine, methionine, phenylalanine, threonine, tryptophan, and valine, all of which are essential; three fatty acid components of fats; sugar or carbohydrate; the fat-soluble vitamins A, D, E, and K; the water-soluble vitamins of the B complex, and vitamin C; the minerals calcium, chlorine, cobalt, copper, fluorine, iodine, iron, magnesium, manganese, phosphorus, potassium, sodium, sulfur and zinc.

Foods Generally Water Systems

From an overall view, plant or animal tissues are generally water systems of carbohydrates, proteins and fats. Contained in the water phase are the water-soluble carbohydrates, proteins, fatty acids, mineral salts, vitamins, physiologically active compounds and pigments. Proteins are held in a colloidal state in the water system, and fat in an emulsion. Dissolved in the fatty phase are the fat-soluble vitamins, physiologically active compounds, and pigments.

Proximate Composition of Foods

The chemical composition of a food is usually described in terms of its content in percentage of carbohydrates, proteins, fats, ash (mineral salts), and water. Important differences between plant and animal tissues, important in food preservation, are found in terms of their composition in this sense. Plant tissues are usually rich sources of carbohydrates; animal tissues are usually rich sources of proteins. For example, an apple may have 16% carbohydrate, 0.2% protein, 0.8% fat, 2.0% ash, and 81% water, while lean muscle may contain 2.0% carbohydrate, 20% protein, 2.0% fat, 2.0% ash, and 74% water.

Small Difference Important

While foods are found to vary substantially in terms of their major and minor nutrients, usually the composition of closely related plants or animals is similar. There are also important differences (e.g., in vitamin content) among various varieties of a fruit or vegetable.

Those differences are now sought and selected to yield foods of improved value to man.

FOODS OF PLANT ORIGIN

The plant life found by man on earth has been classed either as lower or higher from a standpoint of molecular complexity and degree of organization. The higher plants are those commonly seen and which generally produce flowers and seeds.

A number of lower plants, including mushrooms, truffles and algae, are eaten by man. Experience indicates, however, that the useful lower plants, if eaten as a steady diet, measurably lower the satisfaction of eating. On the other hand, such materials can be important additives to our diets and are important to consider in future food production systems.

Dry Fruits and Seeds

The most important higher plant structures used as food are the dry fruits and seeds. These would include all the cereals and small grains, the legumes and nuts. Because they are dry, they store easily and are not difficult to transport. Roots, tubers, bulbs, and other so-called "earth" vegetables are next in importance to us, although their food value is lessened because of the increased moisture content compared to dry grains. The leafy parts of plants contain little stored or reserve food and, although very perishable due to their high moisture content, are eaten as sources of vitamins, minerals and roughage. The fleshy fruits such as peaches and apples are also very perishable, are good sources of carbohydrates and vitamins, and, importantly, add pleasure to eating.

Cereals and Grasses

Cereals are the dry seeds of those members of the grass family grown for their grains and are by far the most important plants eaten by man. Of the cereals, wheat, corn and rice are the most important.

One or more of the cereals is adapted to each type of climate on earth: barley and rye grow in the cool temperate regions; corn and rice in the warmer temperate regions and in the tropics. Cereals have a wide range of requirements of moisture and soil types. They can be grown with little labor and the yield of food is high for the work involved.

Cereals are a principal source of carbohydrates although they contain proteins, fats, some vitamins, and minerals. Cereals in their native dry state are alive. They respire just as we do, giving off carbon dioxide, water and heat. Grain with about 12 to 14% moisture can be stored for several years.

Wheat.—Wheat is the most important annual grass. The mature grain is a germ covered with a starchy quilt. The germ is a good source of vitamins, fat and protein. The germ and starchy cover is packaged in a bran or husk which is also a good source for protein and vitamins. There are 8 important types of cultivated wheats and they grow from 2 to 4 ft in height.

The spring wheats are sown in the spring and harvested in the fall. Winter wheat is sown in the fall and harvested in the early summer. The yields from the winter types are higher. In the United States about half of the wheat crop is hard, red, winter wheat; one-fourth is soft, red, winter wheat; one-fifth is hard, red, spring wheat; one-twentieth is durum wheat; and part of the rest is white wheat. These wheats are the most widely used item in human diets (Fig. 1.2).

Courtesy of Deere and Co.

FIG. 1.2. CEREALS ARE THE MAJOR SOURCE OF FOOD IN THE WORLD

Food yields are high in terms of calories per man hour invested.

Corn.—Indian corn is America's only contribution to the important group of cereals and is the largest of the cereals grown, sometimes reaching 15 ft. Two kinds of flowers are produced—the tassel or male flowers, and the ear or female. Each kernel is a fertilized ovary. The silks are pollen tubes. The ovaries, and consequently the mature kernels, are produced in rows on the cob, which is surrounded by an inedible husk. No wild species of corn has ever been found. Corn is considered the most difficult and spectacular of the isolations of food crops from nature.

Dent corn is the type grown in the largest amounts. The inside of the kernel is somewhat soft and there is an indentation found at the crown of the kernel as the inside dries and shrinks. Corn has been bred to have ears of a type and location that can be severed from the stalk by a mechanical harvester and is an example of the possibilities available in manipulating plants.

Sweet corn is an immature form of corn, of varieties selected for sweetness. When at prime eating quality, the kernels are easily punctured with the thumb nail, yielding a milky exudate of 73 to 76% moisture. At 68 to 70% moisture, the exudate is thick and pulpy. Such corn is chewy and less tasty.

Rice.—Rice is the most important cereal grown in tropical areas and is indispensable as a food for half the population of the earth. About 95% of the world crop is grown in the Orient. The rice plant is an annual grass which produces a number of fine branches terminating in a single grain covered with a husk.

Commercial polished rice has the husk removed. Most of the vitamins in rice are in the husk; hence, they are lost in this practice. Enriched, polished rice is husked rice to which the vitamins and minerals, lost by removing the husk, are returned. Unpolished rice is best for human nutrition, or, as a substitute, polished rice which has been enriched with the lost nutrients.

Legumes

Legumes are second to the cereals as important sources of human food. Legumes are the "meat" of the vegetable world and are close to animal flesh in protein food value. The soybean is one of the oldest known food plants. The seed is the richest in food value of all the vegetables consumed throughout the world. It is used in the fresh, fermented, or dried form.

Field and garden peas are also legumes of high value as food. Field peas are mainly used as livestock feed. Garden peas in an immature tate are eaten fresh, or are preserved by freezing, canning, or drying for later use.

The entire young pods of string or snap beans, and many types of cultivated beans, are edible. The peanut is also a legume and is a good source of protein and oil. Lentils are also legumes.

Nuts

A nut is a one-seeded, one-celled fruit with a hard shell. Hazelnuts, chestnuts and filberts are truly nuts. Other plant tissues may be called nuts, for example: Brazil nuts (seeds), peanuts (legumes), or walnuts, coconuts, pecans, and almonds (dry fruits). We distinguish three groups of nuts: those with high fat content, those with high protein content, and those with high carbohydrate content. High-fat nuts include the Brazil nuts, cashew nuts, coconuts, pecans and walnuts. The high-protein nuts include the almonds, beechnuts and pistachio nuts. High-carbohydrate nuts include acorns and chestnuts.

Roots and Tubers

Plant roots and modified roots are storage organs for plants and also are important to us as food. We enjoy fleshy tap roots such as carrots, beets, radishes and turnips.

Roots are dependent upon plant growth above ground to synthesize the chemical compounds to be stored and, in general, a large top growth usually indicates a large root. Certain plants have large, fleshy, lateral roots, such as are found with sweet potato. One sweet potato plant may have from 6 to 8 large fleshy roots.

Underground Stems.—The white potato is an underground stem and is the most widely grown of the earth vegetables. The potato is not a root, as it has nodes (eyes) and other stem characteristics. From 6 to 8 large potatoes are obtained per plant. The potato was cultivated for centuries by the natives in Peru and was introduced into world commerce following the discovery of the Americas. The fruits produced by the plant are not edible. Yields are shown in Fig. 1.3.

Vegetables

In comparison with the cereals and sugar plants (cane and sugar beets), the vegetable food crops are very numerous and usually limited geographically in production. Many vegetables, for example, are grown in Asia which are unknown in the United States. Vegetable growing is a highly local affair, unlike cereal growing. Tomatoes,

Courtesy of Purdue University

FIG. 1.3. HARVEST OF POTATOES

10 Kilos harvested per square meter of this peat soil.

peppers and eggplant serve to demonstrate this point. The first two are widely eaten in the United States, while eggplant has never achieved wide acceptance. In the Orient the eggplant has the widest acceptance of the three.

Onions represent an important class of bulbs grown as food crops. Members of the cabbage family, including cauliflower, broccoli, Brussels sprouts, kale, collards and kohlrabi are also widely eaten. Turnips, rutabaga, radishes, beets, carrots and parsnips are biennial plants. They store rich supplies of carbohydrates in their roots dur-

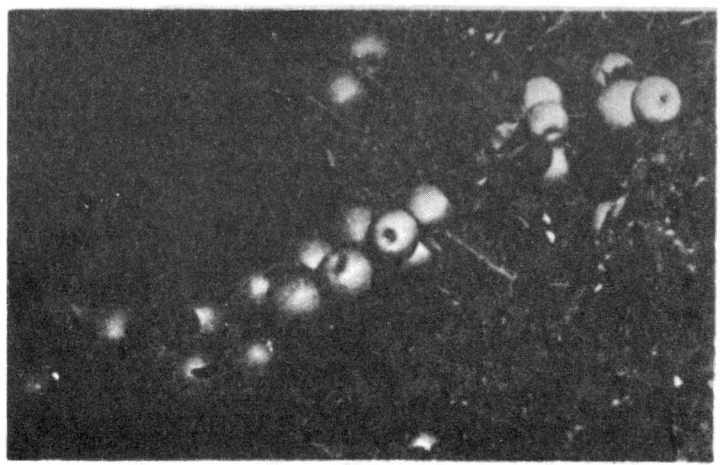

Courtesy of Purdue University

FIG. 1.4. TYPES OF APPLES DESIRED CAN BE CULTURED AND CONTROLLED

Successful apple culture involves selection of functional root stock to which is grafted the desired variety of fruit.

ing the first season to support growth during the second season, when they produce seeds. The salad crops of lettuce, celery, endive, chicory, cress and parsley are important foods in all parts of the world.

Fruits

Fruits are ripened ovaries of a flower. The edible portion is usually the fleshy covering over the seed, although the seeds were noted previously to be the edible nonfloral or vegetative growths of plants.

Fruits and vegetables can also be divided into groups mainly according to their uses. Fruit vegetables (pumpkin, cucumbers, tomatoes) are technically fruits but are eaten as vegetables. In food value and other properties fruit vegetables resemble the other vegetables.

When speaking of fruits, in general, we usually mean tree fruits (Fig. 1.4) or berries. Tree fruits are grouped into those from fruit trees which shed their leaves in the fall (apple, pear, peach) and those which shed their leaves in the spring (citrus fruits). The former are called deciduous trees, and the latter evergreen trees.

Berry fruits include strawberries, blueberries, gooseberries, cranberries, currants, blackberries and raspberries.

Peaches, plums, cherries and apricots are members of a class of fruits which are widely cultivated. Apples and pears belong to another class of useful fruits, as do the orange, lemon and grapefruit.

Composition of Plant Products

The composition of selected important members of the plant kingdom which are eaten as food by man are presented in Table 1.1.

TABLE 1.1

TYPICAL COMPOSITION OF FOODS OF PLANT ORIGIN

Food	Composition—Edible Portion				
	Carbo-hydrate (%)	Protein (%)	Fat (%)	Ash (%)	Water (%)
Cereals					
Wheat flour, white	73.9	10.5	1.9	1.7	12.0
Rice, milled, white	78.9	6.7	0.7	0.7	13.0
Maize (corn) whole grain	72.9	9.5	4.3	1.3	12.0
Earth vegetables					
Potatoes, white	18.9	2.0	0.1	1.0	78.0
Sweet potatoes	27.3	1.3	0.4	1.0	70.0
Vegetables					
Carrots	9.1	1.1	0.2	1.0	88.6
Radishes	4.2	1.1	0.1	0.9	93.7
Asparagus	4.1	2.1	0.2	0.7	92.9
Beans, snap, green	7.6	2.4	0.2	0.7	89.1
Peas, fresh	17.0	6.7	0.4	0.9	75.0
Lettuce	2.8	1.3	0.2	0.9	94.8
Fruits					
Banana	24.0	1.3	0.4	0.8	73.5
Orange	11.3	0.9	0.2	0.5	87.1
Apple	15.0	0.3	0.4	0.3	84.0
Strawberries	8.3	0.8	0.5	0.5	89.9
Melon	6.0	0.6	0.2	0.4	92.8

Source: Food Composition Tables, Food and Agricultural Organization of the U.N., Rome.

FOODS OF ANIMAL ORIGIN

About 10,000 years ago in the Near East the bow was invented, the cow and pig were domesticated along with the dog and horse, pottery making was discovered, and foods were being cooked and boiled in containers.

At least 8000 years ago man inhabited the large river valleys of the Nile, the Tigris-Euphrates and the Indus Rivers. Meadows and shrubs were emerging from drying swamps and man moved in with his selected animals and seeds. The domestication of an animal takes time. The animal must not only accept the process but must show signs of benefiting from it. The new strains of animals resulting from the selection by man had new instincts. All our barnyard animals of today had ancestors roaming wild on earth at that time. They are gone now. Domesticated animals released to the wild have difficulty surviving without us because they have been selected on the basis of utility to man rather than survival. There is a difference worth noting between a tame animal and a domesticated animal. The tame tolerates man; the domesticated serves man.

Domestication of Animals

The domestication of animals was a tremendous achievement. To be made to serve man, the animal must have some degree of intelligence, and by habit tend to live in flocks or herds in a community of sorts. Mammals either serve man or harm him. They supply him with food, clothing and other materials. Early paintings depict the hunt and defense against these large predators. The great interest man has in hunting today demonstrates this continuing relationship. Civilized man now hunts for pleasure and recreation. Live mammals, especially the rare types, are popular in our zoos.

Certain mammals are damaging to our crops and food animals, harbor diseases, and the larger species are capable of combatting us. Wolves, wildcats and bears prey on domestic cattle, sheep, hogs, and poultry. Current day predatory mammals range in size from the shrew to the whale. Each feeds on animals of appropriate size. Some of the diseases transmitted to man from the mammals are rabies, typhoid fever, spotted fever, tularemia, trichinosis, tuberculosis and undulant fever.

In beef cattle, animals were selected for shape (Fig. 1.5) and meat and milk yield. In dairy herds they were selected for high milk production and high butterfat milk. The Hereford and Angus were for meat, the Holstein for high milk yield, the Jersey for high milk-fat production, the Brown Swiss for meat, milk and work.

Milk Drinking and Dairy Products

Well-developed dairy herds were tended about 5000 years ago, and perhaps twice that long. It is believed that first sheep, then goats, then cows, were milked for human food. The milk collected was carried in pouches made from the stomachs of animals, a common and ancient practice. The stomach of a goat or sheep was taken, cleaned, tied at one end, and filled with milk. The stomach of a young goat or sheep has an enzyme called rennin, which we use to make cottage cheese and curd cheeses. A pouch of milk carried in the sun also provided adequate conditions for the growth of micro-organisms. At the end of a week's journey the pouch contained a

Courtesy of Food Processing
FIG. 1.5. BEEF CATTLE WERE SELECTED FOR SHAPE AND MEAT YIELD

ball of cheese in a watery whey. The cheese was found to be edible and, in many instances, tasty.

Milk is transported today from dairy farms to market in tank trucks. Partially filled tanks of milk are subjected to much agitation enroute by truck from farm to factory. It is not unusual to find that, if the milk is allowed to become warm, some has churned . . . butter flakes are seen in the milk! Butter making was thought to be discovered in a somewhat similar process. Churning or violently agitating milk yields butter. The fat in small globules strike one another and coalesce. There is a change from tiny fat globules in water to tiny water globules in fat—the latter we call butter.

At one time skin pouches filled with sour milk were beaten to form butter, and later hollow logs filled with milk were bounced to form butter. It was noted that it was from cream that rose on milk that butter was made. Butter has long been known as a medicine for correcting eye diseases and also as a salve for flesh wounds. We recognize the vitamin A (carotene) or yellow color of butter to be important for good health.

Agriculture Defined.—The domesticated animals and birds provide man with the bulk of the protein in his diet (aside from fish). Animal breeding was and is important to man. However, animal breeding alone was inadequate. It is but one of the factors in agriculture.

Tending animals alone does not yield food enough for man and animals and civilization. The difficulty with tending animals is that they eat themselves out of food. Man followed them constantly to new pastures. There were problems finding food for the animals as well as for man, but eventually man learned to provide food for both. This is known as agriculture, a combination of plant and animal culture, and they are closely related in yielding surplus food for man. Historically, whenever some men were not required to produce food, societies evolved.

Why Man Eats Meat

From the dawn of our beginning we have been meat-eaters when we had a choice. Why is animal tissue prized by us as food? It is because animal flesh closely resembles our own. We should find the balance of the amounts and kinds of nutrients that we need in meat, rather than in plant foods. Animal tissues all through nature have much in common. We can make fewer nutritional mistakes eating fresh meat that we have slightly cooked, than choosing from 100 plants those nutrients we need.

Animal Foods

The head, body and limbs of animals are composed of bone, cartilage, muscle, fatty tissue, skin and nerve. Nearly half the weight of the body is muscle. We eat these body muscles, if they are not too tough. Muscle is about three-fourths water, one-fifth protein, and the remainder fat, carbohydrates, vitamins and minerals. There may be large amounts of fat associated around muscle tissues. Fat tends to accumulate just beneath the skin, around muscles and in the body cavities.

Muscles vary in thickness and length from animal to animal. They are covered and held together with a tough tissue. This is not digestible, and it contributes to the toughness of meat. The tenderness of a piece of meat is directly related to the amount of tough covering tissue the muscles have. Aging meat by storing the carcass in a cool room tenderizes the muscles. This occurs by the degradation of some of this tough tissue by the enzymes in the now dead flesh. Gelatin may be made by boiling this tough covering tissue in water. Cooking meat, then, tenderizes in addition to the sanitary considerations. Some of the toughness of flesh is reduced by cooking. The tougher the meat, the longer it is necessary to cook it in order to produce a tender product (Table 1.2).

TABLE 1.2

INFLUENCE OF HEATING (107°C) ON THE TENDERNESS OF MEAT (BEEF)
AS MEASURED BY SHEAR FORCE

Heating Time (min)	Tenderness (Shear Force—lb)[1]
0	16.0
10	15.2
20	13.5
40	11.2
60	10.0

[1] See Appendix for metric conversion tables.

The protein in flesh is what we seek daily. The edible portions of beef contain approximately 17% protein, regardless of where we get the meat from the animal. Flank has 15% and round steak has 19% protein. Roast beef has 25% protein. During roasting some of the water is evaporated from the meat and some of the fat has dripped. The result is a greater percentage of protein in the weight that remains.

The pork we eat has a greater variation in protein content. Bellies have only 9% while tenderloin has 20%. Most pork we eat in chops, roasts, and hams and shoulders has about 16% protein, slightly less than beef flesh.

Poultry and Eggs

The contribution of birds to man's welfare has been improved by the domestication of birds to yield flesh and eggs. The wild hen of India was a bantam-sized slender bird laying a dozen eggs a year. Through breeding and selection, improved usefulness has been incorporated in the hen. Present-day varieties produce more than 200 eggs a year. In addition to egg production, the production of broilers weighing 3 lb in 10 weeks has become the most efficient means of producing flesh. The conversion of grain to flesh is now at the rate of 2½ lb of grain per pound of chicken. This is to be compared with the conversion of 4 lb of grain per pound of pork and 10 lb of grain per pound of beef.

Fish and Shellfish

Fish and shellfish are the major sources of protein in diets of people living along many coastal areas. Fresh-water and salt-water fish have been major and minor food sources for man. At the time of the Roman Empire live fish were carried in tanks on wagons from ponds and lakes to Rome. There the fish were kept alive until ready to be eaten. Trout are similarly handled in some American markets. Along our coasts we find lobster pounds. Rice and carp are farmed together in the Orient. Oysters are cultivated along several coasts around the world. Fish are as efficient in converting feed to flesh as many of our domesticated animals. Yields of 400 tons per hectare per year are possible (Fig. 1.6).

Courtesy of Brown

FIG. 1.6. CARP FEEDING IN POND CULTURE IN JAPAN

Meat and Fish Equally Nourishing

The flesh of fish and shellfish is nutritionally as effective in human diets as the flesh of animals. In addition, fish livers are excellent sources of the fat-soluble vitamins.

Problem of Raw Fish as Food.—Clams, lobster, mussels, shrimp and cockles contain an enzyme, thiaminase, which can destroy vitamin B-1 if they are eaten raw. Frogs, birds, toads and mammalian tissues do not contain thiaminase. Whitefish, carp, catfish and herring do. Salmon, trout, perch, bluegills, eels, bass, wall-eyed pike, cod, haddock and halibut are thiaminase-free. Carp and clams contain high levels.

Thiaminase also occurs in leaves of many trees, in mustard seed, wheat germ, ground fern and linseeds. Fortunately thiaminase is easily destroyed by heating or cooking.

Insects as Food

Insects are eaten in cooked form. Men eat grasshoppers, beetles, crickets, caterpillars, the pupae of butterflies and moths, termites, ants, and bee larvae.

Animal By-products

Bone forms the storage place for calcium salts in the animal. Bone matrix is similar to that found in cartilage. Cartilage consists of cells imbedded in an organic structure secreted from the cells. Cartilage serves as a web for bone formation and occurs as part of the skeletal structure of the higher animals. The normal bone contains nearly half water and one-fourth fat. The chief minerals are calcium phosphate and carbonates. Nearly 1% of bone is citric acid; 90% of the citric acid in a mammal is found in bone.

A mineral analysis of bone indicates that it is chiefly tricalcium phosphate with smaller portions of calcium carbonate. Bone is a complex of these salts, not a simple structure. Bone is actively metabolized, as is other animal tissue. Using radioactive phosphorus it has been shown that one-third of the ingested phosphorus is eliminated from rat skeletons in a month! About three-fourths of the minerals of a mammal's body consists of calcium and phosphorus. Ninety percent of the calcium and 70% of the phosphorus are deposited in bones and teeth. Throughout life, even our bones are changing . . . nothing remains constant.

Teeth are complex calcified structures in which calcification is differentiated into distinct regions. Dentine is the major tooth component. It is covered by a layer of cementum in the root and a layer of enamel in the exposed part of the tooth. Tooth enamel is the hardest tissue in the animal body. Dentine is softer than enamel but harder than ordinary bone. Cementum is as hard as bone. In order of decreasing hardness we have enamel, dentine, cementum and bone. Enamel is 89% mineral, dentine is 77% and cementum is 70%. Little wonder teeth are found in million-year-old fossils.

Blood collected under sanitary conditions from healthy animals may be made into blood sausage and similar products, or converted into by-products such as albumin and adhesives. Barbarians of 20,000 years ago in Northern Europe rode horses and drank their blood when on the hunt. Freshly drawn blood served as their source of vitamin C. Most fruits are suitable alternative sources.

Composition of Animal Products

The composition of selected members of the animal kingdom which are eaten as food by man is presented in Table 1.3. It is clearly seen that foods of plant sources are generally high in carbohydrate content, while foods of animal sources are rich in proteins. Furthermore, since the chemical composition of closely related plants and closely related animals is similar, it should be possible to develop a technology of food preservation for groups of closely allied food products. This turns out to be the case, as will be shown in later chapters.

Elements of Food Preservation

Except for surfaces exposed to air, soil or water, the internal sections of healthy plant and animal tissues are essentially sterile. By nature, the growth and development of tissues employ asepsis.

The preservation of foods in the first instance concerns itself

TABLE 1.3

GENERAL COMPOSITION OF FOODS OF ANIMAL ORIGIN

Food	Carbo-hydrate (%)	Composition—Edible Portion			
		Protein (%)	Fat (%)	Ash (%)	Water (%)
Meat					
Beef, medium fat	—	17.5	22.0	0.9	60.0
Veal, medium fat	—	18.8	14.0	1.0	66.0
Pork, medium fat	—	11.9	45.0	0.6	42.0
Lamb, medium fat	—	15.7	27.7	0.8	56.0
Horse, medium fat	1.0	20.0	4.0	1.0	74.0
Poultry					
Chicken	—	20.2	12.6	1.0	66.0
Duck	—	16.2	30.0	1.0	52.8
Turkey	—	20.1	20.2	1.0	58.3
Fish					
Non-fatty fillet	—	16.4	0.5	1.3	81.8
Fatty fish fillet	—	20.0	10.0	1.4	68.6
Crustaceans	2.6	14.6	1.7	1.8	79.3
Dried fish	—	60.0	21.0	15.0	4.0
Milk					
Cow, whole	5.0	3.5	3.5	0.7	87.3
Goat, whole	4.5	3.8	4.5	0.8	86.4
Cheese					
Hard, whole milk	2.0	25.0	31.0	5.0	37.0
Soft, partly whole milk	5.0	15.0	7.0	3.0	70.0

Source: Food Composition Tables, Food and Agriculture Organization of the U.N., Rome.

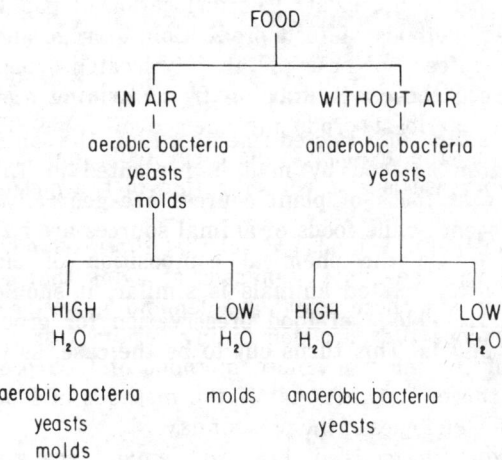

FIG. 1.7. MICROBIAL FOOD SPOILAGE

with controlling the growth of contaminating microorganisms. A schematic diagram of the organisms to be encountered is shown in Fig. 1.7. Mold will grow in a food as dry as grain if air is present and warm humid conditions prevail. Yeasts need slightly more water than mold to grow, but can grow in air or without air in warm humid conditions. Bacteria need more moisture than molds to grow, but can grow in air or without air in warm humid conditions.

There are generally considered to be three types of microorganisms: those with an optimum temperature for growth at 55°C called thermophiles; those with an optimum near 37°C called mesophiles, which include human pathogenic organisms; and those with an optimum growth below 10°C called psychrophiles. Psychrotrophic organisms are those capable of adjusting to temperatures below about 7°C.

NEEDS AND BENEFITS OF INDUSTRIAL FOOD PRESERVATION

A major problem facing us is that most people live in nations with low levels of industrialization, and preserved foods are found as a significant component of diets of populations in the highly industrialized nations. At the present time, most countries are in the process of forcing industrialization and with it is coming further urbanization. As a result, people are leaving the food production areas and are moving into the regions where industrial opportunities and the possibilities for a better life exist. This dislocation of populations means that improved methods of food production, storage and distribution are required to feed not only the already located urban populations, but also to feed those withdrawing from farming occupations who formerly were at least producing their own foods. Tied with this development is the explosive increase in human populations on earth. This increase is also concentrating in the metropolitan areas of the world.

On the other hand, it is also a fact that there are now *more* people in the world with adequate standards of living than ever before in human history, and these people are demanding better quality foods. The kind of foods they are demanding results from the successful integration of the most advanced methods of food production technology with the most functional of the methods of food storage and distribution technologies of the present day.

The high quality foods in greatest demand are also the highly perishable foods. Fortunately most perishable foods can be made stable and acceptable by the judicious application of present technology. With the successful application of commercial food preservation technologies the availability of perishable foods can be extended, thereby contributing usefully to human welfare.

Commercial food preservation improves food supplies in other ways as well. It encourages and/or initiates intensive food production practices and at the same time reduces losses due to spoilage and decay in harvested foods. Together these increase food supplies and eventually lower unit food costs.

With perhaps only 10% of the world's population currently consuming preserved foods regularly as important components in diets, the potential for growth of the food preservation industry is enormous. This growth is clearly recognized at this time and it is urgently needed.

Definition of Terms

The world's food supplies can be divided into three broad categories: perishable, semiperishable and shelf-stable.

Perishable foods are those that deteriorate quickly (such as fresh leafy vegetables and soft fruits) after harvest, or, soon after slaughter (fresh meat, poultry and fish).

Semiperishable foods are those that contain natural inhibitors to spoilage (such as eggs and root vegetables) or those that have received some type of mild preservation treatment which creates greater tolerances to the environmental conditions and abuses during distribution and handling (such as pasteurized milk, lightly smoked fish or pickled vegetables).

Shelf-stable foods are considered nonperishable at room temperature (such as cereal grains and nuts). Some have been made room temperature stable by suitable means (such as in canning) or processed to reduce their moisture content (such as raisins and crackers).

Food storage terms also include the following:

Dry storage means storage without extremes of temperature change, at about 20°C, and humidities below 50%.

Protected from light means to store in a dark place such that light reactions are prevented, as in an aluminum foil pouch or opaque container or in a closed area protected from visible light.

Cool storage means to store at a temperature below about 12°C, as in a refrigerated chamber or underground.

Refrigerated storage means to store at temperatures below those normally available naturally throughout the year, and between about 0° to 8°C.

Freezer storage means to store at temperatures below 0°C and preferably below −23°C.

Freezer compartment in a refrigerator storage means to store at temperatures of about 0°C.

The storage temperature of foods directly influences and controls the storage life of foods. As a rule of thumb, reducing the temperature of a stored food 10°C doubles the storage life. Increasing the temperature 10°C cuts the storage life in half. The temperature of a storage chamber is important in retaining the quality of foods, as well as for extending the period of time during which that quality is maintained.

Shelf-life means the period between the processing and the retail purchase and use of a food product. During this finite period of time the product is in a state of satisfactory quality in terms of nutritional value, taste, texture, appearance and use.

The shelf-life of food products varies but is generally recognized for each particular product by its manufacturer or processor and in general by consumers. This information is an essential factor in determining the kinds of conditions and methods used in distribution of food.

Storage studies are part of each product evaluation program (whether new product, product improvement, or simply a change in type or specification) as we shall see. The processor or manufacturer attempts to provide the longest shelf-life practical, consistent with costs and the pattern of handling and use by distributors, retailers and consumers.

STUDY PLAN

With the preceding as background, we are now in a position to expand the dialogue to include preservation technologies.

In Chapter 2 the health hazards encountered in foods and their prevention are discussed.

In Chapter 3 we will explore the use of cool temperatures to prolong the storage life of fresh foods by influencing their metabolism and that of their contaminating microorganisms. Next, in Chapter 4 we will study the use of controlled atmospheres in preserving these foods. In Chapter 5 the use of freezing temperatures to preserve foods for prolonged periods of time is developed.

In Chapter 6 the discovery and development of technology in canning foods is presented. There is no counterpart of canning in nature.

Removing moisture as a means of preserving foods by drying and desiccation are studied in Chapter 7. This is followed in Chapter 8 by the use of fermentation, and pickling to preserve foods in Chapter 9.

Then, the use of sugar and non-sugar concentrates as a means of food preservation is explored in Chapter 10. In Chapter 11 the use of chemical agents to preserve foods, including the philosophy behind such means, and an understanding of the approaches to follow in gaining government approval for these materials, is presented. In Chapter 12 the rapidly developing associated technology of preserving foods in semi-moist condition is studied and the importance of water activity is explored.

In Chapter 13 the state of knowledge in using ionizing radiation to preserve foods is reviewed, with special emphasis on radiation effects, what they are, and how to measure and control them.

All food products have a useful storage life. This matter is discussed in Chapter 14, including applications in predicting the storage life of products. In Chapter 15 the principles of quality assurance in food manufacturing are explored. Finally, the general method of applying these technologies in new product development is presented in Chapter 16.

The pursuit of additional information in each area presented in this text is encouraged by cited and selected reference citations. At the same time, the literature citations suggest the variety of sources of information available in government publications, in scientific, technical and trade journals, and in books.

REFERENCES

ANGELINE, J.F., and LEONARDS, G.P. 1973. Food additives, some economic considerations. Food Technol. 27, No. 4, 40-50.

ANGELOTTI, R. 1973. FDA regulations promote quality assurance. Food Technol. 29, No. 11, 60-62.

ANON. 1968. National diet heart study. Final Rep. Am. Heart Assoc. N.Y. Circ. 37.

ANON. 1973. Code of Federal regulations. Fed. Regist. 38, No. 143, 20061.

BAUMAN, H.E. 1974. The HACCP concept and microbiological hazard categories. Food Technol. 28, No. 9, 30-34, 70.

BRIGGS, H.M. 1949. Modern Breed of Livestock. Macmillan Co., New York.

BROWN, E. 1977. World Fish Farming Cultivation and Economics. AVI Publishing Co., Westport, Conn.

BULLERMAN, L.B., and OLIVIGNI, F.J. 1974. Mycotoxin producing potential of molds isolated from Cheddar cheese. J. Food Sci. 39, 1166-1168.

BYRAN, F.L. 1974. Microbiological food hazards today—based on epidemiological information. Food Technol. 28, No. 9, 52-66, 84.

CANNON, P.R. 1945. The importance of proteins in relation to infection. J. Am. Med. Assoc. 128, 360-363.

CHRISTENSEN, C.M., NELSON, G.H., MICROCHA, C.J., and BATES, F. 1968. Toxicity to experimental animals of 943 isolates of fungi. Cancer Res. 28, 2293-2295.

CLARK, J.G.D. 1947. Forest clearance and prehistoric farming. Econ. Hist. Rev. 17, 45-51.

CLARK, J.G.D. 1948. The development of fishing in prehistoric Europe. Antiquaries J. 28, 45-85.

CLAUSI, A.S. 1973. Improving the nutritional quality of food. Food Technol. 27, No. 6, 36-40.

DARBY, W.J. 1972. Fulfilling the scientific community's responsibilities for nutrition and food safety. Food Technol. 26, No. 8, 35-37.

DAVENPORT, C.B. 1945. The dietaries of primitive peoples. Am. Anthropol. 47, 60-82.

DESROSIER, N.W. 1961. Attack on Starvation. AVI Publishing Co., Westport, Conn.

DOBZHANSKY, T. 1955. Evolution, Genetics, and Man. John Wiley and Sons, New York.

DUGGAN, R.R., and LIPECOMB, G.Q. 1969. Dietary intake of pesticide chemicals in the United States (II), June 1966-April 1968. Pestic. Monit. J. 2, 153.

FILER, L.J. 1976. Patterns of consumption of food additives. Food Technol. 30, No. 7, 62-70.

FOSTER, E.M. 1972. The need for science in food safety. Food Technol. 26, No. 8, 81-87.

FRANKFORT, H. 1951. The Birth of Civilization in the Near East. Indiana Univ. Press, Bloomington, Ind.

GATES, R.R. 1931. The origin of bread wheats. Nature 127, 325-327.

GAUME, J.G. 1958. Nutrition in space operations. Food Technol. 12, 433-435.

GRAS, N.S.B. 1946. A History of Agriculture. F.S. Crofts, New York.

GUILD, L., DEETHARDT, D., and RUST, E.,II. 1972. Data from meals selected by students. Nutrients in university food service meals. J. Am. Dietet. Assoc. 61, 38.

HALL, R.L. 1975. GRAS—concept and application. Food Technol. 29, No. 1, 48-53.

HOFFMAN, W.E. 1947. Insects as human food. Proc. Entomol. Soc. 42, 233-237.

HOWE, P.E. 1950. Foods of animal origin. J. Am. Med. Assoc. 143, 1337-1342.

IFT EXPERT PANEL AND CPI. 1973. Organic foods. Food Technol. 28, No. 1, 71-74.

IFT EXPERT PANEL AND CPI. 1974A. Nutrition labeling. Food Technol. 28, No. 7, 43-48.

IFT EXPERT PANEL AND CPI. 1974B. Shelf life of foods. Food Technol. 28, No. 8, 45-48.

IFT EXPERT PANEL AND CPI. 1975. Naturally occurring toxicants in foods. Food Technol. 29, No. 3, 67-72.

KAUFFMAN, F.L. 1974. How FDA uses HACCP. Food Technol. 28, No. 9, 51, 84.

KRAMER, A. 1973. Storage retention of nutrients. Food Technol. 28, No. 1, 50-60.

KRAMER, A., and FARQUHAR, J.W. 1976. Testing of time-temperature indicating and defrost devices. Food Technol. 30, No. 2, 50-53, 56.

LACHANCE, P.A., RANADIVE, A.S., and MATAS, J. 1973. Effects of reheating convenience foods. Food Technol. 27, No. 1, 36-38.

LIBBY, W.F. 1951. Radiocarbon dates. Science 114, 291-296.

LIVINGSTON, G.E., ANG, C.Y.W., and CHANG, C.M. 1973. Effects of food service handling. Food Technol. 27, No. 1, 28-34.

MACGILLIVRAY, J.H. 1956. Factors affecting the world's food supplies. World Crops 8, 303-305.

MATCHES, J.R., and LISTON, J. 1968. Low temperature growth of Salmonella. J. Food Sci. 33, No. 6, 641-645.

MIDDLEKAUF, R.D. 1976. 200 years of U.S. food laws: a gordian knot. Food Technol. 30, No. 6, 48-54.

MITCHELL, H.H., and BLOCK, R.J. 1946. Some relationships between the amino acid contents of proteins and their nutritive values for the rat. J. Biol. Chem. 163, 599-620.

MURPHY, E.W., PAGE, L., and WATT, B.K. 1970. Major mineral elements in Type A school lunches. J. Am. Dietet. Assoc. 57, 239.

NATL. ACAD. SCI. 1970. Evaluating the Safety of Food Chemicals. Appendix: Guidelines for estimating toxicologically insignificant levels of chemicals in food. National Academy of Sciences, Washington, D.C.

NAS/NRC. 1970. Evaluating the Safety of Food Chemicals. Food Protection Committee, National Academy of Sciences—National Research Council, Washington, D.C.

NAS/NRC. 1972. GRAS Survey Report. Food Protection Committee, National Academy of Sciences—National Research Council, Washington, D.C.

NAS/NRC. 1973. Subcommittee on review of the GRAS list, Food Protection Committee. A comprehensive survey of industry on the use of food chemicals generally recognized as safe (GRAS). Natl. Tech. Inf. Serv. Rep. PB-221-925 and PB-221-939.

NAS/NRC. 1973. Toxicants Occurring Naturally in Foods, 2nd Edition. Food Protection Committee, National Academy of Sciences—National Research Council, Washington, D.C.

NATL. SCI. FOUND. 1973. President's science advisory committee panel on chemicals and health. Science and Technology Policy Office, Washington, D.C.

ORR, J.B. 1943. Food and People. Pilot Press, London.

PETERSON, M.S. 1963. Factors contributing to the development of today's food industry. In Food Technology the World Over, Vol. 1, M.S. Peterson, and D.K. Tressler (Editors). AVI Publishing Co., Westport, Conn.

PRESIDENT'S SCIENCE ADVISORY COMMITTEE. 1973. Report of the Panel on Chemicals and Health. U.S. Gov. Print. Off., Washington, D.C.

PUBLIC HEALTH SERV. 1965. Division of radiological health, Public Health Service: Radionuclides in institutional diet samples, April-June 1964. Radiol. Health Data 6, 31.

PURVIS, G.A. 1973. What do infants really eat? Nutr. Today 8, No. 5, 28.

SCOTT, P.M. 1973. Mycotoxins in stored grain, feed and other cereal products. In Grain Storage: Part of a System, R.N. Sinha, and W.E. Muir (Editors). AVI Publishing Co., Westport, Conn.

SEBRELL, W.H., JR. 1953. Chemistry and the life span. Chem. Eng. News 27, 3624-3629.

SNYDER, D.G. 1967. The fish protein concentrate story. Food Technol. 21, No. 7, 70-73.

STUNKARD, A.J. 1968. Environment and obesity: recent advances in our understanding of regulation of food intake by man. Fed. Proc. 27, No. 6, 1367-1374.

TANNER, F.W. 1953. Food Borne Infections and Intoxications. Twin City Printing Co., Champaign, Ill.

TRESSLER, D.K., and LEMON, J.M. 1951. Marine Products of Commerce. Reinhold Publishing Corp., New York.

ULRICH, W.F. 1969. Analytical instrumentation—its role in the food industry. Food Prod. Dev. 2, No. 6, 18-25.

VON SYDOW, E. 1971. Flavor—a chemical or psychophysical concept? Part I. Food Technol. 25, No. 1, 40-44.

WODICKA, V.O. 1971. The consumer protection team. Food Technol. 25, No. 10, 29-30.

WOODEN, R.P., and RICHESON, B.R. 1971. Technological forecasting: the Delphi technique. Food Technol. 25, No. 10, 59-62.

WHO. 1967. Joint FAO/WHO Expert committee on food additives. Procedures for investigating intentional and unintentional food additives. WHO Tech. Rep. Ser. 348.

Chapter 2

Nature of Food Hazards

In order to feed ourselves we must act to control deteriorative processes. Decomposition is a natural phenomena and an inescapable problem (Table 2.1). Plant and animal tissues are consumed one way or another by biological forces. There is a contest among people, animals, insects and microorganisms as to which will consume the nutrients first. In attempting to prevent the deterioration of plant and animal tissues, we have a doubly difficult task in that we must not only preserve the food for our own use, but we must also exclude the other forces of nature from it. Although food preservation has been practiced throughout the history of man, it was not until Louis Pasteur that insight was gained into why foods spoil. The last word has yet to be written.

CAUSES OF FOOD SPOILAGE

It is now recognized that the principal causes of spoilage in foods are the growth of microorganisms, the action of naturally occurring enzymes in the foods, chemical reactions, and physical degradation and desiccation. The type of spoilage for a particular food item depends to a great extent on the composition, structure, types of microorganisms involved, and conditions of storage of the food.

Microorganisms have nutrient and environmental requirements, as do other plants. Specific food spoilage is generally associated with certain types of foods and microorganisms. Some of the factors which control the type and extent of microbial food spoilage are the moisture content, the temperature, the oxygen concentration, the nutrients available, the degree of contamination with spoilage organisms, and the presence of growth inhibitors. Generally the con-

TABLE 2.1

USEFUL STORAGE LIFE OF PLANT AND ANIMAL TISSUES

Food Product	Generalized Storage Life (Days) (20°C)
Animal flesh	1-2
Fish	1-2
Poultry	1-2
Fruits	1-7
Dried fruits	360 and more
Leafy vegetables	1-2
Root crops	7-20
Dried seeds	360 and more

trol of one or more of these factors controls microbial spoilage.

Cleanliness is one of the most important deterrents of food spoilage. Microbiological deterioration of a food involves multiplication and growth of billions of cells. It follows then that whatever we can do to reduce the numbers of microorganisms will prolong the life of the product. The larger the initial population of organisms, the more rapidly the food is attacked. The application of good sanitary practices in the handling of food does reduce the incidence of spoilage. Time is another important consideration. The growth of spoilage organisms is a function of time as well as temperature and environment. The reduction of the amount of time elapsed between harvest and consumption is therefore an important consideration in the control of spoilage of foods (Table 2.2).

TABLE 2.2

RELATION BETWEEN SANITATION, BACTERIAL POPULATIONS IN MILK AND ITS STORAGE STABILITY AT DIFFERENT TEMPERATURES

Holding Temperature (°C)	Holding Times		
	0 hr	24 hr	48 hr
	Bacteria per ml		
	Clean Cows and Sterile Utensils Used (avg of 20 samples)		
4	4,295	4,138	4,566
15	4,295	1,587,388	33,011,111
	Dirty Cows and Utensils Used Not Sterile (avg of 30 samples)		
4	136,533	281,646	538,775
15	136,533	24,673,571	639,884,615

With the exception of chemical poisons and parasitic infections, foods which we harvest or gather or slaughter and consume without storage offer little danger to us when properly prepared for consumption. Food infections and food intoxications may occur when we store or pre-prepare or preserve our foods. It takes a combination of time and proper environment for microbial spoilage to occur in our foodstuffs. In the preparation of foods, both factors may be such that serious deterioration of food may occur.

Deter Natural Destructive Forces

Food properly prepared for our tables from freshly harvested, gathered or slaughtered products offers the peak of nutritional reward. Unfortunately it is not possible to mass distribute or mass produce our foods such that this condition exists. We must therefore prevent deterioration in our foodstuffs.

The plant and animal foods we eat must be healthy and contain no living disease-causing organisms. Decay and injury can usually be seen. Healthy plant and animal foods do not harbor disease organisms because healthy, living tissue can resist an infection. This is the first rule to be learned: food must be kept alive and healthy as long as possible. This is particularly important in the case of foods of vegetable origin.

Potatoes, carrots, beets, cabbage and turnips have been stored alive since man first grew them. These foods left alone in nature may live through the winter and bear seeds the second year. Bacteria do not easily invade living, healthy tissues. Cabbage may remain alive while some sugar is left in its leaves for energy; such plant tissues killed by heating or freezing lose their ability to resist invasion by microorganisms.

Fruits differ from the vegetables in one important aspect; they are made up of very old cells, incapable of reproducing themselves. They are not destined by nature to grow; they are meant to deteriorate and supply an appropriate seed bed for the germination of the seed. Fruits are not resistant to injury. On the other hand a potato can heal a wound.

Berries and cherries can be kept for only a few days before they begin to mold. Oranges are carefully handled and usually disinfected before being sent to market. Bananas are picked and shipped from the tropics in a green condition. So long as the banana is not ripe, it can heal a wound and withstand some rough handling during shipment. A ripe banana decomposes rapidly after a slight injury.

Living organisms are composed of carbohydrates, fats, proteins, cellulose (in woody parts of plants), keratin (in the hair and horns of animals), chitin (in the skeletons of insects), lignin (in trees), organic acids (in fruits and vegetables), and many other chemical substances. When conditions are right, substances produced by a living organism may be decomposed by microorganisms. If there were a compound produced by plants and animals which resisted this microbial decomposition, it would have accumulated in the billion years life has been on earth, and would be present in enormous quantities. We do find some exceptions in shell (White Cliffs of Dover), bone, coral, petroleum and various inert materials.

Every kind of organism does not decompose all organic matter produced by living plants and animals. Bacteria specialize in their food habits, just as the animals specialize. Lions eat animals, cows eat only plants, hogs eat both plant and animal food. The anteater lives principally on ants. Some caterpillars will eat only the leaves of one kind of plant and refuse all others. The appetites of bacteria are just as different, and only a few groups are "all-eaters." Most microorganisms are specialists.

The decomposition of dead plant and animal matter is therefore not a simple process, but represents the combined effort of many different types of microorganisms. Watching bacteria grow, we wonder how they can accomplish such a task. They work slowly but continuously. Ultimately all living tissue is degraded to water, carbon dioxide, ammonia, and mineral elements.

The storage of prepared foods in warm areas in the absence of oxygen creates conditions for putrefaction.

The storage of comminuted food in containers open to the air will cause losses in nutritive value. For instance, the storage in the air of any food containing vitamin C after the cells of the tissue have been ruptured will cause a considerable loss of this vitamin. Exposure of milk to light causes loss of its vitamin B-2 (riboflavin) content. Fats and oils in their fresh state, if stored at warm temperatures, will eventually become rancid and harsh tasting, and lose much of their nutritive value. The storage of fresh meat is followed with an immediate loss of its vitamin C unless the meat is refrigerated. Many other chemical reactions can take place in our foods if we do not give them adequate care. Decomposed food is little better for us than no food at all. Neither condition is to be desired or created.

Food Poisoning

Little pleasure is derived from eating decayed food. Our ancestors may not have been easier to please, they just had no choice. When food tissues die, they are invaded by microorganisms. In addition, the food may become infected by insects and pests. Food poisoning has been with us from the beginning. Cases of food poisoning are relatively numerous today and probably have been even more common in the past. Although caused by a variety of agents, most of these cases have been labeled ptomaine poisoning ever since 1870 when the term was coined. The types of food poisoning we encounter are those resulting from chemical poisons, those resulting from poisonous plants or animals, those resulting from toxins elaborated by bacteria, and certain intestinal infections caused by bacteria that are transmitted through food.

There are many chemicals which are poisonous to us when taken by mouth. Some encountered in food are: arsenic (sprays, etc.); certain alkaloids (strychnine); sodium fluoride; and salts of heavy metals such as lead, mercury, cadmium and antimony.

A number of plants and animals are poisonous to us. Cases of food poisoning have been traced to mussels obtained from the Pacific Coast during certain seasons of the year. Some wild mushrooms are commonly known to be poisonous. Ergotism is a poisoning from fungus growth on wheat and rye. Ergot finds use in medicine, however. Snakeroot poisoning may result from drinking milk of cows that have pastured and eaten snakeroot plants. Rhubarb leaf poisoning is due to the oxalic acid content of the leaves. Through years of of trial and error, we have learned to beware of certain plants and animals.

TABLE 2.3

CHARACTERISTICS OF FOOD POISONING CAUSED BY BACTERIA

Symptoms	Start of Symptoms After Eating	Type Poisoning	Cause		
			Eating of Toxin	Eating of Organism	Mortality (%)
Vomiting, nausea, diarrhea, abdominal cramps	1-6 hr or less	Staphylococcus	Yes	Yes	1 or less
Vomiting, nausea, chills, fever, diarrhea	7-72 hr	Salmonella	No	Yes	1 or less
Nausea, colicky pains, diarrhea	5-18 hr	Streptococcus	No	Yes	1 or less
Difficulty swallowing, double vision, difficulty breathing, paralysis	1-2 days	Botulism	Yes	No	70 or more

Food Intoxications

Some bacteria produce a toxin in food prior to being eaten, which causes a poisoning. These are called bacterial intoxications. Bacteria which grow in us and elaborate a toxin after being eaten are called food infections (Table 2.3).

Botulism.—Botulism is a food poisoning caused by the ingestion of the toxin elaborated by *Clostridium botulinum* prior to eating the food. It is soil bacterium, one of those which decomposes cellulose and grows without air. We have all eaten this organism at one time or another. This bacterium itself is non-infectious and is incapable of causing symptoms of poisoning in us. When the organism is allowed to grow under favorable conditions in food (without air) a powerful toxin may be produced. If the toxin is consumed, botulism will result. A spoonful of this toxin might kill a million persons. The toxin formed is destroyed by boiling for 10 min, hence, the recommendation to boil home canned foods for 10 min prior to eating. Botulinus toxin may be produced only under anaerobic conditions. The food responsible for an outbreak of botulism could appear normal, but usually is putrid. It usually smells like a combination of rancid butter and rotten meat. This bacterium has been found in soil all over the world. If we are affected by botulism we have ingested the toxin; the organism itself does no harm (Table 2.4).

TABLE 2.4
TIME-TEMPERATURE INFLUENCES ON SELECTED BACTERIAL TOXINS

Toxin	Temperature (°C)	Time	Toxicity
Cl. botulinum	-78	60 days	Potent
Cl. botulinum	100	10 min	Destroyed
Staph. aureus	-28	60 days	Potent
Staph. aureus	100	60 min	Destroyed

The organism names come from botulus—meaning sausage. The first case reported was in Germany due to eating infected sausage in 1735, and believed then to be a special type of meat poisoning. From 1900 to 1941 there were more than 1000 cases reported in the United States and Canada, of which nearly 700 were fatal. A few deaths a year still occur. Most cases come from home canned vegetables and meat. This point will be discussed later. Death is generally due to respiratory failure, although the patient maintains clear mental reactions up to death, which comes in 3 to 6 days following ingestion of the toxin. Nonfatal cases recover very slowly; the mortality rate is about 70%.

The only known therapeutic agent is an antitoxin. One must know the type of toxin, however, as there are two (Type A and Type B) important to man. Recently a Type E outbreak occurred. Usually a patient does not recognize the symptoms early enough and antitoxins are not widely available.

Staphylococcus.—Staphylococcus food poisoning, like botulism, is produced by a powerful toxin elaborated in food prior to ingestion. This is the most common type of food poisoning. Many cases are not severe and are seldom called to the attention of health authorities, so it is difficult to estimate its occurrence. Probably most of us have had this poisoning. The disease is milder than botulism, symptoms are of short duration, and recovery is rapid and complete. Mortality is extremely low, less than 1%. Most "ptomaines" are actually this type of food poisoning. Only when large outbreaks occur do we learn of it. These are at banquets, wedding parties, picnics and gatherings where food is prepared with inadequate facilities and/or care.

The organism is widely distributed; found on our skin, nose and throat. *Staphylococcus aureus* needs air to grow and a warm, moist environment, such as on a warm ham or turkey covered with a moist, clean towel to keep the meat from drying out. The organism is killed by boiling in water. The poisoning toxin is peculiar in that *it is very heat stable*. It is potent after a half hour of boiling. Vigorous growth is necessary for toxin formation. Most food may be contaminated with this organism, but if held cold no toxin is produced. However, if allowed to warm to room temperature for several hours our foods can be toxic. Any food that will support growth of staphylococci may develop toxin if left at favorable temperatures. Most cases of this poisoning are reported during the summer from cream-filled pastry, eclairs, custard cakes, dairy products, meat salads, salad dressing and mildly cured ham. In nearly all cases the incriminating food has been left at room temperature for several hours.

With this food poisoning (evidencing itself suddenly from 1 to 6 hr after ingestion) nausea, cramps, vomiting and diarrhea appear. Mortality is low. No special treatment is known and recovery is rapid. Several hundred million organisms per particle of food are usually present!

MYCOTOXINS

Toxic metabolites (mycotoxins) produced by filamentous fungi may be carried in trace amounts into human food supplies. Mycotoxins from *Fusarium, Aspergillus, Penicillium* and other genera, from the viewpoint of natural occurrence, may have toxic effects when ingested.

Aflatoxins are secondary metabolites of *Aspergillus flavus*-Link and *A. parasiticus*-Speare and have been shown to be both toxic and carcinogenic in test animals (Bullerman 1974). Bullerman reports that aflatoxins have been produced experimentally in a wide variety of food products, ranging from breads, cheese, meats, nuts and fruit. Aflatoxins are potentially most hazardous to human health. Much research has been done and is currently being done on aflatoxins, but too little is known in overview about other mycotoxins.

As noted previously, ergotism is perhaps the best-known mycotoxicosis (Scott 1973) because of its recorded effects on man as far back as the Middle Ages. Characteristic symptoms are gangrene of the limbs and extremities, convulsions, hallucinations and abortion. The chief causitive fungus, *Claviceps purpurea* (Fr.) Tul., produces sclerotia (ergots) on rye and other cereals that contain toxic alkaloids. With the quality now maintained in cereal production and milling, the fungus is no longer a hazard to human health (Scott 1973). There is a danger of poisoning farm animals. Scott maintains that toxic substances are produced by filamentous fungi more freqently than commonly realized. He states that of 246 fungal isolates from 25 moldy corn samples collected in central Iowa and cultured on rice, 99 yielded extracts lethal to laboratory animals. Water or ether extracts of 24 of the original samples were toxic. In a similar survey in South Africa, 46 out of 228 isolates from cereals and legumes caused death of ducklings when fed as cultures on maize meal. Japanese workers using mice as test animals reported incidences of toxigenic fungi on milled rice (10 out of 82) and various flours (25 out of 88). The majority of the toxin-producing isolates in these three studies belonged to the genera *Aspergillus* and *Penicillium*. A comprehensive survey showed that 284 (54%) of 527 fungi isolated from feeds caused death of experimental animals (mainly rats) when fed as cultures on autoclaved corn or corn-rice mixture. The feed samples were already associated with cases of illness or death of domestic animals. Toxigenic fungal isolates were predominantly of the genera *Fusarium, Aspergillus, Penicillium* and *Alternaria* (Christenson *et al.* 1968).

Experts define natural poisonings of man or animals through foods or feeds containing fungal toxins as mycotoxicoses. The term mycosis is used when invasion of animal or human tissues by a pathogenic fungus occurs. Many examples of mycotoxicosis attributable

to moldy grains have been reported (Scott 1973), but systematic
classification is difficult and isolation of toxigenic fungi is complicated when suspected mycotoxins in foods are involved. The number
of confirmed mycotoxicoses whose causative fungus has been identified, and its toxin characterized, is small. Detection of mycotoxin in
toxic grain has rarely been attempted. Apart from aflatoxins (Table
2.5), only six other specific mycotoxins have so far been found naturally in grains and feeds (Table 2.6).

TABLE 2.5
NATURAL OCCURRENCE OF MYCOTOXINS OTHER THAN AFLATOXINS IN GRAINS

Mycotoxin	Grain	Remarks
Ergotamine	Ryegrass	Caused lameness and gangrenous syndrome in cattle
Zearalenone (F-2)	Corn	
	Hay	14 ppm, sample associated with infertility in dairy cattle
	Feeds	10-200 ppm, suspected cause of illness in animals
F-3 (related to F-2)	Feeds	Associated with infertility and abortion in dairy cattle
Ochratoxin A	Corn	110-150 ppb (1/283 samples)
	Wheat	0.1-1.3 ppm (5 samples)
	Oats	Isolated sample; 460 ppb
	Mixed grain and feed	Up to 11 ppm (3 samples)
Citrinin	Barley and oats	Isolated sample; 23 ppm
Sterigmatocystin	Wheat	Isolated sample

Source: Adapted from Scott (1973).

Food Infections

Intestinal organisms *(Salmonella, Shingella)* can cause food poisoning. Toxin is elaborated after the food is eaten. This group
comprises a large number of closely related organisms found in the
intestines of infected meat animals, insects, rodents and man.

This form of food poisoning is caused by the infection of our body
with these organisms. Then they produce a toxin inside us. It takes
time to have this occur, so the poisoning begins some time after the
food is eaten. This separates this type of poisoning from the previous
two. The onset of the disturbance is not sudden. The mortality is
low. Here again, refrigeration of food is a positive control. The infection may be from diseased animals, poultry, eggs, milk, and from

TABLE 2.6

NATURAL OCCURRENCE OF AFLATOXINS IN GRAINS AND FOOD PRODUCTS

Source	Country	Remarks
Wheat	United States	2/531 samples contained 7 ppb B_1 and 2 ppb G_1
	Ethiopa	Isolated sample; 42 ppb B_1, 5 ppb G_1
	France United Kingdom	47% lots contained B_1, up to 225 ppb; also G_1
Wheat flour	France Germany	28% lots positive; up to 150 ppb B_1
Spaghetti	Canada	Isolated sample; 13 ppb B_1
Corn	United States	2-3% samples positive; up to 25 ppb B_1
	United States	Up to 8733 ppb B_1, samples associated with toxic hepatitis of swine
	United States	0.1-10 ppm caused aflatoxicosis in chickens
	France United Kingdom	26% lots positive: up to 187 ppb B_1
	Senegal	80; 440 ppb (2/14 samples)
Rice	United States	Up to 50 ppb
	Taiwan	226 ppb B_1, sample associated with deaths of 3 children
	Senegal	53% samples positive: 0.03-2 ppm
Sorghum	United States	1% samples positive: 3-19 ppb B_1 and G_1
	Ethiopa	Isolated sample; 75 ppb B_1, 3 ppb G_1
	France	25% lots positive: less than 100 ppb B_1
	Australia	8 ppm B_1, sample associated with abortion
	Senegal	in sows
Oats	United States	Less than 1% samples positive: 6 ppb B_1
	France	42% lots positive: less than 100 ppb B_1, also G_1
Rye	France	19% lots positive: less than 100 ppb B_1
Barley	France	7% lots positive: less than 10 ppb B_1, also G_1
	United Kingdom	Traces
Malt sprouts	Germany	0-5 ppm
Breakfast cereals	United States	Up to 20 ppb B_1 (8 samples), also 50 ppb G_1

Source: Adapted from Scott (1973).

foods contaminated by rodents and other animals and insects. Rigid cleanliness of food preparation and eating areas, and sterilization of utensils and serving dishes are required. These are practices we all must adhere to under all circumstances, otherwise food poisoning may occur.

Trichinosis is a disease caused by a nematode, *Trichinella spiralis*, which lodges in the muscle fibers of hogs. Swine fed uncooked food wastes are more likely to be infected than hogs fed grain. Digestion of infected meat by man sets free the larvae, which penetrate our intestines. There they mature. The fertilized female bores through the wall of the intestine and deposits young trichinae, which pass into the blood stream and are transported over our body, particularly to our muscles, giving rise to soreness symptoms mistaken for other diseases. Fresh pork must be cooked to a temperature of 59°C in all parts to make certain that this organism is killed (Fig. 2.1). Government regulations cover the production and handling of frankfurters and processed pork products. Pork products must be heated sufficiently to kill trichinae cysts whether present or not. All pork is suspect.

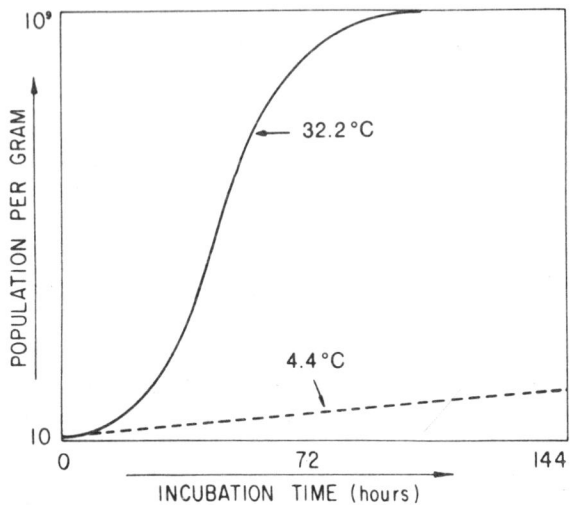

FIG. 2.1.   GROWTH OF MICROORGANISMS—TEMPERATURE CONTROLLED

Food may deteriorate due to improper handling and control. Improper washing, improper preparation, and improper refrigeration of foods may permit the transfer of bacillary dysentery, amoebic dysentery, food infection and poisoning, paratyphoid fever, and typhoid fever.

Using unsanitized glasses, dishes and silverware, and the eating of food exposed to coughing and sneezing persons, may transmit the common cold, diphtheria, encephalitis, measles, mumps, pneumonia, poliomyelitis, scarlet fever, septic sore throat, tuberculosis, and whooping cough, to name a few.

The use of raw milk and unheated milk products may lead to diphtheria, bacillary dysentery, paratyphoid fever, scarlet fever, septic sore throat, tuberculosis, typhoid fever and undulant fever. Milk must be pasteurized.

The use of contaminated drinking water can spread many diseases.

The lack of screening and exposure of food to flies and insects and pests not only leads to disease but there are aesthetic values involved. Improper storage of food and lack of rodent control where food is handled may lead to disaster from diseases and parasites. The common parasites and disease-causing bacteria of man are easily killed by a short exposure to boiling water temperatures.

EPIDEMIOLOGY OF FOOD HAZARDS

The U.S. Center for Disease Control (CDC) in Atlanta, Georgia has identified 1703 outbreaks of food-borne diseases (97,590 cases) in the United States in a 5-year period covering the first part of the 1970s. These cases were of microbial and chemical sources. There is no indication of any lessening in such outbreaks.

About 6% of the outbreaks were attributed to foods from food processing plants. Such a small proportion seems unimportant until one considers the quantities of food processed industrially in 1 hr. One instance of mishandling can jeopardize large numbers of people. One code of ice cream base (Byran 1974) caused 14 outbreaks involving 9000 cases of salmonellosis in 4 states within a period of 13 days. Table 2.7 is an example of actual cases traced and identified by the CDC involving the food preservation industry over a five-year period.

It is clear that hazards of microbial origin are the predominate kind and that poisonings due to the inadvertent contamination of food with chemicals also occurs, though the incidence is lower.

CHEMICALS IN FOODS

If a chemical additive is hazardous to public health, no amount of justification should obscure that fact. However, recent discoveries that some additives to foods, substances having been used for cen-

TABLE 2.7

DISEASE OUTBREAKS RESULTING FROM VARIOUS FOOD INDUSTRIES

Industry	Salmonellosis	Trichinosis	Staphylococcal Intoxication	Botulism	Shigellosis	Vibrio Parahaemolyticus Gastroenteritis	Brucellosis	Scombroid Poisoning	Viral Hepatitis	Chemical Poisonings	Outbreaks Reviewed
Meat processing	6	13	4	1							24
Baking	10		5						1		16
Egg processing	6										6
Canning					6			1		1	8
Fish processing	1				3	1		1			6
Milk products processing	3							1			4
Dietary supplements	3										3
Poultry processing	2										2
Candy manufacturing	1									1	2
Food color manufacturing	1										1
Poi processing						1					1
Salad processing	1										1
Cider processing						1					1
Brewing										1	1
Frozen specialties processing										1	1

Source: Byran (1974).

turies, are of questionable safety have created an emotional response condemning all additives. The consequences from blanket condemnation as presently occurring cannot be in the public interest.

There are a number of chemical substances which have emerged in the main for sound technical and economic reasons. They are used to improve or extend shelf-life, to reduce waste in distribution, to heighten aesthetic qualities, and to improve the nutritional values of foods.

With each passing year, knowledge of the foods we eat improves slightly. At the same time, there is no denying that the food we eat includes emulsifiers, sequestering agents, stabilizers, thickeners, preservatives, antioxidants, acidulants, surfactants, anticaking agents, colorings, flavorings, and nutrient supplements (Table 2.8). These materials have been intensively investigated.

TABLE 2.8

ESTIMATED CONSUMPTION OF MAJOR CLASSES OF FOOD
ADDITIVES IN THE UNITED STATES

Item	Quantity (millions of lb)
Preservatives	40
Antioxidants	20
Sequestrants	1
Surfactants	200
Stabilizers—thickeners	400
Acidulants	200
Leavening agents	150
Food colorants	100
Nutritional supplements	34
Flavoring materials	300
Flavor enhancers	50
Basic taste modifiers	5
Functional protein additives	350
Miscellaneous	100
Total estimate	1950

Market Basket Study

Consumer exposure to both direct and indirect food and color additives can be determined in a variety of ways. The most precise approach is that provided by the so-called USDA Market Basket Analysis. In such studies foods are purchased through usual retail outlets and appropriate meals are prepared.

Another method of study is used by the U.S. Public Health Service, Division of Radiological Health to quantitate intake of radionuclides. Forty-one institutions varying in type from orphanages with severe economic limitations to well-to-do boarding schools were enrolled in this program. Representative dietary samples obtained from institutions on a monthly basis for constructing a 7-day, 21-meal composite were studied. Analyses of these representative diets were made for calcium, potassium, strontium-90 and cesium-137 (Public Health Serv. 1965).

GRAS Studies I and II

Subsequently, in response to a Presidential directive concerning the need for reevaluation of safety of substances "generally recognized as safe" (GRAS), the Food and Drug Administration requested the National Academy of Sciences (NAS) to develop and test procedures that could be used to elicit information from industry on the extent of consumer exposure to GRAS substances.

Survey I, a full scale study of GRAS substance use by the Food Industry, was launched in 1971.

Survey II included substances listed as GRAS in the Code of Federal Regulations (Title 21, Section 121.101, Subpart B), substances considered GRAS by FEMA (Flavor Extract Manufacturers Association), substances considered GRAS on the basis of independent scientific judgement, and substances covered by "no objection" letters.

The questionnaire used in the Phase II survey attempted to elicit the following information regarding a GRAS substance:

(1) The food category in which the GRAS substance was used (i.e., baked goods, breakfast cereals, cheese, fats and oils, meat products, soups, etc). A total of 28 regular food categories and 13 infant food categories was identified and the respondent was asked to fit his specific food into one of these broad categories.

(2) A reason or reasons for its use (i.e., the technical effect to be achieved through its use) was pursued in depth. Was the GRAS substance a nutrient supplement, an emulsifier, a flavoring agent, or was its use based on other technical effects?

(3) The final concentration of the GRAS substance in the food as consumed.

(4) Annual poundage of the GRAS substance sold directly to a consumer, used in preparation of a premix or blend, or used directly in food processing.

(5) Published or unpublished data relative to safety of substance.

Expert Panels

Concurrent surveys of flavoring ingredients and adjuncts were made by the FEMA and other industry groups.

The FEMA expert science panel periodically reviewed the criteria employed to arrive at judgments of GRAS status. In essence, these requirements include: evidence for the identity and purity of the substance; its chemical and pharmacological relation to structurally analogous substances; its presence and level as a naturally occurring constituent of foods; intended use levels; and any pertinent metabolic or toxicologic data. From the accumulated experience in the evaluation of large numbers of chemically related substances certain general principles have evolved, which have established the rationale and aided the process of safety evaluation by the FEMA expert panel.

Official U.S. guidelines for evaluating toxicological insignificance

for food additives employed by FEMA (Hall 1975) were published by the Food Protection Committee of the National Academy of Sciences—National Research Council and state:
For many substances that are functionally effective in food at dietary concentrations above 0.1 ppm, but still much below any reasonable judgment as to their maximum safe level as previously defined, there is need to arrive at estimates of toxicologically insignificant levels. For these substances it is justifiable to employ accumulated scientific experience and to recognize their structural analogy to other chemicals whose metabolism or toxicity is known. Reasoning by analogy may be used to arrive at conclusions of toxicological insignificance. If a substance meets all the following criteria, it may be presumed to be toxicologically insignificant at a level of 1.0 ppm or less in the human diet: (1) The substance in question is of known structure and purity. (2) It is structurally simple. (3) The structure suggests that the substance will be readily handled through known metabolic pathways. (4) It is a member of a closely related group of substances that, without known exception, are or can be presumed to be low in toxicity.

Survey Results

Based on the results of Surveys I and II, and the survey conducted the same year by FEMA, there are now approximately 3890 substances added to food. Of these, 950 are GRAS and not in any FDA regulation and 2080 are regulated food additives. Flavorings, by far the greatest single category in number, comprise 1300 of the 3890 additives. The number is increasing according to Middlekauf (1976).

However, in terms of amount used, the situation changes significantly. Refined sugar is used in the largest quantity—the poundage of sugar used each year being more than that for all others combined. Of course, in each case the usage figure represented is somewhat higher than the amount actually consumed, since portions of each food additive are lost in production processes through evaporation, spillage, etc., and foods are not consumed entirely.

The amount of sugar used is approximately 9 billion kg per year or 46 kg per year per person in the United States. We consume salt at the rate of approximately 6.8 kg per year, corn syrup at 3.8 kg per year, dextrose at 2 kg per year, and the remaining food additives at a total of slightly more than 4.1 kg per year.

According to the report of the President's Science Advisory Committee (1973) on Chemicals and Health, 3.5 kg of that 4.1 kg per year is accounted for by 32 of the most commonly used chemicals,

18 of which are leaving agents and agents for the control of acidity and alkalinity. The rest of the 32 materials includes: flavorings (mustard, pepper, MSG); propellants; carbonating and protective gases (carbon dioxide and nitrogen); and nutrient supplements (calcium salts and sodium caseinate).

The nearly 1900 other direct additives account for only 0.8 kg per year per person. This amounts to an average of 0.015 oz per year per person additive, or less than 0.5 mg per year. According to Middlekauf (1974) 0.5 mg per year is approximately 1 ppb of the daily diet for each additive. He suggests this to be compared with the NAS/NRC Guidelines for Estimating Toxicologically Insignificant Levels of Chemicals in Food (NAS/NRC 1970). It states that certain substances which are structurally simple, readily handled through known metabolic pathways, and in closely related groups of substances with low toxicity, may be considered toxicologically insignificant at a level of 1.00 ppm or less in the human diet.

FEMA undertook a second nationwide survey of the food and flavor industry. Its survey revealed that of the 1249 substances on the FEMA list at the time of the survey, 831 were estimated to be used in total amounts not exceeding 450 kg annually. Moreover, the average maximum use levels in food were below 10 ppm in 228 of these substances. These criteria of total annual usage and minimal levels in foods, together with a safe history of use in food, are regarded by the FEMA expert panel as a basis for "toxicological insignificance" relative to flavorings and extracts, in the absence of any specific evidence or reasonable suspicion (based, for example, on chemical structure) to the contrary.

Filer (1976) summarizes the matter by indicating that reasonably good data is now available on the concentration of many additives contained in the foods we eat. Evaluation of consumer exposure to additives has proven to be difficult. The difficulties relate to: (1) the large number of food additives with potential for use in processed foods for which no use reports exist; (2) the multiplicity of processed foods including main dish meals and ethnic dishes; (3) uncertainties of losses or changes in additives during processing and storage; (4) complexities of dietary patterns including meals away from home and those of target populations such as the young, the aged and the poor and (5) uncertainties in portion size.

The Committee on GRAS Survey III believes even more reliable data on consumer intake of food additives will come from the expanded number of food categories, which will more accurately predict additive use in the future. Food additives are undergoing a close scrutiny. This evaluation will probably continue well into the 1980s. The food industry must continue to demonstrate its concern for the public health by continuing every effort to make products safe and to label them properly.

NATURE'S SEAL OF QUALITY

Every food has a characteristic appearance, odor, taste and feel which we associate with normality. Any deviation is suspect. Changes in the color of foods may be an indication of change in their nutritive value. Raw vegetables which have lost their bright color and are limp and wilted are of lower food value than those with good color and crisp, firm tissue. Meat or poultry which has an off-odor is not fresh. Fish with dull eyes and slimy skin is suspect. The odors from ripe melons and bananas are an indication of their eating conditions.

Our senses tell us much about the quality of our foods. It is our method of determining the seal of quality imprinted by nature on and in our food. We can depend upon these senses providing we have the full complement and we exercise them. The combination of these senses with the information accumulated during the years of experiences of others is adequate to obtain the optimum responses from our foods. But, we recognize that our senses can fail us when food poisoning or nutritive values are considered. We are not able to see poisons in foods. We cannot see that a protein lacks essential amino acids (i.e., corn protein).

Nevertheless, over the years these matters have plagued people. Such matters which recur are susceptible to intellectual penetration. What has been found and what can be done is presented in the following chapters. What remains to be penetrated is also indicated, as best we can, on the basis of present knowledge and insight.

REFERENCES

AMMERMAN, G.R. 1957. Effect of equal lethal heat treatments at various times and temperatures upon selected food constituents. Ph.D. Dissertation, Purdue Univ., Lafayette, Ind.

ANGELINE, J.F., and LEONARDS, G.P. 1973. Food additives, some economic considerations. Food Technol. 27, No. 4, 40-50.

ANGELOTTI, R. 1973. FDA regulations promote quality assurance. Food Technol. 29, No. 11, 60-62.

ANON 1973. Code of Federal regulations. Fed. Regist. 38, No. 143, 20061.

BAUMAN, H.E. 1974. The HACCP concept and microbiological hazard categories. Food Technol. 28, No. 9, 30-34, 70.

BULLERMAN, L.B. 1974. Inhibition of aflatoxin production by cinnamon. J. Food Sci. 39, 1163-1165.

BULLERMAN, L.B., and OLIVIGNI, F.J. 1974. Mycotoxin producing potential of molds isolated from Cheddar cheese. J. Food Sci. 39, 1166-1168.

BYRAN, F.L. 1974. Microbiological food hazards today—based on epidemiological information. Food Technol. 28, No. 9, 52-66, 84.

CHRISTENSEN, C.M, NELSON, G.H., MICROCHA, C.J., and BATES, F. 1968. Toxicity to experimental animals of 943 isolates of fungi. Cancer Res. 28, 2293-2295.

CLAUSI, A.S. 1973. Improving the nutritional quality of food. Food Technol. 27, No. 6, 36-40.

DARBY, W.J. 1972. Fulfilling the scientific community's responsiblities for nutrition and food safety. Food Technol. 26, No. 8, 35-37.

DESROSIER, N.W. 1961. Attack on Starvation. AVI Publishing Co., Westport, Conn.

DUGGAN, R.R., and LIPECOMB, G.Q. 1969. Dietary intake of pesticide chemicals in the United States (II), June 1966-April 1968. Pestic. Monit. J. 2, 153.

FILER, L.J. 1976. Patterns of consumption of food additives. Food Technol. 30, No. 7, 62-70.

FOSTER, E.M. 1972. The need for science in food safety. Food Technol. 26, No. 8, 81-87.

FRANKFORT, H. 1951. The Birth of Civilization in the Near East. Indiana Univ. Press, Bloomington, Ind.

GOLDBLITH, S.A. 1971. Pasteur and truth in labeling: "pro bono publico"— in the best of scientific tradition. Food Technol. 25, No. 3, 32-33.

GRAS, N.S.B. 1946. A History of Agriculture. F.S. Crofts, New York.

GUILD, L., DEETHARDT, D., and RUST, E., II. 1972. Data from meals selected by students. Nutrients in university food service meals. J. Am. Dietet. Assoc. 61, 38.

HALL, R.L. 1975. GRAS—concept and application. Food Technol. 29, No. 1, 48-53.

IFT EXPERT PANEL AND CPI. 1972A. Botulism. Food Technol. 26, No. 10, 63-66.

IFT EXPERT PANEL AND CPI. 1972B. Nitrites, nitrates, and nitrosamines in food—a dilemma. Food Technol. 26, No. 11, 121-124.

IFT EXPERT PANEL AND CPI. 1973. Organic Foods. Food Technol. 28, No. 1, 71-74.

IFT EXPERT PANEL AND CPI. 1974A. Nutrition labeling. Food Technol. 28, No. 7, 43-48.

IFT EXPERT PANEL AND CPI. 1974B. Shelf life of foods. Food Technol. 28, No. 8, 45-48.

IFT EXPERT PANEL AND CPI. 1975A. Naturally occurring toxicants in foods. Food Technol. 29, No. 3, 67-72.

IFT EXPERT PANEL AND CPI. 1975B. Sulfites as food additives. Food Technol. 29, No. 10, 117-120.

ITO, K.A., et al. 1973. Resistance of bacterial spores to hydrogen peroxide. Food Technol. 27, No. 11, 58-66.

JAHNS, F.D., HOWE, J.L., CODURI, R.J., and RAND, A.G., JR. 1976. A rapid visual enzyme test to assess fish freshness. Food Technol. 30, No. 7, 27-30.

JENSEN, L.B. 1954. Microbiology of Meats. Garrard Press, Champaign, Ill.

KAUFFMAN, F.L. 1974. How FDA uses HACCP. Food Technol. 28, No. 9, 51, 84.

KRAMER, A. 1973. Storage retention of nutrients. Food Technol. 28, No. 1, 50-60.

KRAMER, A., and FARQUHAR, J.W. 1976. Testing of time-temperature indicating and defrost devices. Food Technol. 30, No. 2, 50-53, 56.

LACHANCE, P.A., RANADIVE, A.S., and MATAS, J. 1973. Effects of re-heating convenience foods. Food Technol. 27, No. 1, 36-38.

LEUNG, H., MORRIS, H.A., SLOAN, E.A., and LABUZA, T.P. 1976. Development of an intermediate moisture processed cheese food product. Food Technol. 30, No. 7, 42-44.

LIBBY, W.F. 1951. Radiocarbon dates. Science 114, 291-296.

LIVINGSTON, G.E., ANG, C.Y.W., and CHANG, C.M. 1973. Effects of food service handling. Food Technol. 27, No. 1, 28-34.

MACGILLIVARY, J.H. 1956. Factors affecting the world's food supplies. World Crops 8, 303-305.

MATCHES, J.R., and LISTON, J. 1968. Low temperature growth of Salmonella. J. Food Sci. 33, No. 6, 641-645.

MCRADLE, F.J., and DESROSIER, N.W. 1954. A rapid method for determining the pericarp content of sweet corn. Canner 118, 12-15.

MIDDLEKAUF, R.D. 1976. 200 Years of U.S. food laws: a gordian knot. Food Technol. 30, No. 6, 48-54.

MITCHELL, H.H., and BLOCK, R.J. 1946. Some relationships between the amino acid contents of proteins and their nutritive values for the rat. J. Biol. Chem. 163, 599-620.

MURPHY, E.W., PAGE, L., and WATT, B.K. 1970. Major mineral elements in Type A school lunches. J. Am. Dietet. Assoc. 57, 239.

NATL. ACAD. SCI. 1970. Evaluating the Safety of Food Chemicals. Appendix: Guidelines for estimating toxicologically insignificant levels of chemicals in food. National Academy of Sciences, Washington, D.C.

NAS/NRC. 1970. Evaluating the Safety of Food Chemicals. Food Protection Committee, National Academy of Sciences—National Research Council, Washington, D.C.

NAS/NRC. 1972. GRAS Survey Report. Food Protection Committee, National Academy of Sciences—National Research Council, Washington, D.C.

NAS/NRC. 1973. Subcommittee on review of the GRAS list, Food Protection Committee. A comprehensive survey of industry on the use of food chemicals generally recognized as safe (GRAS). Natl. Tech. Inf. Serv. Rep. *PB-221-925* and *PB-221-939.*

NAS/NRC. 1973. Toxicants Occurring Naturally in Foods, 2nd Edition. Food Protection Committee, National Academy of Sciences—National Research Council, Washington, D.C.

NATL. SCI. FOUND. 1973. President's science advisory committee panel on chemicals and health. Science and Technology Policy Office, Washington, D.C.

PETERSON, M.S. 1963. Factors contributing to the development of today's food industry. *In* Food Technology the World Over, Vol. 1, M.S. Peterson, and D.K. Tressler (Editors). AVI Publishing Co., Westport, Conn.

PRESIDENT'S SCIENCE ADVISORY COMMITTEE. 1973. Report of the Panel on Chemicals and Health. U.S. Gov. Print. Off., Washington, D.C.

PUBLIC HEALTH SERV. 1965. Division of radiological health, Public Health Service: Radionuclides in institutional diet samples, April-June 1964. Radiol. Health Data 6, 31.

PURVIS, G.A. 1973. What do infants really eat? Nutr. Today 8, No. 5 28.

SCOTT, P.M. 1973. Mycotoxins in stored grain, feed and other cereal products. *In* Grain Storage—Part of a System. R.N. Sinha, and W.E. Muir (Editors). AVI Publishing Co., Westport, Conn.

SMITH, C.A. JR., and SMITH, J.D. 1975. Quality assurance system meets FDA regulations. Food Technol. 29, No. 11, 64-68.

STEWART, R.A. 1971. Sensory evaluation and quality assurance. Food Technol. 25, No. 4, 103-106.

STOTT, W.T., and BULLERMAN, L.B. 1976. Instability of patulin in cheddar cheese. J. Food Sci. 41, 201-202.

STUNKARD, A.J. 1968. Environment and obesity: recent advances in our understanding of regulation of food intake by man. Fed. Proc. 27, No. 6, 1367-1374.

TANNER, F.W. 1953. Food Borne Infections and Intoxications. Twin City Printing Co., Champaign, Ill.

TRESSLER, D.K., and LEMON, J.M. 1951. Marine Products of Commerce. Reinhold Publishing Corp., New York.

ULRICH, W.F. 1969. Analytical instrumentation—its role in the food industry. Food Prod. Dev. 2, No. 6, 18-25.

VON SYDOW, E. 1971. Flavor—a chemical or psychophysical concept? Part I. Food Technol. 25, No. 1, 40-44.

WODICKA, V.O. 1971. The consumer protection team. Food Technol. 25, No. 10, 29-30.

WOODEN, R.P., and RICHESON, B.R. 1971. Technological forecasting: the Delphi technique. Food Technol. 25, No. 10, 59-62.

WHO. 1967. Joint FAO/WHO Expert committee on food additives. Procedures for investigating intentional and unintentional food additives. WHO Tech. Rep. Ser. 348.

Chapter 3

Principles of
Fresh Food Storage

All foods have a useful storage life (Table 2.1) and this period can be prolonged or shortened depending on the conditions of storage and the physical abuse the food products receive prior to, during and following storage. In overall analysis the cool storage of nearly all foods, fresh or preserved, is beneficial to the retention of quality. In this chapter we will explore the technology of storage of fresh foods. In Chapter 14 we will study the storage of preserved foods.

NATURE OF HARVESTED CROPS

The quality of harvested fruits and vegetables is dependent on the conditions of growth and on postharvest treatments. The climatic factors and cultural practices are complex. Fruits as well as vegetables, which are botanically fruits, form a special class of foods which can maintain an independent existence when detached from the tree or vine. The length of storage is a function of composition, resistance to attack by microorganisms, the external conditions of temperature, and the gases in the environment (Biale 1975).

A mature harvested fruit contains a variety of oxidizable substrates and the molecular machinery required to perform oxidative reactions. Respiration is the major process of concern and its mechanism is essentially the same in fruits as in other plant and animal life. In fruits, as in other materials, biological oxidations involve a number of metabolic pathways in which synthetic and degradative reactions are interdependent (Fig. 3.1).

Ripening in fruit is a critical period of transition from the stages of cell disorganization and death. Ripening means those changes in sensory factors of color, texture and taste which render the fruit accep-

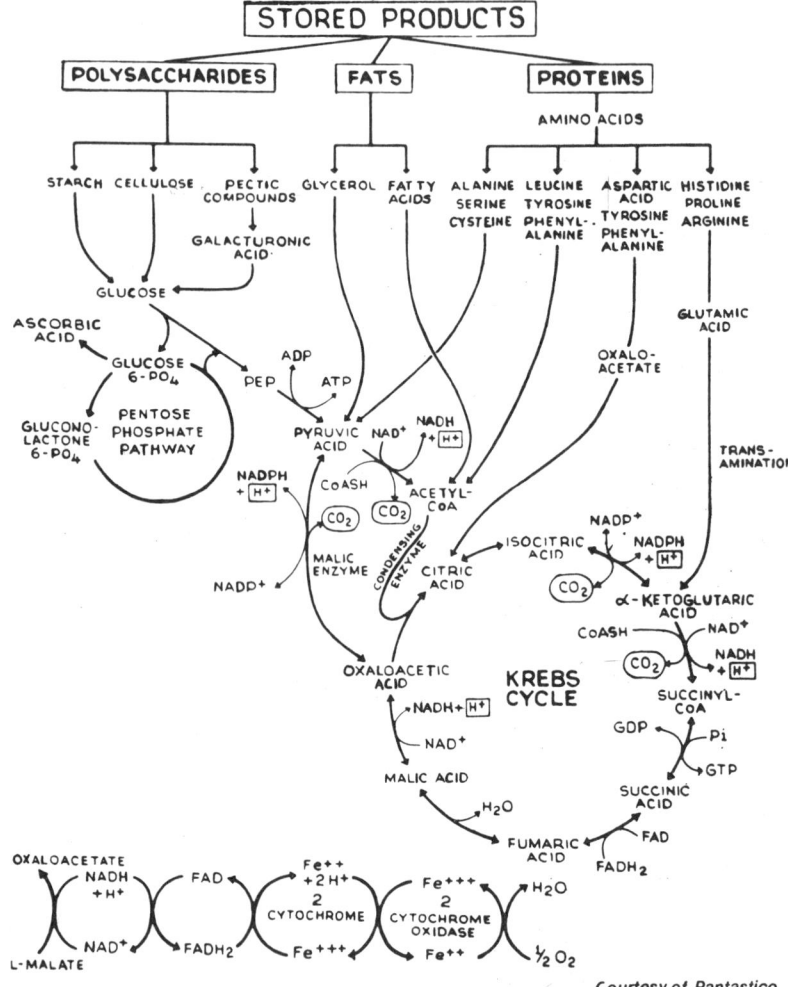

Courtesy of Pantastico

FIG. 3.1   METABOLIC INTERRELATIONS AMONG VARIOUS STORED PRODUCTS

table to eat. Biale (1975) reports these changes may be detected by analyzing transformations in pigments, pectins, carbohydrates, acids, tannins, etc. Fruits of different species differ in the nature of and rate of these changes, but most of them share a respiratory pattern known as the climacteric.

Rhodes (1970) describes the significance of the upsurge in respiration at the end of the maturation phase and preceding breakdown

(Fig. 3.2). After harvest the decline in the rate of oxygen uptake and CO_2 evolution to a low value (preclimacteric minimum) is followed by a sharp rise to a peak, terminating in a post-climacteric stage. The peak to minimum ratio tends to increase with temperature. This ratio varies among fruits. For example, it is much higher for a banana than for an apple. Fruit reaches the eating stage at the climacteric peak or sometime after the peak, depending upon the species and to some extent upon the conditions of storage (Fig. 3.2). Unique changes take place during the sharp rise in respiration from the pre-climacteric minimum to the climacteric peak. The slope of the rise varies with species, maturity, temperature, and the oxygen and carbon dioxide content of the storage chamber.

A partial list of anabolic and catabolic reactions associated with the ripening of fruit is presented in Table 3.1. Examples of depolymerization include the hydrolysis of starch to glucose in the banana and the breaking of polygalacturonide chain of pectins.

A multitude of physiological processes is present in edible plant tissues at the time of harvest. For example, on removal of vegeta-

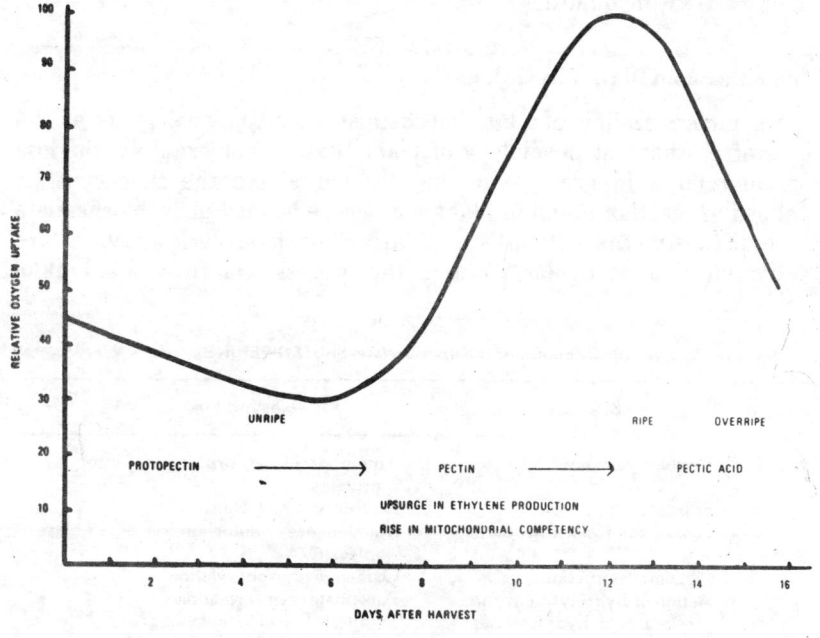

Courtesy of Rhodes

FIG. 3.2 THE CLIMACTERIC PATTERN OF RESPIRATION AND ASSOCIATED CHANGES IN FRUIT RIPENING

tive parts (such as fruits, roots, stems, etc.) from the parent plant, the tissues are deprived of their normal supply of water, minerals and, in some instances, of simple organic molecules (e.g., sugars, hormones) which previously would have been translocated to them from other parts of the plant. Although photosynthetic activity is then negligible, most tissues retain capabilities of a diversity of metabolic reactions which are specific to a given commodity and variety. These are seen in events such as rotting, ripening, sprouting, toughening, yellowing, etc. The kind and intensity of activity in detached plant parts determines to a large extent their storage life.

Seeds, fleshy roots, tubers and bulbs are morphologically and physiologically designed in nature to maintain the tissues in a dormant condition until favorable environmental situations are available for germination or growth. Metabolic activity is low but not completely halted in such tissues.

Fruits, leaves and stems, on the other hand, are physologically conditioned for senescence rather than dormancy. Fleshy fruits are unusual in that the ripening process is followed by the development of optimal eating qualities.

Insights From Plant Physiology

An understanding of plant biochemistry and physiology is at the core of postharvest physiology of plant tissues. For example, the loss of chlorophyll in green vegetables or fruits and the characteristic yellowing reaction found in plant senescence has led plant biochemists to hypothesize that groups of oxidative and hydrolytic enzymes are activated at a particular point in the process. Haard and Salunkhe

TABLE 3.1

PROCESSES ASSOCIATED WITH FRUIT RIPENING

Biodegradation	Biosynthesis
Depolymerization	Amino acid incorporation into proteins
Substrate utilization	Nucleic acid metabolism
Loss of chloroplast structure	Maintenance of mitochondrial integrity
Pigment destruction	Oxidative phosphorylation
Action of hydrolytic enzymes, esterases, dehydrogenases, oxidases, phosphatases, ribonuclease	Phosphate ester formation
	Synthesis linked to metabolic pathways: EMP, HMP, TCA
Ethylene production	

Source: Haard and Salunkhe (1975).

(1975) see degreening as a result of a diminishing rate of chlorophyll synthesis relative to a constant rate of degradation. With knowledge of the mechanism of loss, questions of biological control arise. They ask what cellular loci and messengers are responsible for a given expression of enzyme activity. It is known that storage of green vegetables under conditions which retard respiration (e.g., low temperatures) delays changes such as chlorophyll loss. The link or message is the focus of attention. They find three facets: the link (biochemical control), the machinery (mechanics of quality change), and the environment (temperature, atmosphere, etc.).

Protein synthesis and other anabolic events are closely allied to catabolism during senescence and stress response. Senescence and other types of tissue necrosis are not necessarily the result of a wearing-out or collapse of cellular machinery. There are indications that senescence is a highly ordered and directed process. Biale (1975) reports the coupling of synthetic and degradative processes, along with changes in membrane permeability, causes metabolic shifts associated with ripening. The presence of ethylene plays a central role in promoting fruit ripening. He reports that recent evidence points also to the involvement of other hormones as regulators of fruit ripening. The auxins, gibberellins and cytokinins appear to be native repressors of senescence. On the other hand, abscisins, like ethylene, are native stimuli of senescence-linked events. Chemical messengers, such as hormones, are thus very important components of biochemical control. Evidence indicates that auxins are endogenous antagonists of fruit ripening.

It has also become evident that inorganic ions play a role in the control of these physiological processes (Shear and Faust 1975). For example, the rapid conversion of sugar to starch occurring after sweet corn is harvested may be markedly delayed by pyrophosphate sprays after harvest (Haard and Salunkhe 1975). They believe that preharvest nutrition of the plant and postharvest application of minerals can dramatically influence stress response in harvested commodities and is far-reaching in implication. Texture changes obviously occur during storage. Pulpy fruits undergo extensive softening after harvest. Vegetables, such as beans and asparagus, toughen after harvest. The asparagus spear develops substantial fiber within a few hours after harvest unless controlled. Such changes occur at the cellular level and cell wall metabolism is obviously related to tissue toughening (Sterling 1975). The molecular organization of certain cell wall polysaccharides has been compared to the rheological properties of food systems by Sterling. The role and cellular metabolism of pigments also involve a wide variety of reactions and are involved in color change for a given commodity (Markakis 1975). Texture and color characterists of fruits and vegetables have been found useful as an index of internal storage disorders. In addition to the physiological deterioration, there are losses attributed to spoilage microorganisms. In many instances, physiological deterioration is a necessary prior condition to invasion by microorganisms, as will be discussed later. The toxic metabolities produced by filamentous fungi (mycotoxins) were discussed in Chapter 2.

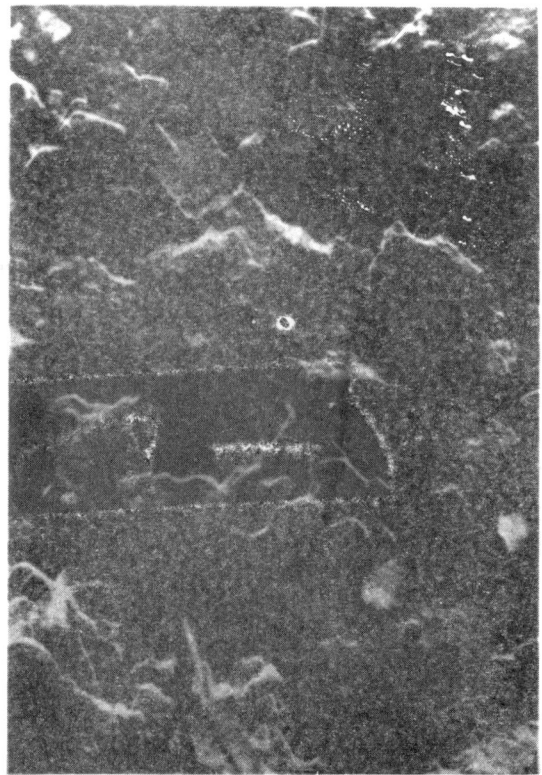

Courtesy of D.S. Skene

FIG. 3.3.   THE WAX PLATES ON THE FRUIT SURFACE OF APPLE (9000x APPROX)

Insight into the ripening process of fruit is gained also from observation of the ultrastructures. Apple surfaces become sticky as they approach ripeness and wax plates appear to soften and merge into one another (Skene 1963) (Fig. 3. 3).

PLANT PRODUCT STORAGE

Fresh fruits and vegetables maintain life processes during storage. So long as they are alive, they are able to resist the growth of spoilage organisms to some degree.

The storage disorders of fruits and vegetables and selected other foods have been studied in detail by the USDA. Excellent publications are available by writing and requesting Handbook No. 66, c/o Superintendent of Documents, Washington, D.C.

TABLE 3.2

TREATMENTS TO CONTROL POSTHARVEST DECAY OF FRESH PRODUCE

Crop	Treatment
Apple	sodium o-phenylphenate
Pear	thiabendazole benomyl
Banana	thiabendazole benomyl
Citrus fruits	sodium carbonate borax sodium o-phenylphenate thiabendazole benomyl 2-aminobutane (sec-butylamine) biphenyl 2,4-dichlorophenoxyacetic acid nitrogen trichloride fumigation
Grape	sulfur dioxide fumigation
Peach	hot water
Nectarine	2,6-dichloro-4-nitroaniline
Cherry	benomyl
Pineapple	sodium o-phenylphenate salicylanilide
Papaya	hot water
Mango	hot water, benomyl
Potato	hypochlorite
Vegetables	chloramines

Source: Eckert (1975).

Decay Control

Antimicrobial agents and senescence inhibitors can be used to delay the onset of spoilage in storage. Two methods of applying these agents are used: (1) a spray or dip in a solution or suspension in water/ wax formulations and (2) fumigations. Treatments currently being used are summarized in Table 3.2. An example of the type of equipment used is shown in Fig. 3.4.

Courtesy of FMC Corp.

FIG. 3.4. APPLYING FUNGISIDE TO CITRUS FRUITS BEFORE STORAGE

Heat Evolved by Living Tissues

Freshly harvested fruits, vegetables, grains and beans are alive These living tissues respire and energy is released in the form of heat The amount of heat released varies with the commodity and increases as the temperature of the storage chamber increases (Table 3.3).

It is also becoming clear that the storage life of living matter, within limits, varies inversely with the rate of respiration. It was postulated and then found that reducing the rate of respiration could prolong storage life.

Temperature of Cold Storage Rooms

Temperature control in storage rooms is most important. Variations (Table 3.4) from desired conditions may be most damaging. These variations can be prevented if the storage rooms are sufficiently insulated, have adequate refrigeration equipment, and the spread between the temperature of the refrigerating coils and the temperature

TABLE 3.3

HEAT EVOLVED IN THE RESPIRATION OF FRUITS AND VEGETABLES

Commodity	Btu[1] per Ton per 24 Hr		
	(0°C)	(4°C)	(15°C)
Apples			
Jonathan or Winesap	300 to 800	590 to 840	2,270 to 3,470
Beans, green	5,500 to 6,160	9,160 to 11,390	32,090 to 44,130
Broccoli	7,450	11,000 to 17,600	33,870 to 50,000
Cabbage	1,200	1,670	4,080
Carrots, topped	2,130	3,470	8,080
Celery	1,620	2,420	8,220
Corn, sweet	6,560	9,390	38,410
Onions	600 to 1,100	1,760 to 1,980[2]	—
Oranges	420 to 1,030	1,300 to 1,560	3,650 to 5,170
Peaches	850 to 1,370	1,440 to 2,030	7,260 to 9,310
Pears, Bartlett	660 to 880	—	8,800 to 13,200
Peas, green	8,160	13,220	39,250
Potatoes	440 to 880	1,100 to 1,760	2,200 to 3,520
Spinach	4,240 to 4,860	7,850 to 11,210	36,920 to 38,000
Strawberries	2,730 to 3,800	3,660 to 6,750	15,460 to 20,280
Sweet potatoes	1,190 to 2,440	1,710 to 3,350	4,280 to 6,300
Tomatoes			
Mature green	580	1,070	6,230
Ripe	1,020	1,250	5,640

Source: U.S. Dep. Agric., Agric. Handb. 66.
Note: See Appendix for metric conversions.
[1] Heat values obtained by multiplying respiration rate in kg of CO_2 per kilo, per hr by 220.
[2] At 10°C.

of the storage room is minimum. In a room with a desired temperature of 10°C, cooled by coils operating at −3°C, the air temperature may vary by a number of degrees. A room kept at 0°C with sufficient coils at a temperature of −3°C should vary in temperature by less than one degree. The difference between the temperature of the refrigerant and the temperature of the storage room is important in maintaining desired humidities.

Metabolism a Function of Temperature

The metabolism of living tissue is a function (Table 3.5) of the temperature of the environment. Living organisms have a temperature which is optimum for growth. Higher temperatures are injurious. Lower temperatures greatly retard metabolism. Low temperatures, near the freezing point of water, are effective in reducing the rate at which respiration occurs. Such temperatures have been found to be valuable in short term preservation of foods. For every 10°C the temperature is lowered, it may be estimated that the rate of a reaction will be halved. Storage of food at temperatures near 0° to 5°C may be anticipated to prolong the period in which foods may be stored. Not only is the respiration rate of such foods as fruits decreased, but the growth of many spoilage microorganisms as well is retarded.

TABLE 3.4

RECOMMENDED STORAGE TEMPERATURES, RELATIVE HUMIDITIES, APPROXIMATE STORAGE LIFE AND AVERAGE FREEZING POINTS OF FRUITS AND NUTS

Commodity	Storage Temperature (°C)	RH (%)	Approx Storage Life	Avg Freezing Point (%)
Apples	-2 to -1	85 to 90	[1]	-2
Apricots	-1 to 0	85 to 90	1 to 2 weeks	-2
Avocados	2 to 8	85 to 90	[1]	-2
Bananas	11 to 15	85 to 90	1 to 3 weeks	[2]
Berries				
Blackberries	-1 to 0	85 to 90	7 to 10 days	-1
Dewberries	-1 to 0	85 to 90	7 to 10 days	-1
Gooseberries	-1 to 0	85 to 90	3 to 4 weeks	-1
Loganberries	-1 to 0	85 to 90	7 to 10 days	-1
Raspberries				
Black	-1 to 0	85 to 90	7 to 10 days	-1
Red	-1 to 0	85 to 90	7 to 10 days	-0
Strawberries	-1 to 0	85 to 90	7 to 10 days	-1
Cherries	-1 to 0	85 to 90	10 to 14 days	[4]
Coconuts	1 to 2	80 to 85	1 to 2 weeks	-3
Cranberries	2 to 4	85 to 90	1 to 3 months	-2
Figs	-1 to 0	85 to 90	10 days	
Grapefruit	see text	85 to 90	[1]	-2
Grapes				
Vinifera	-2 to -1	88 to 92	3 to 6 months	-3
American	-1 to 0	80 to 85	3 to 8 weeks	-2
Lemons	12 to 14	85 to 90	1 to 4 months	-2
Limes	7 to 8	85 to 90	6 to 8 weeks	-1
Mangoes	10	85 to 90	15 to 20 days	-1
Nuts	0 to 2	65 to 70	8 to 12 months	[4]
Olives	7 to 10	85 to 90	4 to 6 weeks	-1
Oranges	see text	85 to 90	8 to 10 weeks	[5]
Papayas	7	85 to 90	15 to 20 days	-1
Peaches	-1 to 1	85 to 90	2 to 4 weeks	-1
Pears	-1 to 1	88 to 92	2 to 7 months	[6]
Persimmons	-1	85 to 90	2 months	-2
Pineapples				
Mature green	10 to 15	85 to 90	3 to 4 weeks	-1
Ripe	4 to 7	85 to 90	2 to 4 weeks	-1
Plums, prune	-1 to 0	85 to 90	3 to 8 weeks	-2
Pomegranates	-1 to 0	85 to 90	2 to 4 months	-2
Quinces	-1 to 0	85 to 90	2 to 3 months	-2

Source: U.S. Dep. Agric., Agric. Handb. 66.
[1] See text for varietal differences.
[2] Green bananas -1°C, ripe bananas -3°C.
[3] Sweet cherries -4°C, sour cherries -2°C.
[4] English walnuts -6°C, pecans -6°C, chestnuts -4°C, filberts -9°C, peanuts, shelled -10°C.
[5] Peel -2°C, flesh -2°C.
[6] Anjou -2°C. Bartlett -1°C.

There are generally considered to be three types of microorganisms: those with an optimum temperaure for growth at 55°C, called thermophiles; those with an optimum at 37°C, called mesophiles, which include most organisms pathogenic for man; and those with an optimum growth below 10°C, called psychrophiles. Temperature control (Table 3.6) then is a positive means of influencing the growth of microorganisms associated with foods, as previously noted.

TABLE 3.5

RATE OF RESPIRATION OF APPLES RELATED TO
TEMPERATURE OF ENVIRONMENT

Variety	Temperature (°C)	Mg CO_2 Evolved, per kg per hr
Grimes Olden	0	4.76
	4	8.34
	16	29.25
	34	67.82
Winesap	0	2.75
	4	5.30
	16	20.45
	34	33.95

Source: Smock and Neubert (1950).

CREATING ENERGY DEFICITS

Ice has been employed since early times to prolong the storage life of foods. Ice was one of the export items from the American Colonies to the tropics. It is the energy deficit of ice which has great utility. However, when all the energy deficit of ice is supplied, water remains; the temperature of the water-food substrate begins to come into equilibrium with the environment. Protective mechanisms (insulation) prolong this process. When the temperature of the water-food substrate reaches that where microorganisms can multiply, the food will deteriorate very rapidly. Ice is therefore used presently where its characteristics are of value. One feature of ice in cooling foods is that ice does not desiccate the food.

Creating Energy Deficits Mechanically

One of the important inventions of man is mechanical refrigeration. A simple diagram of an ammonia system is shown in Fig. 3.5. Both the first and second laws of thermodynamics are operative. Ammonia gas takes up energy as it is expanded. This heat is taken from the

TABLE 3.6

USEFUL TEMPERATURES

Temperature (°C)	Effect
100	Boiling point of water at standard pressure; short exposure kills most vegetative cells
93	Steam-cabinet sterilization of utensils for 5 min at this temperature
76	Hot water sterilization of utensils in 5 min
73	Cooking starts
71	High temperature, short-time pasteurization of milk (15 sec) upper range of growth of thermophiles, ropy bacteria killed
62	Holding pasteurization of milk (30 min)
58	Brucella (anthrax), typhoid-dysentery bacteria, and pathogenic streptococci killed (30 min)
56	Temperature for caustic solution in soaking-type washer for bottles
46	Temperature for water and cleaner for hand washing utensils
38	Temperature of cow's body
37	Temperature of man's body
26	In 12-24 hr bacteria may multiply 3000 times
21	In 12-24 hr bacteria may multiply 700 times
15	In 12-24 hr bacteria may multiply 15 times
12	Average well water temperature in northern belt for U.S.
10	In 12-24 hr bacteria may multiply 5 times
4	In 12-24 hr bacteria may double numbers
-1 to 1	Common cold storage temperaoure
0	Freezing point of water; very slow growth of most microorganisms
-17 to -23	Common frozen food storage
-28 to -34	Common food freezing temperature
-62	All water probably as crystals in frozen foods

atmosphere, or chamber, or environment. The expanded ammonia gas is then compressed. This requires energy to be exerted on the system. The compressed gas is now hot. Heat is removed from the compressed gas by means of running water or circulating air over the tubes containing the hot gas. The gas is liquefied. The cycle is then repeated. The gas is allowed to evaporate under controlled conditions, the gas takes up heat, the hot gas is compressed, the heat is removed, and the gas returns to the liquid state. With such a system accurate temperature control is possible. The refrigeration developed may be made to work directly on the food, or the refrigeration system may be employed to cool a brine which in turn is employed to cool. Many alternatives in refrigerants and systems are possible.

Once mechanical devices capable of imparting motion were invented, machines to power apparatus evolved. The ability to transform electrical energy to mechanical energy yielded such devices. With motors came pumps; pumps led to compressors. With motors, pumps and compressors, the elements of a refrigeration device were available. Applying cumulative knowledge of gases and vapors, with these mechanical devices, mechanical ice-making machines evolved. The process of invention took 300 years.

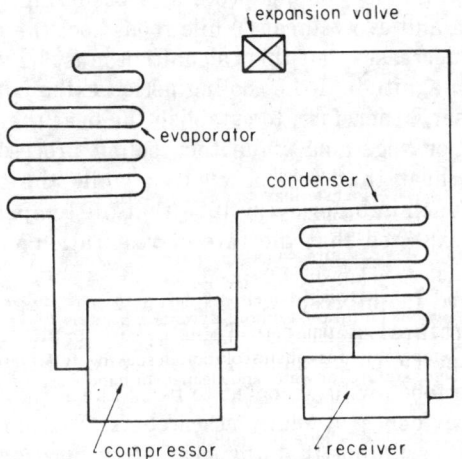

FIG. 3.5.   SCHEMATIC DIAGRAM OF A SIMPLE AMMONIA
REFRIGERATION SYSTEM

In 1595 Galileo made an accurate thermometer. In 1622 Boyle evolved his laws of the relation of volume to pressure in gases. In 1823 Faraday found that he could change ammonia gas to a liquid by applying pressure. In 1824 Carnot described his ideal heat cycle called the Carnot cycle, covering the expansion and compression of gases. In 1834 Perkins invented the compression refrigeration system we use today. In 1875 Linde introduced the ammonia refrigeration system (Fig. 3.6).

DETERMINING REFRIGERATION LOAD NEEDED

In order to establish the refrigeration requirement for a chamber of fruit or vegetable, certain information must be known. We must know the initial temperature of the food, the final storage temperature, the rate of respiration and heat evolved, the specific heat of the food, and the amount of food to be placed in the room.

If we could instantaneously lower the temperature of a food to the storage temperature, the heat load would be obtained by multiplying the specific heat for the food by the number of degrees the temperature will be lowered, by the weight of food. This value is generally reported in terms of the British Thermal Unit (Btu). One Btu is equal to the amount of heat required to raise one pound of water one degree Fahrenheit at or near the water's point of maximum density (1 Btu = 0.252 kg calories per °C).

The Btu value obtained from this calculation is called the sensible heat. In practice, the cooling period is extended and is not instantaneous. Therefore, while the cooling process is occurring, the fruits or vegetables live on and evolve heat. While foods cool the rate at which they evolve heat decreases, but this calcualtion must be included, as it may be of great magnitude if the cooling period extends for a week or more. It is necessary, therefore, to establish the heat that will be evolved by the fruit or vegetable while the cooling process occurs, and also, to add the heat that will be evolved while the food is being stored. As a rule of thumb, for every 10°C that the temperature is lowered, it may be expected that the rate of respiration will be halved, as noted earlier.

The heat evolved in the respiration of fruits and vegetables stored at several temperatures is given in Table 3.4 in terms of Btu per ton per 24 hr. (For metric conversions, see Appendix.) In this table it is assumed that the foods will be cooled to 0°C. These figures are based on experimental evidence by many researchers. The data given have been obtained by assuming that the heat of respiration is produced solely by the oxidation of glucose to carbon dioxide and water. Because of little other evidence, and the fact that this system is workable, these values are useful in establishing the refrigeration load for storage of commodities.

Courtesy of Vilter Corp.

FIG. 3.6. REFRIGERATION PLANT FOR COLD STORAGE CHAMBER

Compressors in a row service the storage chambers.

Specific Heat of Foods

The specific heat of food is required in calculating the refrigeration load. The specific heat for a food can be estimated from the equation:

$$\text{Specific heat} = 0.008 \,(\text{percent } H_2O \text{ in food}) + 0.20$$

For example, apples have 85% moisture. The specific heat of apples can be estimated to be $0.008 \times 85 + 0.20$, or 0.88. (Actual values are given in Table 3.3.)

Cold Injury to Fruits and Vegetables

Fruits and vegetables are susceptible to cold injury at above freez-

TABLE 3.7

COMMODITIES SUSCEPTIBLE TO COLD INJURY WHEN STORED AT
ONLY MODERATELY LOW TEMPERATURES

Commodity	Approx Lowest Safe Temperature (°C)	Character of Injury When Stored Between 0°C and Safe Temperature
Apples		
Certain varieties	1-2	Internal browning, soggy breakdown
Avocados	7	Internal browning
Bananas		
Green or ripe	13	Dull color when ripened
Beans (snap)	7-10	Pitting increasing on removal, russeting on removal
Cranberries	1	Low-temperature breakdown
Cucumbers	7	Pitting, water-soaked spots, decay
Eggplants	7	Pitting or bronzing, increasing on removal
Grapefruit	7	Scald, pitting, waterby breakdown, internal browning
Lemons	12-14	Internal discoloration, pitting
Limes	7	Pitting
Mangoes	10	Internal discoloration
Melons		
Cantaloups	7	Pitting, surface decay
Honey Dew	4-10	Pitting, surface decay
Casaba	4-10	Pitting, surface decay
Crenshaw and Persian	4-10	Pitting, surface decay
Watermelons	2	Pitting, objectionable flavor
Okra	4	Discoloration, water-soaked areas, pitting, decay
Olives, fresh	7	Internal browning
Oranges, California	1-2	Rind disorders
Papayas	7	Breakdown
Peppers, sweet	7	Pitting, discoloration near calyx
Pineapples		
Mature-green	7	Dull green when ripened
Potatoes		
Chippewa and Sebago	4	Mahogany browning
Squash, winter	10-12	Decay
Sweet potatoes	12	Decay, pitting, internal discoloration
Tomatoes		
Mature-green	12	Poor color when ripe, tendency to decay rapidly
Ripe	10	Breakdown

Source: U.S. Dept. Agric., Agric. Handb. 66.

ing temperatures. There is a wide variation in fruits and vegetables in their injury due to freezing. Some foods are injured with a slight exposure to freezing temperatures; other foods can be frozen and thawed several times without permanent injury. The freezing point of foods is not enough in itself to explain this problem. Tomatoes and parsnips, as examples, freeze at −1°C. Parsnips can be left in the soil over winter with no apparent damage. The first frost ruins tomatoes. Green tomatoes held at 2°C for three days do not ripen properly if warmed. Some varieties of potatoes may turn brown internally if held at less than 4°C (Table 3.7). Their sugar content also increases. In any event living tissues must be kept living if their food values and eating qualities are to be maintained by cold storage practices.

ANIMAL PRODUCT STORAGE

Microorganisms causing spoilage of fresh meat, poultry, eggs and most dairy products have optimum temperatures for growth at 20° to 35°C (Table 3.8). To obtain the maximum refrigerated shelf-life of these foods, microorganisms which grow best at room temperature as well as those that can also grow at refrigeration temperatures must be controlled. This general observation also conveys insight into potential sources of contamination and what to expect from sanitation practices and refrigeration of the environment and processing equipment. While Pseudomonads are the major cause of spoilage for most fresh meats, poultry, eggs and bulk tank raw milk, other bacteria may be the dominant spoilage organism (Table 3.9).

Meat

The two most critical factors influencing spoilage rate are temperature and initial level of contamination. Research on beef, poultry and pork show that small changes in temperature in the range of 0° to 7°C have an effect on refrigerated shelf-life and on the types of bacteria growing on fresh meat (Tompkin 1973). Coli-aerogenes and micrococci are found at very low levels in spoiled beef held at 4°C. These groups increase to moderate levels at 9°C and become a significant portion of the spoilage population at 15°C. Pseudomonads dominate the spoilage flora at all temperatures studied. With chicken meat, Pseudomonads dominate at 1°C and become of less signifiance as holding temperature increased. Acinetobacter and members of the family Enterobacteriaceae are of little consequence at 1°C but increased significantly with the higher temperatures of 10° and 15°C (Tables 3.10 and 3.11).

TABLE 3.8

REPORTED GENERATION TIMES FOR SOME PSYCHROTROPHS AT 0° to 32°C

	Generations Time (hr)						
	0°	2°-2.5°	4°-5°	10°	14°-15°	20°	30°-32°
Clostridium hastiforme	—	—	73	25.5	14.5	—	—
Clostridium 61	17	—	—	9	5	3.5	—
Bacillus W25	23	—	8.5	6	3	2.5	—
Bacillus W16B	24	—	11.5	7	2.8	2.5	—
Bacillus coagulans	—	24-30	—	—	—	—	—
Bacillus (Group A)	30	13-19	7.5-14	4-4.5	—	—	—
Pseudomonas 92	26.6	—	11.7	5.4	2.3	1.7	0.90
Pseudomonas 69	29.1	—	14.7	6.5	2.7	1.9	0.75
Pseudomonas fluorescens	—	—	8.2	3.5	2	1.5	1.2
Pseudomonas fluorescens (P-200)	35	13.7	—	4	2	1.5	0.85
Pseudomonas fluorescens	30.2	—	6.7	—	—	1.4	—
Pseudomonas fragi	11.3	7.7	5	2.6	—	1.1	—
Pseudomonas (21-3c)	20	9.2	7.3	3.2	2	1.3	0.77
Pseudomonas (1-3b)	10.3	8	7	2.7	2.3	1.6	0.93
Aerobacter aerogenes (Ps48)	37.7	—	12.2	4.1	2.2	1.3	0.8
Gram negative rod							
Aerobic	21.2	—	4.3	—	—	1.4	—
Anaerobic	23.4	—	9.8	—	—	0.6	—
Gram negative rod	20	—	7.6 (6°C)	1.9 (12°C)	—	—	0.78
Achromobacter #7	—	7.1	5.3	2.5	1.8	1.4	1.6
Achromobacter #438	—	7.8	6	2.5	1.8	1.3	1.5
Achromobacter #5	—	—	7.4	3.1	1.8	1.4	—
Pseudomonas #451	—	13.8	9.7	4.0	2.0	1.2	0.8

Source: Tompkin (1973).

TABLE 3.9

DISTRIBUTION OF 3880 SAMPLES OF MANUFACTURING
GRADE MILK ACCORDING TO BACTERIAL LEVELS

Standard Plate Count/ml	Canned Milk (%)	Bulk Milk (%)
100,000	22.3	41.5
100,000-1,000,000	30.9	30.1
1,000,000-3,000,000	11.1	10.3
3,000,000	35.7	18.1

Source: Tompkin (1973).

TABLE 3.10

RELATIONSHIP BETWEEN TEMPERATURE AND
LEVELS OF COLI-AEROGENES AND
MICROCOCCI IN SPOILED BEEF

	No./g at Time of Spoilage		
	4°C	9°C	15°C
Coli-aerogenes	2×10^4	1.5×10^5	1.6×10^7
Micrococci	3×10^3	6.3×10^4	1.9×10^6

Source: Tompkin (1973).

TABLE 3.11

RELATIONSHIP BETWEEN TEMPERATURE AND
TYPES OF BACTERIA CAUSING SPOILAGE OF CHICKEN MEAT

	Percentage		
	1°C	10°C	15°C
Pseudomonas sp.	90	37	15
Acinetobacter	7	26	34
Enterobacteriaceae	3	15	27
Streptococci	—	6	8
Aeromonas	—	4	6
Others	—	12	10

Source: Tompkin (1973).

Milk

The importance of temperature as well as time in holding milk cannot be overstated. Manufacturing-grade milk received into a plant with bacterial counts in the low millions requires little time to spoil further. This is especially true for milk received at 7°C or more, where the generation time may be only 8 hr or less compared to 12 hr at 5°C and 16 hr at 2°C. This is due to the psychrotrophic nature of the majority of bacteria in the milk.

In general, coliforms do not rapidly multiply in adequately refrigerated milk. Psychrotrophic coli-aerogenes bacteria may constitute 5 to 20% of the psychrotrophic microflora of farm bulk tank milk at time of retailing (Tompkin 1973). They may increase only 100- or 1000-fold in 3 days at 3° to 5°C, yet rarely attain dominance in milk at temperatures of less than 8° to 10°C.

Eggs

The spoilage pattern of eggs is also dependent upon the degree of refrigeration. Tests on raw albumen for 3 weeks at 5°, 10°, 15° and 20°C reveal that *Pseudomonas* sp. dominate at 5°C and decrease in significance as the holding temperature approaches 20°C (Table 3.12). Conversely, *Enterobacter liquifaciens* was not detected at 5°C and increased in significance as the holding temperature was raised to 20°C. In whole eggs held at 10°, 20° and 30°C, Pseudomonads dominate at 10°C and are of lesser importance as the temperature increases to 30°C. It is clear that temperature markedly influences the rate of spoilage of animal products and the types of bacteria. It is logical to assume that there may be a seasonal effect relative to the types of bacteria in the environment. Probably less than 1% of the microorganisms contaminating meat carcasses as they leave the killing floor and enter the cooler are of the psychrophilic type. Refrigerated processing areas may act as a selective breeding ground for the psychrotrophic bacteria most likely to cause spoilage of the refrigerated packaged product. One must also consider that environmental temperature can influence the types and numbers of bacteria present in a processing area.

The limiting factor for refrigerated shelf-life has been recontamination after pasteurization with psychrotrophs. Tompkin (1973) found that during recent years it has been profitable to use aseptic packaging conditions to exclude these psychrotrophs from specialty dairy products (e.g., whipping cream, half-and-half, nondairy creamer, chocolate milk). Aseptic packaging has extended refrigerated shelf-life of 6 weeks to 6 months.

The spoilage pattern for fresh meats, poultry, eggs and raw milk is, very similar. The degree of refrigeration not only markedly influences refrigerated shelf-life but also the types of bacteria causing spoilage. Advantage may be gained by minimizing contamination with the bacteria which occur in refrigerated processing areas which can cause spoilage of perishable refrigerated foods.

TABLE 3.12

RELATIONSHIP BETWEEN TEMPERATURE AND
TYPES OF BACTERIA CAUSING SPOILAGE OF ALBUMEN

| | Percentage | | | |
	5°C	10°C	15°C	20°C
Pseudomonas sp.	89	71	79	23
Enterobacteria[1]	—	10	14	77
Not identified	—	19	7	—

Source: Tompkin (1973).
[1] Probably *Enterobacter liquefaciens*.

General

Meats must be refrigerated at all stages from butchering to eating. If we desire to hold meat for a week we must rapidly get its temperature down below 4°C, otherwise the meat will begin to spoil. Most commercial meat we eat fresh is consumed within a week or ten days from the time it is slaughtered. Beef is sometimes "aged" for 4 or even 6 weeks. If we want to keep meat longer, we must use the more drastic methods of preserving (freezing, canning, irradiating, drying, curing).

The heat present in the freshly killed animal must be removed rapidly to avoid decomposition and retard losses in body weight. The period required to remove this heat is less than 24 hr. The temperature of the carcasses entering the cooler will be about 37°C. The thickest section of the carcass should reach a temperature of 0° to −1°C. Beef (Fig. 3.7) at this temperature can be stored for 8 days, lamb for 6, and veal for 6, before entering retail channels.

Calculation of Refrigeration Load

Meat.—The animal carcass body heat to be removed is sensible heat. Dead tissues do not respire. Calculation of refrigeration needs for meat chilling is then a multiplication of the specific heat of the flesh by the weight of meat to be cooled, multiplied by the temper-

TABLE 3.13

BACON FLAVOR SCORES AS INFLUENCED BY TEMPERATURE
AND TIME OF STORAGE

Storage Time (Weeks)	Temperature		
	(-2°C)	(2°C)	(7°C)
0	2.0[1]	2.0	2.0
2	2.0	2.0	1.6
4	2.0	1.6	1.2
6	1.6	1.2	1.0

Source: Brown and Schmucker (1960).
[1] Flavor scoring: Above average, 3.0; average, 2.0; below average, 1.0.

ature drop and divided by the conversion factor. This will yield the
tons of refrigeration needed to chill the carcasses. It is assumed that
this cooling will be accomplished in 24 hr. Heat losses must be considered for the storage room, lights, etc., in establishing the total
refrigeration needs for the storage chambers.

Courtesy of American Meat Industn

FIG. 3.7.   AGING BEEF CARCASSES

Aging improves the quality of beef. The subsidence of rigor mortis is essential. Further aging
tenderizes the flesh.

For quick marketing of fresh meat, temperatures near 0°C are used in storage rooms. For delayed marketing, meat is frozen and held at −17°C or lower (Table 3.13).

The humidity of the storage room is a factor requiring attention. The carcass will lose nearly 2% of its weight in the chilling process, due to loss of moisture. Vapors rising from a warm animal carcass are principally water vapor. If the relative humidity of the storage room is above 90% the meat will mold. If the relative humidity is less than 90% the carcass will lose excessive weight.

Mold and slime formation on chilled carcasses can be retarded by using ultraviolet lights in the storage chamber. Ultraviolet light at a wavelength of 2700A has germicidal power. If the carcass is exposed thoroughly to the germicidal rays, the meat will not mold. However, the exterior fat may turn rancid sooner than would be the case if it is not exposed to ultraviolet light.

Fish.—Fresh fish is more perishable than meat. Storage of fish in ice slows the process, but the flesh becomes soft and flabby. The bright color of the skin dulls. The sweet fresh fish flavor is lost. Unpleasant, strong odors develop. When caught at sea, fish is packed quickly into crushed ice, and iced again along the line from the sea to our tables. Soft fish like the whiting will not keep long under any circumstance. If cod is kept at 0°C it can be held for three weeks. For inland markets much of our fish is frozen and distributed in that condition.

Shellfish may be kept for a week if cooled to ice temperatures. Shrimp, lobster and crab are very perishable, keeping in storage for a few days at the most. Lobsters are often held alive in sea water. Immersion in fresh water causes their quick death.

Egg Storage.—Eggs should be stored at the lowest temperature possible yet not permit the interior of the egg to solidify. If the interior solidifies the resulting expansion may cause shells to crack. This results in difficulties as eggs with firm, thick albumen solidify at a higher temperature than thin-albumen type eggs. Shell characteristics also affect the storage quality of eggs, as thick shells withstand the solidification better than thin-shelled eggs. It is solidify at a higher temperature than thin-albumen type eggs. Shell characteristics also affect the storage quality of eggs, as thick shells withstand the solidification better than thin-shelled eggs. It is considered that −1°C is the ideal storage temperature for eggs. The room must be maintained at constant temperature for best results. A relative humidity betweε.ı 82 and 85% is generally considered optimum for eggs. Lower humidities result in desiccation, and therefore an enlargement of the air cell, which is undesirable (Table 3.14). Eggs may pick up odors during storage and therefore should not be stored in the same room with other commodities.

Recommendations for cold storage of selected foods are presented in Tables 3.15 and 3.16.

TABLE 3.14

QUALITY LOSSES IN SHELL EGGS ON STORAGE
AT DIFFERENT TEMPERATURES. LOSSES IN MOISTURE
AND THINNING OF WHITES DECREASE EGG QUALITY

Storage (Months)	Grade A Eggs Remaining (%)		
	-1°C	7°C	15°C
0	100	100	100
3	80	50	0
6	70	0	0

TABLE 3.15

RECOMMENDED COLD STORAGE TEMPERATURES FOR SELECTED FOODS

Food	Temperature (°C)
Dairy Products	
Butter[1]	1
Cheese (cream)	-1 to 2
Cheese (soft)	0 to 2
Condensed milk (bulk)	1 to 4
Cream (fresh)	0 to 1
Margarine	0 to 1
Milk (fresh)	0 to 1
Milk (evaporated)	2 to 4
Storage of Fish and Fish Products	
Fresh fish	0
Fish (dried)	1 to 4
Oysters	0 to 1
Storage of Miscellaneous Products	
Beer	0 to 4
Chocolate	3 to 5
Cider	0 to 2
Honey	4 to 7
Maple syrup	4 to 7
Molasses	4 to 7
Olive oil	1 to 4
Sauerkraut	1 to 3
Wine	4 to 7

[1] If butter is to be stored for longer than 2 or 3 weeks, it should be held at -17° to -23°C.

TABLE 3.16

RECOMMENDED STORAGE TEMPERATURES, RELATIVE HUMIDITIES,
APPROXIMATE STORAGE LIFE AND FREEZING POINTS OF VEGETABLES

Commodity	Storage Temperature (°C)	RH (%)	Approx Storage Life	Avg Freezing Point °C
Artichokes, Jerusalem	0	90 to 95	2 to 5 months	-2
Asparagus	0	90 to 95	3 to 4 weeks	-1
Beans				
Green	7	85 to 90	8 to 10 days	-1
Lima—unshelled	0	90 to 95	2 to 4 weeks	-1
shelled	0	90 to 95	15 days	-1
Beets				
Topped	0	90 to 95	1 to 3 months	-2
Not topped	0	90 to 95	10 to 14 days	-2
Broccoli, Italian	0	90 to 95	7 to 10 days	-1
Brussels sprouts	0	90 to 95	3 to 4 weeks	1
Cabbage	0	90 to 95	3 to 4 months	-1
Carrots				
Topped	0	90 to 95	4 to 5 months	-1
Not topped	0	90 to 95	10 to 14 days	-1
Cauliflower	0	90 to 95	2 to 3 weeks	-1
Celery	-1 to 0	90 to 95	2 to 4 months	-1
Corn, green	-1 to 0	90 to 95	4 to 8 days	-1
Cucumbers	7 to 10	90 to 95	10 to 14 days	-1
Eggplants	7 to 10	90 to 95	10 days	-1
Endive	0	90 to 95	2 to 3 weeks	-1
Garlic	0	70 to 75	6 to 8 months	-3
Horseradish	0	90 to 95	10 to 12 months	-3
Leeks, green	0	85 to 90	1 to 3 months	-1
Lettuce	0	90 to 95	2 to 3 weeks	-1
Melons (ripe)				
Watermelon	2 to 4	80 to 85	1 to 2 weeks	1
Muskmelon	4 to 10	80 to 85	10 to 14 days	2
Honeydew	4 to 10	80 to 85	2 to 4 weeks	3
Casaba, Persian	4 to 10	80 to 85	4 to 6 weeks	.
Mushrooms	0	85 to 90	5 days	-1
Okra	10	85 to 95	2 weeks	-1
Olives	7 to 10	85 to 90	4 to 6 weeks	-1
Onions	0	70 to 75	6 to 8 months	-1
Peas, green	0	85 to 90	1 to 2 weeks	-1
Peppers, green	7	85 to 90	8 to 10 days	-1
Potatoes	3 to 7	85 to 90	6 to 9 months	-1
Pumpkins	10 to 12	70 to 75	2 to 6 months	-1
Rhubarb	0	90 to 95	2 to 3 weeks	-3
Rutabagas	0	90 to 95	2 to 4 months	-1
Spinach	0	90 to 95	10 to 14 days	-1
Squashes				
Summer	4 to 10	85 to 95	2 to 3 weeks	1
Winter	10 to 12	70 to 75	4 to 6 months	-1
Sweet potatoes	12 to 15	80 to 85	4 to 6 months	-1
Tomatoes				
Ripe	4 to 10	85 to 90	7 to 10 days	-1
Mature green	12 to 21	80 to 85	3 to 5 weeks	-1

Source: U.S. Dep. Agric., Agric. Handb. 66.
[1]Rind (-1°C) flesh (-1°C).
[2]Rind (-2°C) flesh (-1°C).
[3]Rind (-1°C) flesh (-1°C).

EFFECT OF COLD STORAGE ON QUALITY

An unrefrigerated fruit or vegetable usually spoils rapidly and soon has little food value for man. If similar fruits and vegetables are held temporarily in cool storage, life processes are retarded, but the net result is a longer period in which the food is acceptable for man to eat (Tables 3.17 and 3.18). Some of the life activity is used by storing the tissues. It is not to be expected that a fruit after seven months storage will be identical to a freshly harvested fruit. If the temperature and humidity conditions optimum for the storage of a fruit or vegetable are adhered to, there will be ample time for cold-stored commodities to be marketed through usual channels. Very perishable fruits and vegetables, those having a short period during which they keep in cold storage, must be consumed shortly after being removed from storage rooms (Tables 3.19, 3.20 and 3.21). It cannot be expected that cold storage will make extremely perishable foods nonperishable.

Occasionally it is necessary to refrigerate fruits and vegetables in common storage chambers. There can be substantial cross-transfer of odors. Apples should not be stored with celery, cabbage, potatoes or onions. Celery and onions damage each food's quality. Citrus fruits take up most strong odors. Apple and citrus odors are readily transferred to dairy products. Eggs stored with fish or certain vegetables lead to off-flavored eggs.

Preserving Foods in a Microenvironment

It is possible to preserve our foodstuffs if we alter the environment. As an example, the cold storage chamber is a modified environment

TABLE 3.17

LOSSES IN VITAMIN C IN SELECTED VEGETABLES ON COLD STORAGE

Produce	Storage Conditions		Losses (%)
	Days	Temp (°C)	
Asparagus	1	1	5
	7	0	50
Broccoli	1	7	20
	4	7	35
Green beans	1	7	10
	4	7	20
Spinach	2	0	5
	3	1	5

TABLE 3.18

CHANGES IN CAROTENE CONTENT OF SWEET POTATOES
STORED AT 10°C FOR FOUR MONTHS

Variety	At Harvest mg per 100 g Wet Basis	After Storage
Maryland Golden	4.7	5.9
Nance Gold	5.4	5.7
Heartogold	6.7	7.4

in which we place foods. The storage chamber is therefore a large package. The concept of microenvironment is the realm of individual packages for foods. Prepackaging perishable foods is an application of controlled environment (Fig. 3.8).

The ability to control moisture losses inexpensively in fresh meats made major improvements in the economics of distribution and marketing of meats. This development had far-reaching implications, even to the relocation of livestock yards and meat processing plants further away from points of consumption.

Food packaging is an area of specialization in itself. A package has many functions. Four of the most important considerations of a package are the protection of the food, the economy of the package, the convienence of the package, and its appearance.

There are a number of factors to be considered for a package.

(1) Product properties: Tendency of the food to gain or lose moisture; free oil or fat content of the food; tendency of food to lose volatile flavors or pick up foreign odors; caking tendencies at different temperatures and different moisture contents; susceptibility of food to spoilage by light; susceptibility of food to spoilage by atmospheric oxygen; susceptibility of food to infestation by insects; particle size of food and sifting considerations.

(2) The conditions of environment which alter the package and contents must receive attention. These include: relative humidity of storage chamber, temperature, ventilation, pressure, warehousing and transportation problems.

(3) The packaging material must lend itself to being used with processing equipment. The packaging material must have specifications regarding tensile strength, tearing resistance, softness, ability to make dead folds, moisture content, thickness, heat sealing abilities, gluing requirements, vapor transmission factors, and a host of other considerations.

TABLE 3.19

RELATION OF TEMPERATURE TO FIRMNESS OF BARTLETT PEARS
IN COLD STORAGE (MEASURED BY PRESSURE TEST)

Temperature (°C)	0	2	4	8
1	25[1]	19	12	9
4	25	7	4	2

[1] Pressure test: pounds.
Note: See Appendix for metric conversions.

TABLE 3.20

THIAMIN, RIBOFLAVIN, AND NIACIN CHANGES IN
COLD STORED PORK AND PORK PRODUCTS

Product	Days Storage (3°C)	Retention		
		Thiamin (%)	Riboflavin (%)	Niacin (%)
Pork Loin	11	97	100	72
Ground pork	14	94	104	97
Cured pork	14	95	105	90

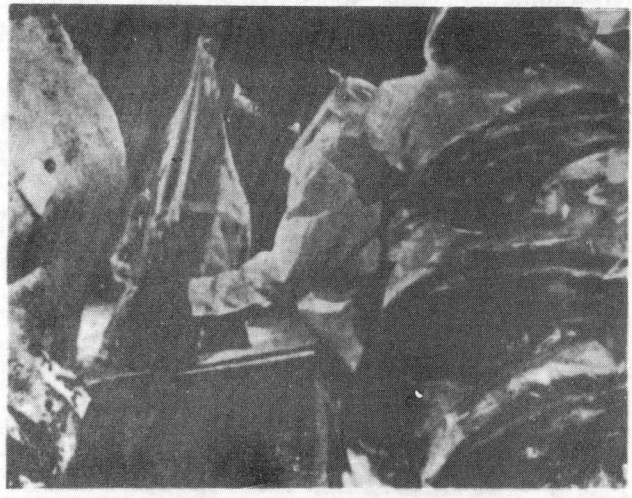

Courtesy of Food Processing

FIG. 3.8.   PREVENTING MOISTURE LOSSES DURING THE SHIPMENT
OF BEEF

Packaging meat decreases moisture losses during shipment, but care must
be exercised in controlling microbial attack.

TABLE 3.21

LOSS OF SWEETNESS IN SWEET CORN DURING STORAGE

Storage	Loss in Total Sugars (%)	
(hr)	0°C	20°C
24	8.1	25.6
48	14.5	45.7
72	18.0	55.5
96	22.0	62.1

Packaging Materials Tests Which May Be Performed.—Weight per unit of area, weight per 100 units, caliper or thickness, specific gravity or density, bursting strength, tearing resistance, tensile strength, folding endurance, stiffness, moldability, elongation, porosity, transmission of air and gases, transmission of water vapor, transmission of organic vapors, oil and grease resistance, water absorption (penetration), moisture content, wax content, surface wax, melting and softening range, pressure and heat blocking, sealing strength, stability (heat and light), resistance to chemicals, printability, gloss, smoothness, opacity, transparency, and color are criteria of quality for packaging materials (Fig. 3.9).

Formed Container Tests.—Compression tests, standard method or offset method, drop test, impact and vibration test (Fig. 3.10) may be used to determine the quality of the formed containers.

Knowing the specifications for a food package, the check list for three types of packaging materials (Table 3.22) provides a guide to making a decision as to which material to select.

U.S. Dept. of Agriculture Photo

FIG. 3.9.   PALLET OF LOAD OF VENTILATED FULL-
TELESCOPE-TYPE FIBERBOARD CARTONS. SIMILAR
CARTONS ARE USED FOR SHIPPING MANY KINDS OF
PRODUCE

Note stacking pattern to allow air circulation.

TABLE 3.22

PARTIAL CHECK LIST FOR FLEXIBLE PACKAGING FILMS

Qualities	Film I	Film II	Film III
Water-vapor transmission rate	Fair	Excellent	Good
Grease resistance	Poor	Excellent	Excellent
Resistance to passage of O_2 and CO_2	Good	Excellent	Excellent
Resistance to passage of organic vapors	Fair	Excellent	Excellent
Strength	Good	Fair	Excellent
Dimensional stability	Good	Good	Good
Stiffness	Poor	Good	Poor
Tear fold	Poor	Excellent	Poor
Type of seals	Heat	Good	Heat
Strength of seals	Excellent	Fair	Excellent
Appearance of film	Good	Good	Good
Printability	Fair	Good	Fair
Price	Satisfactory	Satisfactory	Too high
Decision	Reject	Accept	Reject

Courtesy of QM Food and Container Institute

FIG. 3.10.   COMPRESSION TEST ON CONTAINERS

Containers are designed to withstand abuse during shipment and storage.

Disorders of Stored Foods

Our fruits, vegetables, eggs, meats and dairy products may be harvested, gathered or slaughtered some time prior to going into refrigerated storage. Food deterioration may have already begun. A food will not improve in quality if harvested, gathered or slaughtered in a decayed condition. Only sound foods should be given the attention required for successful cold storage. Even under optimum conditions, storage in a cool chamber only retards food deterioration (Fig. 3.11).

A number of diseases and disorders of foods occur in cold storage chambers. A description of such conditions and control measures

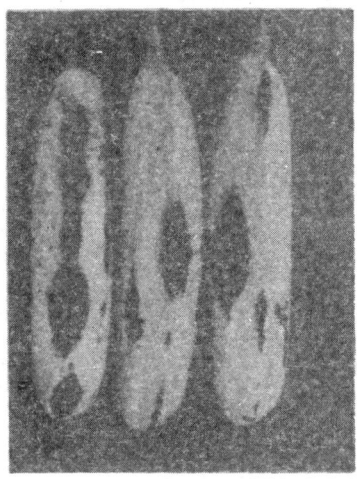

Courtesy of Eckert et al.

FIG. 3.11.   (TOP) CROWN ROT OF BANANA HAND; (BOTTOM) WOUND ON BANANA FINGERS CAUSED BY *GLOEOSPORIUM MUSARIUM*

that have been found for these conditions are described in the selected references, given below.

STORAGE OF GRAINS

Experts agree that grain stores better and more cheaply than other major foods. Nevertheless, storage is not an end in itself. Storage is usually repeated in transporting grain from producer to processor to consumer. As a minimum, grain must be stored from one harvest to the next; and prudence demands additional carryover as insurance against a subsequent crop of poor quality. Storage of grain occurs on farms, at collection points serving a number of farms, and at terminals where grain is processed or moved forward. The characteristics of the system are that grain is accumulated in larger and larger lots.

Once grain has been processed the procedure is reversed. Wheat flour goes to bakeries and bread to consumer outlets, etc. At each step the product is distributed in smaller and smaller amounts. Quality factors in grain that can be readily controlled are: variety (i.e., inherited qualities), soundness, admixtures, and moisture content. Variety has little effect on storage per se; within a given class of grain all varieties tend to store equally well. By contrast, the remaining three quality factors do affect storage. Grain stores better when it is sound, clean and especially when it is dry (Anderson 1973).

Moisture presents a special problem. High moisture content increases storage hazards. Fortunately moisture can be determined accurately and grain can be dried without damage to its quality. Anderson reports that grain can be damaged during drying. The milling and baking qualities of wheat, the milling quality of corn, the malting quality of barley or the germination of seed grains can be damaged when the kernels become too hot during drying. Drying in traditional equipment costs less per unit of moisture removed with newer techniques. Grain can be dried more rapidly at high temperatures, precisely the conditions that damage the quality of grain. A trade-off occurs between costs and quality, which requires careful attention.

Technological changes are occuring steadily in the grain industry. Examples from the milling and baking industries are found in α-amylase activity and protein content. The importance of controlling α-amylase is seen when mechanical work during mixing dough is used to reduce the fermentation period in this system. Bread of acceptable quality can be made from flour of lower protein content. When the protein content is reduced, the moisture content of the bread is more difficult to control. This difficulty can be offset by mechanically damaging the starch granules during milling so that the starch will

carry additional moisture. However, damaged starch is more susceptible to attack by α-amylase, with consequent deterioration of the
crumb texture of the loaf.

The α-amylase in grain increases rapidly if the grain becomes wet
after harvest and starts to sprout. It can increase before visual
evidence of sprouting is apparent. The first step in germination is the
development and activation of enzymes. Their function is to transform the contents of the kernel, making them available to nourish
form the contents of the kernel, making them available to nourish
the new plant until the emerging roots and shoots take over the task.
This first step occurs within the kernel and cannot be observed
(Anderson 1973). Considerable control of α-amylase can be exercised
by visual examination for signs of incipient sprouting or even
weathering.

Protein content of flour can be controlled by blending wheats of appropriate protein contents. The primary aim during storage is simply
to prevent deterioration in quality. This is done indirectly through
control of moisture and air movements and by preventing attack
by microorganisms, insects and pests.

In most countries, mills, bakeries, macaroni factories, malting
houses, breweries and feed plants generally have their own laboratories and use a wide variety of tests to control processes and
products. Examples of such analyses for fresh and 7-year-old wheats
are shown in Table 3.23.

Sinha and Muir (1973) have summarized existing knowledge of
grain storage. They report that the choice of the site for the granary,
its design, and the materials used in its construction to a great extent
determine whether certain harmful organisms, including birds and
rodents, will be significant pests. They found that structural requirements for grain storage generally vary according to climate, crop
type, and dominant pest species of a country or geographical area.
They found that most building structures usually reduce pest infestation if they minimize heat uptake from the environment and
maximize heat and moisture loss from storage to environment.

Bulk Storage

Sinha (1973) describes a grain bulk or storage unit as a man-made
ecological system in which living organisms and their nonliving
environment interact. Deterioration results from interactions among
physical, chemical and biological variables. Since grain occupies
about 60% of the physical space in a bulk, the most important
living organism is the grain itself. However, the grain and the grain
bulk have several physical (e.g., porous nature) and biological (e.g.,
respiration) attributes, the importance of which depends mainly on

TABLE 3.23

CHEMICAL QUALITY DATA FOR FRESH UNHEATED AND 7-YEAR OLD
WHEAT FROM A WHEAT BULK

Property	Fresh	Aged
Wheat		
Grade	2.0	2.0
Bushel wt (kg)	28.5	28.0
Flour yield (%)	74.4	74.2
Protein (%)	12.8	13.1
Fat acidity (mg)	13.4	29.0
Flour		
Wet gluten (%)	38.0	36.6
Ash (%)	0.49	0.51
Color, units	1.2	1.9
Diastatic activity (mg)	193	180
Baking absorption (%)	58.4	59.1
Gassing power (mm)	365	335
Loaf Volume (method)		
AACC, zero bromate (cc)	660	725
AACC (cc)	800	777
Remix (cc)	805	872
Remix, zero bromate (cc)	690	785
Extensogram		
Length (cm)	23	19
Height, BU (5 cm)	225	345
Height, BU (max)	470	565
Area (cm^2)	134	137

Source: Sinha (1973).

the surrounding medium.

The environment of the grain includes physical variables (such as temperature, carbon dioxide, oxygen, moisture) and an array of organic compounds which are the by-products of biological activity in the bulk.

The major biological variables other than grain would include: microorganisms, such as fungi, actinomycetes and bacteria; arthropods, such as insects and mites; and vertebrates, such as rodents and birds (Fig. 3.12). An ecological kinship develops, supported by the complex processes which result in deterioration of the quality of the grain. Although this deterioration is usually slow, rapid and complete loss of the grain can occur if certain combinations of variables are maintained in an undisturbed bulk.

Experts agree that the main physical properties whose interplay influences deterioration of the grain are: the flow properties and tendency towards layering; intergranular spacing or porosity; absorption, adsorption, and desorption capacities for various vapors and gases (sorption); heat conductivities; and thermal capacity.

It is now generally agreed that above 70 to 75% RH, small

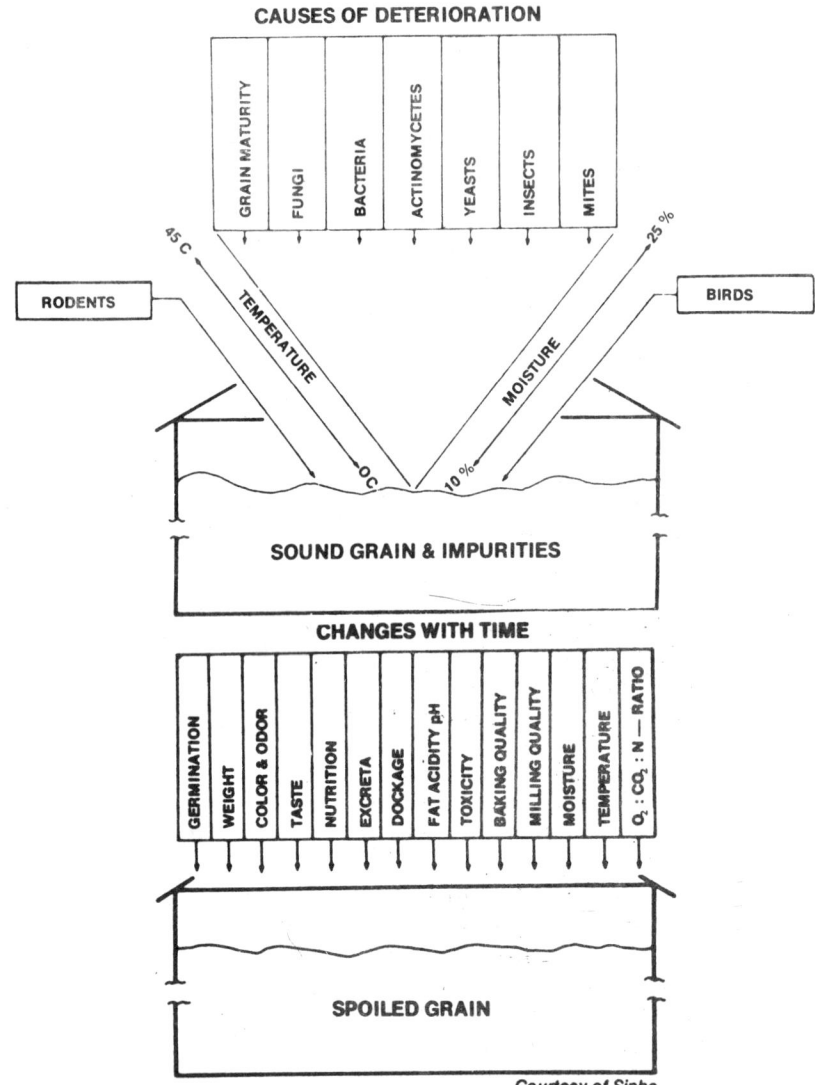

FIG. 3.12.   A DIAGRAMATIC REPRESENTATION OF INTERRELATIONS AMONG THE GRAIN
BULK, ORGANISMS, AND THEIR ABIOTIC ENVIRONMENT IN THE SPOILAGE OF STORED
GRAIN

increases in the relative humidity of the atmosphere can result in a
large rise in grain moisture content. Charts showing moisture content/
relative humidity equilibrium relationships of various grains have
been published by Smith (1969) and others.

Temperature

The atmospheric temperature of grain and the intergranular air temperature are crucial for safe and prolonged storage. Heat from external sources penetrates slowly into the grain bulk. Daily temperature fluctuations rarely affect the grain below a few centimeters from the surface. Reports indicate that at about 300 cm (8 to 10 ft) depth the effects of the summer and winter temperature cycles are minimal and, if at all noticeable, are delayed by 2 to 3 months. The metabolic heat produced exclusively by dry grain is about 1×10^{-7} cal $\sec^{-1}$ cm^{-3} and by damp grain approximately 1.3×10^{-5} cal $\sec^{-1}$ cm^{-3} (Muir 1970). The amount of heat produced by fungi, insects and other organisms invading the grain is considerably higher. Careful measurement and control of temperature in grain bulks is important for proper storage.

In considering the storage of grain, the important points to be remembered are: (1) mites do not develop below 5°C, nor insects below 15°C; (2) most storage fungi do not develop below 0°C; and (3) the effect of temperature on an organism is correlated with the amount of moisture present, because rise in temperature corresponds with decrease in the relative amount of moisture in the atmosphere (Sinha 1973).

Moisture

Moisture content is critical in controlling the development of bacteria, actinomycetes, yeasts, fungi, mites, and insects in stored grain, as noted previously. The important points to remember for safe storage are: (1) moisture contents of grain below 13% arrest the growth of most microorganisms and mites; (2) moisture contents below 10% limit development of most stored grain insect pests; and (3) moisture contents within a grain bulk are rarely uniformly distributed and are changeable from season to season and from one climatic zone to another. Routine measurement of moisture within a grain bulk is required.

Weather

Weather obviously influences stored grain bulks. However, diurnal and yearly variations in temperature affect the deeper regions of large grain bulks slowly. Changes in barometric pressure affect the volume of intergranular air. With pressure changes, the grain bulk "breathes" by taking in fresh air and releasing intergranular air. An external annual pressure range of 950 to 1050 mb might affect an aeration of about 1/10 of the surface of the bulk (Howe 1965). Interrelationships among climatic variables, grain bulks and insect outbreaks have been shown.

Respiration

The respiration of grain and the microflora it contains are also important in understanding deterioration. Both the grain and its microbial associates respire by the same physiological principle; the latter plays a greater role in the deterioration of the grain (Christensen and Kaufmann 1969). Respiration in grain bulks, or metabolism to produce energy, occurs either in the presence or in the absence of oxygen. In aerobic respiration, complete oxidation yields carbon dioxide, water and energy (647 Kcal), whereas in anaerobic respiration hexose is incompletely decomposed forming carbon dioxide, ethyl alcohol and energy. The resulting effects are the loss of weight, grain in the moisture content, rise in the level of carbon dioxide in the air, and a rise in the temperature of the grain.

Chemical Aspects

Oxygen is the most important variable influencing grain storage because fungi, mites and insects all require free oxygen for their development. Grain can be stored with minimum quality loss if oxygen is excluded (as in air-tight storage) or manipulated by the structural modification of the granary.

Various studies show that the quality of stored wheat in various regions of the world vary considerably. Part of the variation is attributed to storage conditions. Chemical changes in wheat (about 12% moisture content) originating in Canada and stored for 8 years under storage conditions in Britain were found to have both crude and salt-soluble protein virtually unchanged. Total fat increased slightly and fatty acid increased significantly. The physical characteristics of the dough showed little change during the period (Sinha 1973). He also reported on changes in the chemical quality (including baking and milling) of rice stored for up to 100 years in Japan. The grain was still edible; fat, glucose and dextrin had decreased; vitamin B complex was reduced to 8%; and lipase activity fell to 67% of that of the newly harvested rice.

Cereal grains and oilseeds release many identifiable volatile components, giving rise to characteristic aromas of food prepared from them. Recent gas chromatographic studies of maize, rye, wheat, rapeseed, and other cereals reveal that the collected vapors from

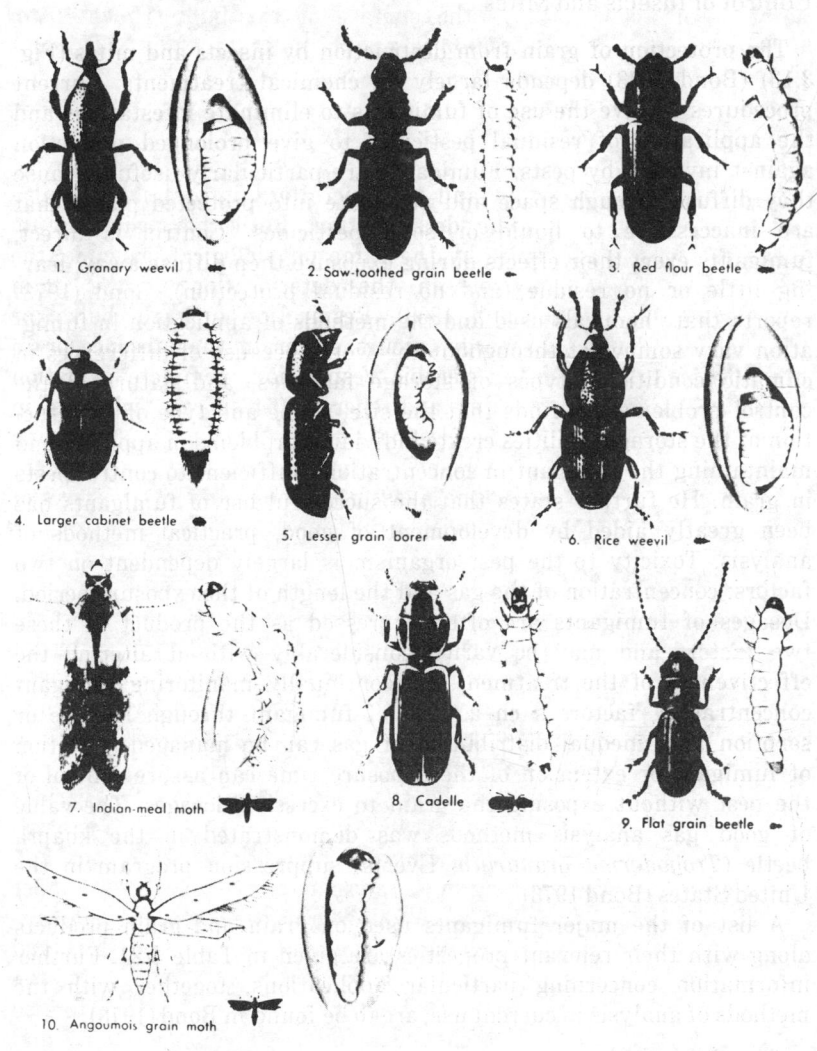

1. Granary weevil
2. Saw-toothed grain beetle
3. Red flour beetle
4. Larger cabinet beetle
5. Lesser grain borer
6. Rice weevil
7. Indian-meal moth
8. Cadelle
9. Flat grain beetle
10. Angoumois grain moth

Courtesy of Extension Entomologists,
North Central States, USDA

FIG. 3.13.   PRINCIPAL STORED GRAIN INSECTS

grain samples infested by various species of storage insects can be identified.

Control of Insects and Mites

The protection of grain from destruction by insects and mites (Fig. 3.13) (Bond 1973) depends largely on chemical treatments. Current procedures involve the use of fumigants to eliminate infestations and the application of residual pesticides to give prolonged protection against invasion by pests. Fumigants are particularly useful because they diffuse through space and penetrate into protected places that are inaccessible to liquid or solid pesticides. Control is direct; fumigants exert their effects during exposure then diffuse away, leaving little or no residue (and no residual protection). Bond (1973) reports that chemicals used and the methods of application in fumigation vary somewhat throughout the world because of differences in climatic conditions, types of storage facilities, and nature of the control problems. He finds that the size, shape and type of construction of the storage facilities create individual problems in applying and maintaining the fumigant in concentrations sufficient to control pests in grain. He further states that the successful use of fumigants has been greatly aided by development of good, practical methods of analysis. Toxicity to the pest organism is largely dependent on two factors: concentration of the gas and the length of the exposure period. Dosages of fumigants are often expressed as the product of these two factors and may be varied considerably without altering the effectiveness of the treatment. By continually monitoring fumigant concentration, factors such as loss of fumigant through leakage or sorption and unequal distribution of gas can be managed. Addition of fumigant or extension of the exposure time can assure control of the pest without exposing the grain to excessive dosages. The value of good gas analysis methods was demonstrated in the khapra beetle (*Trogoderma granarium* Everts) suppression program in the United States (Bond 1973).

A list of the major fumigants used on grain and grain products along with their relevant properties are given in Table 3.24. Further information concerning particular applications, together with the methods of analysis in current use, are to be found in Bond (1973).

SUMMARY

It has been half a century since the first publications of information on postharvest handling of perishable commodities were written by members of the USDA. With the laboratory equipment then available they studied the chemistry of food materials in storage and interpreted their findings in procedures for maintaining the quality of fresh foods.

TABLE 3.24

PRINCIPAL FUMIGANTS USED ON GRAIN AND GRAIN PRODUCTS

Name	Molecular Weight	Boiling Point (°C)	Flammability by Vol in Air (%)	Remarks
Methyl bromide	94.95	3.6	Nonflammable	General fumigant
Phosphine	34.04	-87.4	Highly flammable	General fumigant; good penetrating ability
Ethylene oxide	44.05	10.7	3-80	Controls microorganisms but affects seed germination
Hydrogen cyanide	27.03	26.0	6-41	General fumigant; safe on seeds; poor penetrating ability
Ethylene dibromide	187.88	131.0	Nonflammable	Mainly spot fumigant admixed with other fumigants
Carbon disulphide	76.13	46.3	1.25-44	Usually mixed with nonflammable compounds
Chloropicrin	164.39	112.0	Nonflammable	Safe with seeds; mixed with methyl bromide as warning agent
Acrylonitrile	53.06	77.0	3-17	Spot fumigant often mixed with carbon tetrachloride
Ethylene dichloride	98.97	83.0	~16	Usually mixed with carbon tetrachloride
Carbon tetrachloride	153.84	77.0	Nonflammable	Used chiefly in mixtures with flammable compounds to reduce fire hazard and aid distribution
Dichlorvos	221	120/14 mm	Nonflammable	Does not penetrate materials; good only in open space; very toxic to most stored product insects

Source: Bond (1973).

There are now sophisticated analytic instruments available to expand those early studies. With these new tools to study cellular and intercellular activities, there is great promise for improving procedures for maintaining and prolonging the storage stability of fresh foods, while at the same time retaining those desirable qualities, nutritional values and acceptability of fresh foods as never before possible. Great strides have been made; yet they only tend to stimulate further activity and investigation.

A further step has been taken which promises to aid future development. This step is the use of modified atmospheres to prolong the storage life of fresh foods, which is the subject of the next chapter.

REFERENCES

ASHRAE 1967. Refrigeration Data Book. American Society of Heating, Refrigeration and Air-conditioning Engineers, New York.

ASHRAE 1976. Conversion to System Internationale (SI) Units. In Handbook and Product Director, 1976 Systems. American Society of Heating, Refrigeration and Air-conditioning Engineers, New York.

ANDERSON, J.A. 1973. Problems in controlling quality in grain. In Grain Storage: Part of a System, R.N. Sinha, and W.E. Muir (Editors). AVI Publishing Co., Westport, Conn.

BIALE, J.B. 1975. Synthetic and degradative processes in fruit ripening. In Postharvest Biology and Handling of Fruits and Vegetables, N.F. Haard, and D.K. Salunkhe (Editors). AVI Publishing Co., Westport, Conn.

BOND, E.J. 1973. Chemical control of stored grain insects and mites. In Grain Storage: Part of a System. R.N. Sinha, and W.E. Muir (Editors). AVI Publishing Co., Westport, Conn.

BROWN, W.L., and SCMUCKER, M.L. 1960. The influence of low level irradiation, antibiotic treatment, storage temperature and vacuum packing on flavor, bacterial changes and cured bacon. Food Technol. 14, 92-93.

CHRISTENSEN, L.M., and KAUFMAN, H.H. 1969. Grain Storage. Univ. of Minnesota Press, Minneapolis.

DE FIGUEIREDO, M.P. 1971. Quality assurance of liquid eggs. Food Technol. 25, No. 7, 70-76.

DENNIS, C.A.R. 1973. Health hazards in grain storage. In Grain Storage: Part of a System, R.N. Sinha, and W.E. Muir (Editors). AVI Publishing Co., Westport, Conn.

ECKERT, J.W. 1975. Diseases of tropical crops and their control. In Postharvest Physiology, Handling and Utilization of Tropical and Subtropical Fruits and Vegetables, E.B. Pantastico (Editor). AVI Publishing Co., Westport, Conn.

FILER, L.J. 1976. Patterns of consumption of food additives. Food Technol. 30, No. 6, 62-70.

FOSTER, E.M. 1972. The need for science in food safety. Food Technol. 26, No. 8, 81-87.

HAARD, N.F., and SALUNKHE, D.K. 1975. Symposium: Postharvest Biology and Handling of Fruits and Vegetables. AVI Publishing Co., Westport, Conn.

HALL, H.E. 1971. The significance of Escherichia coli associated with nut meats. Food Technol. 25, No. 3, 34-36.

HOWE, R.W. 1965. A summary of estimate of optimal and minimal conditions for population increases of some stored product insects. J. Stored Prod. Res. 1, 177-184.

HUFFMAN, D.L., DAVIS, K.A., MARPLE, D.B., and MC GUIRE, J.A. 1975 Effect of gas atmospheres on microbial growth, color and pH of beef. J. Food Sci. 40, 1229-1233.

IFT EXPERT PANEL AND CPI. 1975A. Naturally occurring toxicants in foods. Food Technol. 29, No. 3, 67-72.

IFT EXPERT PANEL AND CPI. 1975B. Suliftes as food additives. Food Technol. 29, No. 10, 117-120.

ITO, K.A. et al. 1973. Resistance of bacterial spores to hydrogen peroxide. Food Technol. 27, No. 11, 58-66.

JAHNS, F.D., HOWE, J.L., CODURI, R.J., and RAND, A.G. 1976. A rapid visual enzyme test to assess fish freshness. Food Technol. 30, No. 7, 27-30.

JAMES, D.E. 1971. Managing raw materials and good manufacturing. Food Technol. 25, No. 10, 30-31.

KRAMER, A. 1973. Storage retention of nutrients. Food Technol. 28, No. 1, 50-60.

KRAMER, A., and FARQUHAR, J.W. 1976. Testing of time-temperature indicating and defrost devices. Food Technol. 30, No. 2, 50-53, 56.

KROCHTA, J.M., TILLIN, S.J., and WHITEHAND, L.C. 1975. Ascorbic acid content of tomatoes damaged by mechanical harvesting. Food Technol. 29, No. 7, 28-30, 38.

KWOLEK, W.F., and BOOKWALTER, G.N 1971. Predicting storage stability from time-temperature data. Food Technol. 25, No. 10, 51-57, 63.

MARKAKIS, P. 1975. Anthocyanin pigments in food. In Postharvest Biology and Handling of Fruits and Vegetables, N.F. Haard, and D.K. Salunkhe (Editors). AVI Publishing Co., Westport, Conn.

MIDDLEKAUF, R.D. 1976. 200 Years of U.S. food laws: a gordian knot. Food Technol. 30, No. 6, 48-54.

MITSUDA, H., KAWAI, F., and YAMAMOTA, A. 1972. Underwater and underground storage of cereal grains. Food Technol. 26, No. 3, 50-56.

MUIR, W.E. 1970. Temperatures in grain bins. Can. Agric. Eng. 12, 21-24.

PANTASTICO, E.B. 1975. Postharvest Physiology, Handling and Utilization of Tropical and Subtropical Fruits and Vegetables. AVI Publishing Co., Westport, Conn.

PHAN, C.T. 1975. Respiration and respiration climateric. In Postharvest Physiology, Handling and Utilization of Tropical and Subtropical Fruits and Vegetables. E.B. Pantastico (Editor). AVI Publishing Co., Westport, Conn.

RHODES, M.J.C. 1970. The climacteric and ripening of fruits. In The Biochemistry of Fruits and Their Products, Vol. I, A.C. Hulme (Editor). Academic Press, New York.

SCOTT, P.M. 1973. Mycotoxins in stored grains, feeds and other cereal products. *In* Grain Storage: Part of a System, R.N. Sinha, and W.E. Muir (Editors) AVI Publishing Co., Westport, Conn.

SHEAR, C.B., and FAUST, M. 1975. Preharvest nutrition and postharvest physiology of apples. *In* Postharvest Biology and Handling of Fruits and Vegetables, N.F. Haard, and D.K. Salunkhe (Editors). AVI Publishing Co., Westport, Conn.

SINHA, R.N. 1973. Interrelationships of physical, chemical and biological variables in deterioration of stored grain. *In* Grain Storage: Part of a System, R. N. Sinha, and W.E. Muir (Editors). AVI Publishing Co., Westport, Conn.

SINHA, R.N., LISCOMBE, E.A., and WALLACE, H.H. 1962. Infestations of mites, insects and microorganisms in large wheat bulks after prolonged storage. Can. Entomol. *94*, 542-555.

SINHA, R.N., and MUIR, W.E. 1973. Grain Storage: Part of a System. AVI Publishing Co., Westport, Conn.

SKENE, D.S. 1963. The fine structure of apples, pears and plum fruit surfaces, their changes during ripening, and their response to polishing. Am. Bot. 27, 581-585.

SMITH, C.V. 1969. Meterology and grain storage. W.M.O. Tech. Note *101*.

SMOCK, R.M., and NEUBERT, A.M. 1950. Apples and Apple Products. Interscience Publishers, New York.

SPLINTER, W.E. 1974. Harvesting, handling and storage of grain. *In* Wheat: Production and Utilization, G.E. Inglett (Editor). AVI Publishing Co., Westport, Conn.

STERLING, C. 1975. Anatomy of toughness in plant tissue. *In* Postharvest Biology and Handling of Fruits and Vegetables, N.F. Haard, and D.K. Salunkhe, (Editors). AVI Publishing Co., Westport, Conn.

TOMPKIN, R.B. 1973. Refrigeration temperatures. Food Technol. *27*, No. 12, 54-58.

U.S. DEP. AGRIC. 1970. Market Diseases of Fruits and Vegetables. U.S. Dep. Agric., Agric. *66*.

U.S. DEP. AGRIC. 1972. U.S. Standards and Grades for Fruits and Vegetables. Superintendent of Documents, Washington, D.C.

Chapter 4

Principles of Refrigerated Gas Storage of Foods

There are clear benefits from the cool temperature storage of foods. As this thesis evolved there was another development which paralleled it. While slow in maturing, it also had significant potential. The area dealt with the gas storage of fresh foods. The matter had appeared in quite an unexpected system—refrigerated dough. In this chapter we will explore the nature and direction of the technology of gas storage as it is evolving.

GAS PACKED REFRIGERATED DOUGH

Cooking is the art of preparing foods by heating until they have changed in flavor, appearance, tenderness and chemical composition. Baking is a form of cooking that is carried out in an oven. Baked bread is not only one of the most ancient foods manufactured by man, it is also the food most widely eaten in the world. The actual baking process is really the last and most important step in the production of bakery products. Through the agency of heat, an unpalatable dough mass is transformed into a light, porous, easily digestible and appetizing product.

Freshly baked products can be produced and distributed effectively. However, the storage life of soft baked goods is generally less than a week. During this time, products stale and lose desirable tastes and textures. The appealing aroma of baked goods is slowly lost after the products are removed from the ovens. For best eating, baked products are generally consumed soon after baking.

Refrigerated Dough

An ingenious means of providing consumers with ready-to-bake dough products evolved in the past few decades. This product line is called refrigerated dough. The essential technology involves the preparation of a dough, packaging it in a sealed container capable of holding about 5 atm of gas, and causing an initial controlled release of carbon dioxide. Under refrigerated conditions, slightly above freezing, such products can be effectively produced and distributed, and when baked yield excellent products.

Matz (1968) reported that in such commercial refrigerated dough practices, there is ordinarily used, as the leavening agent or gas generating component, a slowly acting sodium acid pyrophosphate of the approximate formula $Na_2H_2P_2O_7$ in combination with sodium bicarbonate. All ingredients are mixed under very rigidly controlled temperature conditions. (Note: dough temperature out of mixer, 10° to 15°C.) The resulting dough is rolled out and the dough is then sheeted and cut into blanks, such as discs about 5 cm in diameter by 1.2 cm thick. The cut-outs are dusted with rice flour or oiled to prevent sticking together. They are then stacked and packed in a suitable can. These cans are dough-tight but not gas-tight. As a result, air and carbon dioxide may and do escape so that the dough reaches and blocks the gas outlets. Within about 1.5 to 3 hr after the package is sealed, for instance, the biscuits will have so expanded as to fill the container and close the original vents for gas and the internal pressure of carbon dioxide generated by the leavening materials will have risen to around 1 atm. Pressure within the can will be maintained over a period of 8 weeks or so if the biscuit dough and cans are normal and the storage temperature is between 7° to 12°C.

One of the difficulties in the manufacture of doughs using phosphate leavening agents has been the formation of visible phosphate crystals (disodium phosphate dodecahydrate). This crystal formation occurs at storage temperatures below 10°C, and is frequently most abundant at about 7°C. This crystallization is prevalent in the canned refrigerated doughs because of their extended storage, including the normal period of transportation and storage in the stores and homes. These visible crystals cause consumer rejection of the product because of their glass-like appearance and an uneven coloration of the baked goods which develops upon baking.

It is found that phosphate leavened cereal doughs can be prepared,

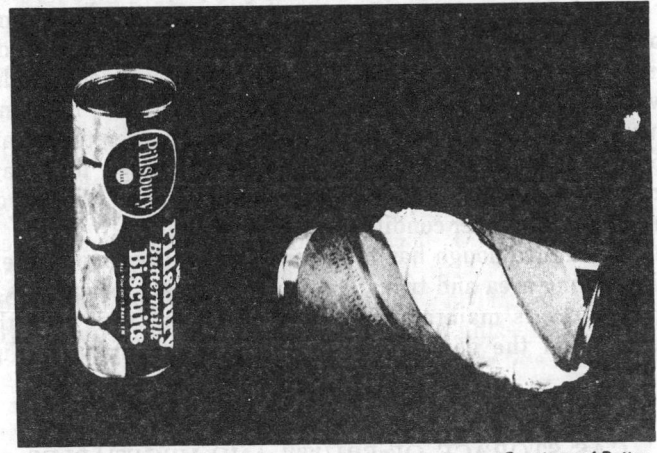

Courtesy of Potter

FIG. 4.1.   REFRIGERATED BISCUIT CONTAINER

which can be kept at their normal storage temperatures (i.e., between about 7° and 12°C) for periods up to 12 weeks or more, and which are free from visible phosphate crystal formation. Such a dough is made in the conventional manner, by mixing the ingredients such as flour, shortening, flavoring, water, the usual minor ingredients, and the chemical leavening agent, and then allowing the dough to develop, and shaping the dough as desired.

The dough is then placed in the consumer container and sealed, then allowed to proof rapidly so that dough temperature reaches 14° to 20°C as quickly as possible. Proofing involves the reaction of the chemical leaveners to provide sufficient carbon dioxide to cause the dough to expand and fill the container so as to close the gas vents.

The canned dough is cooled in a conditioning area, such as controlled temperature room, at a temperature above its freezing point but below 0°C. Normally the freezing point of such doughs is about −6°C. The time and temperature required to condition the dough will vary dependent upon can dimensions, net weight of the dough, formation of the dough, and refrigeration conditions used. However, it is important that the temperature of the conditioning room not be below the freezing point of the dough. The necessary time and temperature can be determined by withdrawing sample cans at intervals, opening them, and noting the temperature and condition of the dough. When microscopic crystals of disodium phosphate dodecahydrate are observed dispersed substantially uniformly throughout the dough, the dough is conditioned and will not exhibit visible phosphate crystals even after 12 weeks of storage. Once the time has been obtained for a given product under given cooling conditions, the same time of treatment can be used for subsequent production. Conventional refrigerated doughs can be conditioned by being held at about $-7°$ to $-5°C$ for about 48 hr. With a higher temperature it is recommended that the holding time be increased to 4 days. At temperatures of about $-1°C$ a longer holding time of about 3 days is required to obtain proper conditioning of the dough.

After the canned dough has been conditioned, the cans are removed from the cooling area and transferred to the usual storage area where the temperature is maintained at $4°$ to $10°C$. The cans may then be transported and the dough used in the ordinary manner (Fig 4.1).

GAS STORAGE OF FRUITS AND VEGETABLES

CA Defined

Controlled atmosphere (CA) storage refers to the composition of the atmosphere altered from that of air in respect to the proportions of O_2 and/or CO_2. The proportions are controlled; O_2 usually is lowered and CO_2 increased; nitrogen acts as an inert "filler;" other gases may be added in low concentrations.

Modified atmosphere (MA) storage is similar in principle to CA storage, except control of gas concentrations is less precise. Respiratory CO_2 or CO_2 derived from dry ice accumulates and O_2 decreases.

Lipton (1975) believes that the modification of the O_2 and/or CO_2 concentration in the atmosphere surrounding fresh produce is justified if the vegetable or fruit will be more valuable after CA storage than after a similar storage period in the air. CA is used most commonly to slow ripening of fruits, but appropriate mixtures of O_2 and CO_2 also can retard the spread of certain diseases and lower the incidence of some disorders. This is usually not apparent when storage time is brief and/or when storage temperature is optimal.

The use of CA can also prevent desirable ripening, induce severe physiological disorders and cause an increase in decay when misused (Lipton 1975). Each kind of vegetable and fruit has its own specific, unpredictable tolerance for atmosphere modification.

Controlled atmosphere storage has been commonly used to delay ripening of fruits, retard the spread of disease (Table 4.1), lower the incidence of storage disorders and inhibit toughening and yellowing (Lipton 1975). An extension of this technique is the use of packaging

TABLE 4.1

DECAY OF STRAWBERRIES AS INFLUENCED BY CO_2 CONCENTRATION
AND DURATION OF STORAGE

Storage Condition	Percentage Decayed Berries at CO_2 Concentration (%)			
	0	10	20	30[1]
3 or 5 days at 5°C	11.4	4.5	1.7	1.3
plus 1 day at 15°C in air	35.4	8.5	4.7	4.0
plus 2 days at 15°C in air	64.4	26.2	10.8	8.3

Source: Lipton (1975).
[1] Persistent off-flavors developed in 30% CO_2.

films to develop a microcontrolled environment in retail packages
(Henig 1975). Modification of the storage environment by suitable
packaging can provide storage benefits which exceed those observed
with refrigeration and controlled atmosphere.

Produce Package System

A package of apples can be a dynamic system in which two processes, respiration and permeation, occur simultaneously (Henig 1975).
There is an uptake of O_2 by the apples and evolution of CO_2, C_2H_4,
H_2O and other volatiles, and, at the same time, specific restricted
exchange of these gases through the packaging film. Variables that
affect respiration are: weight of apples, stage of maturity, membrane
permeability, temperature, O_2 and CO_2 partial pressures, ethylene
concentration, light, etc. Variables affecting gas exchanges into and
out of the package are: structure of the packaging film, thickness,
area, temperature, O_2 and CO_2 concentrations (Fig. 4.2 and
Table 4.2).

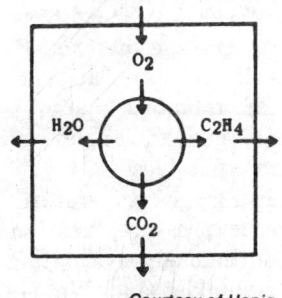

Courtesy of Henig

FIG. 4.2.   PRODUCE PACKAGE SYSTEM

TABLE 4.2

PERMEABILITY OF EIGHT TYPES OF PACKAGING FILMS
TO O_2 AND CO_2

Film	Permeability (cc/mil/1 sq m/24 hr at 1 atm, 20 °C, 0% rh)	
	O_2	CO_2
Saran	8—26	52—150
Cellophane (lacquered)	15—77	15—95
Polyester	52—130	180—390
Vinyl (PVC)	77—7,500	770—55,000
Pliofilm	130—1,300	520—5,200
Polypropylene (unoriented)	1,300—6,400	7,700—21,000
Polystyrene (oriented)	2,600—7,700	10,000—26,000
Polyethylene (low-density)	3,900—13,000	7,700—77,000

Source: Pantastico (1975).

It has been demonstrated that steady state conditions are established within such an intact packaging system; equilibrium concentrations of O_2 and CO_2 prevail, and the respiration rate is equal to the rate of gas exchange. Any change in the system variables will affect the equilibrium or the time to establish steady state conditions. The packaging of fresh food in polymeric films is now frequently used in retail stores and about 40% of the produce is now distributed to retail stores in consumer packages. This packaging is designed for consumer appeal. Better use to regulate ripening and, thus, to prolong useful storage life is predicted by many people.

SUBATMOSPHERIC STORAGE

The storage life of several fruits and vegetables can also be increased significantly by reduced pressure under refrigeration, subsequently decreasing respiration and evacuation of ethylene given out by the produce. Salunkhe and Wu (1973) and others have shown that subatmospheric low pressure storage, a form of controlled atmosphere with reduced atmospheric pressure at a given temperature, has extended storage life of tomatoes, potatoes, bananas, avocados, mangoes, cherries, limes, guavas, apples, peaches, pears, apricots and many other fruits and vegetables.

A food is maintained at a given temperature in a sealed container at a constant subatmospheric pressure, ventilated with air saturated with water vapor by continuously evacuating the container with a sealed vacuum pump. Salunkhe and Wu (1973) report two major consequences that result from the storage of fresh produce under subatmospheric pressure: (1) oxygen supply to the food is reduced,

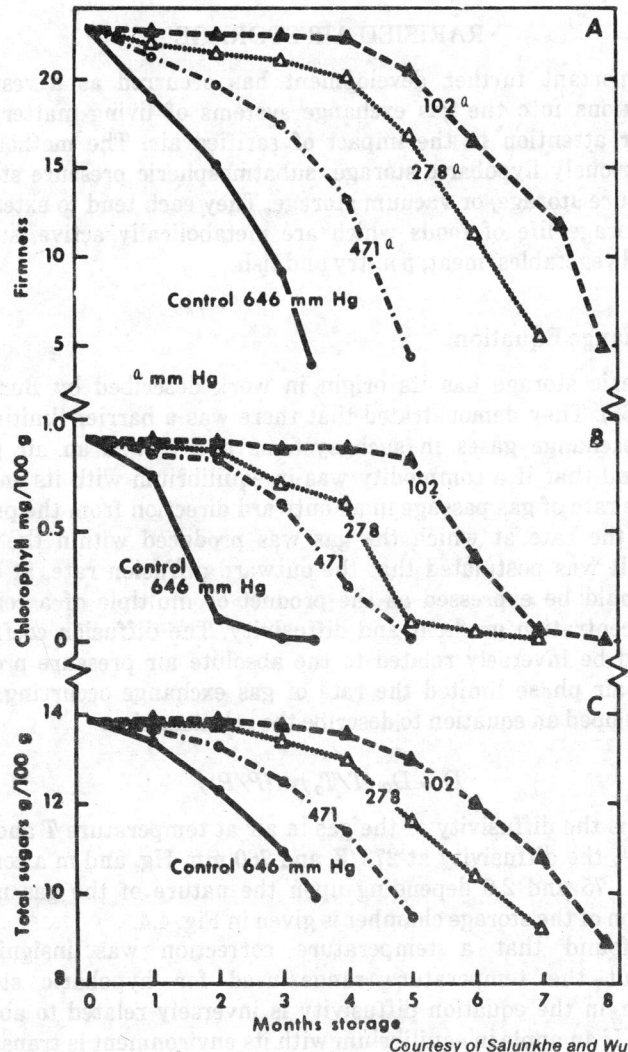

FIG. 4.3. EFFECTS OF SUBATMOSPHERIC PRESSURE STORAGE ON (A) FIRMNESS, (B) CHLOROPHYLL, AND (C) TOTAL SUGARS OF PEARS

thus reducing the respiratory rate; and (2) ethylene and other gases released by the food are evacuated, consequently, ripening and aging processes of the produce are inhibited. Examples are shown in Fig. 4.3.

RARIFIED AIR STORAGE

An important further development has occurred as a result of investigations into the gas exchange systems of living matter, with particular attention to the impact of rarified air. The methods are called variously hypobaric storage, subatmospheric pressure storage, low pressure storage, or vacuum storage. They each tend to extend the useful storage life of foods which are metabolically active, such as fruits and vegetables, meat, poultry and fish.

Gas Exchange Equation

Hypobaric storage has its origin in work described by Burg and Burg (1966). They demonstrated that there was a barrier limiting the rate of exchange gases in such systems and it was an air phase. They found that if a commodity was in equilibrium with its environment, the rate of gas passage in an outward direction from the product equalled the rate at which the gas was produced within the living product. It was postulated that the outward diffusion rate, at equilibrium, could be expressed as the product or multiple of a constant with concentration gradient and diffusivity. The diffusion coefficient *(D)* would be inversely related to the absolute air pressure provided that the air phase limited the rate of gas exchange occurring. They then developed an equation to describe the relationships:

$$D = D_0 \, (T/T_0) \, \text{m} \, (P/P_0)$$

where D is the diffusivity of the gas in air at temperature T and pressure P, D_0 the diffusivity at 273°K and 760 mm Hg, and m a constant between 1.75 and 2.0 depending upon the nature of the gas mix. A description of the storage chamber is given in Fig. 4.4.

They found that a temperature correction was insignificant throughout the temperature range used for hypobaric storage. Therefore, in the equation diffusivity is inversely related to absolute pressure. If an apple in equilibrium with its environment is transferred from atmospheric pressure to rarified air having an absolute pressure of 1/10 atmosphere, when a new equilibrium is established any volatile will be produced and escape at the same (initial) rate. Now, however, the concentration gradient driving the gas in an outward direction adjusts to a lower value, 1/10 of the initial, because diffusivity increases ten-fold.

Burg (1975) found that the low oxygen partial pressure created by the rarified environment can, and often does, reduce the rate of synthesis of volatiles, either by limiting energy availability or by

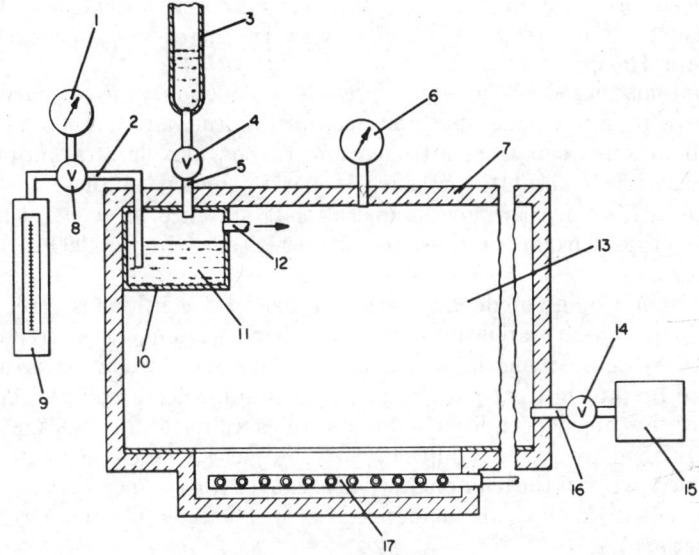

FIG. 4.4. BASIC HYPOBARIC STORAGE APPARATUS

(1) Vacuum gauge—reading pressure downstream of diaphragm vacuum regulator; (2) Conduit carrying in-coming rarified air; (3) Water reservoir containing make-up water; (5) Conduit to admit make-up water; (6) Temperature gauge; (7) Insulated walls; (8) Diaphragm vacuum regulator; (9) Air-flow meter; (10) Humidifier; (11) Water; (12) Conduit admitting air from humidifier to storage chamber; (13) Vacuum storage chamber; (14) Valve to throttle vacuum pump; (15) Vacuum pump; (16) Conduit to vacuum pump; (17) Refrigeration coolant tubes.

preventing an essential oxidative reaction peculiar to the biogensis of the specific volatile. Alternatively, if the synthesis of a volatile is under feedback control dependent upon concentration, lowering the endogenous concentration can increase the rate of formation of the volatile. Similarly, anaerobiosis could favor its production if the volatile is a highly reduced compound. In the case of carbon dioxide he found that its removal inhibited the "dark" fixation reactions, altering the organic acid pool in such a way that respiration, and hence the production of carbon dioxide, decreased. When such living matter is transferred from hypobaric to atmospheric conditions it resumes emanation of volatiles at a normal rate unless it has entered a state of semidormancy or has been otherwise changed by injury, disease, etc.

In practice Burg (1975) describes hypobaric storage as placing a commodity in a flowing stream of air substantially saturated with water (RH = 80 to 100%), at a reduced pressure (4 to 400 mm Hg absolute), and at a controlled temperature ($-2°$ to $15°C$). The process is distinguished from other methods of storage employing slightly reduced pressure in that the hypobaric system operates at lower pressure than previously, which was restricted to approximately 650 mm Hg absolute.

Systems based on a slight pressure reduction aim primarily at improving air through-put, or facilitating the addition of various combinations of oxygen, nitrogen and carbon dioxide at atmospheric pressure. The absolute pressure is not important in these systems (Burg 1975) since the lowest operational pressure is about equal to the average barometric pressure at an altitude of only 4000 ft above sea level.

A second point of departure is the use of a continuous stream of nearly water-saturated air in the hypobaric system, in contrast to a sealed system or one in which a slight pressure reduction is maintained by intermittent pumping and a resultant decrease in humidity. Under hypobaric conditions vapors released into the storage area are flushed away. In addition, noxious gases, which normally are retained within the commodity, are caused to escape because their outward diffusivity is enhanced proportionate to the pressure reduction.

At atmospheric pressure air changes remove the vapors in the air surrounding stored matter, but cannot remove them from within the commodity where they are produced. Consequently, even when atmospheric air surrounding a commodity is changed often enough to prevent accumulation of these vapors, substantial concentrations persist within the tissue contrary to the principle that frequent air changes prevent accumulation of undesired gaseous by-products (Burg 1975). The first application of the hypobaric system was for storage of various nonripe fruits. In these studies it was found that firmness was maintained, internal or external mold development, decay or discoloration were controlled, and poststorage shelf-life was unimpaired. A comparison of storage lives of various products under conventional and rarified air storage is given in Table 4.3.

GAS ATMOSPHERE STORAGE OF MEATS

Based on the foregoing it becomes clear that modified atmospheres should also be applicable to meat storage. For example, as shown earlier, the common spoilage organisms of fresh meat are aerobic microorganisms of the types *Pseudomomonas* and *Achromobacter*. Inhibiting the growth of these organisms ought to be possible by drastically reducing oxygen concentrations in suitably packaged products. Huffman *et al.* (1975) report that depletion of oxygen will limit growth of pseudomonads on meat in a gas atmosphere of 75%

TABLE 4.3

COMPARATIVE STORAGE LIFE OF PRODUCE STORED
IN REFRIGERATION AND UNDER HYPOBARIC CONDITIONS

Variety	Storage Life (Days)	
	Cold	Hypobaric
Ripe, fully mature fruit		
Pineapple (field ripe)	9-12	40
Grapefruit	30-40	90-120
Strawberry, Florida 90		
and Tioga	5-7	21-28
Cherry, sweet	14	60-90
Vegetables		
Green pepper	16-18	50
Cucumber	10-14	41
Bean, pole	10-13	30
Onion, green	2-3	15
Corn	4-8	21
Lettuce, iceberg	14	40-50
Nonripe, fully mature fruit		
Tomato (mature green)	14-21	60-100
Tomato ("breaker")	10-12	28-42
Banana, Valery	10-14	90-150
Avocado, Lula	23-30	90-100
Lime, Tahiti	14-35	60-90
Apples (general varieties)	60-90	300
Pear, Bartlett	45-60	300

Source: Berg (1975).

N_2 and 25% air. A gas atmosphere containing 20% CO_2 was found to be effective in inhibiting the growth of slime-producing bacteria at high storage temperatures and high relative humidity.

Bacterial Counts

Packaged beef steaks stored at $-1°C$ in CO_2 have significantly lower aerobic bacterial counts for 16 through 27 days postslaughter than steaks stored in N_2, O_2 or air (Fig. 4.5). For the first 20 days postslaughter there are small differences in aerobic growth on steaks stored in CO_2 or in the gas mixture; however, a significant difference appears at both 23-day and 27-day periods. The aerobic counts for pork chops stored in high CO_2 atmospheres are consistently lower than for those stored in the gas mixture containing approximately 70% N_2, 25% CO_2 and 5% O_2. Anerobic counts are usually low (less than 10 organisms per cm^2) with increase during 27 days of storage under the stated conditions.

pH of Tissue

Changes in pH for all gaseous atmospheres follow the typical pattern of stored fresh meat, increasing gradually from pH 5.5 to

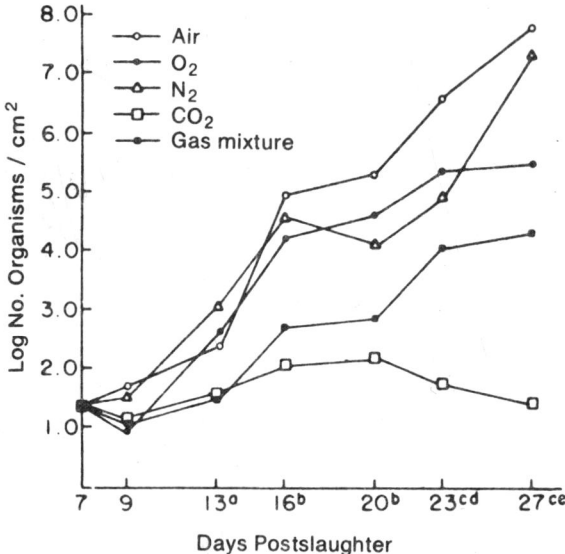

Courtesy of Huffman et al.

FIG. 4.5. AEROBIC BACTERIAL PLATE COUNTS OBTAINED FROM
THE SURFACE OF BEEF RIBEYE SAMPLES STORED AT 1°C IN THE
DARK FOR VARIOUS PERIODS OF TIME UNDER VARYING
GASEOUS ATMOSPHERES

approximately pH 5.8 after 27 days of storage at 1.1°C. Thereafter a
slight depression in pH occurs under CO_2 or gas mixtures (pH 5.75
compared to the O_2 stored meat of 5.80 and the air and N_2 stored
meats with a pH of 5.85).

The depression of pH caused by the CO_2 atmosphere and the gas
mixture has been proposed as one of the mechanisms by which CO_2
inhibits microbial growth. The decrease in pH caused by CO_2 is so
slight (approx 0.1 unit) that it is unlikely that pH depression alone
accounts for the lower bacterial counts (Huffman et al. 1975).

They summarize present knowledge and report that a gas
mixture of approximately 70% N_2, 25% CO_2 and 5% O_2 will cause
an apparent CO_2 inhibition of aerobes. Anaerobic counts are reduced
by both a high CO_2 atmosphere and a gas mixture of approximately
70% N_2, 25% CO_2 and 5% O_2 at 14 days postmortem when
compared to air. At 28 days postmortem there are significantly lower
counts for samples stored in CO_2 than for those stored in air. This
growth inhibition is believed to be an influence of CO_2 on specific
enzymes of the bacteria and, to lesser extent, the lowering of pH by
the CO_2.

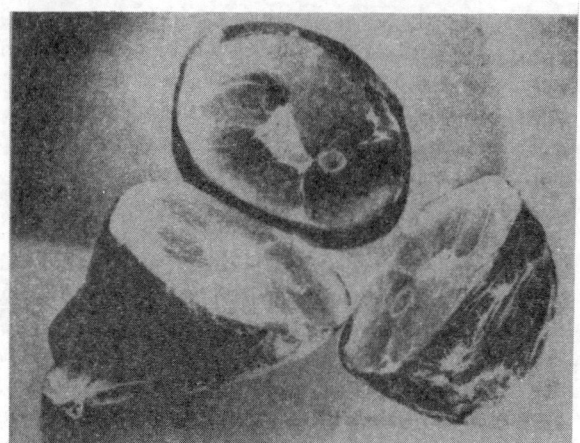

FIG. 4.6. SHRINK-PACKAGED CUT HAM

Meat Color

Fresh meat color is related to the amount of heme pigment myoglobin and to its biochemical state. The autoxidation of myoglobin is greatest when myoglobin is half-saturated with oxygen, according to reports. Thus, formation of oxidized metmyoglobin is not increased in elevated carbon dioxide atmospheres provided the oxygen partial pressure is maintained above a limiting value of about 5%. Packaged meat samples stored in O_2 have better color than samples stored in air, N_2, CO_2 and the gas mixtures. Meat stored in air had better color 13 days postslaughter than samples stored in N_2, CO_2 and gas mixtures. Huffman et al. (1975) report that by subjective observation samples stored in N_2 have the least desirable color, and all samples had undesirable color after 23 days postslaughter.

A CO_2 concentration higher than 25% caused discoloration in cut-up meats. If undesirable color problems could be overcome, CO_2 might be more widely used to inhibit microbial growth on fresh meat. Color is the major factor in acceptability of fresh meat at the retail counter. Therefore, a process that maintains color while inhibiting bacterial growth would permit centralized packaging of fresh retail cuts and create a major change in how meats are distributed and marketed. The existing commercial method is now applicable only to cured meats (Fig. 4.6).

GAS STORAGE OF GRAINS, SEEDS AND FLOUR

Cereals and beans are the major sources of proteins and calories for human nutrition. There is great need to increase production of these foods and reduce their loss or waste during transit and storage. Preservation to stabilize the supply is hindered by storage problems. The most desirable temperature for maintaining the quality of grain is about 13°C. As storage temperature or humidity rises, the stored food loses quality, as previously discussed.

High land and construction costs often are deterrents to the building of temperature-controlled warehouses. In addition, the maintenance cost for temperature-controlled warehouses is higher than that for ordinary warehouses. Consequently, the possibility of storing grain in naturally cold places warrants attention. Appropriate places satisfying the temperature requirements are underwater locations (lakes, ponds or man-made pools at the seacoast) and underground locations (caves or abandoned mines). Such storage locations require special packaging. Packaging materials suitable for underwater and underground storage of cereals must be: resistant to structural attack by water; resistant to physical shock; airproof; and moistureproof (Fig. 4.7, 4.8 and 4.9).

Conventional jute or paper bags do not satisfy these requirements. On the other hand, plastic films developed for food packaging are acceptable. Laminated films found satisfactory for the purpose (Mitsuda et al. 1972) were composed of oriented nylon film for high physical strength, aluminum foil or polyvinylidene chloride film for gas- and water vapor-impermeability, and polyethylene film for heat-sealability.

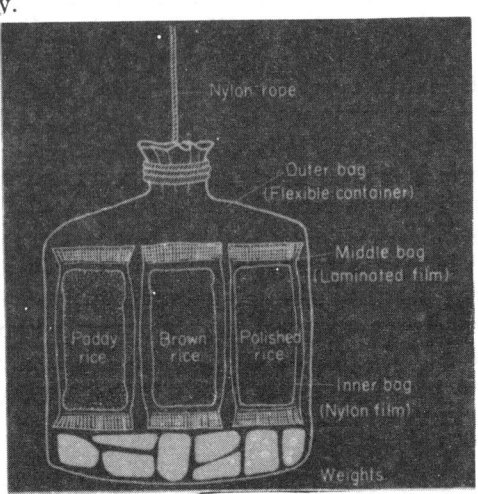

Courtesy of Mitsuda et al.

FIG. 4.7.   PACKAGING FOR UNDERWATER STORAGE
OF CEREALS

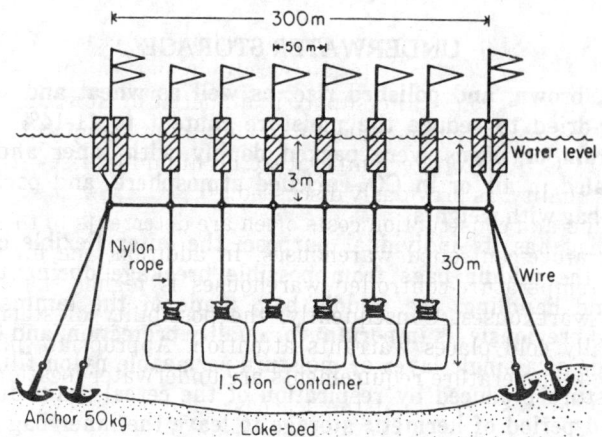

Courtesy of Mitsuda et al.

FIG. 4.8.   EXPERIMENTAL SYSTEM FOR UNDERWATER STORAGE
OF CEREALS

Courtesy of Mitsuda et al.

FIG. 4.9.   SAMPLING IS ACCOMPLISHED BY RAISING THE
CONTAINER WITH NYLON ROPE

UNDERWATER STORAGE

Paddy, brown, and polished rice, as well as wheat and soybeans, were air-dried to reduce the moisture content to 11-14%. Then 60 kg of grain or beans were packed doubly with inner and middle bags, sealed in air or in CO_2-enriched atmosphere, and packaged in an outer bag with weights.

Each bag has its individual purpose: the outer flexible container protects the middle bags from possible breakage during transportation and handling; the middle bag, made of the laminated film described previously, is important for quality protection; and the inner bag, made of a single layer of moisture-permeable nylon film, allows the moisture produced by respiration of the cereal grain during the prolonged period of hermetic storage to leave the inner bag and condense between the inner and middle bags. Analyses of the rice stored under water and rice stored under atmospheric conditions (Mitsuda et al. 1972) were:

(1) Moisture content: An increase in moisture content of the rice stored under water was less than 0.5% after one year, whereas the moisture content of the rice stored in open air gradually increased.

(2) Chemical composition: The change in free fatty acid content is less with the rice stored under water than with the rice stored under atmospheric conditions. Changes in the content of vitamin B-1, reducing sugar, water-soluble nitrogen, and other substances were also found to be less for the rice stored under water. The changes observed in underwater storage were within the ranges found in storage in temperature-controlled warehouses.

(3) Biological activity: The germinative capacity of paddy rice stored under water was maintained, whereas that of rice stored under atmospheric conditions diminished. Considerably higher biological activities (e.g., catalase and peroxidase activities) are found in grain stored under water than under natural conditions. The original freshness of the grain is retained even after prolonged underwater storage (Fig. 4.10).

(4) Palatability: The stable flavor and rheological characteristics show that deterioration proceeded at a considerably lower rate under water than under atmospheric conditions. The appearance, aroma, taste and overall quality of cooked rice prepared from grain stored under water were similar to rice stored at controlled low temperatures. The atmosphere en-

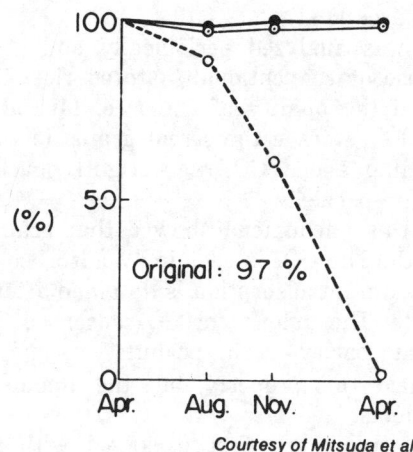

Courtesy of Mitsuda et al.

FIG. 4.10. GERMINATIVE CAPACITY OF PADDY RICE IS MAINTAINED DURING UNDERWATER STORAGE

-o---o- Natural conditions

-o——o- Under water, air group

-•——•- Under water, CO_2 group

riched with CO_2 was more effective than air in preventing changes in chemical composition and in retarding the development of stale flavor.

Results with Wheat and Beans

Wheat and soybean tests by Mitsuda et al. (1972) yielded results similar to those obtained with rice. Amylograms of flour prepared from wheat stored under various conditions exhibited a higher maximum viscosity in the heating curve than the original sample, while the samples from underwater storage showed the lowest values. These results indicate that changes in rheological properties of wheat during underwater storage follow a different process from that of naturally stored wheat. This was also true of the changes in free fatty acid content of soybeans.

UNDERGROUND STORAGE

Underground storage of brown rice in abandoned mines has also received attention. Special attention to avoid possible ravages from field rodents is needed. At 570 m deep in a gallery of a mine, the air temperature ranges from 9° to 11.5°C through the year and the relative humidity is over 90%.

The stored rice was analyzed periodically and the results were compared with those of conventionally stored rice. Laminated film packaging preserved the quality of the rice. Mitsuda *et al.* (1972) also noticed that after storage the cereal grains in the plastic laminated bags containing added CO_2 were tightly packed, as though they had been vacuum-packed.

Investigation of this phenomena showed that grain can adsorb a large amount of CO_2. The CO_2 is seemingly adsorbed on the surface of cereal grains. Maximum adsorption is obtained after 24 hr and the process is reversible. The amount of CO_2 adsorbed depends on the temperature. Wheat, barley, corn, peanuts, sesame, soybeans and their flours also have this property, but the amount of adsorption varies with the products.

In this method cereals are first packaged with an appropriate volume of CO_2 in a bag made of plastic film of low gas permeability and then sealed. At this stage CO_2 surrounds the cereals and expands the bag. The bag gradually decreases in volume because of the adsorption of the CO_2 by the cereals. This technique is expected to displace the usual vacuum or shrink packaging in these products. Packaging of wheat flour, for example, is readily achieved without high vacuum treatments. Packaging selected products by this technique preserves the quality of the product and also protects against aerobic microbial infection and effectively reduces the storage space required.

SUMMARY

Controlled atmosphere storage of foods has evolved as a new technology in preserving fresh foods. Such storage has become important in delaying the ripening of fruits, controlling the development of toughness in vegetables, reducing the incidence of spoilage in stored products as well as retarding the spread of diseases of fruits and vegetables. The technology is now being extended to other foods, including meats and cereal grains.

The gas packing of various products in selected packaging systems holds promise of and is stimulating further developments. The possibilities of preserving fresh foods were quite limited a decade or two ago. Gas storage warrants further study (Fig. 4.11 and 4.12).

Courtesy of TransFresh Corp.

FIG. 4.11.   REFRIGERATED VAN BEING CHARGED WITH A
PREPARED MIXTURE OF GASES

Courtesy of TransFresh Corp.

FIG. 4.12.   MOBILE INJECTION EQUIPMENT FOR ATMOS-
PHERE MODIFICATION IN PRODUCE SHIPMENTS

REFERENCES

ADAMS, J.R., and HUFFMAN, D.L. 1972. Effect of controlled gas atmospheres and temperature on quality of packaged pork. J. Food Sci. *37*, 869.

BROWN, W.L., and SCHMUCKER, M.L. 1960. The influence of low level gamma irradiation, antibiotic treatment, storage temperature, and vacuum packaging on flavor and bacterial changes in cured bacon. Food Technol. *14*, 92-93.

BURG, S.P. 1967. Method of storing fruit. U.S. Pat. 3,333,967.

BURG, S.P. 1967. Molecular requirements for ethylene action. Plant Physiol. *42*, 144-152.

BURG, S.P. 1975. Hypobaric storage and transportation of fresh fruits and vegetables. *In* Postharvest Biology and Handling of Fruits and Vegetables, N.F. Haard, and D.K. Salunkhe (Editors). AVI Publishing Co., Westport, Conn.

BURG, S.P., and BURG, E.A. 1965. Ethylene action and the ripening of fruits. Science *148*, 1190-1191.

BURG, S.P., and BURG, E.A. 1966. Fruit storage at subatmospheric pressure. Science *153*, 314-315.

BURG, S.P., and HENTSCHEL, W. 1974. Low pressure storage of metabolically active matter with open cycle refrigeration. U.S. Pat. 3,810,508.

CAIN, R.F. 1967. Changes during processing and storage of fruits and vegetables. Food Technol. *21*, No. 7, 60-62.

CECIL, S.R. 1969. Storage of foods in shelters. Ga. Agric. Exp. Stn. Bull. *156*.

CECIL, S.R., and WOODROOF, J.G. 1962. Long term storage of military rations. Ga. Agric. Exp. Stn. Bull. *25*.

DAUN, H.K., SOLBERY, M., FRANKE, W., and GILBERT, S. 1971. Effect of oxygen-enriched atmospheres on storage quality of fresh packaged meat. J. Food Sci. *36*, 1011.

HARRIS, C.M., and HARVEY, J.M. 1973. Quality and decay of California strawberries stored in CO_2 enriched atmospheres. Plant Dis. Rep. *57*, 44-46.

HENIG, Y.S. 1972. Computer analysis of the variables affecting respiration and quality of packaged produce in polymeric films. Ph.D. Dissertation, Rutgers Univ., New Brunswick, New Jersey.

HENIG, Y.S. 1975. Storage stability and quality of produce packed in polymeric films. *In* Postharvest Biology and Handling of Fruits and Vegetables, N.F. Haard, and D.H. Salunkhe (Editors). AVI Publishing Co., Westport, Conn.

HUFFMAN, D.L., DAVIS, K.A., MARPLE, D.B., and MCGUIRE, J.A. 1975. Effect of gas atmospheres on microbial growth, color and pH of beef. J. Food Sci. *40*, 1229-1233.

LIPTON, W.J. 1960. Effect of atmosphere composition on quality. U.S. Dep. Agric., Agric. Res. Serv. Mark. Res. Rep. *428*.

LIPTON, W.J. 1967. Some effects of low-oxygen atmospheres on potato tubers. Am. Potato J. *44*, 292-299.

LIPTON, W.J. 1975. Controlled atmospheres for fresh vegetables and fruits. *In* Postharvest Biology and Handling of Fruits and Vegetables, N.F. Haard, and D.K. Salunkhe (Editors). AVI Publishing Co., Westport, Conn.

LIPTON, W.J., and HARRIS, C.M. 1974. Controlled atmosphere effects and market quality of stored broccoli. J. Am. Soc. Hortic. Sci. *99*, 200-205.

MATZ, S.A. 1968. Refrigerated dough product. U.S. Pat. 3,397,064.

MITSUDA, H., KAWAI, F., and YAMAMOTA, A. 1972. Underwater and underground storage of cereal grains. Food Technol. 26, No. 3, 50-56.

PANTASTICO, E.B. 1975. Postharvest Physiology, Handling and Utilization of Tropical and Subtropical Fruits and Vegetables. AVI Publishing Co., Westport, Conn.

RYALL, A.L., and LIPTON, W.J. 1972. Handling, Transportation and Storage of Fruits and Vegetables, Vol. 1. AVI Publishing Co., Westport, Conn.

SALUNKHE, D.K., and WU, M.T. 1973. Effects of subatmospheric pressure storage on ripening and associated chemical changes of certain deciduous fruits. J. Am. Soc. Hortic. Sci. 98, 113-117.

SALUNKHE, D.K., and WU, M.T. 1975. Subatmospheric storage of fruits and vegetables. In Postharvest Biology and Handling of Fruits and Vegetables, N.F. Haard, and D.K. Salunkhe (Editors), AVI Publishing Co., Westport, Conn.

STRINGER, W.C., BILSKIE, M.E., and NAUMANN, H.D. 1969. Microbial profile of fresh beef. Food Technol. 23, No. 1, 97-102.

TOLLE, W.E. 1969. Hypobaric storage of mature green tomatoes. U.S. Dep. Agric., Agric. Res. Serv. Mark. Res. Rep. 842.

WANG, S.S., HAARD, N.F., and DIMARCO, G.R. 1971. Chlorophyll degradation during controlled atmosphere storage of asparagus. J. Food Sci. 36, 657-661.

WELLS, F.E., SPENCER, J.V., and STADELMAN, W.J. 1958. Effect of packaging materials and techniques on shelf life of fresh poultry meat. Food Technol. 12, 425-428.

WU, M.T., JADHAV, S.J., and SALUNKHE, D.K. 1972. Effects of subatmospheric pressure storage of ripening of tomato fruits. J. Food Sci. 37, 952-953.

WU, M.T., and SALUNKHE, D.K. 1972. Fungistatic effects of subatmospheric pressures. Experimentia 28, 866-868.

Chapter 5

Principles of Food Freezing

DEVELOPMENT OF A FROZEN FOOD INDUSTRY

Freezing temperatures, once feared by mankind, have been turned to great advantage by man's inquiry into the phenomena. While ice-salt systems were used to freeze foods in the mid 1800s, and patents for freezing fish, for example, were granted in 1861 to Enoch Piper in Maine and even earlier to H. Benjamin in England in 1842, the invention of mechanical refrigeration in the late 1800s provided the base for subsequent commercial exploitation of the process. Frozen foods have become important items of commerce (90% of Iceland's export trade is frozen fish) and important in food preparation for dinner tables (Tables 5.1 and 5.2).

Clarence Birdseye fathered this revolution as a technologist by developing quick freezing processes and equipment, and by successfully promoting consumer units of frozen foods. He overcame tremendous obstacles. In the 1920s there were few mechanical refrigerators in homes in the United States.

In the 1930s, as facilities for food freezing and retail distribution developed across the United States, frozen foods began to find their place in commerce. Yet, it was not until 1940 that they became important competitors of other consumer-type preserved foods. While Clarence Birdseye was a prime mover industrially, the frozen food industry had support in the scientific aspects of the development by men such as Dr. Donald K. Tressler at Cornell, and Dr. C.R. Fellers at the then Massachusetts State College.

The present day finds competition between all methods of food preservation and the competition is being resolved by consumer purchases. Those foods best preserved by freezing are largely frozen. Those foods highly acceptable as canned products continue as highly

successful consumer goods. The economic struggle for survival between fresh commodities, canned foods and frozen foods in a free
market evidences itself in better foods at lower prices for consumers.

TABLE 5.1

ESTIMATES OF FROZEN FOOD
PER CAPITA CONSUMPTION

Country	Kilos per Capita[1]
U.S.A.	37-38
Sweden	20.8
Denmark	17.8
Great Britain	13
Switzerland	12.3
Australia	12
Norway	10.4
New Zealand	9
France	7 [2]
Finland	7
Netherlands	5.5
W. Germany	5.3
Belgium	5
Hungary	4
Austria	4.0
Italy	1.44
Spain	1.20

Source: Williams (1976).
Note: Most estimates include poultry but not ice cream.
[1] 1 Kilo equals 2.2 lb.
[2] Includes both surgelé and congelé.

THE FREEZING POINT OF FOODS

Living cells contain much water, often two-thirds or more of their
weight. In this medium there are organic and inorganic substances,
including salts and sugars and acids in aqueous solutions, and more
complex organic molecules, such as proteins which are colloidal
suspension. To some extent gases are also dissolved in the watery
solution.

The physical, chemical and biological changes occurring during the
freezing and subsequent thawing of foods are complex and not completely understood. Nevertheless it is useful to study the nature of
these changes which have been recognized in order to design a successful freezing process for a food.

The freezing point of a liquid is that temperature at which the
liquid is in equilibrium with the solid. A solution with a vapor
pressure lower than that of a pure solvent will not be in equilibrium
with the solid solvent at its normal freezing point. The system must
be cooled to that temperature at which the solution and the solid
solvent have the same vapor pressure. The freezing point of a solu-

TABLE 5.2

FUTURE OF FROZEN FOODS - 1980's; PROJECTED SALES COMPARED TO 1970

Category	Increase (%)
Top ten frozens categories	
Cakes and Pastries	+198.9
Pies	+137.7
Breaded Shrimp	+ 68.3
Toppings	+ 67.5
Seafood	+ 66.3
Grapefruit Juice	+ 64.3
Meat	+ 60.4
Fruit Drinks	+ 52.9
Precooked Foods	+ 52.7
Ice Cream	+ 51.5

Category	$ Volume Increase (%)
The dollar sales increases in grocery store frozen food business	
Precooked Foods	+30.3
Poultry	+11.9
Ice Cream	+11.5
Vegetables	+11.4
Orange Juice	+ 9.7
Cakes and Pastries	+ 8.3
Pies	+ 4.6
Seafood (excluding breaded shrimp and sticks)	+ 3.3
Meat	+ 2.9
Toppings	+ 1.7

Source: Williams (1976).

tion is lower than that of a pure solvent. The freezing point of food is lower than that of pure water.

When a liquid evaporates the escaping molecules exert a pressure known as the vapor pressure. The total pressure of a system will be equal to the sum of the partial pressures of the system. The addition of a nonvolatile solute (sugar) to water lowers the vapor pressure of the water solution of sugar, and the freezing point of the water solution will be lower than that of pure water (Table 5.3).

Because of the high content of water in most foods, most of them freeze solidly at temperatures between 0° and −3°C (Fig. 5.1). The temperature of the food undergoing freezing remains relatively constant until the food is mostly frozen, after which time the temperature approaches that of the freezing medium. Quick freezing has been defined, by those who adhere to rapid crystallization theory, as that process where the temperature of the food passes through the zone

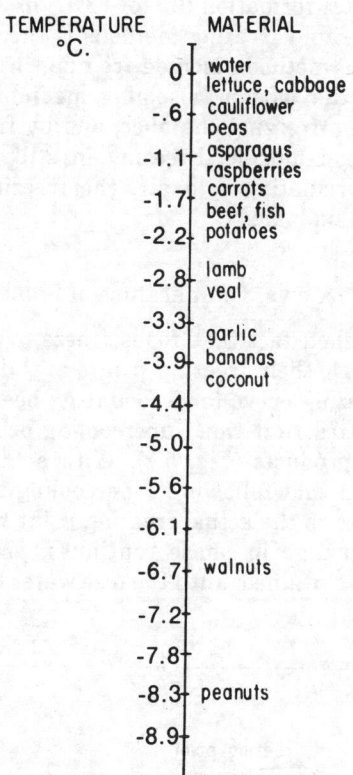

FIG. 5.1. FREEZING POINT OF SELECTED FOODS

TABLE 5.3

INCREASING THE CONCENTRATION OF SOLUTES IN
SOLUTION LOWERS THE FREEZING POINT

Calcium Chloride in Water (sp gr)	Freezing Point (°C)
1.00	0
1.05	-2
1.10	-6
1.15	-12
1.20	-21
1.25	-32
1.30	-41

of maximum ice crystal formation (0° to −3°C) in 30 min or less. The basic principle of all rapid freezing methods is the speedy removal of heat from food. These methods include freezing in cold air blasts, by direct immersion of the food in a cooling medium, by contact with refrigerated plates in a freezing chamber, and by freezing with liquid air, nitrogen or carbon dioxide. Freezing in still air is the poorest method of all. By circulating cold air, the freezing rate is greatly accelerated, as will be explained.

Percentage Water Frozen vs. Temperature of Food and Its Quality

It is a well established fact that foods freeze over a wide range of temperatures, although their freezing points are identifiable. Careful evaluation of the freezing curve for a food (e.g., beef) under controlled conditions demonstrates first that supercooling occurs, and that this is characteristic for products (Fig. 5.2). With a thin section of food tissue it can be shown that following supercooling the temperature of the cooled section rises to the actual freezing point when the change in phase occurs. This change in phase continues, providing a temperature differential is maintained, until the free water becomes ice.

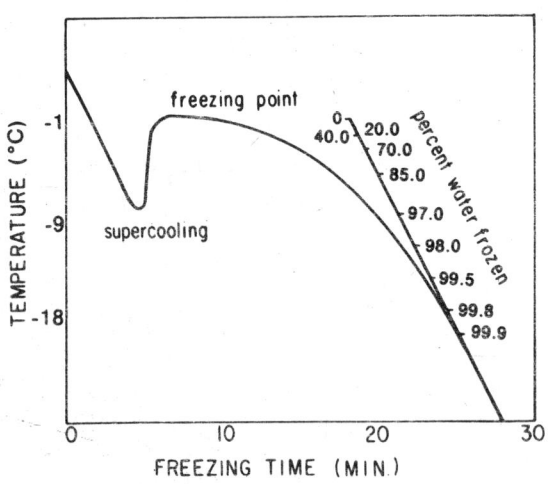

FIG. 5.2.   FREEZING CURVE FOR THIN SECTIONS OF BEEF

Supercooling usually occurs in freezing. The degree of ice crystal formation may be estimated from the slope of the cooling curve once freezing begins.

It is a fact that the water in foods exists in two (or more) states. The terms generally used are bound water and free water. Some accept the definition of bound water as that which does not freeze at −20°C. Free water on the other hand exhibits the physical and chemical properties of liquid water, and freezes according to its condition of solution. In fish flesh researchers have found that even at −33°C there remains some noncrystalline water. True, the amount is infinitesimal. Figure 5.2 shows the percentage of water frozen in a sample of food at different temperatures. A satisfactory definition of bound water is difficult to make, but good evidence exists for its justification. Bound water exists in proportion to the free water content rather than the solid material of the system. There may be a shift due to freezing in the amount of bound water (Table 5.4).

Reducing the amount of free water in a food can therefore be expected to improve the quality of the frozen food. The more complete the change from free water to a more stable state, the better is the retention of quality in the frozen food.

Flavor changes, color changes, nutrient losses and texture losses

TABLE 5.4

THE DEGREE OF WATER FREEZING OVER THE FREEZING RANGE FOR SAMPLES OF BEEF, VEAL, CARROT AND PEACH. SAMPLES WERE FROZEN AT A CONSTANT THERMAL HEAD OF 7°C

Moisture Content (%)	Temp (°C)	Degree Frozen	Moisture Content (%)	Temp (°C)	Degree Frozen
	Beef			Veal	
73.6	-2	20.0	72.6	-1	68.6
	-3	42.8		-2	63.3
	-4	69.2		-3	76.7
	-7	88.8		-6	90.4
	-9	94.6		-9	94.6
	-12	97.3		-12	97.0
	-15	98.6		-15	98.3
	-18	99.5		-17	98.8
	-20	100.0		-20	100.0
	Carrot			Peach	
83.0	-3	11.2	85.2	-2	20.0
	-4	63.8		-3	63.3
	-7	89.4		-6	86.5
	-9	96.0		-9	93.6
	-12	98.8		-12	97.0
	-15	99.6		-15	98.9
	-16	100.0		-17	1
				-18	100.0

[1] Too close to the value of 100.0 for practical measurement.

occur relatively rapidly above −9°C (as compared to −18°C or lower). The lower the temperature, the slower the rate of loss of ascorbic acid. Further, some products deteriorate more rapidly under fluctuating temperatures. Fresh fish flesh is gelatinous. If its temperature is lowered rapidly to −40°C,when thawed the flesh loses little tissue fluid. On the other hand, if slowly frozen the flesh may, when thawed, lose much as "drip." There is a point of electrolyte concentration attained in the tissues which may cause irreversible changes in colloidal structure.

Dehydro-freezing is a combination process of dehydration followed by freezing. As indicated previously, the processes which reduce free water in foods do not damage the quality of the frozen product, providing the removal of water itself does not cause deleterious changes in the food substrate.

Size of Ice Crystals Formed

Under standard conditions the temperature of water must fall below 0°C before ice crystals form. When forming, the temperature of the ice-water slurry returns to 0°C. If ice crystals are allowed to form slowly, relatively large crystals are produced (Fig. 5.3). If the water is made to freeze rapidly, due to rapid heat removal from the system, the ice formed will have a fine texture (Fig. 5.3). Looking at the ice formed, in the first instance there are large, sharp needle-like structures, and in the latter instance, there are many more and smaller ice crystals. If a fine textured ice is partially allowed to melt and refreeze and the process repeated several times, the ice crystals will change from small to large.

Ice cream made simply by placing it in an ice cube tray of a small home refrigerator will develop large ice crystals. The ice cream will have a "coarse" texture, unlike the usual rapidly frozen commercial product which has a velvet-smooth ice crystal system. This phenomenon is equally demonstrated with strawberries. Fruit placed in such a tray and frozen slowly will be lacking characteristic texture when thawed. Piercing ice crystals form, puncturing tissue cells. When thawed the cells spew their contents; the berry is flabby and loses its form.

According to the crystal damage theory, ice crystal growth impairs food quality generally. Slow freezing permits ice crystal growth. The cells of meat, poultry, fish, shellfish, fruits and vegetables all contain jelly-like protoplasm. To fix the original jelly-like mass the rate of freezing must be such that minute crystals form uniformly throughout the tissues. If such quick frozen tissue is thawed immediately,

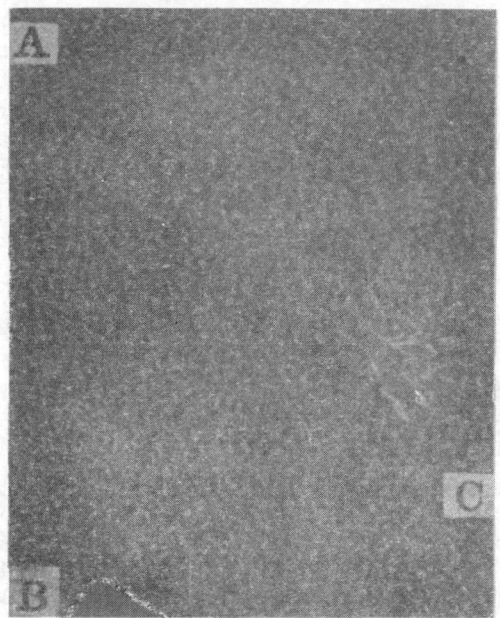

Courtesy of Tressler and Evers

FIG. 5.3. CROSSSECTION OF ASPARAGUS SPEARS FROZEN BY THREE METHODS
A—Immersion freezing; B—Contact freezing; C—Still air freezing.

the water is reabsorbed within the tissues as the ice crystals melt. If the product is frozen slowly, or fluctuating conditions of temperature during storage permit ice crystal growth, cells become punctured and the thawed tissues cannot return to their original jelly-like state. Some of the fluid resulting from the thawing does not become reabsorbed and appears as free liquid.

However, as noted previously, ice crystal growth is but one of the factors influencing the quality of frozen foods. As freezing progresses through a substrate, there is an increasing concentration of electrolytes which may cause irreversible changes in colloidal structures. It is not uncommon to observe coagulated protein in thawed milk, for example, or to find curdled, thawed cream sauces. Damage to the structure of tissues can result from ice crystal growth. Both phenomena occur simultaneously; each is important.

Volume Changes During Freezing

In a freezing environment an unprotected bottle of milk freezes. The milk expands on freezing; the volume of milk increases. The cap sits on top of a protruding cylinder of frozen milk. If the milk bottle has a solidly held cap of metal, one that would resist being pushed from the bottle, then it would burst. The liquid is not compressible and neither is the ice. As the volume increases within the bottle during freezing, either the bottle bursts or the cap is forced free. When freezing food in a rigid container, opportunity for such expansion must be considered. However, not all food products expand during freezing. Strawberry jam does not increase in volume when frozen. The frozen sugary solution occupies the same height in the jar frozen as unfrozen—and it may even occupy less. Water alone can be expected to increase about 10% in volume on freezing. With high sugar concentrations in water, the expansion may be nil and actually a decrease in volume can occur. Strawberries packed in sugar as ordinarily frozen will expand slightly on freezing.

Refrigeration Requirements in Freezing Foods

Hot and cold are relative concepts, requiring a reference point. The term ice carries information. In its common usage it refers to the solid state of water molecules. Water freezes below 0°C. To keep water molecules in the solid state, it is necessary to have them in an environment which permits the solid state to exist.

In colder climates during the winter this presents little difficulty. In warmer climates, it is necessary to create an artificial environment in order to obtain, and maintain, water in a solid state. A condition must first be established such that heat can be removed from liquid water and a change in state permitted to occur. Next, the frozen material must be protected (insulated) to prevent it from acquiring heat, raising the temperature, and thawing the ice formed. There are two distinct problem areas: the problem of bringing a sample to a frozen condition and the problem of maintaining the frozen material in a suitable solid state. In order to establish the refrigeration requirements to achieve this desired state, it is necessary to consider both aspects of the problem.

Establishing the Refrigeration Requirements to Freeze Food.—The temperature at which a food will freeze, under standard conditions, is dependent upon the concentration of solutes in the water phase. The average freezing points for selected foods are given in Fig. 5.1. A temperature of −2°C is usable as the average freezing point of foods in general.

To freeze a food it is first necessary to bring the temperature of the mass down to the freezing point. The work required to accomplish this is shown in the equation:

$$H_1 = (S_L)\,(W)\,(T_i - T_f)$$

where H_1 = kgc[1] required to lower the temperature of the food
from the initial temperature (T_i) to the temperature at which the food freezes (T_f)

S_L = the specific heat of the food above the freezing point of the food (Table 5.5)

W = the weight of the food mass in kilos.

To freeze a food it is necessary to remove the heat of fusion, once the food mass has been brought to the freezing point. The work required to accomplish this is shown as:

$$H_2 = (H_f)\,(W)$$

where H_2 = kgc required to change liquid food at freezing point, to solid state, at freezing point

H_f = heat of fusion (Table 5.5)

W = weight of food in kilos.

TABLE 5.5

SPECIFIC HEAT OF FOODS

Food	Above Freezing	Below Freezing	Latent Heat of Fusion (Btu per lb)[1]
Fish, fresh	0.76	0.41	101
Oysters	0.90	0.46	125
Bacon	0.55	9.31	30
Beef	0.66	0.38	94
Liver	0.72	0.40	94
Pork	0.60	0.38	66
Poultry	0.80	0.41	99
Apples	0.71	0.39	92
Berries	0.89	0.46	125
Peas	0.80	0.42	108
Carrots	0.87	0.45	120
Beans, string	0.92	0.47	128
Asparagus	0.95	0.44	134
Eggs	0.76	0.40	98
Milk	0.90	0.46	124
Butter	0.30	0.24	18

[1] 1 Btu = 0.252 kg calories.

[1] Btu = British Thermal Unit, or the amount of heat required to raise the temperature of 1 lb of water 1°F. A more precise definition is 1/180th the heat required to raise the temperature of 1 lb of water from 0° to 100°C at standard pressure; 1 Btu = 0.252 kg calories.

Once food is frozen, the temperature of the frozen mass must be lowered to the frozen storage temperature. Commercial frozen storage temperature in the United States is usually −18°C, although lower temperatures are often desirable. Work is required to bring the temperature of the frozen mass, at its freezing point, to the temperature of the storage chamber. This can be calculated as:

$$H_3 = (S_s) (W) (T_f - T_s)$$

where H_3 = kgc required to lower the temperature of the mass (W) from the freezing point (T_f) to the storage temperature (T_s)

S_s = the specific heat of the frozen material (Table 5.5).

The refrigeration requirement to freeze food and bring its temperature to frozen storage temperatures is a sum of the values H_1, H_2 and H_3, and is reported usually in terms of the Btu load or kgc load.

The refrigeration requirement (R_f) solely for the frozen food mass is generally reported in terms of tons of refrigeration needed. The value is obtained by changing the kgc load (H_{fs}) into equivalent tons of refrigeration.

The next step is to establish the refrigeration requirement for the chamber to hold the frozen food in a solid state at a suitable given temperature. For purposes of this discussion it is assumed that the temperature of the frozen food chamber is lower than its environment.

In order to maintain a frozen mass of foods at a desired temperature, it is necessary to insulate the food and create an artificial environment. A suitable chamber must be designed in order to create a desired microenvironment. In the case of a frozen food storage chamber, it is necessary to prevent the transfer of heat from surroundings to the frozen food. There are several considerations. First there is no perfect insulation system; there will be heat losses through the chamber walls. Second, to be functional, there must be an access to the frozen chamber; there will be heat losses due to opening and closing the frozen food storage chamber. Third, there will be numerous other losses, including heat given to the chamber by electric lights or running electric motors in the chamber, and, there is heat given off by persons working in the microenvironment. These can be estimated as follows:

The refrigeration requirements to maintain an empty frozen food storage chamber at a desired storage temperature may be calculated. There is work required in maintaining a chamber at a given temperature. The work required in terms of kgc requirements is dependent upon the temperature at which the chamber is to be maintained, the temperature of the outside air, the surface area of the storage chamber, the amount and kind of insulation material applied to the chamber.

The kgc load to maintain a chamber at a given temperature for 24 hr can be calculated from the equation:

$$H_C = \frac{(K)\,(24)\,(S_a)\,(T_1 - T_2)}{I}$$

where H_c = kgc losses per 24 hr in the frozen storage chamber
K = thermal conductivity of the insulation material
S_a = surface area of the outer wall of the storage chamber
T_1 = temperature outside the chamber
T_2 = temperature inside the chamber
I = thickness of the insulation material.

The thermal conductivity (K) and density of commonly used insulating materials are shown in Table 5.6.

TABLE 5.6
THERMAL CONDUCTIVITY AND DENSITY OF SELECTED INSULATING MATERIALS

Material	K[1] (Btu per in.-thick per sq ft of Surface per hr per °F)	Density[1] (lb per cu ft)
Air	0.175	0.08
Asbestos paper and air spaces	0.50	8.80
Balsa wood	0.35	7.5
Celotex	0.30	13.8
Corkboard	0.28	6.9
Glass wool	0.27	1.5
Pure wool (lamb)	0.26	5.0
Masonite	0.33	15.0
Wood (oak)	1.00	38.0
Wood (white pine)	0.79	32.0
Tar roofing	0.71	55.0

[1]See Appendix for metric conversions.

Miscellaneous Heat Loads in Maintaining and Operating a Frozen Food Chamber.—In addition to work required in terms of kgc for both the food mass and the storage chamber, there are heat loads in lighting the chamber, in running motors, in having people operating within the chamber, and the opening and closing of the chamber doors. The latter consideration is one involving the changes of air due to entrances and exits from the chamber. The following values are useful in establishing the miscellaneous heat loads:

Electric lights = 3.42 Btu per hr per watt (13.57 kgc)
Electric motors = 3000 Btu per hr per hp (11,905 kgc)
Working man (within chamber) = 750 Btu per man hour (2975 kgc)

Average air changes per 24 hr are dependent upon the size of the chamber and its use, which may be estimated from Table 5.7.

The average kgc losses due to air changes within a chamber are related to the temperature and relative humidity of the exchanged air.

The heat load contributed by men (H_m) working within the storage chamber can be calculated as:

$$H_m = 2975 \text{ (No. of man hours in chamber) kgc}$$

TABLE 5.7
ESTIMATED AVERAGE AIR CHANGES PER DAY IN CHAMBER

Room Vol (cu m)	Number of Changes (per 24 hr)
9	29
18	20
36	13.5
72	9.3
144	6.3
288	4.3
576	2.9
1,152	2.0
2,304	1.4
4,608	0.8

The heat load to operate an empty frozen food chamber is therefore the sum of the individual heat loads contributing to the operation of the chamber. This requirement is therefore:

Estimated empty chamber heat load = $H_c + H_e + H_m + H_a$

where H_c = kgc requirement to maintain empty chamber at desired temperature

H_e = kgc requirement to operate electrical equipment within the chamber

H_m = kgc load contributed by men working within the chamber

H_a = kgc load added by changing air within the chamber during normal operating circumstances.

The refrigeration requirement for the operation of the chamber (R_c), without any consideration for the food, is therefore obtained using the equation:

$$R_c = H_c + H_e + H_m + H_a$$

The refrigeration requirements (R) in freezing foods and maintaining foods in frozen storage can be calculated from the equation:

$$R = R_f + R_c$$

The refrigeration need for freezing foods is a sum of the requirements for lowering the temperature of the food to its freezing point, freezing the food at that temperature, and lowering the temperature of the frozen food to the temperature of the storage chamber. To this must be added the refrigeration needs in maintaining the temperature of the empty storage chamber, with its complement of lighting and electrical equipment, and under use by man.

A safety factor, in order of magnitude of 10%, is usually added to the calculated refrigeration requirement for an installation. Also, the refrigeration equipment should not operate 24 hr a day, but in the order of 18 hr a day. The equipment needs are then calculated on the basis of having the compressor capable of handling a 24-hr load in 18 hr. This results in an adjustment upwards of the calculated tons of refrigeration required by a factor of 25%, and applies an additional factor of safety.

Freezing in Air

There are two types of air systems for food freezing—still air and forced air. Still air freezing is accomplished by placing packaged or loose foods in suitable freezing rooms. Still air freezing is the cheapest and slowest method. Products remain in the freezing chamber until frozen. The length of time required to freeze the food is dependent upon the temperature of the freezing chamber, the type of food being frozen, the temperature of the food as it enters the freezer, the type, size and shape of the food package (Tables 5.8-5.10) and the arrangement of the packages (size of piles, aisle space, etc.) in the freezer.

Individual strawberries on trays placed in a −32°C room will freeze in 5 hr or less. A 55-gal. barrel of strawberries will take several days

TABLE 5.8

RELATION OF AIR VELOCITIES ON TIME REQUIRED TO
FREEZE CAKE DOUGHNUTS AT -18°C

Air Velocity (m/min)	Freezing Time (min)
0	42
50	18
150	11.5
350	9.5

TABLE 5.9

THE LOWER THE FREEZER TEMPERATURE, THE SHORTER THE TIME
REQUIRED TO FREEZE FOODS. EXAMPLE: CAKE-TYPE DOUGHNUTS

Freezer Temperature (°C)	Freezing Time (min)
18	42
23	23
28	21

TABLE 5.10

RELATION OF AIR VELOCITIES ON FREEZING RATE OF 5 KG BLOCKS
OF FISH FILLETS. AIR TEMPERATURE HELD AT -28°C

Air Velocity (m/min)	Freezing Time (hr)
17	7.5
700	5.5
1000	4.0

to freeze in a still air zero degree room.

The freezing time for a given package of food can be drastically reduced by installing fans in the freezing chamber. Very cold air moved at high speeds results in more rapid freezing. For quick freezing, cold air blasts in an insulated tunnel have been found functional. Tunnel freezing is a commonly used system in the United States, and production rates are available commercially for continuously quick freezing 4000 and more consumer packages per hour.

Dehydration is a serious factor to be considered when freezing unpackaged foods (i.e., individual peas, berries or fish) by either still or blast air methods. Packaging foods prior to freezing has obvious advantages in controlling this dehydration, but has the disadvantage of insulating the food.

Freezing by Indirect Contact with Refrigerants

The early ice-making machines consisted of a metal cannister submerged in a refrigerated brine. Food may be frozen by being placed in contact with a metal surface which is cooled by a refrigerant. Food also may be packed in a can and immersed in a refrigerant. Food may be packaged in paperboard box and the box placed in contact with a refrigerated metal plate. The refrigerated metal plates may be moving in the form of a belt, or be stationary. The refrigerated brine may be still or in turbulent motion.

The Birdseye multiplate freezer (Fig. 5.4) consists of a series of superimposed, hollow metal plates actuated by hydraulic pressure lifts in such a manner that they may be separated to receive packaged food between the plates, then brought together in intimate contact

Courtesy of Tressler et al.

FIG. 5.4.   BIRDSEYE MULTIPLATE FROSTER

Intimate contact between food packages and freezing plate is obtained in
this widely used system.

with the packages. The entire series of plates is enclosed in an insulated cabinet. The freezing time commercially for a 5 cm thick consumer package of meat or fish is in the order of an hour and a half. Continuous freezing systems are commercially available (Fig. 5.5).

Courtesy of Tressler et al.

FIG. 5.5.   FLUIDIZED PEAS ON LEWIS IQF FREEZER BELT

This system features minimum fluidization with maximum heat transfer between air and product.

Direct Immersion Freezing

Direct immersion of a food particle in a liquid refrigerant offers the most rapid method of food freezing (Fig. 5.6). Liquids are good heat conductors, when compared to air or gases. Sodium chloride and sugar solutions have been used as low temperature heat exchanger systems. Food products can be frozen quickly and the contact is intimate between food and refrigerant. High heat exchanges rates can be obtained using turbulent flow techniques.

Food particles or packages can be frozen in liquid baths (Fig. 5.5), in sprays and in fog systems. Individual fruits and vegetables can be frozen in a matter of minutes using suitable brines and temperatures (Fig. 5.6).

Courtesy of Tressler et al.

FIG. 5.6.   THE CRYO-QUICK LIQUID NITROGEN FREEZING SYSTEM

Individual food pieces can be frozen in a matter of minutes using suitable brines and temperatures.

Freezerburn

As mentioned previously, all frozen foods must be packaged to protect them from dehydration (Fig. 5.4) by sublimation during air freezing and in all conditions of frozen storage. Freezerburn irreversibly alters the color, texture, flavor and nutritive value of frozen foods. Roast beef acquires the appearance of light brown paper due to freezerburn. Adequate packaging prevents and controls freezerburn.

Packaging Requirements for Frozen Foods

The snow which accumulates on the surface of a freezer plate or a coil in a frozen storage room comes from moisture vapor from the air in the chamber. The moisture vapor in the atmosphere of the chamber attempts to reach an equilibrium with the materials within the chamber, as well as with the chamber itself. The vapor however is condensed onto the freezing coil or plate, accumulating as snow. As moisture is removed from circulation the moist materials within the chamber yield more water vapor, attempting to satisfy the vapor pressure deficit within the chamber. There is a steady removal of water vapor from the air in the form of ice to the coils; and there will be a constant loss of water in the form of vapor from unprotected materials in the chamber. If the material is food, then the food develops freezerburn.

In order to protect the food, a barrier must be placed around the food to eliminate the constant yielding of moisture vapor to the coils or plates. In addition to the direct protection the package offers the food from desiccation, there is another problem which the package solves. Aside from the loss of moisture vapor, the unprotected food is subject to oxidation and contamination from the atmosphere within the chamber. Substantial deterioration in quality of unprotected food in a frozen storage chamber occurs.

The package must be functional in protecting the food, must lend to mechanical handling (Fig. 5.7), must be economical of space and must be practical in application insofar as costs are concerned. Wooden, metal, glass, paper and plastic materials have been used successfully

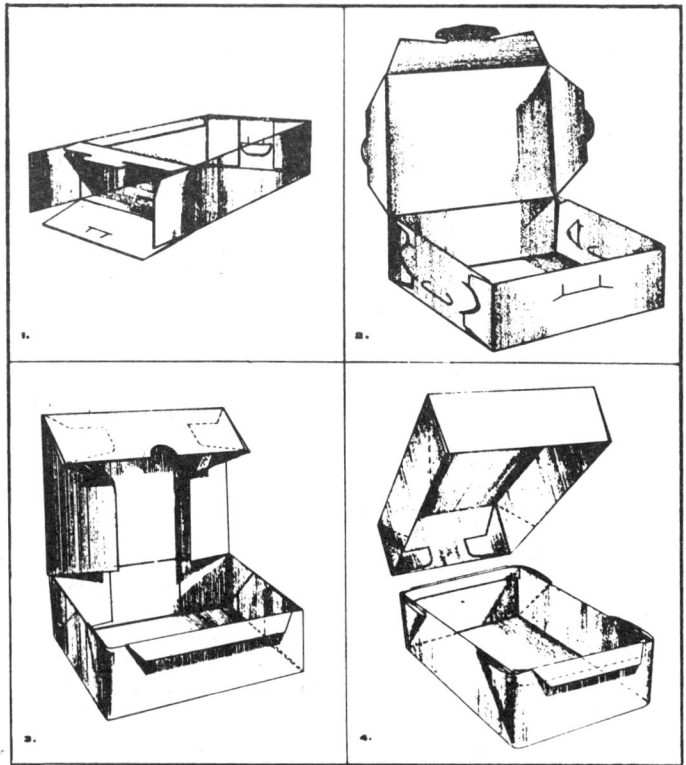

Courtesy of Tressler and Evers

FIG. 5.7.   LINE DRAWINGS OF TYPICAL CARTONS USED FOR FROZEN FOODS

as frozen food containers.

The simplest protective coating possible for frozen foods is a glaze (coating of ice). This has been employed for decades and consists of an ice coating over the frozen commodity. The glaze must be replaced periodically. Glazing has been widely practiced in the fishing industries.

The most complicated packaging problems involve the processing and distribution of consumer-type, individual serving, units of frozen foods (Fig. 5.8).

Courtesy of Tressler and Evers

FIG. 5.8.   CARTON FORMING, FILLING AND CLOSING MACHINES

Carton forming, filling and closing are performed mechanically with great speed.

Influence of Freezing on Microorganisms

Microorganisms may be classed by their optimum temperatures for growth. Most microorganisms do not grow at temperatures below 0°C. However, it is known that there are some yeasts that can grow at temperatures as low as −9°C in nonfrozen substrates. In general yeasts and molds are capable of growing at temperatures much lower than those supporting the growth of bacteria. Slow freezing is damaging to a microbial population. The most susceptible forms of microorganisms are the vegetative cells. Spores are not usually injured by freezing (Table 5.11). As an example, the changes in microflora during the processing of peas is shown in Table 5.12. Raw peas contain a significant number of organisms. Blanching and washing can reduce the population from millions to hundreds per gram. In the course of being processed and packaged, the population tends to increase, depending of course on the elapsed time and the temperature of the peas. The packaged frozen peas may contain less than half the bacteria present in the package just prior to freezing. A greater percentage reduction is not unusual.

The survival of pathogenic bacteria in frozen foods during a period of six months at −17°C can be expected to follow the general patterns shown in Table 5.12. The immediate effect of freezing is seen in a sharp drop in population numbers, followed by a much less drastic killing effect. Repeated freezing and thawing of a bacterial culture in vegetative growth has drastic killing effects.

The temperature at which frozen food is thawed also has a marked influence on the growth of microorganisms.

TABLE 5.11

EFFECT OF FROZEN STORAGE AT -17°C ON SURVIVAL OF *Staph. aureus* IN BROTH AND FISH HOMOGENATES

Media	0	Cell Counts After Days of Storage 40	130	393
Broth	170,000	5,000	1,800	3,500
Fish homogenate	170,000	51,000	22,500	17,700

TABLE 5.12

CHANGES IN MICROBIAL POPULATIONS DURING PROCESSING AND FREEZING OF PEAS

Stage of Process	Avg Microbial Count/Gram
Before blancher	85,000
After blancher	1,000
In reels—end of flume	5,800
On inspection belt	14,000
Elevator to filler	43,000
In filler hopper	49,000
After packaging	83,000

Influence of Freezing on Proteins

While there is little change in the nutritive value of protein due to freezing, it is obviously possible to denature protein by such a treatment. This may be seen in the curdling of proteinaceous materials, especially during repeated freezing and thawing. While the biological value of denatured protein, as a food for man, need not necessarily differ from native protein, the appearance and quality of the food material may be seriously altered by such treatments. Proteolysis may occur in animal tissues during frozen storage if enzymes are not inactivated.

Influence of Freezing on Enzymes

Enzyme activity is temperature dependent. The activity has a pH optimum and is influenced by the concentration of substrate. The activity of an enzyme or a system of enzymes can be destroyed at

TABLE 5.13

TIME IN MINUTES FOR THE INACTIVATION OF PEROXIDASE IN BEANS
ACCORDING TO VARIETIES

| Variety | Inactivation of Peroxidase (min) | |
	(87 °C)	(93 °C)
Tendergreen	4.30	2.68
Wade	5.40	2.87
Topcrop	5.80	3.05
Seminole	5.94	2.84
King Green	5.95	3.08
Tenderlong 15	5.07	2.52

temperatures near 93°C. Enzymes retain some activity at temperatures as low as −73°C, although reaction rates are extremely low at that temperature (Table 5.13).

Animal enzyme systems tend to have optimum reaction rates at temperatures near 37°C. Plant enzyme systems tend to have temperature optima at lightly lower temperatures.

Freezing stops microbiological activity. Enzyme activity is only retarded by freezing temperatures. Enzyme control is most easily obtained by a short heat treatment (blanching) prior to freezing and storage (Table 5.13).

Enzyme activity is stimulated in the supercooled area of the freezing curve over counterpart temperatures in the frozen tissue. The velocity of enzyme reactions is greater in supercooled water than in crystallized water at the same temperature. From Fig. 5.2 it is ap-

parent that there is much noncrystalline water at temperatures above
−9°C. Hence there is sound scientific basis to a commonly held view
that temperatures above −9°C yet below freezing point for food
(−2°C, for example) permit severe damage to the quality of foods, not
only in appearance, but in losses of nutrients. Long-term storage at
−6°C yields unacceptable foods.

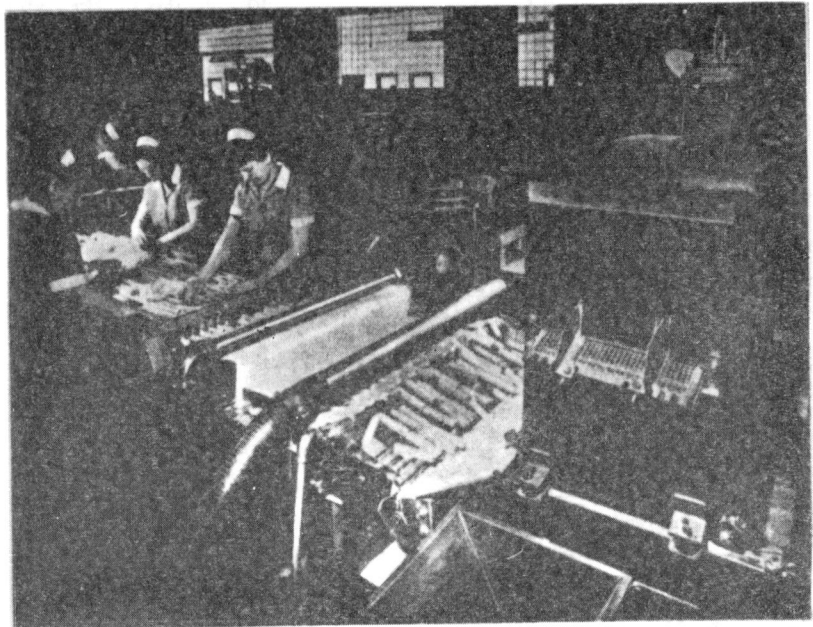

Courtesy of Foxboro Co.

FIG. 5.9.   FREEZING PERMITS PRESERVATION OF PREPARED FATTY FOODS, FOR EXAMPLE,
FRIED FISH STICKS

Forming and coating operations shown.

Influence of Freezing on Fats

Oxidative deteriorations of fats and oils are not uncommon in
frozen foods rich in these nutrients. Fatty fish deterioration is notable.
Fats in frozen fish tissue tend to become rancid quicker than the
fats in frozen animal tissues. Plant tissues are least susceptible. In
the case of meat, pork fat may become rancid after six months storage at zero, while beef fat may retain good quality after two years of
storage at that temperature. At −34°C rancidity development in frozen
fatty tissues is greatly reduced. Rancid fats tend to have lower nutritive values than fresh, sweet fats (Fig. 5.9).

Emulsions of oil in water or water in oil may become destabilized by freezing. This may be serious in prepared precooked frozen foods and foods products. Because fat and oil deteriorations are, in general, temperature dependent, freezing preservation offers maximum potentials in preserving many fatty foods.

Influence of Freezing on Vitamins

The freezing process itself is not destructive of a nutrient. In fact, the lower the temperature of a food, the better will be the retention of nutrients. However, involved in the freezing preservation of food is the preparation and processing of the commodities. During the processing steps nutrient losses may occur. Losses of vitamins occur throughout processing operations; for example, during blanching and washing, trimming and grinding. Exposure of tissues to the atmosphere results in vitamin losses due to oxidation. In general, vitamin C losses will occur when tissues are ruptured and exposed to the air. During storage in the frozen state, vitamin C losses continue. The higher the storage temperature, the greater the destruction of the nutrient. Greater losses are found with vitamin C than with other vitamins in frozen foods. Blanching to inactivate enzymes is important in protecting not only the vitamins but the quality of frozen foods in general (Table 5.14; Fig. 5.10).

TABLE 5.14

LOSSES IN VITAMIN C IN STRAWBERRIES (4-1) DURING
FROZEN STORAGE AT 0°C

Storage Time (Days)	Ascorbic Acid, (mg/100 g)
0	31
60	30.5
120	29
240	26
480	20
720	14

Adding ascorbic acid to fruits prior to freezing is used commercially to protect their quality.

Vitamin B-1 is heat sensitive, and is destroyed in part during blanching to inactivate enzymes. Slight further loss is found during frozen storage at temperatures below zero, in fruits, vegetables, meat and poultry. Vitamin B-2 levels of frozen foods may be lowered during preparation for freezing, but little or no destruction of this nutrient occurs during frozen storage.

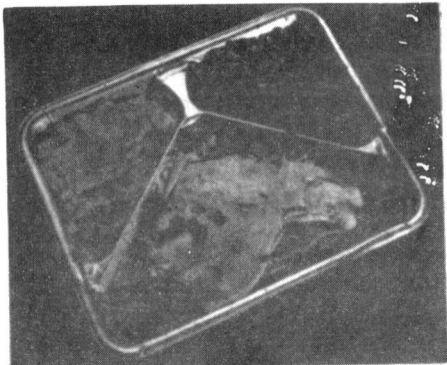

Courtesy of Quick Frozen Foods

FIG. 5.10. FROZEN TURKEY DINNER

A compartmentalized tray of sliced turkey with gravy, green peas and mashed potatoes ready for serving.

Of the fat-soluble vitamins, the carotene precursors of vitamin A are altered little by the freezing of foods although storage losses occur. Blanching of plant tissues improves the storage stability of carotenes.

The storage of foods in a frozen state without package protection leads to oxidation and destruction of many nutrients, including vitamins.

Influence of Freezing on Parasites

Freezing foods has advantages in destroying certain parasites. The best example of this beneficial action is found in *Trichinella spiralis* destruction by freezing. Bringing the temperature of infected food down to −18°C or lower kills all stages of the organism. The recommended treatments to ensure destruction is given in Table 5.15 and

TABLE 5.15

CONDITIONS OF FREEZING FOR THE DESTRUCTION OF TRICHINA

Temperature (°C)	Group 1 (Days)[1]	Group 2 (Days)[2]
15	20	30
23	10	20
29	6	12

[1] Group 1 comprises the meat or product in separate pieces not exceeding 6 in in thickness.
[2] Group 2 comprises the meat or product in pieces or layers or within containers, the thickness of which exceeds 6 but not 27 in.

is a time and temperature relationship. Frozen foods are not suitable for the growth of parasites, and insect infestations obviously cannot occur.

Thawing Damage to Frozen Foods

Refreezing a thawed mass of food may result in important quality changes. These are measurable to some extent.

For beef (Table 5.16) the refreezing curve is displaced upwards over the curve for fresh flesh. Freezing continues at a higher temperature in the refrozen food, although in the final stages of freezing the two curves coincide. In overall view there has been little change in the actual total amount of water frozen, although the freezing point of the thawed flesh is higher.

For lamb, refreezing thawed flesh does not demonstrate the same alteration in pattern of freezing. Thawing and freezing lamb several times does not appear to alter the freezing point of the food mass as is the case with beef. There is no notable change in the freezing curve, with the exception of a slight shift in the level of supercooling. Lamb is characteristic in having negligible losses in fluid on thawing, and also has little structural damage by freezing. On the other hand, beef, as well as carrots, fish and peaches, evidence some freezing damage. Fluid loss from tissues may occur during thawing and altered freezing points with thawed tissues can be demonstrated.

TABLE 5.16

THE DEGREE OF WATER FREEZING AT -15°C INTERVALS OVER THE FREEZING RANGE FOR BEEF AND CARROT FOR AN INITIAL AND A SUBSEQUENT FREEZING OF THE SAME SAMPLE. FREEZING WAS DONE UNDER A CONSTANT THERMAL HEAD OF -6°C

Beef[1]			Carrot[2]		
Temp (°C)	First Freezing Degree Frozen	Second Freezing Degree Frozen	Temp (°C)	First Freezing Degree Frozen	Second Freezing Degree Frozen
-1[3]	0	21	-2[3]	0	13.0
-2	20	64.5	-3	11.2	75.8
-3	42.8	69.2	-4	63.8	77.5
-4	69.2	77.0	-6	89.4	92.2
-6	88.8	81.6	-9	96.0	96.5
-9	94.6	95.4	-12	98.8	98.8
-12	97.3	97.3	-15	99.6	99.6
-15	98.6	98.6	-16	100.0	100.0
-17	99.5	99.5			
-19	100.0	100.0			

[1] Initial temperatures of freezing.
[2] Beef round, U.S. Choice, moisture content 73.6%.
[3] California Bunching, topped, moisture 83.0%

The changes in the freezing point of thawed frozen foods may well be a measure of the changes which occur in freezing plant and animal tissues. Animal tissues appear less prone to freezing and thawing damage than are fruit and vegetable tissues. Fruits are particularly sensitive. To obtain the best performance from frozen foods, storage temperatures should be maintained constant and no higher than $-18°C$. Food is preserved better at $-23°C$ and best at $-34°C$. Food is significantly altered if stored at a higher temperature (Table 5.16).

If frozen concentrated orange juice is processed quickly and carefully, but allowed to thaw and freeze several times during distribution, the thawed juice when consumed will little resemble the juice of fresh oranges in odor, taste or appearance.

Repeated freezing and thawing are detrimental to frozen fruits, vegetables, prepared foods, ice cream, fish and poultry.

Soon after beeves, lambs, hens or fish are killed rigor mortis sets in. As the rigor subsides, due to the action of enzymes in the flesh, a slow decomposition begins. This action in beef may be beneficial. In fish it is objectionable. Meat is tenderized by this autolytic process. Ageing of beef is this controlled decomposition, resulting in more tender meat. With poultry the disappearance of rigor mortis itself yields more tender meat, but ageing (as is done with beef) does not improve the chicken flesh.

Fish and poultry are best quick frozen. Thawed frozen fish has more loss of tissue fluids than meat or poultry. Drip loss can be reduced if fish is cooked quickly before thawing completely.

Nutrient losses occur in the drip from frozen meat. This is shown in Table 5.17 for beef.

TABLE 5.17

SOME NUTRIENT LOSSES IN THE DRIP FROM FROZEN BEEF
THAWED OVERNIGHT AT ROOM TEMPERATURE

Nutrient	Losses (%)
Pantothenic acid	33
Niacin	14.5
Folic acid	8

FREEZING OF BAKERY PRODUCTS

Quick freezing is an almost ideal method of preserving nearly all baked products. Bread frozen and maintained at −22°C or below retains its freshness for many months. Some kinds of frozen cakes also retain their freshness for long periods. At present some bakeries freeze bread and then ship it a considerable distance; at destination, the bread is thawed for sale.

Cakes, cookies, shortcakes, waffles and pancakes are also frozen and marketed frozen. Unbaked doughs and batters of many kinds are also distributed this way. The replacement of the corner bake shop is rapidly occurring. A part of the reason is the great advance being made in bakery technology combined with the advances in freezing (Table 5.18).

TABLE 5.18

ADVANCES IN FREEZING EFFICIENCIES WITH BAKERY PRODUCTS

Product	Freezing Medium		
	Plate Freezer (-30°C)	Cabinet Freezer (-18°C)	Immersion in Liquid Nitrogen
Pound cake (454g)	2½ hr	7 hr	56 sec
Bread (454g)	2¾ hr	8 hr	88 sec
Danish roll	22 min	45 min	4 sec
Iced cake	1¼ hr	3 hr	20 sec

Note: Freezing time 38°C to -18°C.

Packaging

To prevent dehydration and condensation from damaging products, the baked goods must be packaged in a carton or container which is completely impervious to moisture.

Storage Life of Frozen Bread

In contrast to fresh bread, which stales in less than a week, frozen bread becomes stale very slowly. The lower the storage temperature the more slowly it becomes stale. Tressler *et al.* (1968) reported that bread quick frozen immediately after baking and held one year at −18°C was equivalent in freshness to bread held two days at 20°C. In order to retain freshness as well as this commercially, the storage temperature should be −30°C or below, which is not the present practice.

Tressler also reported that the defrosting conditions should be such that no moisture condenses on the cold bread. In still air the rate of defrosting is very slow. At temperatures just below 0°C staling is rapid. Therefore, it is important to thaw the bread and bring it rapidly to room temperature or slightly above. This can be done best with air or low humidity which circulates freely and uniformly around the bread.

Cookies and Cakes

Most cookies contain little moisture and can be frozen and thawed several times without losing their quality. In general, the lower the storage temperature, the longer they can be kept without noticeable deterioration. Cookies will retain their palatability for at least six months at −18°C, and longer at lower temperatures. Cakes are now successfully frozen.

FROZEN DAIRY FOODS

Frozen dairy products (Arbuckle 1972) began as early as 54 A.D. Wines and fruit juices were cooled with ice and snow at the Court of Nero, Emperor of Rome.

Ice cream probably came to America with the early English colonists. A letter written in the year 1700 by a guest of Governor Bladen of Maryland reported that he had been served ice cream. The development of the ice cream industry including important processing methods, ingredient formulations, and merchandizing methods are given in *Ice Cream* by Arbuckle (1972) from which the following has been abstracted. The development of refrigeration (reviewed earlier) played an important part in ice cream making, as the cooling of wines and fruit juices led the way to the freezing of similar liquids and finally to the freezing of milk and cream.

THE ICE CREAM INDUSTRY

The frozen dairy foods industry grew slowly in the United States until after 1900. The annual increase has been quite rapid since that date. During the past 25 years the number of plants manufacturing frozen dairy foods has decreased over 60% but the average production per plant has increased 6-fold.

Basic Ingredients

When commercial ice cream was being introduced in this country, the ingredients were cream, fluid milk, sugar and stabilizer. Later condensed milk, nonfat dry milk and butter became popular ice cream ingredients.

A wide range of choice of ingredients for ice cream is now available from various sources. These ingredients may be grouped as (a) dairy products and (b) nondairy products. The dairy products group is most important as they furnish the basic ingredients of milkfat and milk-solids-not-fat (MSNF) which have essential roles in good ice cream. Some dairy products provide fat, other MSNF, others supply both fat and MSNF, and still others supply bulk to the mix.

The nondairy product group includes sweetener solids, stabilizers and emulsifiers, egg products, flavorings, special products and water.

The basic ingredients in frozen dairy foods are milkfat, MSNF, sweetener solids, stabilizers, emulsifiers and flavoring (Table 5.19).

TABLE 5.19

TYPICAL FORMULA FOR ICE CREAM

Component	Kg	Fat	MSNF	Sugar	Stabilizer	Total Solids
Cream 40%	300.0	120.0	16.2			136.2
Condensed skim milk (27%)	247.6		66.9			66.9
Skim milk	299.4		26.9			26.9
Sugar	150.0			150.0		150.0
Stabilizer	3.0				3.0	3.0
Total	1000.0	120.0	110.0	150.0	3.0	383.0

Source: Arbuckle (1972).

Manufacture of Ice Cream

Ice cream means the pure, clean, frozen products made from a combination of milk products, sugar, dextrose, corn syrup in dry or liquid form, water, with or without egg or egg products, with harmless flavoring and with or without harmless coloring, and with or without added stabilizer or emulsifier composed of wholesome edible material. It shall contain not more than 0.5% by weight of stabilizer and not more than 0.2% by weight of emulsifier, not less than 10% by weight of milkfat and not less than 20% by weight of total milk solids; except when fruit, nuts, cocoa, chocolate, maple syrup, cakes, or confection are used for the purpose of flavoring, then such reduction in milkfat and in total milk solids as is due to the addition of such flavors shall be permitted, but in no such case shall it contain less than 8% by weight of milkfat, nor less than 16% by weight of total milk solids. In no case shall any ice cream weigh less than 2.04 kg per 3.78 liters or contain less than 0.73 kg of total food solids per 3.78 liters.

The Mix

The mix consists of all ingredients with the exception of flavorings, fruits, and nuts. The amount of ingredients needed is accurately calculated and is carefully compounded to give the proper composition and balance of fat, solids-not-fat, sugar, and stabilizer. Only the highest quality products should be used. The use of inferior products will result in an inferior ice cream and reduced sales.

The properties of the formulated mix should be such that it has the proper viscosity, stability, and handling properties and such that the finished ice cream will meet the conditions which prevail in the plant where it is to be produced.

A typical mix formula for ice cream of a good average composition mix of 12% butterfat, 11% MSNF, 15% sugar, 0.3% stabilizer and 38.3% total solids (Table 5.19).

The basic steps of production in manufacturing ice cream are composing and blending the mix, pasteurization, homogenization, cooking, ageing, flavoring, freezing, packaging, hardening and storage.

The diagram shown in Fig. 5.11 presents a flow chart of the typical processes used in the manufacture of different frozen dairy foods.

The first step of processing is composing the mix. This procedure may range in scope from the small batch operation where each ingredient is weighed or measured and added, to the large pushbutton oper-

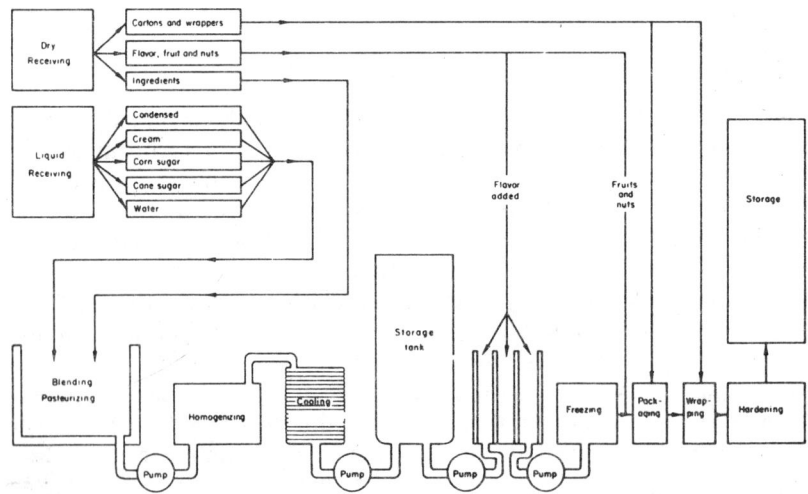

Courtesy of Arbuckle

FIG. 5.11. FLOW CHART FOR ICE CREAM PROCESSING

ation where many of the ingredients are metered into the batch. The common procedure is to: (1) add liquid materials (cream, milk, or other liquid milk products) to mix vat or pasteurizer; (2) apply heat (optional) and then add dry solids such as egg yolk, gelatin, etc. Mixing dry products with three parts of sugar and adding to the mix will aid in their dispersion; (3) add sugar when the mix reaches approximately 49°C; and (4) use caution to ensure that all materials are dissolved before pasteurization temperature is reached.

Pasteurization

The pasteurization process consists of heating products for approved temperature and time (Table 5.20).

Pasteurization (1) renders the mix free of harmful bacteria, (2) brings into solution and aids in blending the ingredients of the mix, (3) improves flavor, (4) improves keeping quality, and (5) produces a more uniform product. There is a trend toward the higher temperature processes.

TABLE 5.20

HEAT TREATMENTS FOR ICE CREAM MIX

Method	Temperature (°C)	Time
Holding method	71	30 min
High temperature-short time	79	25 sec
Vacreation	90	for instant (about 3 sec)
Ultra high temperature	115	for instant

Source: Arbuckle (1972).

When the batch pasteurization method is used the mix is pasteurized by heating to 68° or 71°C and held at that temperature for 30 min, thus killing all pathogenic types of bacteria and all or nearly all other objectionable organisms. The required time and temperature of pasteurization varies in different localities depending on state and city laws and ordinances. For that reason, temperature of pasteurization is carefully controlled and recorded.

When the continuous method of pasteurization is used the high-temperature, short-time (HTST) treatment is used. Mix processing arrangements vary greatly with the HTST treatment and in some arrangements the hot mix is homogenized before pasteurization is accomplished. Immediately following pasteurization, the hot ice cream mix is passed through the high-pressure pump of the homogenizer at pressures that range from 1500 to 3000 psi; this machine breaks the particles of butter fat into very small globules.

Homogenization

The purpose of homogenization is to produce a homogeneous mix. The hot mix is pumped from the pasteurizer through the homogenizer. The optimum homogenization temperature may range as high as 82°C. This process reduces the size of fat globules.

The advantages of homogenization are: (1) thoroughly blends the ingredients of the mix; (2) breaks up and disperses the fat globules, thus preventing churning of the fat during freezing; (3) improves the texture and palatability of the ice cream; (4) makes possible the use of different ingredients; (5) reduces ageing and aids in obtaining overrun; and (6) produces a more uniform product.

Homogenization reduces the size of fat globules to less than 2μ ($1\mu = \frac{1}{25,000}$ in.).

The correct amount of pressure to apply for a given mix is influenced by the type of homogenizer, temperature of mix (low temperature-lower pressure), acidity of mix (high acid-lower pressure), and composition of mix (high fat, stabilizer, and solids require low pressure to prevent excessive viscosity).

Cooling

The smooth mix then flows to a cooler where the product is cooled as rapidly as possible in order to prevent bacterial growth. The cooler chills the mix to a temperature of 4°C or colder. After chilling, the mix may go directly to the freezers, or it may go to small flavor tanks where liquid flavorings like vanilla or chocolate are added, or it may go to so-called ageing tanks. These tanks (flavor or ageing) are insulated to maintain the temperature of the mix at 4°C or lower. If the mix is aged, it is held from 4 to 24 hr. The ageing step is believed to be necessary if gelatin is used as the stabilizer. By ageing the mix, the gelatin has time to set and better accomplish the purpose for which it was added. Most authorities agree that a 4-hr ageing is ample, but many plants prepare a mix one day and hold it overnight for freezing the next day. Most vegetable stabilizers set up immediately upon being cooled and if one of them is used, there appears to be less advantage in ageing.

Changes that take place during ageing include: (1) combination of stabilizer with water of the mix; (2) the fat solidifies; (3) the proteins may change slightly; (4) increase in viscosity; and (5) mix ingredients may become more stable.

Under present operating conditions, an ageing period of 3 or 4 hr is satisfactory. Prolonged ageing beyond 7 days may result in abnormal product properties.

Freezing

Freezing the mix is one of the most important steps in the making of ice cream. The freezing process should be accomplished as rapidly as possible to ensure small ice crystals and a smooth texture in either the batch freezer or the continuous freezer. The function of the freezing process is: (1) to freeze a portion of the water of the mix and (2) to incorporate air into the mix. There are four phases of the freezing process: (1) lowering the temperature from the ageing temperature (4°C) to the freezing point of the mix; (2) freezing a portion of the water of the mix; (3) incorporating air into the mix; and (4) hardening the ice cream after it is drawn from the freezer.

Freezing involves refrigerating the mix in a freezer cylinder which is surrounded by sub-zero ammonia or brine, as today most plants use the continuous freezer. The cylindrical freezer is provided with blades which scrape the freezing mix from the refrigerated metal walls of the machine. During the whipping, air is forced into the mix increasing the volume of the frozen ice cream. Without this overrun ice cream would be an almost inedible hard frozen mass.

The temperatures at which the mix starts to freeze varies with the percentage of total solids, but for the average formula that temperature is approximately −2.8°C. When the ice cream is drawn from the freezer, its temperature will usually range from −3.7° to −6.7°C.

The freezing time and temperature is affected by the type of freezer used: batch freezer, the freezing time to 90% overrun approximate is about 7 min and the drawing temperature is about −4° to −3°C; continuous freezer, the freezing time to 90% overrun approximate is about 24 sec and the drawing temperature is about −6° to −5.6°C; counter freezer, the freezing time to 90% overrun approximate is about 10 min and the drawing temperature is about −3°C; and for the soft serve freezer, the freezing time to 90% overrun approximate is about 3 min and the drawing temperature is about −8° to −7°C. About half the water of the mix is frozen in the freezer and most of the remaining water is frozen in the hardening room.

If fruit is to be added to the mix, the soft ice cream coming out of the freezer is run through a flavor feeder machine which adds the fruit. The soft ice cream then goes to a packaging machine where it may be packed in bulk containers or small packages for retail sale.

Hardening

From the packaging machine the product goes to the hardening room where the freezing process is completed. In the hardening room

(Fig 5.12) the packaged soft ice cream becomes firm within 24 hr. Hardening rooms are maintained at a temperature of −23° to −34°C, either with or without forced air circulation. The ice cream is now ready for storage or delivery.

Courtesy of Delvale Dairies

FIG. 5.12. ICE CREAM HARDENING ROOM

In the hardening room, the soft ice cream from the freezers becomes firm within 24 hr. Unit air cooler and ducts are shown which maintain this room temperature of -29°C.

HAZARD ANALYSIS

Frozen food processors use various control measures to ensure the production of safe and wholesome frozen foods. Hazard analysis differs from such efforts in that it provides a systems approach (Peterson and Gunnerson 1975) for estimating the risk in producing a frozen food product. The assessment of all possible hazards, the elimination of avoidable hazards, and setting limits for those hazards not possible to eliminate are the basis of a rational systematic technique.

Hazard Categories

The analysis of hazards, according to experts, is based on three general hazard characteristics: (1) A frozen product contains a "sensitive" ingredient or ingredients assumed to be potential sources of contamination under normal circumstances. (2) The manfacturing process employed does not contain a controlled step that effectively destroys harmful bacteria. (3) There is substantial potential for the abuse of frozen products in distribution or in handling by consumers that could render the product harmful when eaten due to the result of microbial growth.

The combination of these factors is used to classify frozen foods as to the consumer's risk. If all three hazard characteristics are present in a product, it would have a hazard classification of + + +. If a hazard is absent a 0 would designate this, such that: 0+ + = no sensitive ingredient; +0+ = product pasteurized; and 000 = no hazard involved.

The hazard categories in order of decreasing risk are:

Category I: A special category of nonsterile products designed and intended for consumption by infants, the aged or the infirm.

Category II: Food products subject to all three general hazard characteristics (+ + +).

Category III: Food products subject to two general hazard characteristics (+0+, + +0 or 0+ +).

Category IV: Food products subject to one general hazard characteristic (+00, 0+0 or 00+).

Category V: Food products subject to none of the general hazard characteristics (000).

Examples of Hazards

Frozen foods range in complexity from a single-component, single-product process to a multicomponent series of processes and assembly operations. The following examples illustrate the scope and application of hazard analysis:

Single-component, Single-product Processes: The production of frozen corn, peas, green beans, broccoli and other vegetables are in Category III, a single-component, single-product, straight-line process.

Multicomponent Blended Products: A straight-line process for frozen soft-filled bakery products, essentially a single product with a complex mixture of ingredients.

Multicomponent Products with Add-on: The use of fish for battering, breading and frying operations is an example of a multicomponent product. The "add-on" comes when a sauce is added to the seafood.

Multicomponent, Multiprocess Assembly Operations: Frozen prepared chicken dinners are component prepared foods. Examples of hazard categories for ingredients are:

Butter-margarine—Category III
Cheese—Category II *(Salmonella, Staphyloccocus)*
Egg products—Category II *(Salmonella)*
Flour—Category III
Milk—Category I
Milk solids—Category II *(Salmonella)*
Potatoes—Category III
Seasonings—Category III
Seasonings, spices—Category IV
Shortening—Category IV
Vegetables—Category III

It is important to emphasize that it is not possible to create by inspection either quality or safety in frozen foods. High-quality, safe, wholesome frozen foods are created by determined, knowledgeable, careful, well-organized and well-implemented programs designed to anticipate and prevent problems (Fig. 5.13).

Courtesy of Greer Division of
Joy Manufacturing

FIG. 5.13.   CONVEYING PIES TO A MULTI-TRAY FREEZER

The upper level of pies is on the way to a blast freezer; the pies
below are leaving the freezer for packaging.

Thaw Indicators

The quality of frozen foods depends on the temperature of the product not only while it is in frozen storage but also during distribution. It becomes important to monitor the temperature of frozen foods. There are three types of indicators available.

Defrost: Those indicators which react with a color change when a preselected temperature is reached. The temperature can be any desired temperature ($-10°$, $-5°$, $0°C$, etc.). This type sometimes delays visibility for a period of minutes or hours. Some types require being above the selected temperature for a given period of time.

Such devices may have little value in monitoring ordinary frozen foods because most are not measurably changed by a single exposure for a short time to elevated temperature. Materials that can be made unusable by exposures, either high or low, are examples of where the use of such indicators can be meaningful.

Time-Temperature Integrators: They start to react as soon as they are activated at a preselected combination of time and temperature. For use with frozen foods, manufacturers make the combination of time and temperature related to the rate of disappearance of quality in the food.

Time-temperature integrators change color gradually and develop a quantitative time-temperature equivalent. They do not indicate whether it has been greatly exceeded. The use of a series of integrators, each having a different response capacity, is recommended.

Time-Temperature Integrator/Indicators: These have a graduated scale. They start to react as soon as the activating temperature is reached and at a rate proportional to the storage temperature. Byrne (1976) reports this is the most promising of the available types of devices.

Time-temperature integrators and time-temperature integrator/indicators show accumulated effects of repeated exposures to varying temperatures. As temperature indicators become available, it is becoming apparent they can be useful in promoting good quality.

SUMMARY

A technology of food preservation by freezing has evolved in just a few decades, which supports and sustains a major new industry—the frozen food industry. Its success was contingent upon the development and successful operation of a network of satellite frozen food factories, warehouses and distribution centers, and frozen food cabinets in supermarkets and homes.

Courtesy of Hussmann Refrigeration, Inc.

FIG. 5.14.   FULLY-INDEXED FROZEN FOOD DISPLAY CABINET

Convenience foods are now evolving, taking advantage of the new industry's resources, and perhaps to a point unbelievable to even the most optimistic of the early pioneers in the field. Yet, the industry obviously is only in its infancy and is localized in its effectiveness from a worldwide point of view.

The reader's attention is directed at this point to Chapter 14 where the influence of long term storage of foods at sub-zero temperatures is discussed in more detail.

REFERENCES

AM. FROZEN FOOD INST. 1971A. Good commercial guidelines of sanitation for the frozen vegetable industry. Tech. Serv. Bull. 71. American Frozen Food Institute, Washington, D.C.

AM. FROZEN FOOD INST. 1971B. Good commercial guidelines of sanitation for frozen soft filled bakery products. Tech. Serv. Bull. 74. American Frozen Food Institute, Washington, D.C.

AM. FROZEN FOOD INST. 1971C. Good commercial guidelines of sanitation for the potato product industry. Tech. Serv. Bull. 75. American Frozen Food Institute, Washington, D.C.

AM. FROZEN FOOD INST. 1973. Good commercial guidelines of sanitation for frozen prepared fish and shellfish products. Tech. Serv. Bull. 80. American Frozen Food Institute, Washington, D.C.

ASHRAE. 1969. Refrigerating Data Book. American Society of Heating, Refrigeration and Air-conditioning Engineers, New York.

ARBUCKLE, W.S. 1972. Ice Cream, 2nd Edition. AVI Publishing Co., Westport, Conn.

ARBUCKLE, W.S. 1976. Ice Cream Service Handbook. AVI Publishing Co., Westport, Conn.

BAUMAN, H.E. 1974. The HACCP concept and microbiological hazard categories. Food Technol. 28, No. 9, 30-34, 70.

BECHTEL, W.G., and KULP, K. 1960. Freezing, defrosting, and frozen preservation of cake doughnuts and yeast raised doughnuts. Food Technol. 14, 391-394.

BOCKIAN, A.H., and AREF, M. 1958. Some effects of sweeteners of frozen fruits used for preserve manufacture. Food Technol. 12, 393-397.

BYRAN, F.L. 1974. Microbiological food hazards today—based on epidemiological information. Food Technol. 28, No. 9, 52-66, 84.

BYRNE, C.H. 1976. Temperature indicators—state of the art. Food Technol. 30, No. 6, 66-68.

CORLETT, D.A., JR. 1973. Freeze processing: Prepared foods, seafood, onion and potato products. Presented to FDA Training Course on Hazard Analysis in a Critical Control Point System for Inspection of Food Processors, Chicago, July and August.

DESROSIER, N.W. and TRESSLER, D.K. 1977. Fundamentals of Food Freezing. AVI Publishing Co., Westport, Conn.

DOAN, F.J. 1952. Frozen concentrated milk. Food Technol. 6, 402-404.

FARLEY, J.F. 1958. A Study of the Degree of Freezing of Foods. M.S. Dissertation, Purdue Univ., Lafayette, Ind.

GUADAGNI, D.G., DOWNES, N.J., SANCHUCK, D.W., and SUMIKO, S. 1961. Effect of temperature of stability of commercially frozen bulk pack fruits—strawberries, raspberries, and blackberries. Food Technol. 15, 207-209.

GUADAGNI, D.G., and KELLY, S.H. 1958. Time-temperature tolerance of frozen foods. XIV. Ascorbic acid and its oxidation products as a measure of temperature history of frozen strawberries. Food Technol. 12, 641-644.

JACOBS, M.B. 1951. The Chemistry and Technology of Food and Food Products. Interscience Publishers, New York.

JAHNS, F.D., HOWE, J.L., CODURI, R.J., and RAND, A.G. 1976. A rapid visual enzyme test to assess fish freshness. Food Technol. 30, No. 7, 27-30.

KAUFFMAN, F.L. 1974. How FDA uses HACCP. Food Technol. 28, No. 9, 51, 84.

KRAMER, A. 1973. Storage retention of nutrients. Food Technol. 28, No. 1, 50-60.
KRAMER, A., and FARQUHAR, J.W. 1976. Testing of time-temperature indicating and defrost devices. Food Technol. 30, No. 2, 50-53, 56.
LACHANCE, P.A., RANADIVE, A.S., and MATAS, J. 1973. Effects of reheating convenience foods. Food Technol. 27, No. 1, 36-38.
LEUNG, H., MORRIS, H.A., SLOAN, E.A., and LABUZA, T.P. 1976. Development of an intermediate moisture processed cheese food product. Food Technol. 30, No. 7, 42-44.
LIBBY, W.F. 1951. Radiocarbon dates. Science 114, 291-296.
LINDQUIST, F.E., DIETRICK, W.C., and BOGS, M.M. 1950. Effect of storage temperature on the quality of frozen peas. Food Technol. 4, 5-9.
LIVINGSTON, G.E., ANG, C.Y.W., and CHANG, C.M. 1973. Effects of food service handling. Food Technol. 27, No. 1, 28-34.
MATCHES, J.R., and LISTON, J. 1968. Low temperature growth of Salmonella. J.Food Sci. 33, No. 6, 641-645.
MATZ, S.A. 1972. Bakery Technology and Engineering, 2nd Edition. AVI Publishing Co., Westport, Conn.
MCARDLE, F.J., and DESROSIER, N.W. 1954. A rapid method for determining the pericarp content of sweet corn. Canner 118, 12-15.
MUNDT, O.J., ORVIN, J., and MCCARTY, I.E. 1960. Factors affecting the blanching of green beans. Food Technol. 14, 309-311.
MUNTER, A.M., BYRNE, C.H., and DYKSTRA, K.G. 1953. A survey of times and temperatures in the transportation, storage, and distribution of frozen foods. Food Technol. 7, 356-366.
NATL. ACAD. SCI. 1969. Classification of food products according to risk. An evaluation of the Salmonella problem. NAS—NRC Publ. 1683.
NICKERSON, T.A. 1954. Lactose crystallization in ice cream. Control of crystal size by seeding. J. Dairy Sci. 37, 1099-1105.
PEARSON, A.M., et al. 1951. Vitamin losses in drip obtained upon defrosting frozen meat. Food Res. 16, 85-87.
PENCE, J.W., et al. 1955. Studies on the preservation of bread by freezing. Food Technol. 9, 495-499.
PETERSON, A.C., FANELLI, M.J., and GUNDERSON, M.F. 1968. Microbiological problems. In The Freezing Preservation of Foods, 4th Edition, Vol. 4, D.K. Tressler, W.B. Van Arsdale, and M.J. Copley (Editors). AVI Publishing Co., Westport, Conn.
PETERSON, A.C., and GUNDERSON, M.F. 1968. Microbiology of frozen foods. In The Freezing Preservation of Foods, 4th Edition, Vol. 2, D.K. Tressler, W.B. Van Arsdale, and M.J. Copley (Editors). AVI Publishing Co., Westport, Conn.
PETERSON, A.C., GUNNERSON, R.E. 1975. Microbiological critical control points in frozen foods. Food Technol. 28, No. 9, 37-44.
RAJ, H., and LISTON, J. 1961. Survival of bacteria of public health significance in frozen sea foods. Food Technol. 15, 429-434.
SINNAMON, H.I. 1968. Effect of prior freezing on dehydration and rehydration of apple half segments. Food Technol. 22, No. 2, 101-103.
SMITH, C.A., JR., and SMITH, J.D. 1975. Quality assurance system meets FDA regulations. Food Technol. 29, No. 11, 64-68.
SPLITTSTOESSER, D.F., WETTERGREEN, W.P., and PEDERSON, C.S. 1961. Control of micro-organisms during preparation of vegetables for freezing. Food Technol. 15, 332-334.

TRESSLER, D.K., and EVERS, C.F.   1957.   The Freezing Preservation of Foods. AVI Publishing Co., Westport, Conn. (Out of print).

TRESSLER, D.K., VAN ARSDEL, W.B., and COPLEY, M.J.   1968.   The Freezing Preservation of Foods, 4th Edition, Vol. 1, 2, 3 and 4.   AVI Publishing Co., Westport, Conn.

TURNBOW, G.D., TRACY, P.H., and RAFFETTO, L.A.   1949.   The Ice Cream Industry.   John Wiley and Sons, New York.

U.S. DEP. AGRIC.   1967.   Market Diseases of Fruits and Vegetables.   U.S. Dep. Agric., Agric. Handb. 66.

VAN ARSDEL, W.B.   1957.   The time-temperature tolerance of frozen foods. Food Technol. 11, 28-33.

WATSON, P.P., and LEIGHTON, A.   1927.   Some observations on the freezing points of various cheeses.   J.Dairy Sci. 10, 331-334.

WEIL, B.H., and STERNE, F.   1948.   Literature Search on Preservation of Foods by Freezing.   Georgia School of Technology, Atlanta.

WILLIAMS, E.W.   1976.   A 1976 report on the frozen food industry.   Int. Quick Frozen Foods 17, No 5, 16.

WINTER, F.M., YORK, G.K., and El-NAKHAL, H.   1971.   Quick-counting method for estimating the number of viable microbes on food and food processing equipment.   Appl. Microbiol.22, 89.

WINTER, J.D., KUSTRULID, A., NOBLE, I., and ROSS, S.R.   1952.   The effect of fluctuating storage temperature on quality of stored frozen foods. Food Technol. 6, 311-318.

WLADYKA, B.J., and DAWSON, L.E.   1968.   Essential amino acid composition of chicken meat and drip after 30 and 90 days of frozen storage.   J. Food Sci. 33, No. 5, 1453-1455.

ZABIK, M.F. and FIGA, J.E.   1968.   Comparison of frozen, foam dried, freeze dried and spray dried eggs.   Food Technol. 22, No. 9, 119-125.

Chapter 6

Principles of Food
Preservation by Canning

THE ART OF "APPERTIZING"

France in the late 1790s was at war and having difficulty feeding its people. Napoleon's fighting forces had a diet of putrid meat and other items of poor quality. The foods available couldn't be stored or transported except in a dry state. Recognizing an important problem, a prize was announced offering 12,000 francs and fame to anyone inventing a useful method of food preservation.

Nicolas Appert, a French confectioner working in a simple kitchen, observed that food heated in sealed containers was preserved if the container was not reopened or the seal did not leak. He modestly called the process "the art of Appertizing." Appert received the award from Napoleon after spending ten years proving his discovery.

It should be appreciated that the cause of spoilage of food was unknown. The great scientists of the day were summoned to evaluate Appert's process and offer explanations for its apparent success. The conclusion reached was that the process was successful because in some mysterious and magical fashion air combined with food in a sealed container, preventing putrefaction. This was quite incorrect. Nevertheless, the canning process was discovered and practiced for the next 50 years with some success, but in the darkness of ignorance.

Appert began work on his process in 1795. Peter Durand received patents in England in 1810 for glass and metal containers for packaging foods to be canned. The tin-plated metal containers were called "canisters" from which the term "can" is assumed to be derived. Early metal containers (Fig. 6.1) were bulky, crude and difficult to seal. By 1823 a can with a hole in the top was invented, allowing the food to be heated in boiling water baths with the hole covered with a loose

Courtesy of American Can Co.

FIG. 6.1.   CAN MAKING IN THE EARLY 1800S IN EUROPE

lid. The lid was soldered into place after the heat treatment. Hole-in-top cans are in use presently for canned evaporated milk, although the cans are sealed prior to heating.

By 1824 Appert had developed schedules for processing some 50 different canned foods. Meats and stews processed by Appert were carried by Sir Edward Perry in 1824 in his search for a northwest passage to India. Several cans of food from this voyage were obtained from the National Maritime Museum in London in 1938 and opened. The food was found nontoxic for animals. Interestingly there were isolated from these canned products bacteria which had been dormant for at least 114 years. Given proper environment and substrate, they grew!

In the 1820s canning plants appeared in the United States in Boston and New York. By 1830 sweet corn was being processed in Maine. By 1840 canneries began appearing throughout the United States.

Temperature vs. Pressure

In 1851 Chevalier-Appert invented an autoclave which lessened the danger involved in the operation of steam pressure vessels. It was recognized that some foods could be processed for shorter times if higher temperatures were available. It was learned that the temperature of boiling water could be increased by adding salt. Demands

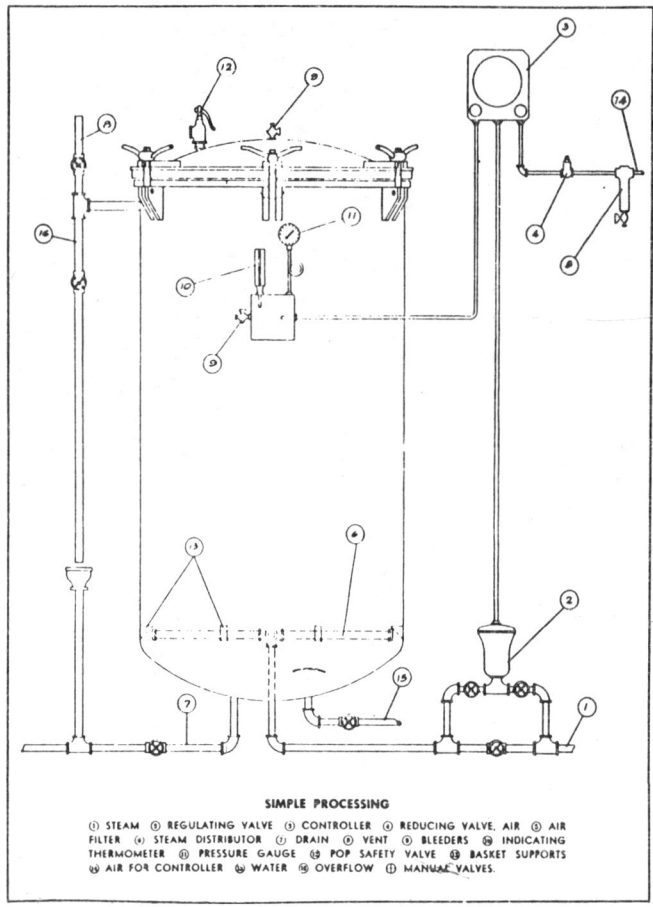

SIMPLE PROCESSING

① STEAM ⑤ REGULATING VALVE ③ CONTROLLER ④ REDUCING VALVE, AIR ⑤ AIR FILTER ⑥ STEAM DISTRIBUTOR ⑦ DRAIN ⑧ VENT ⑨ BLEEDERS ⑩ INDICATING THERMOMETER ⑪ PRESSURE GAUGE ⑫ POP SAFETY VALVE ⑬ BASKET SUPPORTS ⑭ AIR FOR CONTROLLER ⑮ WATER ⑯ OVERFLOW ⑰ MANUAL VALVES.

Courtesy of Continental Can Co.

FIG. 6.2-A. A SIMPLE STEAM PROCESSING SYSTEM COMMONLY USED

High pressure vessel which permits increased vapor pressure, hence increased temperature.

for greater production in factories could be met if the cooking times for foods could be reduced. For instance, the boiling water bath cooking of canned meats could be reduced from 6 hr to perhaps ½ hr by cooking the cans in a water-calcium chloride solution. Production could be increased thereby from some 2000 to 20,000 cans per day. Losses due to failure of containers were large. No pressure was applied to the cooking vessels. Commercial cans were unable to withstand the internal pressures developed by heating to 115°C.

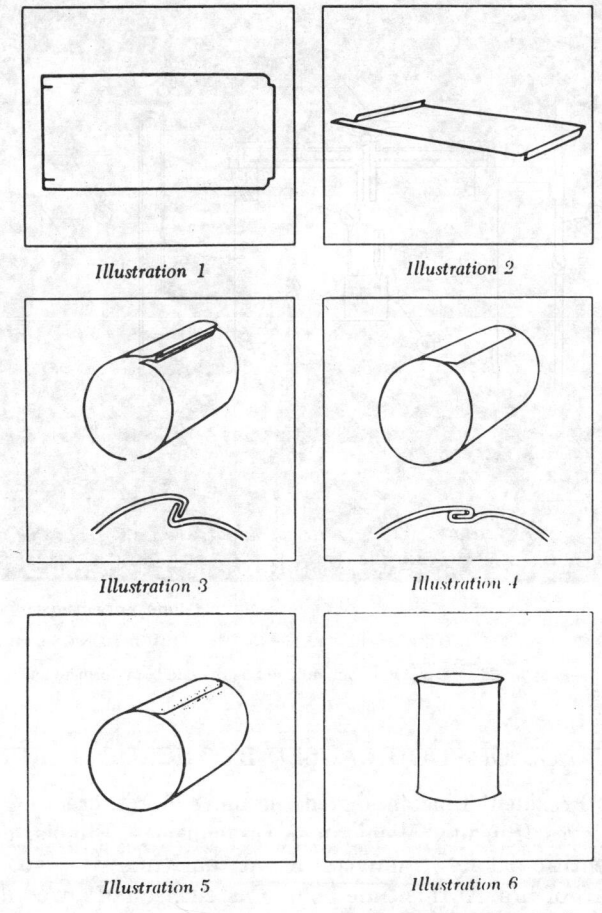

Illustration 1

Illustration 2

Illustration 3

Illustration 4

Illustration 5

Illustration 6

Courtesy of American Can Co.

FIG. 6.2-B. FABRICATION OF SANITARY CAN

Steps in the forming of cans. One end of can is sealed into place by the
can manufacturer, the other end by the food processor.
(1) Body blanks are notched; (2) hooked; (3) hooked blanks formed
around bodymaker; (4) hooked blank is flattened to form side seam; (5)
side seam solder applied to outer surface; and (6) ends of body are curled
outwardly by special form to make the "flange."

The temperature at which water will boil is dependent upon the
pressure. Using a pressure vessel it was possible to achieve temperatures in the vicinity of 115°C. However, these retorts were still
dangerous to operate. (Present counterpart retort design is shown in
Fig. 6.2-A and can fabrication in Fig. 6.2-B and 6.2-C).

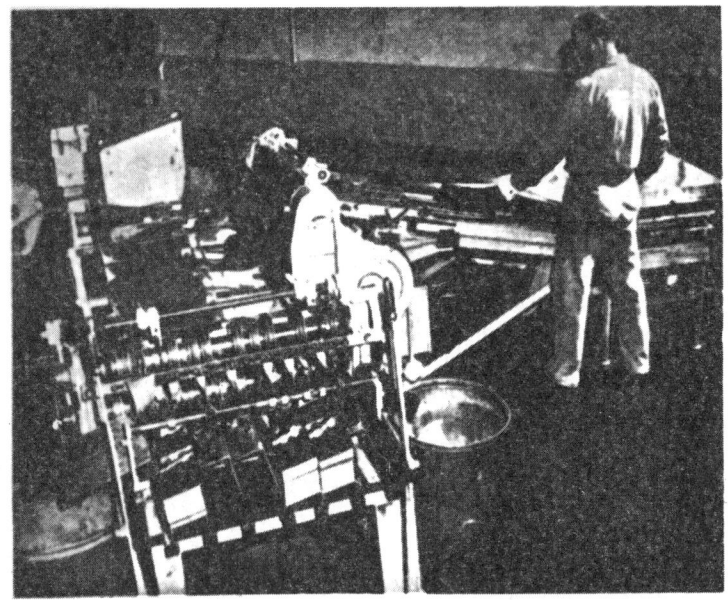

Courtesy of American Can Co.

FIG. 6.2-C.   CUTTING MACHINE—A STEP IN THE FABRICATION OF CANS

Sheets of tinned steel plate are cut, notched, and fed to body-forming unit.

SPOILAGE OF FOOD CAUSED BY MICROORGANISMS

In 1862 President Lincoln signed the Morrill Act, creating the land grant colleges (Purdue, Michigan, Massachusetts, Illinois, etc.). The great scientific debate in universities at that time was "spontaneous generation" of life. At this time Louis Pasteur, son of a well-decorated officer in Napoleon's army, became interested in the problems of the great wine and beer industries of France which were threatened with ruin; their products were diseased and souring from "spontaneous generation" of life in bottles and kegs.

To the Academy of Sciences in France in 1864, Pasteur reported that he had found the cause of the disease of wine and beer to be a microscopic vegetation. When given favorable conditions this vegetation grew and spoiled the products. However, boiled wine sealed from contamination in jars with even cotton plugs would not sour. In fact, it was possible to isolate this microscopic vegetation from the cotton plugs! It was this microscopic growth which spoiled foods, and it was necessary for such organisms to gain entrance to heated foods if they were to spoil! Here was an explanation for the success of Appert more than half a century before. The concept of heat treating foods to inactivate pathogenic organisms is termed appropriately "pasteurization" today.

It is interesting to note that magnifying lenses were used by Bacon in the late 1200s, but had never been focused on a drop of water until the 1600s by Leeuwenhoek. He had noted microscopic growth which he named "animalcules," but they were only a curiosity in water to him. Two more centuries elapsed before this information was organized and synthesized into an explanation for "spontaneous generation" of life.

Appert had established that containers of food must be carefully sealed and heated. Cleanliness was important to his process, although he did not know that microorganisms were the agents of spoilage. Pasteur established several important principles. Most changes in wine depended on the development in it of microorganisms which were themselves the spirits of disease. Germs were brought by air, ingredients, machinery and even by people. Whenever wine contained no living organisms, the material remained undiseased. Some of Pasteur's flasks remain, and are presumably still sterile today.

Heat Resistance of Microorganisms Important in Canning

There are two important genera of bacteria which form spores. Both genera are rod forms: one (Bacillus) is aerobic and the other (Clostridium) is anaerobic. When a rod is about to sporulate a tiny refractile granule appears in the cell. The granule enlarges, becomes glassy and transparent, and resists the penetration of various chemical substances. All of the protoplasm of the rod seems to condense into the granule, or young spore, in a hard dehydrated, resistant state. The empty cell membrane of the bacterium may separate off, like the hull of a seed, leaving the spore as a free, round or oval body. Actually a spore is an end product of a series of enzymatic processes. There is no unanimity of opinion either of spore function in nature or of the factors concerned in spore formation.

Since no multiplication takes place as a result of the vegetative cell-spore-vegetative cell cycle, few bacteriologists accept the concept of the spore as a cell set apart for reproduction. Instead, various explanations of the biological nature and function of bacterial spores have been advanced. These include: the teleological interpretation of the spore as a resistant structure produced to enable the organism to survive an unfavorable environment; the idea that the spore is a normal resting state (a form of hibernation); the notion that spores are stages in a development cycle of certain organisms, or a provision for the rearrangement of nuclear material. It is interesting to note that the protein of the vegetative cell and the protein of the spore are antigenically different.

Spores appear to be formed by healthy cells facing starvation. Certain chemical agents (glutamic acid) may inhibit the development of spores. No doubt sporulation consists of a sequence of integrated biochemical reactions. The sequence can be interrupted at certain susceptible stages.

The literature on the subject of the heat resistance of bacteria contains many contradictions and discrepancies from the records of the earliest workers to those of the present day. This lack of uniformity has been due in part to factors of unknown nature. Until the factors operative in the thermal resistance of bacteria are understood, it will not be possible to control by other than empirical means the processes which require for their success the destruction of bacteria.

Heat may be applied in two ways for the destruction of bacteria. Oven heat may be considered as dry heat, used in the sterilization of glassware. Other materials are heated when moist or in the presence of moisture; this is commonly termed moist heat. Dry cells exhibit no life functions; their enzyme systems are not active. Cell protein does not coagulate in the absence of moisture.

The gradual increase in the death rate of bacteria exposed to dry heat is indicative of an oxidation process.

Whereas death by dry heat is reported as an oxidative process, death by moist heat is thought to be due to the coagulation of the protein in the cell. The order of death by moist heat is logarithmic in nature. The explanation of bacterial death as caused by the inactivation of bacterial enzymes cannot be correct. A suspension containing 99% dead cells has 80% of its catalase active. Since the order of death by moist heat is logarithmic in nature, death must be brought about by the destruction of a single molecule. This change is termed a lethal mutation. To a food technologist, death of a bacterium is described by its inability to reproduce. Heat inactivates or coagulates a single mechanism (gene?) preventing reproduction. The decreasing enzyme content of dead bacteria is the consequence of inhibited growth and probably not the cause. Replacement of the enzyme molecules becomes impossible; the enzyme content slowly decreases.

Regardless of the explanation of death of bacterial spores, the logarithmic order of this death permits the computation of death points, rates or times, independent of any explanation. The death rates or times permit the comparison of the heat resistance of one species at different temperatures or of different species at the same temperatures. It is also possible to describe in quantitative terms the effect of environmental factors upon the heat resistance of the bacteria.

Originally the standard method of establishing the heat tolerance of different species of bacteria was the thermal death point, i.e., the lowest temperature at which the organism is killed in 10 min. This method cannot give comparable results unless conditions such as the age of the culture, the concentration of cells, the pH value of the medium, and the incubation temperature are standardized. Food technologists concerned with processing canned foods have adopted the thermal death time, keeping the temperature constant and varying the times of heating. The thermal death time is the shortest time required at a given temperature to kill the bacteria present.

It is necessary to know the time and temperature required to adequately sterilize canned foods. This procedure involves not only the destruction of spores by moist heat, but also the rate of heat penetration and heat conductivity of containers and their contents. The heat resistance of an organism is designated by the c value (the number of minutes required to destroy the organism at 121°C) and the z value (the number of degrees centigrade required for the thermal death time curve to traverse one logarithmic cycle). These two values establish and describe the thermal death time curve, and are a quantitative measure of the heat resistance of the spores over a range of temperatures (Fig. 6.3).

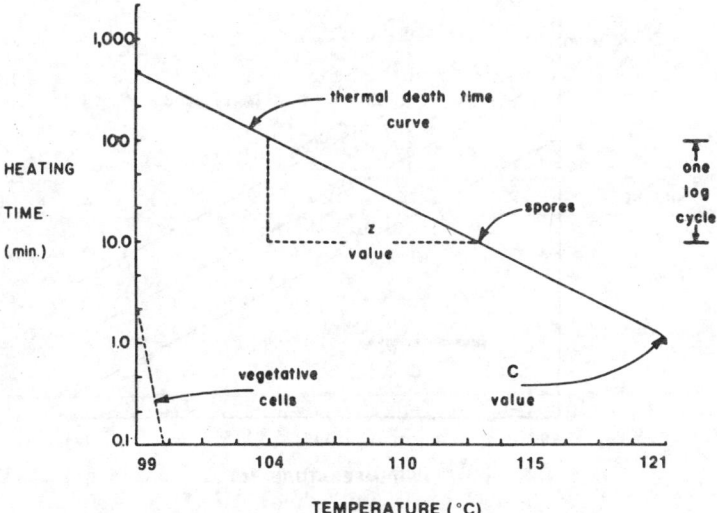

FIG. 6.3.   TYPICAL THERMAL DEATH TIME CURVES FOR SPORES AND VEGETATIVE CELLS OF HEAT RESISTANT ORGANISMS

Spores are heat resistant, vegetative cells are generally not.

It has been recognized that spores of different species, and of strains of the same species, exhibit marked differences in heat resistance, but little or nothing is known in explanation. Some workers have believed that there might be a difference in heat resistance among the vegetative cells, which was transmitted to the spores. Comparing the heat resistance of vegetative cells and spores of a number of bacteria, considerable differences in the spore resistances are found among organisms. Differences in vegetative cell heat resistance is in some instances associated with high spore resistance. Other cultures of vegetative cells produce spores of low resistance. There is evidently no significant relationship between the heat resistance of the vegetative cell and that of the spore produced therefrom. As noted previously, even the proteins of the vegetative cell and spore differ for a species.

Some researchers reason that the spores of a strain are all of the same heat resistance. Others suspect that in a given spore suspension there are a predominant number of spores of relatively low heat resistance, a smaller number with greater heat resistance, and a still smaller number of very heat resistant spores. However, subcultures from heat resistant selections do not yield survivors of uniformly high heat resistance over the parent strain.

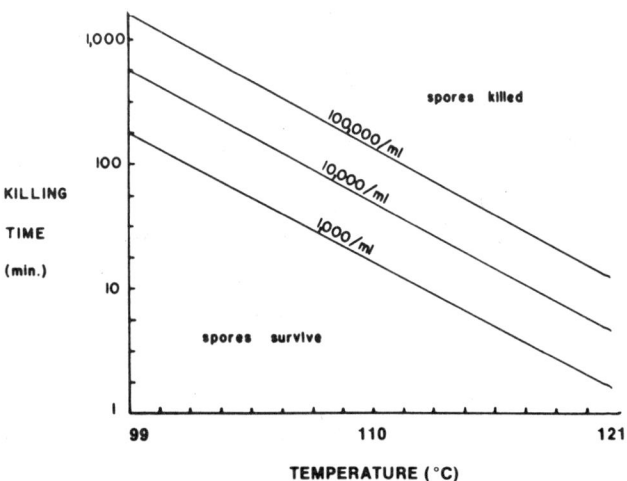

FIG. 6.4. GREATER THE SPORE CONCENTRATION THE MORE HEAT RESISTANT THE SUSPENSION

Heat resistance of a culture, measured by the presence of survivors, related to the concentration of organisms.

FACTORS INFLUENCING THE
HEAT RESISTANCE OF SPORES

Concentration.—The heat resistance of a suspension of bacterial spores is related to the number of organisms present (Fig. 6.4). The greater the number of spores per milliliter, the higher resistance of the suspension.

Environmental Factors.—The resistance of bacterial spores is not a fixed property, but one which under ordinary conditions may tend to be relatively constant. The extent of change in resistance is determined largely by the physical and chemical forces which operate from outside the spore cell. Aside from purely theoretical interest, a better understanding of the cause of heat resistance of spores is of fundamental importance to the canning industry. There are relatively few types of spore-forming organisms especially endowed with heat resistant properties, but these account for most of the spoilage potential in canning. Spore heredity, the environment in which grown, and a combination of these factors must play some part in the production of highly heat resistant spores.

Different yields of spore crops can be determined in various media. This may be demonstrated by plate count or by direct microscopic count. There is little information indicating a relationship between the physiological factors influencing spore formation and the heat resistance of the spores produced. The reaction (pH value) of the medium in which spores are produced has apparently little influence on their heat resistance.

Continuous drying seems to enhance the resistance of spores, but this is irregular in effect. Freezing tends to weaken spores. The following data for an aerobic spore-forming organism isolated from spoiled canned milk is noteworthy (Curran 1935):

Heat Resistance at 121°C

Spore Treatment	Survival in Minutes
Wetted	5
Alternately wetted and dried	6
Dried	7
Frozen	2

Spores formed and aged in soil are found to be more heat resistant than those formed and aged in broth or agar. Natural environmental conditions are evidently more conducive to the development of heat resistant spores than conditions prevailing in artificial cultures. The prolonged action of metabolic wastes from cells appears to decrease the heat resistance of spores.

Bacteria exposed to subleathal heat are more exacting in their nutrient and temperature requirements than undamaged bacteria. The composition of recovery media in which organisms are placed after heating may have considerable effect on the apparent thermal destruction time for the organisms. Depending on the choice of media, heat treated bacteria may be found to be dead in one and alive in another.

Thermophilic bacteria, which form spores in artificial media, produce spores of comparable heat resistance to those formed on equipment and machinery in canning plants.

Spores obtained from soil extractions and remixed with sterile soil are less heat resistant than those heated in the soil directly. The higher natural resistance of spores in soil may be due to some physicochemical influence of the soil and not to any differences between the soil and cultured spores themselves.

Anthrax spores remain viable and virulent in naturally contaminated water for as many as 18 years, while artificial cultures remain in this condition for perhaps 5 months. Soil organisms on corn may remain viable on naturally contaminated tissue for at least 7 years, while the artifically cultured die in 3 months. Artificial media apparently weakens cultures of organisms.

If a culture is to be kept alive for a long period it is apparently desirable to have a medium which permits only a limited growth, limiting metabolic by-products, than media which permit produse growth. *B. tuberculosis* growing on a relatively poor medium may be kept viable for several years while growth on enriched media has viable organisms for only a few weeks. The preserving influence of natural environments may be a similar phenomena.

CATEGORIES OF FOODS FOR CANNING

It is possible to classify foods to be canned (Fig. 6.5) on the basis of acidity and pH value. Plant tissue (except fruits and berries) and animal tissue (including meat, fish and dairy products) are classed as low acid foods. Manufactured items with several ingredients may fall into the medium acid group. Fruits are in the acid group. Berries, fermented products and certain citrus products fall into the high acid group as do jams and jellies. Few foods are basic in reaction if considered in their best quality.

Important Food Groups

Low Acid Foods.—Meat, fish, poultry, dairy products and vegetable foods of man generally fall into a pH range of 5.0 to 6.8. This

TYPICAL COMMERCIAL CANNING OPERATIONS

Harvesting

Receiving raw product

Soaking and Washing

Sorting and grading

Blanching

Peeling and Coring

Filling

Exhausting

Sealing

Processing

Cooling

Labeling

Warehousing and Packing

Courtesy of American Can Co.

FIG. 6.5. SKETCH DEMONSTRATING TYPICAL CANNING OPERATION

large group is commonly referred to as the low acid group, and in some cases these foods are even referred to as non-acid foods. While they are relatively non-acid, they do fall in the acid range of pH values.

Manufactured food items such as soups and spaghetti products, as well as figs and pimientos, fall into what is called the medium acid food group. These foods have pH values between 4.5 and 5.0.

Foods with pH values greater than 4.5 require relatively severe heat treatments. The lower limit of growth of an important food poisoning organism, *Clostridium botulinum*, is at a pH value of 4.5. Inasmuch as a millionth of a gram of the toxin produced by this organism will kill a man, certain precautions are indicated. All foods capable of sustaining the growth of this organism are processed on the assumption that the organism is present and must be destroyed. Foods could be classified into two groups, depending on whether this organism can grow or not.

Acid Foods.—Foods with pH values between 4.5 and 3.7 are called acid foods. Fruits such as peaches, pears, oranges, apricots and tomatoes fall into this class. Potato salad made with vinegar is also in this group.

High Acid Foods.—Next in order of increasing acidity are the berries, pickle products and fermented foods. The pH values range from 3.7 down to 2.3. An example of this high acid group is cranberry juice.

Another important group of foods is one termed high acid-high solids. Jams and jellies are in this classification.

Alkaline Foods.—Foods with pH values in the basic range are few. Old eggs, aged seafood, soda crackers and lye hominy may have pH values higher than 7.0. Lye hominy is the only food item canned which is normally over 7.0. The degree of alkalinity is dependent upon manufacturing procedures. If all the lye is washed from the treated kernels of corn, it would be expected that its pH would be slightly less than 7.0.

A summary of the acidity classification of foods is presented in Table 6.1.

Microorganisms Associated with the Food Groups

Most microorganisms if actively growing (vegetative state) are readily killed by exposure to temperatures near the boiling point of water. Bacterial spores are more heat resistant than vegetative cells. As mentioned previously, bacteria can be classified according to their temperature requirements for growth. Bacteria of the soil, water, air and body growing at room temperature or slightly higher are called mesophiles; their range is between 21°C and 43°C. Some water and soil type bacteria grow best at temperatures ranging from 2° to 10°C and are called psychrophiles. There are bacteria of soil, water and air which grow best at temperatures from 49° to 77°C and are called thermophiles. These ranges are significant in canning.

It is important to distinguish between organisms capable of growing at moderately high temperatures (65°C), the thermophilic group, and those capable of resisting the effect of high temperatures, the thermoduric group. Mesophilic organisms can be thermoduric due to their spores, as can the spores of thermophilic bacteria.

Typical genera of microorganisms commonly associated with the

TABLE 6.1

CLASSIFICATION OF CANNED FOODS ON BASIS OF PROCESSING REQUIREMENTS

Acidity Classification	pH Value	Food Item	Food Groups	Spoilage Agents	Heat and Processing Requirements
Low acid	7.0	Lye hominy; Ripe olives, crabmeat, eggs, oysters, milk, corn, duck, chicken, codfish, beef, sardines	Meat Fish Milk Poultry	Mesophilic spore-forming anaerobic bacteria	High temperature processing (115° to 121°C)
	6.0	Corned beef, lima beans, peas, carrots, beets, asparagus, potatoes	Vegetables	Thermophiles Naturally occurring enzymes in certain processes	
Medium acid	5.0	Figs, tomato soup	Soup		
	4.5	Ravioli, pimientos	Manufactured foods	Lower limit for growth of *Cl. botulinum*	
Acid		Potato salad Tomatoes, pears, apricots, peaches, oranges	Fruits	Non-spore forming aciduric bacteria	Boiling water processing (100°C)
	3.7	Sauerkraut Pineapple, apple, strawberry, grapefruit		Acidic spore-forming bacteria Natural occurring enzymes	
High acid	3.0	Pickles Relish Jam-jelly Cranberry juice Lemon juice Lime juice	High acid foods High acid-high solids foods Very acid foods	Yeasts Molds	
	2.0				

TABLE 6.2

COMMON SPOILAGE ORGANISMS OF FOODS

Food	Organisms Commonly Found in Spoiled Food
Milk and milk products	Streptococci, Lactobacilli, Microbacterium, Achromobacter, Pseudomonas and Flavobacterium, Bacilli
Fresh meat	Achromobacter, Pseudomonas and Flavobacterium, Micrococci, Cladosporium, Thamnidium
Poultry	Achromobacter, Pseudomonas and Flavobacterium, Micrococci, Penicillium
Smoked cured meats	Micrococci, Lactobacilli, Streptococci, Debaryomyces, Penicillium
Fish, shrimp	Achromobacter, Pseudomonas and Flavobacterium, Micrococci
Shellfish	Achromobacter, Pseudomonas and Flavobacterium, Micrococci
Eggs	Pseudomonas, Cladosporium, Penicillium, Sporotrichum
Vegetables	Penicillium, Rhizopus, Lactobacilli, Achromobacter, Pseudomonas and Flavobacterium
Fruits and juices	Saccharomyces, Torulopis, Botrytis, Penicillium, Rhizopus, Acetobacter, Lactobacilli

spoilage of important foods are given in Table 6.2. The thermophiles of importance to the food industries are presented in Table 6.3 along with other information pertinent to the canning process.

Foods have associated microfloras; certain organisms are associated with particular food groups. These organisms gain entrance to the food during the canning operation either from the soil, from ingredients or from equipment. On the basis of the acidity classification of foods, it is possible to make general statements relative to the spoilage organisms encountered of importance to the success of the canning process (Table 6.3).

Microorganisms of Low Acid Foods.—In foods with a pH value greater than 4.5, mesophilic spore-forming anaerobic bacteria are important (Table 6.1). *Clostridium botulinum* is a soil-borne mesophilic spore-forming, anaerobic bacterium. Another is known as Putrefactive Anaerobe (P.A.) No. 3679, a *Clostridium sporogenes* type, common in soil. The latter is more heat resistant than the former and is used to evaluate many heat processing schedules. If heating is adequate to kill the spores of P.A. No. 3679, the process ensures the destruction also of *Clostridium botulinum*.

In addition to the mesophilic spore-forming bacteria there are also thermophilic spore-forming organisms, which are very heat resistant. In fact, they may be more heat resistant than the mesophiles. Processes designed to kill all thermophilic spore-forming bacteria may also result in canned foods far overcooked and degraded in nutritional value. These organisms therefore are controlled through sanitation and by strict control of ingredients, which may be highly contamin-

TABLE 6.3
THERMOPHILES OF IMPORTANCE TO THE FOOD INDUSTRIES

Name	Economic Importance	Heat-Resistant Spores	Growth Temperatures Optimum (°C)	Range (°C)	Oxygen Requirements
Streptococcus thermophilus	Grow during pasteurization of milk; ripening agent in Swiss cheese	None	48	25–60	Facultative
Lactobacillus bulgaricus	Bulgaricus milk; lactic acid manufacture	None	49	25–60	Facultative
Lactobacillus thermophilus	Grow during pasteurization of milk	None	55	30–65	Facultative
Lactobacillus delbruckii	Acidification of brewery mash; lactic acid manufacture	None	45	21–60	Facultative
Bacillus calidolactis	Coagulates milk held at high temperatures	Yes	55–65	45–75	Facultative
Bacillus thermoacidurans	Flat sour spoilage of tomato juice	Yes	45	26–60	Facultative
Bacillus stearothermophilus	Flat sour spoilage of canned foods	Yes	50	45–76	Facultative
Clostridium thermo-saccharolyticum	Hard swells of canned foods	Yes	55–61	43–71	Anaerobic
Clostridium nigrificans	Sulfide-stinkers of canned foods	Yes	55	26–70	Anaerobic

ated. For example, it is nearly impossible to sterilize chocolate with moist heat, due to the high fat content of the chocolate which apparently entraps (thermophilic) organisms, causing death by dry heat conditions rather than by moist heat. Canned products (i.e., chocolate milk) containing such ingredients may be especially difficult to sterilize.

Microorganisms of Acid Foods.—In the acid food grouping the troublesome organisms are aciduric bacteria of no special heat resistant qualities. Bacteria, yeasts and molds are capable of spoiling these foods. The lack of growth of *Cl. botulinum* in acid foods is reflected in their low heat processing requirements. A few mesophilic anaerobic spore-forming organisms (i.e., *Cl. pasteurianum*) may cause spoilage, but in acid foods the organism has relatively low heat resistance. *Bacillus thermoacidurans* is an exception worth noting. It is the flat sour spoilage organism of tomato juice. Canned tomatoes generally are not found spoiled by this organism, although it could spoil this product. Thermophilic flat sour spoilage is due to an inoculation of food by equipment or ingredients and is related to the sanitary condition of a plant. Flat sour spoilage of home canned tomato juice is not common because equipment in homes is easily cleaned. In flat sour spoilage, as the name implies, acid is produced without gas. One difficulty then is that containers do not appear to be spoiled until open, either by the canner, buyer, or worse, by a consumer.

It is a general axiom of the canning industry never to taste spoiled low or medium acid foods, due to the threat of spoilage by *Cl. botulinum*. With acid and high acid foods the spoiled products may be distasteful, but there is reasonably little cause for alarm in their being tasted. It is good practice to give a can of spoiled food due respect in any case. Fortunately most canned food spoilage is associated with the production of gas, bulging the container. There may be visible signs of decomposition in the canned food itself.

Microorganisms of High Acid Foods.—Aciduric bacteria, yeasts and molds are the troublesome organisms in this group. Their heat resistance is generally low. In this group too, the natural enzymes present in food may be as heat resistant as microorganisms. Heat processes for pickles must give due consideration to the destruction of the natural enzymes of cucumbers. It is no less difficult to destroy enzymes than bacteria, yeast, or molds in pickles.

Influence of Food Ingredients on Heat Resistance of Spores

Acids and pH Value of Heating Medium.—Of the many factors

which influence the heat resistance of spores, the pH values of the heating medium has profound effects. For most spore-forming bacteria maximum resistance generally occurs in the region of neutrality. Bacterial spores are not heat resistant at low pH values. For foods with pH values higher than 5.0 apparently factors other than pH are important in the resistance of spores. For instance, the heat resistance of spores of *Cl. botulinum* in fish products with a range in pH from 5.2 to 6.8 is approximately the same. At a pH lower than 5.0, a marked reduction in resistance occurs (Fig. 6.6). This effect is utilized in the processing of certain vegetables and other low acid foods which do not withstand sterilization under usual canning conditions. The liquors in which these foods are packed are acidified, with the result that the resistance of the contaminating organisms is lowered.

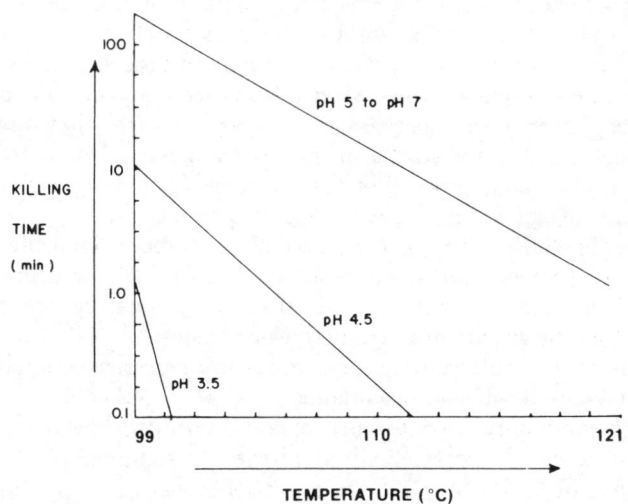

FIG. 6.6.   INFLUENCE OF pH OF HEATING MEDIUM ON HEAT
RESISTANCE OF SPORES

The more acid the substrate, the less heat resistance in spore suspensions.

Altering the pH of tomato juice with citric, lactic or acetic acid greatly alters the heat resistance of *B. thermoacidurans*. If the same percentage of acid is added, they differ in their degree of effectiveness in lowering the heat tolerance of the organisms in the order of lactic, citric and acetic. If based on the pH, the order would be acetic, lactic and citric. Evidently the undisassociated acid molecule is important in this phenomena. Differences in the heat resistance of bacterial spores in different foods cannot be explained solely on the basis of the pH value alone.

Organism—P.A. No. 3679

Substrate	pH Value	Number of Minutes to Kill Organism at 121°C
Asparagus	5.4	3.3
Peas	5.4	3.0
Spinach	5.4	2.6
Milk	6.3	2.6

The causes of these differences are not known.

Sugar.—Longer heating times are required to kill spores as the concentration of sugar in solution increases. Heating yeasts and molds in increasing concentrations of sugar increases their tolerance to heat. Small sugar concentration differences do not evidence this protective effect with spores. Some researchers feel that sugar solutions increase the resistance of spores by a partial dehydration of the cell protoplasm, protecting the proteins from coagulation. Heat coagulation of egg albumin may be retarded with sugar.

Inorganic Salts.—The concentration of sodium chloride in solution may protect spore heat resistance (up to 4%) or decrease the heat resistance (8% or more). Salt is effective, it must be remembered, in inhibiting the growth of putrefactive organisms.

Increasing the salt content of tomato juice decreases slightly the heat resistance of flat sour organisms.

Agents commonly used to cure meats have little influence on the heat resistance of spores. Sodium nitrate or sodium nitrite in concentrations of 0.17% in meat are reported to be ineffective in reducing the heat resistance of spores.

The phosphate level of soil and media influences the heat resistance of bacterial spores present. Phosphate ions are important in spore formation, spore germination and the heat resistance of spores.

Effect of Starch, Protein, Spices and Fat.—It is interesting to note that starch in media permits the growth of greater numbers than will grow without starch. Starch is effective in adsorption of inhibitory substance, including C_{18} unsaturated fatty acids, although starch does not influence the heat resistance of spores, per se. Proteinaceous materials offer some protection to spores against heat.

The essential oils of many spices and flavoring materials (i.e., mustard, clove, onion, pepper, garlic) markedly influence the heat resistance of bacterial spores. In the presence of any lethal substance, bacterial spores may be expected to be of reduced heat resistance as would yeasts and molds. It is noteworthy that many vegetable materials contain substances which lower the heat tolerance of bacterial spores. This may account for some of the differences at least in the variation in heat resistance of a spore suspension in different foods of nearly identical chemical composition.

Spice oils employed in foods as flavoring agents may have preservative qualities. Some are effective in reducing the heat tolerance of spores.

Fats and oils are hindrances in attempting to kill bacterial spores with moist heat. Destruction of bacteria and spores in oil resembles the conditions of dry heat sterilization. Yeasts may be very difficult to kill in salad dressing for example, due to the organism being entrapped in the oil phase. Vegetative, non-heat-resistant bacteria have been isolated from canned fish after high temperature heat treatments. The spores of *Cl. botulinum* survive beyond all reasonable expectation when heated in an oil suspension.

HEAT RESISTANCE OF ENZYMES IN FOOD

Energy of Activation.—Current chemical theory indicates that a reactant in an enzymatic reaction must be activated. This activation requires energy. The energy required to activate the reacting molecules is called the energy of activation. The function of an enzyme is to bring about the reaction with a lower energy of activation:

Reaction	Catalyst	Cal/Mole Required
H_2O_2 decomposition	None	18,000
	Colloidal platinum	11,700
	Liver catalase	5,500
Casein hydrolysis	HCl	20,600
	Trypsin	12,000
Sucrose inversion	Hydrogen ions	26,000
	Yeast invertase	11,500
Ethyl butyrate hydrolysis	Hydrogen ions	13,200
	Pancreatic lipase	4,200

This decrease in the energy of activation due to the catalyst results in increased rates of reaction because a portion of the molecules is sufficiently activated to decompose into products alone.

Enzyme Inactivation with Heat.—The heat inactivation of enzymes is associated with an alteration of the surfaces of the molecules, breaking bonds and opening rings in the protein molecule, with dissociation and loss of structure.

Increasing the temperature of an enzyme-substrate system increases the velocity of reaction catalyzed by the enzyme. Using the equation:

$$\text{Temperature coefficient} = \frac{\text{velocity at } T^\circ + 10\ C}{\text{velocity at } T}$$

The coefficient is usually in between 1.4 and 2.0. This meant that for every 10 C increase in temperature, a doubling of the rate of reaction is obtained (a coefficient of 2).

Temperature coefficients for enzyme-substrate reactions decrease with increasing temperatures. A high temperature causes the relative increase in the rate of reaction to become smaller. This is due in part to the destruction of the enzyme itself, as it is proteinaceous in nature and subject to heat damage.

The point of optimum reaction occurs between the temperature at which there is an accelerated reaction due to temperature and a minimum destruction of the enzyme by heat. At the temperature optimum the temperature coefficient will be 1.0. This is not a specific point; it varies depending upon the length of time the reaction is run, impurities, etc.

Nearly all enzymes are irreversibly destroyed in a few minutes by heating to 79°C. The effect of heat on the rate of coagulation of protein and the effect of heat on the rate of inactivation of enzymes are two phenomena which have high energies of inactivation, and may be due to similar chemical reactions. The heat inactivation energies for several enzymes are:

Enzyme	Heat Inactivation (cal/mole)
Catalase (blood)	45,000
Amylase (malt)	42,500
Lipase (pancreatic)	46,000
Bromelin	76,000
Sucrase	100,000
Trypsin	41,000

Reactivation of Enzymes After Heating.—Although substantial attention has been given to the destruction of enzymes in other methods of food preservation (freezing, dehydration) relatively little has been given in canning foods. The assumption has been that the heat process designed to kill microorganisms is sufficient to inactivate all enzymes. While it is true that heating to 79 C will inactivate many enzymes, it is only recently that studies have been undertaken to evaluate the heat resistance of enzymes in canned foods. Enzymes play a role in the deterioration of canned acid and high acid foods. Also, enzymes are in some instances reactivated after heating (i.e. peroxidase). This problem has developed from studies of extremely high temperature processing (121° to 149°C flash-heat treatments). Microorganisms are heat inactivated but indications are that some enzymes do survive such treatments. The peroxidase in pickles is able to withstand heating to 85°C. The heat destruction of this enzyme is increased by the addition of vinegar. Heavy sugar solutions are protective to enzymes to heat inactivation in pears and peaches. The enzymes of the tomato are not altered in heat resistance by the small amount of salt added. Some suggestions that the enzymes of canned tomatoes remain active after the canning process are available. Peroxidase systems of turnip and cabbage have been found to be reactivated after heating.

Thermal destruction curves obtained for ascorbic acid oxidase and peroxidase in acid foods indicate that standard methods may be used in evaluating the heat inactivation of enzymes.

The internal temperature of fruit and vegetables in the high acid food group, which receive relatively low heat processes in canning, may not rise sufficiently high to inactivate enzymes normally present internally in these tissues. The pectinesterase in canned grapefruit juice is active after an adequate process from the standpoint of microbial spoilage has been administered.

Enzymes—A Chemical Index of Efficiency.—In some instances the inactivation of enzymes may be used as an index of the degree of heating of foods. For instance, the pasteurization of milk can be evaluated by its phosphatase enzyme activity. The destruction of phosphatase in milk coincides with the heat treatment designed to kill *B. tuberculosis* and other human pathogenic organisms. An evaluation of milk for this enzyme indicates the minimum degree of heat treatment.

The peroxidase system in fruits may be useful in evaluating the relative efficiency of acid food canning processes. Unless the enzymes are destroyed, they will continue to function in the container, causing deterioration. So too, if the presence of one enzyme is found, what conclusions are possible relative to the hundreds of others which may be functional, but for which there are no methods of evaluation?

HEAT PENETRATION INTO FOOD
CONTAINERS AND CONTENTS

Disregarding for a moment the acidity of foods and the spoilage

HEATING FOOD IN A CAN

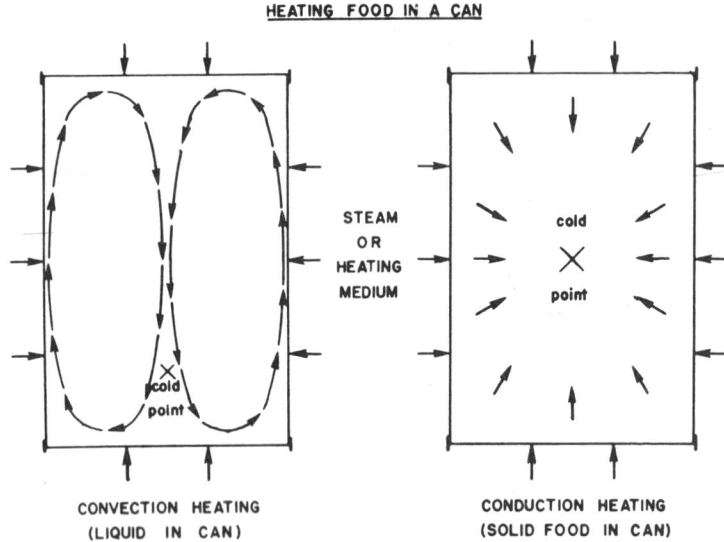

CONVECTION HEATING
(LIQUID IN CAN)

CONDUCTION HEATING
(SOLID FOOD IN CAN)

FIG. 6.7.   COLD POINT OF HEATING FOR CONVECTION AND CONDUCTION TYPE
PRODUCTS

organisms related with these foods, it is necessary to consider the penetration of heat into containers being processed.

Heat.—Heat is a form of energy measured in terms of calories or British thermal units. It cannot be defined in terms of the fundamental dimensions of distances, mass and time. As noted previously, we speak of the temperature of a body but have no clear concept of what kind of a measurement temperature is. Temperature is measured in terms of itself; notions of hot and cold are relative.

Propagation of Heat.—There are three ways to propagate heat energy: convection, conduction and radiation. Convection heating means bodily transfer of heated substances, i.e., molecules. Conduction heating means heat is transferred by molecular activity through one substance to another. Radiation heating is a transfer of heat energy in the same manner as light, and with the same velocity. Heat transfer by convection must be accompanied by some conduction heating. Conduction heating is very slow compared to the usual cases of convection heating (Fig 6.7).

The second law of thermodynamics states that heat energy flows only in one direction, from hot to cold bodies. The difference between a hot and a cold body is a matter of energy. If a hot and cold body are allowed to come to equilibrium, the hot body will cool and the cold body will warm.

The sterilization value of steam depends largely on the transfer of the heat of vaporization to the object upon which steam condenses. The difference between moist heat and dry heat are readily experienced by placing one hand in an oven and the other accidentally in steam, both at 121°C.

Steam is the commonly used sterilizing agent in the canning industry. Steam under pressure has the following characteristics:

Lb Steam Pressure Per Sq In.*	Temperature (°C)
Boiling water vapor (760 mm Hg)	100
1	101
5	108
10	115
15	121

* See Appendix for Metric Conversion.

Heat Penetration Characteristics of Canned Foods.—When a can of food is sealed at 98°C and placed in a steam pressure vessel which is brought to 121°C, the steam chamber is the reservoir of high heat energy and the can of food is the reservoir of lower heat energy. Heat then is transferred from the hot body to the cold. The mechanism of heat transfer in canned food during such thermal processing may be divided into several rather definite classes. To a certain extent it is possible to place food into heat transfer classes by knowing their physical characteristics (Fig. 6.8). The heat is transferred by conduction from the steam to the can, and from the can to the contents. The can contents will either heat by developing convection currents or by conduction. In some instances food will heat first by one method

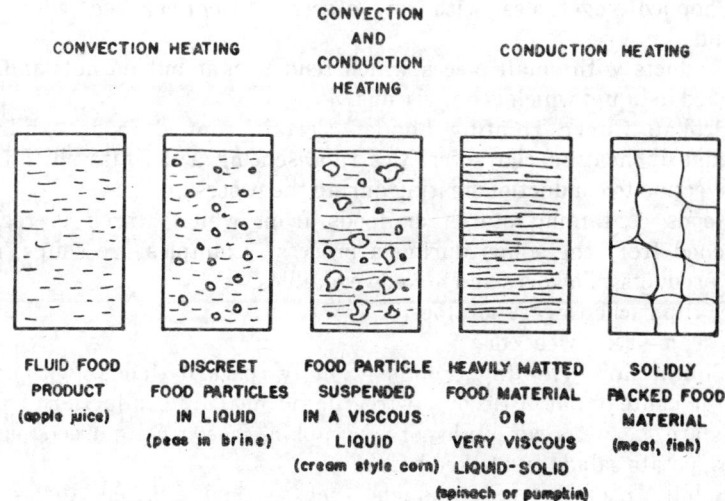

| CONVECTION HEATING | CONVECTION AND CONDUCTION HEATING | CONDUCTION HEATING |

FLUID FOOD PRODUCT (apple juice) DISCREET FOOD PARTICLES IN LIQUID (peas in brine) FOOD PARTICLE SUSPENDED IN A VISCOUS LIQUID (cream style corn) HEAVILY MATTED FOOD MATERIAL OR VERY VISCOUS LIQUID-SOLID (spinach or pumpkin) SOLIDLY PACKED FOOD MATERIAL (meat, fish)

FIG. 6.8.   HEAT TRANSFER CLASSES OF FOOD PRODUCTS

then by the other. For convection heating, heat is initially conducted from the can to molecules inside the can. From this point on, food heats by having the energized molecules (now expanded and lighter in density) rise, while heavier cooler molecules descend. Canned foods heating by convection have a better opportunity of surviving the process in better condition than those requiring heat to be transferred by conduction, which is slower. Heat penetration studies with most food products have been covered. Insofar as possible these products have been classified according to the mechanism of heat transfer. Obviously these considerations are altered by deviations in packaging procedures.

Rapid Convection Heating Foods.—Most fruit and vegetable juices. Pulpiness or gelling slows heating.

Broths and thin soup. Small quantities of starch either added or leached from solid ingredients retards heating.

Fruits packed in water or syrup with large pieces present. Heat penetrates slowly within the pieces.

Evaporated milk heats by convection in agitated retorts. Milk does not withstand ordinary still processing successfully.

Meat and fish products packed in a brine, if the small pieces are not solidly packed into the containers.

Vegetables packed in brine or water, with pieces as above. Exceptions include corn and leafy vegetables (spinach, greens).

Slow Convection Heating Foods.—Small pieces of fruit, vegetable, meat or fish products packed in free liquid.

Chopped vegetables with low starch content packed with free liquid.

Products with small pieces which tend to mat but do not, and are packed in liquid which is not viscous.

Broken Curve Heating Foods.—Certain canned foods exhibit a change in heating characteristics representing a definite shift from convection to conduction heating during the process.

Foods containing starch or foods from which starch is readily leached from the solids during a process. Examples are soups, noodle products, chowders and mixed vegetables.

Syrup packed sweet potatoes.

Cream-style sweet corn.

Conduction Heating Foods.—Solidly packed foods with high water content but little or no free liquid heat by conduction. Examples are: heavy cream-style corn, pumpkin, most thick pureed vegetables, potato salads and baked beans.

Solidly packed fruit products such as jams, baked apples and sliced fruit.

Vegetable, meat, fish products in thick cream sauce, creamed potatoes and chicken-a-la-king.

Concentrated soups of many types.

Meat and vegetable mixtures in thick sauce: chile con carne, chop suey and meat stews.

Solidly packed starch products: spaghetti, chicken and noodles, etc.

Solidly packed meat and fish products: ham, corned beef, chicken loaf, minced clams, codfish, sandwich spreads and spiced ham.

Meat and cereal mixtures such as some meat loaf products.

Measuring the Heat Penetration into Canned Foods.—While thermometers can be used to follow certain heating characteristics of foods, the most satisfactory method involves the use of thermocouples. A thermocouple is formed when two dissimilar metal wires are fused together at the ends. If the ends of these wires are placed at different temperatures a measurable voltage is developed, which is related to the temperature difference between the two ends or thermocouple junctions. By attaching a suitable measuring device (potentiometer) to the thermocouple, it is possible to calibrate it and follow the temperature changes inside a can which itself is being heated in a retort under steam pressure. Examples of thermocouple measuring systems are presented in Fig. 6.9. A commonly used thermocouple

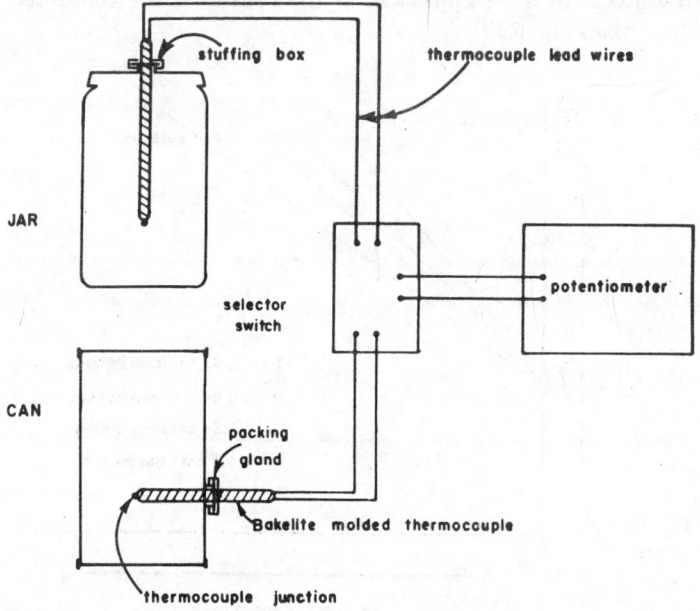

FIG. 6.9.   THERMOCOUPLE INSERTED INTO CONTAINER THROUGH STUFFING BOX

system is composed of copper-constantan wires and a potentiometer, reading directly in degrees centigrade. Recording potentiometers are also available.

Molded bakelite thermocouple assemblies are manufactured which have insulated wires and great utility in studying the heating of canned foods. Thermocouples may be introduced into glass jars by soldering a stuffing box to the lid and boring a hole of correct dimensions through the lid. The thermocouple is then adjusted through the stuffing box to any desired position inside the jar. For metal cans, the stuffing box may be soldered at the desired height to the side of the can body, through which a hole is drilled. The thermocouple is then adjusted to the desired position inside the can through the stuffing box.

Prior to use, thermocouples should be calibrated against a standard thermometer throughout the operating range of temperatures important to a study.

Cold Point Determinations.—All points within a container being heated are not at the same temperature. The zone of slowest heating is called the cold point of a container. It is that zone which is most difficult to sterilize due to the lag in heating. With products heating mainly by convection the cold point is on the vertical axis, near the bottom of the containers. Products heating by conduction have the point of slowest heating approaching the center of the container, on the vertical axis (Fig. 6.10).

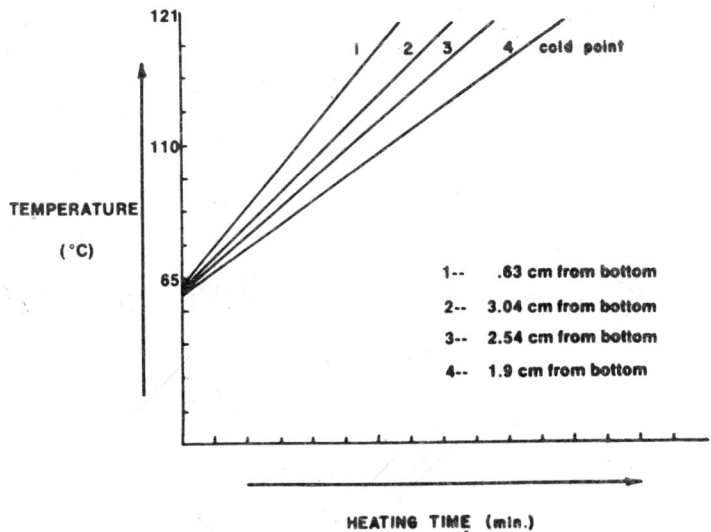

FIG. 6.10.   COLD POINT DETERMINATION—CONVECTION TYPE PRODUCT

To determine the cold point, stuffing boxes are soldered to cans, starting 1.27 cm from the bottom of the first can. A stuffing box is soldered 1.9 cm from the bottom of the second can, 2.54 cm from the bottom of the third can, etc. Product is prepared and a uniform fill is maintained in all containers. Cans are sealed and thermocouples are fitted to each container. Cans then are available with thermocouples located every 0.63 cm from the bottom of cans, starting at 1.27 cm. The cans are placed into a retort and it is brought to temperature with due precaution to remove air. Temperatures are recorded every minute manually. If a recording potentiometer is used, the time-temperature values will be plotted on strip charts. These data are then plotted on semi-log graph paper, which yields a straight line relationship between time and temperature, with minor deviations. An example of such data is shown in Fig. 6.10 for a convection heating product. All subsequent heat penetration studies in this instance would be made at the cold point, locating thermocouple junctions at a position 1.9 cm from the bottom of the cans, for this product, under these circumstances. Varying the filling weight alters the heating characteristics. For example, in pork and beans, changing the amount of beans per can and the amount and composition of sauce changes the heating characteristics (Fig. 6.11).

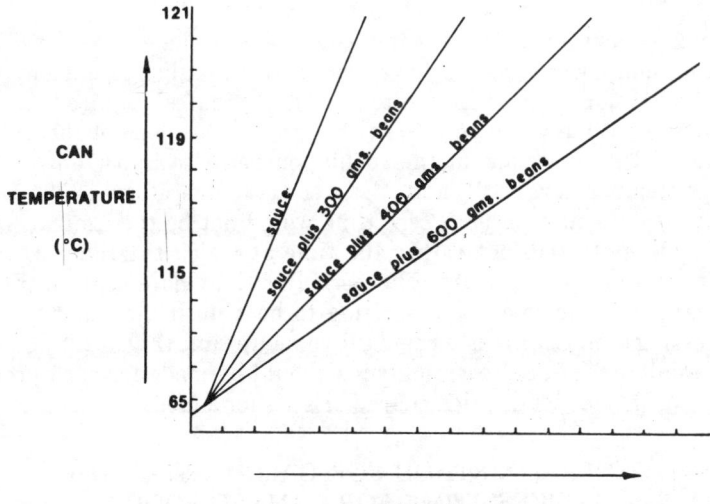

TIME (min.)

FIG. 6.11.   INFLUENCE OF VARYING WEIGHTS OF BEANS ON HEATING CURVE FOR PORK AND BEANS (NO. 2 CANS)

Important consideration in quality control.

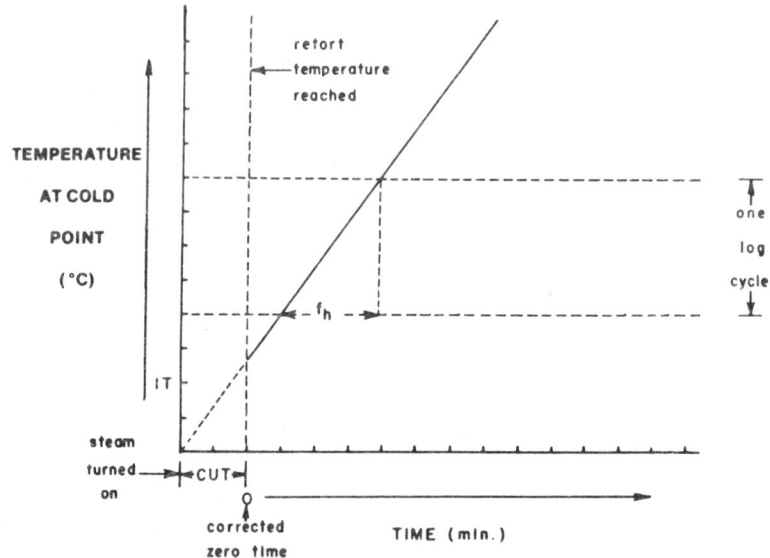

FIG. 6.12.   HEATING CURVE CHARACTERISTICS OF CANNED FOODS

Cooling curves are also established in the same manner. These heating and cooling curves have slopes. For canning technology purposes the slopes are represented as f_h (minutes required for the curve to traverse one log cycle on the graph), the slope of the heating curve, and f_c, the slope of the cooling curve. Also important are the initial temperature (IT) and the retort temperature (RT). Other information is also needed. Zero time would not be the time at which the heating started, but rather the time at which the retort reached the processing temperature. The period of time elapsed from placing the cans into the retort and starting to heat until the retort reaches processing temperature is termed the come-up-time (Fig. 6.12).

A minimum of 12 processing runs should be made for each product to establish heat penetration characteristics for a food.

GENERAL METHOD FOR CALCULATING THE
PROCESS TIME FOR CANNED FOODS

With information relative to the heat resistance of spoilage organisms to be destroyed in canning and the heating characteristics for the food in question, the information necessary to calculate the proces-

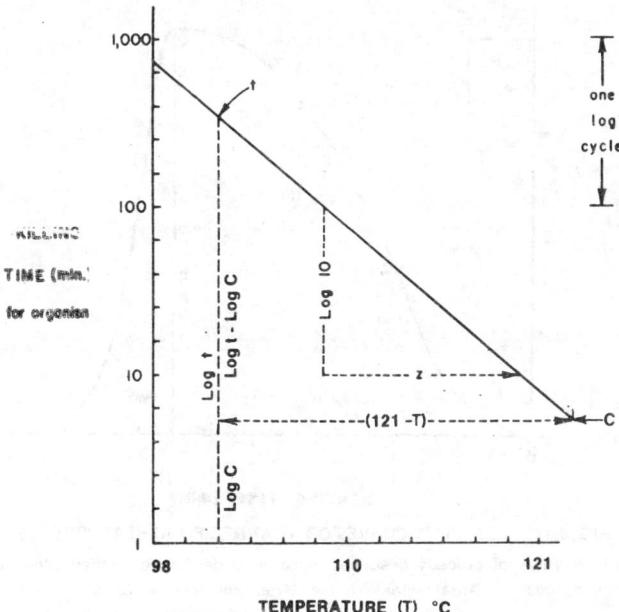

FIG. 6.13. THEORY OF GENERAL METHOD OF HEAT PROCESS
EVALUATION FOR CANNED FOODS

Integration of heat penetration into container with heat resistance
of microorganisms. t = killing time in minutes at temperature (T°C);
C = killing time at 121°C; z = slope of killing curve; T = temperature under consideration (°C).

sing time for the product is available. Each time-temperature interval
during the heating and cooling of the containers has a lethal effect on
food spoilage organisms, if the temperatures are above the maximum
for growth for the organisms. By correlating the killing effects of
these high temperatures with the heating rate of the food, the length
of time theoretically required to destroy any specific bacterial spores
present in the container of food may be calculated for any given temperature. The calculated length of process will be the actual process
necessary, providing all the conditions are given accurate control.
Bigelow and co-workers in 1920 devised a method for calculating
processing times called the *General Method* for process time determinations (Fig. 6.12-6.14).

The rate of destruction of an organism per minute at any given temperature *(T)* in a process is the reciprocal of the time in minutes *(t)*
required to destroy the organisms at that temperature. From the thermal death time curve in Fig. 6.12 a simple geometric relationship

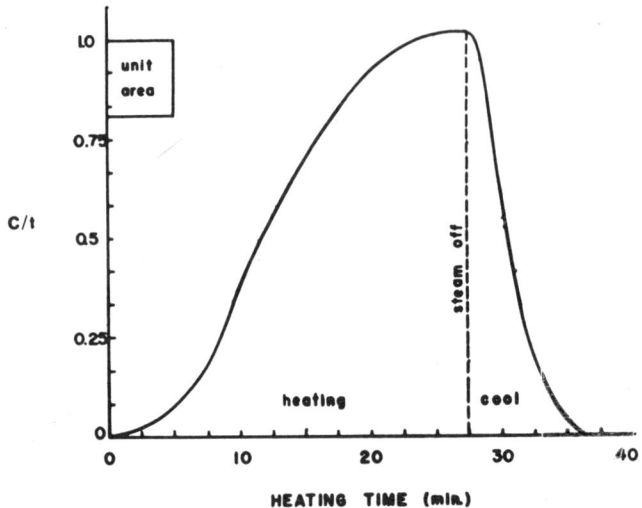

FIG. 6.14.. LETHALITY CURVE FOR A CANNED FOOD HEAT PROCESS

Lethal value of process described by area under curve. Lethal value of
this process = Area beneath curve. Area/unit area = C_0 value; when
C_0 = C = 100 % sterilization process.

exists between the sides of similar right triangles, and may be expressed by the following equations:

$$\frac{\log t - \log C}{\log 10} = \frac{121 - T}{z}$$

$\log 10 = 1$, therefore,

$$\frac{\log t}{C} = \frac{121 - T}{z}$$

and $t/C = \text{antilog} \dfrac{121 - T}{z}$, or $t = C \text{ antilog} \dfrac{121 - T}{z}$

where:

    z = slope of thermal death time curve in °C
    C = minutes to destroy the organism at 121°C
    T = temperatures under consideration (°C)
    t/C = the time to destroy the organism at temperature (T) if $C = 1$
    C/t = lethal rate at T.

From the thermal death time curve, C and z are known. At any temperature (T) it is possible to solve the above equation for t/C, from which the reciprocal C/t can be determined. The lethal rate at any temperature can be calculated then once the C and z value are known.

During the processing of a container of food, the temperature inside the container increases to a maximum and then decreases during cooling.

For each temperature, at definite 1 min intervals, the lethal rate (C/t) for that temperature may be calculated. A curve is formed by connecting each of these values when plotted on linear graph paper against time (Fig. 6.13). The area under this lethality curve represents the total lethal value of the process, and may be measured by means of a planimeter, counting the squares, or cutting the area out, and weighing it. It is now necessary to find the size of the unit sterilization area. By unit sterilization area is meant the area which, if enclosed beneath the lethality curve of a process, represents complete sterilization when the value of C equals one. To define this area a convenient point is arbitrarily chosen on the vertical scale. This represents the height of the unit area. The breadth is found by dividing this arbitrary value into one, which by definition represents the magnitude of the area in terms of lethal rate and time. Since the C values usually required for adequate sterilization are usually greater than one, an adequate process must yield a lethality curve which will enclose an area equal to the C value times the unit area. An adequate process is one which yields a sterilization value (C_O value) equal to the C value of the organism in the product under consideration.

To determine the exact process for a product, a series of three or more heat penetration runs are made, each at a different process time. When plotted on linear graph paper, C_O value versus process time in minutes, the plotted points fall on a straight line. Process times with any desired sterilization value may be interpolated from this graph; the precise value is obtained where $C_O = C = 100\%$ lethality.

Another method of establishing a safe heat sterilization process (C_O requirement) is by use of the D value. This unit is defined as that combination of time and temperature to bring about a 90% reduction in spoilage organisms (bacterial spores). A reduction to the extent of twelve D values is employed in the canning of most foods with pH values higher than 4.5. For acid foods (below 4.5) the following schedule is recommended by the National Canners Association:

pH Value	Number of D Values for Successful Canning
below 3.9	1.0
from 3.9 to 4.3	10.0
from 4.3 to 4.5	20.0
above 4.5	12.0

Formula Method of Process Time Calculation.—The method described above is the *General Method* (Bigelow *et al.* 1920) of process time determination. It is possible to determine the process time for a product by formula (Ball's method) and by Nomograph (Olsen and Steven's method). These methods are fully described in the literature.

Ball's method for the determination of the processing time for a product which has a straight line semi-logarithmic heating curve is accomplished by the solution of the equation:

$$B_B = f_h (\log jI - \log g)$$

where, B_B = the process times in minutes at the retort temperature

f_h = slope of the heat penetration curve

jI = correction factor obtained by extending the heating curve to intersect the time at which the process begins

g = value in degrees below retort temperature where the straight line portion of the heating curve intersects the time at which the heating process ends.

The slope of the heating curve and the point where the extension of the heating curve intersects the time at which the process begins are obtained from heat penetration studies. The formula method value g is obtained by finding the number of degrees which exist between the retort temperature and the maximum temperature, which must be attained by the cold point of heating in the container during the minimum process which will destroy the spores of the organism the process is intending to kill. The C and z values from the thermal death time curves are considered in the g value.

Calculation of the thermal process adequate for a product is the solution of the above equation by substituting the appropriate values, and solving for length of process (B_B). Great versatility is found in the formula method. The features relative to thermal processing of canned foods which may be calculated by Ball's method are given in Table 6.4. It is necessary to refer to the original publications by Ball for graphs and tables used to obtain appropriate values for data in order to solve mathematically the heat process for a canned product. Readers are referred to the original manuscripts for complete information on the *Formula Method.*

Nomogram Method of Process Time Determination.—A method described by Olson and Stevens is a means of solving the processing equations developed by Ball by nomographic methods. These nomograms greatly reduce the time required to reach a solution to a prob-

TABLE 6.4

TYPES OF PROBLEMS ON THERMAL PROCESSING OF CANNED FOOD
SOLVED BY FORMULA METHOD

Class 1—Simple process
Group 1—The heating curve is a simple logarithmic curve
PROBLEMS:
I. Calculation of length of process
II. Calculation of a process equivalent to a given process at a different retort temperature
III. Calculation of effect of change in initial temperature upon the time necessary for sterilization
IV. Calculation of time necessary to reach a given temperature at center of can
V. Calculation of temperature attained at center of can in a given length of time
VI. Calculation of the amount of lethal heat at center of can up to a given time, expressed as percentage of heat necessary to sterilize
Group 2—There is a break in the heating curve
PROBLEMS:
VII. Calculations of length of process
VIII. Calculation of a process equivalent to a given process at a different retort temperature
IX. Calculation of effect of change in initial temperature upon the time necessary for sterilization
X. Calculation of temperature attained at center of can in a given length of time
XI. Calculation of time necessary to reach a given temperature at center of can
XII. Calculation of the amount of lethal heat at center of can up to a given time, expressed as percentage of the heat necessary to sterilize
Class 2—Divided Process
Group 1—The heating curve is a simple logarithmic curve
PROBLEMS:
XIII. Calculation of length of process
XIV. Calculation of divided process equivalent to a given simple process at any retort temperature
XV. Calculation of a divided process equivalent to a given divided process at different retort temperatures
XVI. Calculation of effect of change in initial temperature upon the time necessary for sterilization
XVII. Calculation of temperature attained at center of can in a given length of time
XVIII. Calculation of time necessary to reach a given temperature at center of can
XIX. Calculation of the amount of lethal heat at center of can up to a given time, expressed as percentage of the heat necessary to sterilize
Group 2—There is a break in the heating curve
PROBLEMS:
XX. Calculation of length of process
XXI. Calculation of a divided process equivalent to a given divided process at different retort temperatures and different initial temperatures

Source: Ball (1928).

lem on thermal processing of a product, providing that the thermal death time data and heat penetration data are available. The *nomogram method* is quicker than either the *General Method* or Ball's *Formula Method*. Nomograms are available in published literature and are not reproduced herein. Readers are referred to the original works.

Differences Between the *General Method* and Other Methods of Process Time Determination.—The *General Method* of determining the heat process required for a canned food always gives results in terms of total heating time, and considers the time required to raise the retort to the processing temperature. Providing this does not vary grossly, the *General Method* will yield accurate times for a product under conditions of product composition and bacterial load which the process was intended to consider.

The *Formula* and *Nomogram Methods* require that adjustments in come-up-time (CUT) be established. Ball has established that 42% of the CUT of a retort is usable in terms of the processing temperature. With a 10 min CUT, 4.2 min of this is equivalent in value to actual retort temperature; 4.2 min must be subtracted from the calculated processing time. For example, if the calculated process time was 60 min at 116°C and a 10 min come up time is necessary, the corrected process will be 60 minus 4.2 or 55.8 min. Once the retort reaches 116°C, 55.8 min of holding at that temperature are required to yield an adequate lethal process. The total heating time, on the other hand, from the time the steam is turned on will be 55.8 min plus 10 min CUT, or 65.8 min. In practice this number would be rounded off probably to 70 min. There is then a safety factor involved by rounding off the number of minutes of a process, which is the practice.

INOCULATED PACK STUDIES

In order to ensure that the calculated processing time for a product is adequately established, it is desirable to prepare inoculated packs. The product is prepared and filled into containers. An inoculum of spores of the spoilage organism, important in the food group in which the product falls, is placed at approximately the cold point in the containers. With viscous foods, such as strained pumpkin, the inoculum will remain somewhat at the position placed. For convection heating foods, the inoculum will be carried in the convection currents formed during heating the containers. For solid packed foods, such as potatoes, the inoculum should be injected with needle ¼ in. into the flesh of a potato at the cold point. Excessive processing would be required to kill spores in the center of a 2-in. diameter potato. The inner surfaces are assumed to be sterile.

The concentration of organisms inoculated per container is important. For P.A. No. 3679, 10,000 spores are commonly inoculated per No. 2 can or pint jar. This will vary with the size of container. For relatively non-heat-resistant organisms, 100,000 may be inoculated per container. The longest period of survival of the organisms in the heated containers will be related to the number inoculated.

Assuming that the calculated process is 60 min at 116°C for a food product, a processing schedule will be chosen which brackets this calculated schedule. A minimum of 24 inoculated and 12 control containers will be processed at 50, 55, 60, 65, and 70 min at 116°C. These containers will be coded, and incubated according to the organism's optimum for growth. Carefully kept records of spoilage are important. Microscopic examination of smears from spoiled cans is useful. At the end of four weeks all containers not evidencing spoilage are subcultured. From the inoculated pack studies, the safety of the calculated heat treatment will be obvious. Cans heated for 50 min and 55 min will either all spoil or most of the 50-min treatments will spoil and a few of the 55-min treatments will spoil. None of the 60-min treatments spoil. Under this condition the 60-min calculated process will be adequate. The control containers in the inoculated pack studies probably will not be spoiled at the 55-min process treatments, and even the 50-min treatments may not be spoiled.

It should be apparent that under this condition there is probably a margin of safety in the established process. This may be demonstrated by the following example of sweet corn packed in glass 0.5-liter jars. Suppose sweet corn packs are inoculated with 10, 100, 1000, and 10,000 spores of P.A. No. 3679 and processed for varying lengths of time from 20 to 60 min at 116°C. Control containers were also prepared and processed along with the inoculated jars. The results of such study show that control sweet corn spoils at a level equal to 10 spores and 100 spores per container. Jars inoculated with 1000 spores spoil at a level higher than comparable controls. Containers with 10,000 spores inoculated at the cold point spoil at 50 min of heating, while the containers with 1000 spores spoil only up to 45 min. Control containers viewed as a unit spoil only up to a 35-min process. From such study came the concept that an inoculum of 10,000 spores of P.A. No. 3679 would offer a margin of safety in establishing the safe process for sweet corn if the corn is not grossly contaminated.

Adequacy of Heat Processes

There are two considerations relative to safe processing schedules developed: one relative to the heat resistance of spoilage microorganisms, and the other relative to the heat penetration characteristics of the food in the containers.

Microbiological Considerations.—As shown by the inoculated pack studies a spore load of 10,000 heat resistant organisms is probably greater than the natural contamination in corn. Providing the ingredients and sanitary conditions in the factory are satisfactory, it is unlikely that such a spore load will be present in food processed. Therefore, there is a safety factor involved in the spore concentration employed in establishing the thermal death time characteristics of the spoilage organisms.

Heat Penetration Considerations.—In establishing the heat penetration characteristics of a food product, the slowest heating and the fastest cooling curves are used to calculate the lethal effect of the process.

Statistical Evaluation of Heat Processes.—When process time is plotted against its lethal value, the relationship between the two variables, if considering a range of processes, is generally linear. In view of this, a linear regression line may be fitted to the lethal value-process time data for each product and the standard error of estimate computed.

The line of regression is defined by the regression equation, $y = a + bx$, in which x equals the process time; y, lethal value of process; a, the constant locating the line vertically; and b, the slope of the line. A parallel line is constructed at a distance of 2.6 times the standard error of estimate below the computed regression line. Assuming normal distribution, the probability of an individual container yielding a C_0 value falling below the lower line is only .005.

The process times determined for products by the three methods (i.e., slowest heating-fastest cooling composite data from heat penetration studies, inoculated pack studies, and the statistical evaluation) contain certain margins of safety. Under certain sanitary conditions they may be excessive. Under poor processing conditions they may be inadequate!

SPOILAGE OF CANNED FOODS

The ends of normal cans of food with a vacuum are slightly concave. Ends which are bulged may be caused by microbial, chemical or physical actions. A hard swell is one which resists being pushed back to a normal position. The ends of a soft-swelled can may be forced back slightly but will not resume a normal condition. A springer swell is one which is bulged but which may be forced back into normal position causing the opposite end to bulge by hitting against a solid object. The opposite end flips into the bulged end. Cans may progress through the flipper, springer, soft swell and hard swell stages. The next step is to have the can explode.

The same types of spoilage occur in either cans or jars. For simplicity, can spoilage will be discussed.

Underprocessed cans permit the survival of microorganisms which grow and cause spoilage after the process. The growth may produce gas and acid, or acid alone. All cans spoiling from the survival and growth of microorganisms are underprocessed. This may come about by having an extremely large load of bacteria, or the process time may be inadequate.

Unless grossly understerilized, spoiled cans will contain one organism type. In low and medium acid food this will be a spore-forming organism. In acid products the organisms may be yeasts, molds, aciduric spore-forming or non-spore-forming bacteria. Non-aciduric spore formers may be present. They will grow poorly if at all at this acidity.

When a mixed culture containing one or more non-heat-resistant organisms is present in spoiled containers, they were either grossly underprocessed or became infected after heating.

In flat sour types of spoilage the contamination should be expected from the equipment in the plant or the ingredients. Some inoculation has taken place.

When anaerobic spore-forming organisms are causes of spoilage, it is usual that the contamination comes from the raw material. It is unlikely that conditions in a plant are favorable for inoculation of anaerobes into products.

Leaking containers may be contaminated by cooling water. This spoilage may be found to be due to cocci, non-spore-forming rods, and non-heat-resistant rods. Yeasts and molds may be present. In acid foods the substrate will permit few organisms to grow. The principal source of contamination in leaking containers is the cooling water. Chlorination of cooling waters is a recommended practice. The ratio of spoilage in air cooled cans to that of cans cooled in water may be as high as ten times in the latter over the former.

Bacterial spores are more resistant to the effects of chlorine than vegetative forms. Occasionally spore-forming organisms will appear in leaky cans due to the selective action of the chlorine treatment of water in cooling canals. Also, if the chlorine is not allowed to come into contact with spores for at least several minutes, the spores may survive and be drawn into the leaking container.

Solid products such as meat which have spoiled may have center

FIG. 6.15.   FEATHERING-DETINNING OF THE INNER SURFACE OF SANITARY CAN
This is particularly a problem with acid foods.

portions still sterile. The growth may be centered outside the product or on the surface. Internal tissues of plants and animals normally do not contain organisms if the tissues are not diseased.

Microbial Spoilage

Flat Sour.—Spoilage of canned foods need not be accompanied by bulged ends. Flat sour spoilage, as the term implies, is a condition of high acid formation unaccompanied by gas production. Thermophilic bacteria are characteristic in the production of such spoilage. In flat sour spoilage, either the cans have been under-sterilized or the cans have leaked.

Acid and Gas.—Biological spoilage is commonly evidenced by the production of acid and gas by the spoilage organisms. P.A. No. 3679, for instance, produces such a condition. Generally the mesophiles will produce acid and gas in containers in which they grow. Their presence will be indicated by swelled containers and decomposed foods.

Chemical Swells.—Generally the cans of spoiled food will have extended ends. Gas is formed inside the containers forcing the ends. There are several causes of swells in addition to biological causes.

(1) Chemical swells—resulting from the production of gas by action of the can contents on the container. Hydrogen gas is generally produced.

(2) Chemical swells—resulting from decomposition of products liberating carbon dioxide, i.e., molasses, malt extracts, syrups. No bacteria or viable organisms are found.

Normal imperfections exist in the tin coating of cans. Scratches or internal damage can cause small areas of iron to be exposed. When the tin and iron are in contact with a substrate with a high organic acid content, an electrocouple is formed. The corrosion is more complicated than that of tin or iron alone. The principal factor influencing the corrosion of tinplate is the polarity of the metal in the couple. The polarity is governed by the presence of an oxide film on the metal surfaces and the ability of the electrolyte to remove the tin ions as a complex. When the oxide film is dissolved, if the electrolyte contains anions such as citrate or oxalate (with which tin forms a stable complex), tin becomes anodic and the attack is confined to the tin surface, while the iron base is protected. If the stable complex is not formed, tin remains cathodic and the iron is attacked and perforated.

Oxygen depolarizes and therefore is important in the corrosion process. Hydrogen gas should be evolved when displaced by the action of acid on the metal. This action is very slow at the tin surface unless oxygen or some other depolarizer is present. If hydrogen is not evolved it exerts a back pressure so to speak, and thus opposes further solution of the iron. The rate of attack is slow in the absence of oxygen. Anthocyanin pigments of fruits may act as depolarizers. Red fruits are prone to perforate cans (Fig. 6.15).

Lacquered cans may be more easily perforated than plain cans, due to the fact that areas of exposed iron are not afforded cathodic protection or the protection of dissolved tin. Imperfections in the lacqured surface tend to concentrate this chemical activity to small areas, and perforation may be rapidly accomplished (Fig. 6.16).

Lids on glass containers may be attacked chemically by some foods.

Physically Induced Swells.—Overfilling cans at low temperatures may cause permanent bulging to cans by heating. Expansion of the

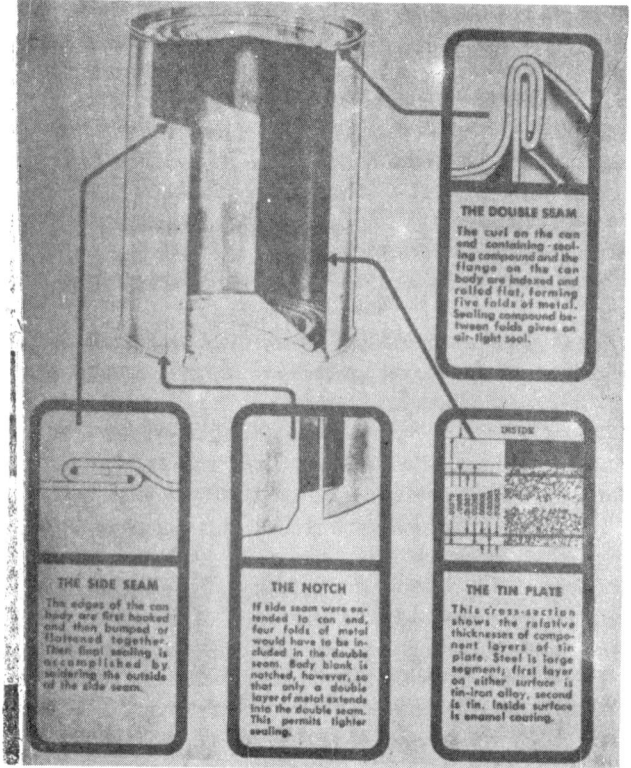

Courtesy of American Can Co.

FIG. 6.16.   ARCHITECTURE OF THE ENAMELED SANITARY TIN CAN

Enamels are useful in protecting food and container.

solids and liquid of the container may permanently distort it.

Foods packed with low vacuums may bulge when placed at high altitudes where there is lower atmospheric pressure.

Freezing food after being canned may cause physical damage to the contents and cans. Freezing may damage the texture and appearance of canned foods. There may be damage to the container if the food contains large amounts of water.

Glass packed foods may be injured by light. Bleaching of products, development of light-struck oil flavors, skunky flavor in beer, and loss of certain nutrients due to light catalyzed reactions may occur. Such spoilage is not a physical spoilage but a photochemical reaction.

There is substantial physical damage to food products due to shipping canned foods by rail, truck or plane. Foods may be degraded in texture noticeably by the agitation of containers and rough handling in general.

STORAGE OF CANNED FOODS

If the canning processes have been successful, the containers should be in a condition where biological spoilage will not occur. Thermophilic organisms may be present, but unless temperature conditions in the storage chamber are excessive such spoilage is unlikely. However, while the biological forces may not be operative, chemical reactions are not eliminated. Chemical reactions bring about many changes in canned foods during storage. The temperature of the storage is directly related to the storage life of the products. If it is considered that 10°C is a highly desirable storage temperature, increasing the temperature of 20°C probably will halve the storage life of the commodities, and raising the temperature of the storage room to 30°C will halve the storage life over that at 20°C. If the color is degraded at 10°C it will take one-fourth as long to arrive at an equal stage of degradation. Chemical reactions taking place during storage of canned foods will affect the flavor, color, texture and nutritive value of the foods.

TABLE 6.5

INFLUENCE OF STORAGE TEMPERATURE ON RANCIDITY VALUES,
THIAMIN RETENTION AND CONDITION OF INTERIOR CAN SURFACES
OF CANNED FRANKFURTERS AND BEANS IN STORAGE

Storage		Rancidity Values			
Temperature (°C)	Time (Months)	Peroxides, (M-Mols/Kg)	Free Fatty Acids (%) Oleic	Thiamin (Mg/ 100 G)	Can Rating[1]
—	0	5.0	3.8	0.059	8.6
3	6	3.6	6.0	0.042	6.6
	12	2.4	3.0	0.027	5.9
	24	1.1	6.9	0.015	4.8
21	6	4.7	4.1	0.057	7.7
	12	2.6	2.8	0.050	6.7
	36	1.8	2.0	0.036	6.4
	72	0.0	3.5		5.2
8	6	3.9	4.0		8.3
	12	3.6	1.8		7.6
	36	2.1	1.4	0.046	6.9
0,-17,-28	6	4.2	4.3	0.064	8.6
	12	3.3	2.4		8.5
	36	2.8	1.3	0.052	7.1
	72	0.0	1.6		6.3

Source: Cecil and Woodroof (1962).
[1] On a scale of 9 (no corrosion) to 1 (interior surface completely corroded).

Internal corrosion of the cans will follow the same pattern. The nutritive value of foods is maintained satisfactorily if the storage conditions are not excessively high. As an example, changes in the thiamin value of canned foods on storage for animal and vegetable products are indicated in Table 6.5.

In order to keep chemical changes to a minimum, temperatures of storage rooms for foods should be held just above the freezing point of the canned product. Summer temperature in warehouses may exceed 38°C in Southern United States and will fall below 0°C in the winter months in the North. It is desirable to cool warehouses in the South and heat in the winter in the North.

While freezing temperatures will not damage food nutritive values, unsightly products result from the freezing action.

Glass packed foods should be protected from light. Light catalyzed reactions include bleaching of color, destruction of vitamins, and flavor deteriorations.

External Corrosion of Cans

The presence of moisture on the surfaces of cans leads to rust formation. Rusting conditions may result from "sweating" of containers when moisture from air condenses on cans when their temperature is below the dew point of the air. When the relative humidity is high and the temperature of the cans is low, condensation may be expected. Warehousing at dry atmospheric conditions and constant low temperatures is important to prevent deterioration of cans and products. Proper air circulation, heating or cooling, and ventilation around stacks of cased products with adequate temperature control, reduce danger to corrosion of cans.

Sweating may also occur when canned foods are shipped from cool warehouses to warm and humid storage areas.

An important consideration in controlling the corrosion of cans of food relates to the temperature of cans after being cooled. If the temperature is reduced much below 40°C there may not be sufficient heat to vaporize the moisture on the cans, and these will remain moist for prolonged periods if placed in a wet condition into cartons in warehouses. Cans must be dry when stored. As a general rule, a can which does not feel slightly warm to the touch is cooler than 40°C.

Storage of canned foods near oceans requires that precautions be taken to reduce the potential of corrosion due to the action of salt in moisture in the atmosphere. Dry, ventilated warehouses which avoid entrance of air currents from an ocean are required.

Coding the Pack

It is important that canned foods be coded when placed into warehouses. In the event of spoilage, lots may be isolated in blocks from the main product. In addition, products of poor quality may be isolated and controlled.

INFLUENCE OF CANNING ON THE QUALITY OF FOOD

The general scheme of commercial canning may be depicted as follows: receive raw products; prepare product (wash, sort, peel, trim, chop, bone, etc.); fill food into containers; exhaust filled containers; seal lids to filled containers; heat process; cool containers; and finally store canned foods.

Unfortunately the application of sufficient heat to destroy food spoilage microorganisms and enzymes also results in some undesirable changes in the foods. There are alterations in the color, flavor, texture and nutritive value of foods in canning.

The prompt dispatch of raw perishable foods through the canning operations is required if high quality products are to result. Any decomposition in the product will be detrimental to the processed foods. There are losses in quality that may occur throughout the canning process. Proper attention to procedures is required. If products are blanched at too high a temperature, if products become contaminated and decomposed, if partially prepared foods are exposed to air and high temperatures for long periods of time, it may be expected that the quality of the processed foods will be poor. Under standard operating procedures in commercial canning plants, the following alations in quality of foods may be anticipated in greater or lesser degree, depending upon plant operation.

Color

Heating foods changes their physical and chemical qualities. In some instances the changes are desirable, making meat more tender. In other cases the changes are detrimental, destroying color characteristics (Table 6.6).

Foods that have been altered by heat can be expected to have altered abilities to reflect, scatter and transmit light. If this occurs there will be determinable changes in the color of foods.

Heating pure pigments causes color characteristics of the pigments to be altered. Heating pigments in complex substrates, such as a canned food, results in degradation of the natural color characteristics.

TABLE 6.6

INFLUENCE OF EQUAL LETHAL HEAT TREATMENTS AT VARIOUS
TEMPERATURES ON THE OPTICAL DENSITY OF BEET JUICE

Heat Treatment (°C)	Optical Density (530 mμ)
Unheated control	1.90
110	0.78
115	0.74
121	0.55
126	0.56
129	0.76

The degradation of color may be enhanced by the action of metals (loss of color of red fruits in tin containers).

The heat impairment of the lycopene (red pigment) of tomato juice is of interest, particularly the influence of equal lethal heat treatments at various temperatures. High temperature exposures for short times, as expected, result in less color degradation than equivalent processes at lower temperatures for longer times.

In addition to the destruction of pigments by heating, colored products may be formed by heating foods. Refluxing reducing sugars with amino acids produces a brown color, the color of maple syrup for example. Caramelization is a process resulting from heating polyhydroxycarbonyl compounds (sugars, polyhydroxycarboxylic acids) to high temperatures in the absence of amino compounds. Oxidative browning may occur in foods heated in the presence of oxygen, i.e., browning of tomato paste when concentrated in open kettle processes. There are enzymatic browning reactions which are inhibited by heating. For example, the browning of a cut slice of apple exposed to the air does not occur in canned apples.

Flavor and Texture

The flavor of a food is a dual response of odor and taste. When combined with the feel (consistency and texture) of foods in the mouth, the consumer is able to distinguish one food from another. Heating may be expected to degrade both flavor constituents and the physical character of foods. The degree of change is related to the sensitivity of the food to heat. Unfortunately the objective methods for evaluation of flavors of foods are lacking. Subjective evaluations are subject to the usual failings of fatigue, memory, etc.

Assuming equal lethal heat treatments, high temperature-short time exposures to heat are less destructive upon flavor and texture than low temperature-long time processes, within limits. Prolonged heating causes gelatin to degrade and lose its gelling powers. Starch will also lose its thickening power when given severe heat treatments. Pectin requires heating at low temperatures to thicken foods, but elevated temperatures or prolonged heat destroys its gelling power.

While heat may be damaging to food quality, the flavor of some products such as pork and beans are improved by heating longer than necessary to sterilize products. Meat is improved in flavor by cooking.

Protein

Denaturation of protein may be brought about by heat in the presence of moisture. When so denatured, the configuration of the native protein molecule is lost and specific immunological properties which distinguish most proteins are diminished. The activity of enzymes disappears when heated. There is an increase in the free sulfhydryl groups (Table 6.7), a change in availability of the protein to enzymatic hydrolysis, and an increase in the viscosity of the protein solution. After denaturation, proteins undergo further alteration known as coagulation or flocculation, and finally precipitation. Coagulation is a process involving the joining together of adjacent protein molecules by means of side-chain hydrogen bonds.

For a process with a C_o value of 4.0, there is a reduction in the rate of trypsin reaction with heated casein compared to unheated, regardless of the temperature of process. Between 104° and 121°C equivalent processes inflict no apparent differences in the rate of enzyme reaction with the heated protein.

There is evidence that heat impairs the nutritional value of protein, without altering the amino acid content as determined chemically. Failure of proteolytic enzymes to digest heated protein as readily as

TABLE 6.7

INFLUENCE OF EQUAL LETHAL HEAT TREATMENTS AT VARIOUS TEMPERATURES ON THE FREE SULFHYDRYL GROUPS IN A TWO PERCENT EGG ALBUMIN SOLUTION

Heat Treatment (°C)	Mg SH/Ml
Control	7.3
104	19.0
110	20.0
115	18.0
121	19.7

unheated may be the explanation for animals thriving less well on highly heated protein than on slightly heated protein.

However, heat processes of equal lethal value usually used in canning do not alter the viscosity of protein solutions significantly.

Fat and Oil

Fats are subject to two main types of rancidification: hydrolytic and oxidative. Enzymatic hydrolysis is characterized by the production of free fatty acids. Oxidative rancidity is an autocatalytic chemical reaction with atmospheric oxygen characterized by the production of peroxides.

Heat has profound influences on both types of deterioration of fats and oils.

Lipases produced by microorganisms are destroyed by the heat process. Oxidative rancidity is accelerated by heat, metallic ions and light. The rate of oxidation of fat is doubled for each increase in temperature of 10°C. The presence of metallic ions (cans) and light (jars) may increase the reaction. Moisture accelerates oxidative rancidity. Fats heated in the presence of oxygen have lowered melting points, lower iodine numbers and increased acidity. Heating is used to accelerate aging of fats in stability tests.

Flavor reversion occurs in unsaturated fats and oils; heat accelerates the reactions.

Fats are stable to moist heat in the absence of oxygen. Under these conditions, fats and oils in canned foods remain relatively unchanged by canning processing temperatures.

Fat and oil heated to high temperatures (204°C) have decreased nutritive values.

Carbohydrates

Sugars and starches are degraded by prolonged heating at high temperatures. Browning-type reactions (of organic acid, amino acids and reducing sugars) are produced by heating under moist conditions. The production of caramel color is an example, too, of the degradation possible with heat. Caramelization of carbohydrates in sweet corn in No. 10 containers is evidence of heat damage.

Vitamins

The vitamins are divided into two groups: those soluble in oil and those soluble in water.

Water-soluble Vitamins.—Thiamin, riboflavin and ascorbic acid have been studied extensively relative to their heat destruction in canning. Thiamin is heat labile; its loss in canning may be substantial. Acid foods retain a higher percentage of thiamin due to their lower processing requirements and increased stability of thiamin in acid foods. Peas may lose 50% of their thiamin during canning preservation. Losses have been found as high as 80% in canned lima beans and corn, but very little loss in canned carrots. Canned meats may lose in the vicinity of two-thirds their thiamin content.

Riboflavin is stable to heat, but light sensitive. Glass packed foods may lose more riboflavin than tinned foods. It is not uncommon to find analyses for riboflavin which report more than 100% of this vitamin present. In these instances more riboflavin is made available by heating for evaluation by present chemical and microbiological assays. Browning reaction end products fluoresce and may confuse results of some assays.

Ascorbic acid is destroyed by heating at low temperatures for long periods of time. This may be due to factors other than heat alone. High temperature-short time processes destroy little ascorbic acid if there is low oxygen tension. The destruction of this vitamin is accelerated by oxygen, copper ions and ascorbic acid oxidase enzyme. Tin protects vitamin C in solution.

High temperatures for short time exposures are generally less destructive of the water-soluble vitamins than lower temperatures for longer periods.

Fat-soluble Vitamins.—Vitamin A is relatively heat stable. If heating occurs in the presence of oxygen, appreciable losses occur. If air is excluded, heating to 115°C has little effect on vitamin A. Prolonged storage of heated products at high room temperatures will cause losses of this vitamin, however.

Vitamin D has been shown to be moderately heat stable and resistant to oxidation. However, heat and oxygen together cause rapid destruction. Vitamin D added to boiling oil is rapidly lost. If added to oil at low temperatures, the vitamin is relatively stable.

Vitamin E is stable to heat in the absence of oxygen, but heating causes rapid destruction of vitamin E in the presence of oxygen. Heating in this condition may yield almost complete destruction of the nutrient.

MISCONCEPTIONS RELATING TO CANNED FOODS

The term ptomaine is derived from the Greek and means a dead body. Ptomaines are produced in putrefying meat and other proteinaceous foods. These are substances which belong to the group of compounds known as amines. They result mainly from the decarboxylation of amino acids. A typical decomposition is:

$$CH_3CHNH_2COOH \rightarrow CH_3CH_2NH_2 + CO_2$$
alanine → ethylamine + carbon dioxide

Alanine loses carbon dioxide yielding the ptomaine ethylamine. Other common ptomaines in decomposed flesh are cadaverine (from lysine) and putrescine (from arginine). Foods containing these materials are too decomposed to be held in the stomach of modern man.

There are widely held notions relating to the safety of opened canned food containers. One is that opened cans of food rapidly become poisonous. The notion does not include evaporated milk, evidently, as it is common to punch holes in these cans and hold them at room temperature without ill effects. In truth, such products should be refrigerated, as should all perishable foods. Canned foods will spoil more rapidly than fresh foods once an inoculation has occurred in the opened canned products.

Certain foods should be removed from opened containers, not from the aspect of danger to health but from the loss of quality of the food. Pigmented foods may bleach in opened cans, and corrosion of tinplate is accelerated by oxygen from the air.

In many circumstances opened cans are the best container for the food, and may be the most sterile containers available in a household. The food and container are at least not hazards to public health unless contaminated. A dish or a pan is likely to be a source of bacteria which will find the food a suitable environment for growth. In any event, opened canned foods should be treated as perishable. The container in which the food is held is of minor significance if it is clean.

Another common notion is that the tinned flavor sometimes obtained from cans is toxic. Unless the amount of tin consumed is very large, it has no harmful effects. Tin has been shown to actually protect vitamin C in foods.

IMPROVEMENTS IN CANNING TECHNOLOGY

Generally for each 10°C increase in temperature, beyond the maximum conditions for growth, a ten-fold increase in the destruction of organisms occurs. Such an increase in temperature results in a doubling of the rate of chemical reactions responsible for product deteriorations. Therefore, heat treatments (of equal lethal influence on bacteria) at high temperatures for short periods of exposure should

Courtesy of Foxboro Co.

FIG. 6.17. STEAM PRESSURE CANNING RETORTS WITH INSTRUMENTATION
FOR CONTROLLING AND RECORDING PROCESSES

Instruments which control and record processes are shown in background.

effect greater retention of the natural quality characteristics of products than equivalent low temperature heating for longer periods of time. Research data, canner experience and consumer acceptance of processed foods have demonstrated this to be true.

With still-retort heating of canned foods (Fig. 6.17), processing temperatures of 115° to 121°C represent the upper limit of heat intensity which yields acceptable finished products. Heating at higher temperatures in still retorts results in deterioration of the quality of the processed foods, particularly conduction heating products.

While the first successful closed, still retort was invented in 1874, a retort was invented in 1884 which could agitate cans. Studies on the manufacture of canned condensed milk (containing added sugar) led to the possibility of canning evaporated milk without the addition of sugar or preservatives. Evaporated milk of acceptable stability cannot be obtained in still retorts. Agitation increases the rate of heat transfer from container to product by renewing the surface in contact with the hot container; the "burned" and "cooked" flavors are reduced in intensity. Agitation prevents the complete coagulation of milk proteins and prevents milk solids from adhering to the walls of cans.

Agitated cookers are used in the canning industry for many products at present. Modern units consist of three sections (Fig. 6.18): the preheater, the cooker and the cooler. Each unit contains a revolving reel

Courtesy of Food Machinery and Chemistry Corp.

FIG. 6.18.   FMC CONTINUOUS PRESSURE COOKER AND COOLER

Continuous high-speed processing of canned foods.

and a spiral track. The cans, placed in individual compartments, are rotated during part of the revolution by movement of the reel, and at the same time are guided by tracks from the inlet to the outlet continuously.

There are many patents concerning agitating cookers. Some deal with reciprocal movement of cans, others have reciprocating movements combined with revolving about an axis in cookers. Cans are held in trays in many of these cookers.

Ball considered the ideal method for food canning to be sterilization of the product prior to filling and aseptically filling sterile cans with

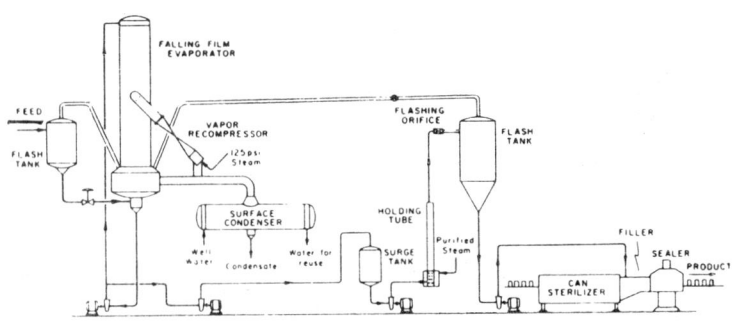

Courtesy of Food Processing

FIG. 6.19.   ASEPTIC CANNING LINE DIAGRAM FOR CONCENTRATED MILK PROCESS

Milk is concentrated and sterilized outside the container, then filled into sterile containers in sterile atmosphere.

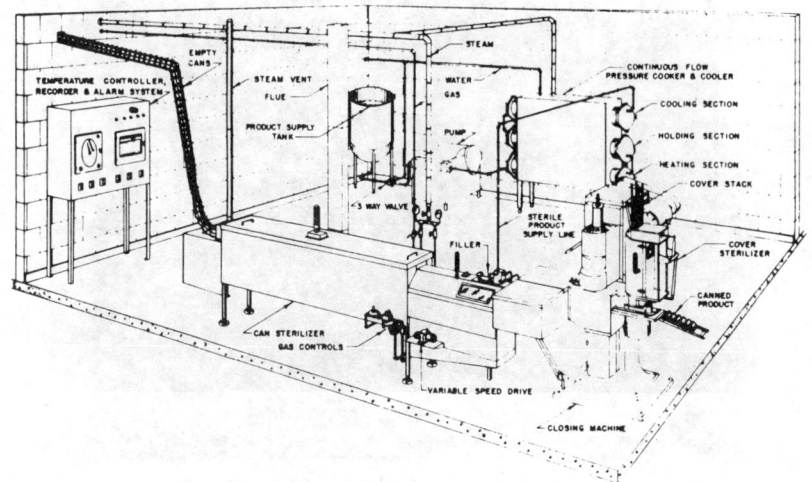

Courtesy of James Dole Engineering Co.

FIG. 6.20.   PLANT LAYOUT COMBINING QUICK FOOD STERILIZATION EQUIPMENT AND
ASEPTIC CANNING LINE

the microbe-free product. Martin described a practical application of
this procedure in 1948 (Fig. 6.19).

It involves the following steps: (1) sterilization of the product by
flash heating and cooling in a tubular type heat-exchange system; (2)
sterilization of the containers and covers with superheated steam; (3)
aseptic filling of the relatively cool, sterile product into the sterile
containers; and (4) application of sterile covers to the filled containers
and sealing the cans in an atmosphere of either saturated or superheated steam (Fig. 6.20 and 6.21).

This process has been shown to be advantageous for heavy or
viscous products which are adversely affected by sterilization in a
sealed container. This method has been demonstrated to be practical.
Another advantage is that the procedure may be used as a continuous
operation, although it is not applicable to solid foods.

The following types of agitation in processing vacuum packed vegetables have been reported: (1) continuous end-over-end rotation;
(2) continuous axial rotation; and (3) intermittent axial rotation.

All three types of agitation increased the rate of heat penetration
over that attained by still processing. In studies involving end-over-
end rotation, it is observed that the rate of heat penetration is only
slightly influenced by the speed of rotation in the range of 10 to 40
rpm. On the basis of information obtained from experimentation a

Courtesy of James Dole Engineering Co.

FIG. 6.21.   ASEPTIC CANNING LINE

substantial reduction in agitating process times over still retort treatments is possible.

The accuracy of results obtained from heat penetration studies have been checked with inoculated packed of peas and corn, containing 200,000 to 250,000 spores of P.A. No. 3679 per can. Some of the results obtained are:

Product	Process	Temperature, (°C)	Calculated Sterilization Value, (C_0)	Minimum Time of Processing Which Did Not Give Swells in Inoculated Packs (min)
Peas	End-over-end	121	7.8	9
Peas	Intermittent axial	121	9.2	12
Peas	Still	115	8.5	45
Corn	End-over-end	121	8.0	12
Corn	Intermittent axial	121	9.2	14
Corn	Still	121	10.9	40

On the basis of these results, it may be concluded that heat penetration measurements are reliable as a basis for calculating agitated can processes, and that substantial reductions in processing times may be effected. The Continental Can Company developed an experimental agitating cooker in which reciprocal and rotational movements could be attained at speeds varying from 0 to 360 rpm and at steam pressures corresponding to 149 C.

The first such studies were made with 300 × 314 cans filled with water to 0.79 cm headspace, with a retort temperature of 126 C and with variable speed of 0 to 200 rotations or reciprocations per minute. When cans were maintained horizontally and subjected to 120 reciprocations a minute through a 2.5-cm stroke, agitation reduced the time from 4.3 to 1.4 min to heat the contents of the can to 1°C below retort temperature in the range of 15° to 126°C. A 5-cm stroke increased the rate of heat penetration but was considered to be impractical for a commercial operation. Results of experiments in which the axes of rotation were inside or at the sides of the cans indicated that the end-over-end type of agitation was superior. A reduction in processing time is obtained when cans are maintained bottom-down and perpendicular to the axis of rotation. End-over-end rotation at the proper speeds is superior to reciprocation, especially if the amount of mechanical work is taken into consideration.

The resultant of forces when the cans were subject to the simultaneous action of both gravity and centrifugation has been studied. By selecting a proper speed, the headspace void can be made to pass through the liquid at various levels. When the speed is such that the centrifugal force equals the weight of the liquid contents, the headspace void passes through approximately the center of the can. When this occurs, maximum turbulence results with the greatest rate of heat penetration (see data following).

The speed of rotation must be decreased for viscous products to allow the headspace volume to cross through the center of the can.

In experiments with end-over-end rotation of large-sized cans, placing the can adjacent to the axis of rotation allows maximum agitation, particularly at moderate speeds. The best speed for a 603 × 700 (No. 10) can with a path radius of 8.9 cm is considered to be 100 rpm depending on the product, of course.

Application of these studies to practical canning results in a decrease in heating time for the end-over-end methods when compared to conventional still processes.

Results are quoted to indicate the improvement possible:

Product	Can Size	Process-Agitated (126°C) (Min)	(°C)	Conventional (Min)	(°C)
Peas	307 × 409	4.90	126	35	115
Carrots	307 × 409	3.40	126	30	115
Beets, sliced	307 × 409	4.10	126	30	115
Asparagus spears	307 × 409	4.50	132	16	120
Asparagus, cuts and tips	307 × 409	4.00	132	15	120
Cabbage	307 × 409	2.75	132	40	115
Asparagus, spears					
brine packed	307 × 409	5.20	126	50	115
brine packed	603 × 700	10.00	126	80	115
vacuum packed	307 × 306	5.00	126	35	121
Mushroom soup	603 × 700	19.00	126	—	—
Evaporated milk	300 × 314	2.25	93	18	115

Another advantage to agitation retorting is that higher temperatures may be used with less danger of overcooking.

Using the thiamin content as an index of the chemical changes taking place during processing, greater retention is found in agitated than in still processes.

Agitation retorting permits successful canning in large-sized containers for low acid foods. No. 10 cans of cream-style sweet corn are caramelized by still retort processes. Agitation processes for corn and most foods result in improved quality retention.

Substantial improvements in color and texture of products that normally become degraded with prolonged heat treatments are possible with agitation retorts. Spaghetti and meat cooked in No. 10 cans by agitating retort have "home cooked" color and texture qualities. Another benefit is found for small canning operations in processing relatively unexploited heat-sensitive food products. Field tests with canned whole milk indicate the product to be substantially as good as freshly pasteurized milk, and it will keep at least ten times as long without refrigeration.

LETHALITY GUIDES FOR COMMERCIAL STERILITY

Stumbo *et al.* (1975) have described commercial sterility, in reference to low-acid canned foods, as the freeing of a product of not only the most heat resistant bacterium of public health significance (i.e., *C. botulinum*) but also of other more heat resistant bacteria that might cause significant economic spoilage losses. Therefore, sterility values have been computed as bases for establishing processes that should normally result in commercially sterile low-acid canned foods. They present a wide range of very useful data. See their published summary for further development of the subject.

Stumbo also developed an equation for calculating Decimal Reduction Times (*D* values):

$$D = \frac{U}{\log a - \log b}$$

where

D = death rate in minutes at the temperature of exposure
U = heating time in minutes (corrected for lag)
a = initial number of spores (inoculum per container × number of containers)
b = number of positive TDT cans or tubes at time U.

Perkins *et al.* (1975) have reported on a study of D values in several substrates (Table 6.8) from which they calculated z values.

TABLE 6.8

CLASSICAL CLOSTRIDIUM BOTULINUM RESISTANCE (19A, 23A, 97A)
IN M/15 PHOSPHATE BUFFER AT pH 7.0 VS 213B
IN PHOSPHATE BUFFER AND FORMULATED SAMPLES

	D_{100}	D_{110}	D_{115}	Extrapolated D_{121}
Phosphate buffer;				
Esty and Meyer (1922) (19A, 23A, 97A)	30.15	7.51	1.67	0.26
CIFR[1] (1973) (213B)	21.90	5.51	1.17	0.17
Formulated meat and vegetable sample pH 6.1	41.20	9.26	1.21	0.11
Formulated poultry sample pH 6.1	42.84	6.25	0.76	0.05
Formulated seafood sample pH 5.9	38.32	6.33	0.69	0.05

Source: Perkins et al. (1975).
[1] Campbell Institute for Food Research.

They believe that if the C_0 concept is to continue in its present wide use, values considerably greater than 2.78 must, in many cases, be considered in establishing safe and adequate processes. There are published processes, for example, based on C_0 values significantly less than 2.78. The assumption on which these low lethalities are based should be reviewed with the potential of a z of 8 in mind.

A more realistic estimate of the hazard potential in cans of product underprocessed by a few degrees or minutes will also be possible if the assumption is made that any contaminating anerobes may have a z value of 8.

Although a lower z value might seem to suggest a lesser heat resistance, the opposite is the case with respect to a given C value and heating times at temperatures below 121°C. A z value of less than 10 means that the spores are more resistant at lower temperatures relative to 121°C than would be predicted on the basis of a 10°C thermal death time curve.

Optimizing Thermal Processes

One of the more recent advances has been the development of means to predict the retention of nutrients in thermal processing. The kinetic parameters describing the thermal stability of nutrients, however, have generally not been available. The most common method of reporting the effect of heat processing on nutrients has been simply to express the nutrient content after processing as a percentage of the initial amount present.

Lund (1973) reports that in determining optimum thermal processes for retention of nutrients, it is important to recognize that the ultimate thermal process must provide a lethality for the microorganisms equal to the existing thermal process.

Teixeira *et al.* (1969) and Hayakawa (1969) developed procedures to predict and optimize a thermal process for nutrient retention. Teixeira's method requires a computer program and z and D_r values for the heat-labile nutrient. Hayakawa's method does not require a computer program but does require extensive interpolation to determine values for six dimensionless groups.

The optimum thermal process is dependent on both the z and D values of the nutrient. Retention of a nutrient with a large z value will be favored by a high temperature-short time process. Lund reports that for thiamin in pureed green beans, the optimum processing temperature was calculated to be 120° C, which is currently extensively used in the canning industry.

These data suggest that optimization for one nutrient does not imply optimization for all nutrients. Consequently, meaningful nutrient optimization of the thermal process should be based on the most important nutrient in the food. To optimize nutrient retention it is important that meaningful data be developed on the thermal destruction of all nutrients under the variety of conditions found in foods. Also, methods for predicting nutrient retention in convection-heated foods are needed. The thermal process should be optimized for the retention of nutrients as well as for destruction of microorganisms.

RETORT POUCHES

There are a variety of preformed food containers. All have a closure seam or mechanism of some type: a tight, double seam for cans; a physical clinch for jar lids; a sewn seam for staple commodity bags; or a fusion weld of two compatible polymeric film surfaces for flexible packaging. Seal performance has been recognized as vital with rigid food containers. Flexible package seals have not had to perform in accordance with as strict and definitive criteria as they do now that flexible packages are being used for thermoprocessed foods (Lampi *et al.* 1976) (Fig. 6.22 and 6.23).

The requirements for retort pouches are: a leaker rate as low as that for the metal can; seals that must withstand 120°C or higher thermoprocessing temperatures; packages that must be durable through the entire distribution system; and extended shelf-life requirements. Attention to the formation of seals and their performance is required.

Courtesy of QM. Food and Container Institute

FIG. 6.22. FLEXIBLE CONTAINERS FOR CANNED FOODS BEING DEVELOPED

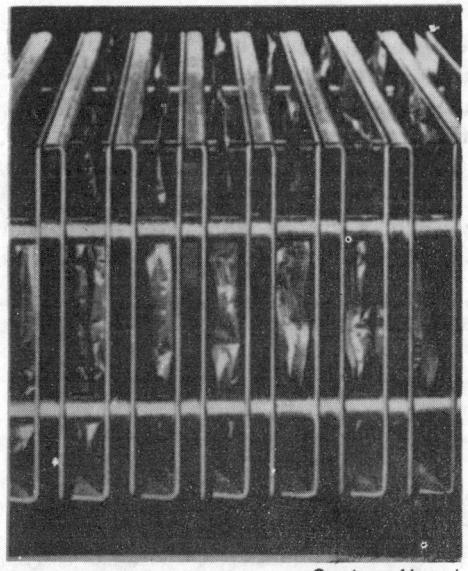

Courtesy of Lampi

FIG. 6.23. VERTICAL-SLOT RACK TO HOLD
FLEXIBLE POUCHES DURING RETORTING

TABLE 6.9

RESISTANCE OF RETORT POUCHES AND METAL CANS
TO A ROUGH HANDLING CYCLE

Product	Package	No. Tested	No. Failed	% Failed
Fluid foods				
	Cans	1,440	32	2.22
	Flexible material No. 1	1,440	30	2.08
	Flexible material No. 2	720	5	0.70
Semi-solid foods				
	Cans	720	4	0.56
	Flexible material No. 1	720	2	0.28
	Flexible material No. 2	720	4	0.56

Source: Lampi et al. (1976).

Testing a Good Seal

The performance standard for the retort pouch has been the sanitary can and its historically documented satisfactory performance. The results of abuse tests are shown in Table 6.9 for fluid and semi-solid foods. Although not restricted to seal failures alone, these data present a tentative performance criterion, suitable for military needs, of 2% or fewer leakers (as detected by microbiological testing).

Four criteria and testing techniques—fusion testing, burst testing, tensile testing, and visual examination—are summarized in Table 6.10.

From the packaging materials standpoint the pouch in a carton is

TABLE 6.10

CRITERIA FOR DEFINING A HIGH-PERFORMANCE SEAL FOR RETORT POUCHES

Test Variable	Criteria
Fusion	Must exist
Internal pressure	Level = 20 psig (1.38 X 10⁵ Pa) Hold time = 30 sec Maximum seal yield = 1/16 in (1.6 mm) Restrained pouch thickness = 1/2 in (13 mm)
Tensile	Level = 12 lb/in (2.1 kg/10 mm) Crosshead spead = 10 in/min (25.4 cm/min) Sample width = 1/2 in (13 mm)
Visual	No visible aberrations

Source: Lampi et al. (1976).

more expensive than the metal can with paper label. The retort pouch overall is less energy-intensive and, therefore, less expensive in that regard than other forms of packaging. As the retort pouch develops in use, more and better equipment, materials and processes will lower costs. A major advance in food packaging will have occurred.

HAZARD ANALYSIS

The hazard categories described in Chapter 5 for frozen foods are useful to review for canned foods. Similarly, those points in process of potential public health hazard must be controlled.

FDA regulations describe the proper methods and equipment to be used in the United States in the processing of low-acid canned foods, and the critical factors which must be adhered to in order to assure adequate processing of these products. All these are primarily involved in monitoring conditions or equipment which, if not functioning properly or within proper operating tolerances, could create matters of public health significance. Therefore, it is necessary to determine what the critical control points are and to make and keep records of those points and areas.

Determining Critical Points

The critical control points in a canning operation vary with the product packed, the container in which it is packed, the retort to process the filled and sealed containers, and the cooling and storage systems used. To find the critical control points, it is necessary to determine the consequences of an event occurring and the possible events which might affect the final product.

It is useful to examine a formulated food product filled into cans and retorted in a continuous cooker. First, spices, sugar, starch, salt, water and other necessary food ingredients are blended in a mixer, sent to filler, filled into cans and capped with a lid. The closed container is retorted, and the processed product is cooled and then sent to a warehouse for labeling and casing.

Ito (1974) of the National Canners Association proposes the following questions to accomplish this (starting with the ingredients, then examining the critical areas associated with the process).

Ingredients: Has the list of ingredients changed since the process was originally designed? Is the ratio of the ingredients the same? Is a different kind of starch being used?

Mixing: Are the ingredients blended properly? Are there any lumps,

Courtesy of Food Processing

FIG. 6.24.   HIGH-SPEED FILLING AND SEALING LINE FOR GLASS CONTAINERS

Glass containers are as functional as tin cans in high-speed operations.

omissions or over-additions? Is the consistency of the product critical? If so, what is the consistency? How should it be measured? What kind of equipment? At what temperature? Can the consistency be adjusted? If yes, how? Under what conditions? Is the consistency recorded? At what point in the line? What are the limits, the maximum and the minimum?

Filling: Does the filler dispose the proper ratio of solids to liquids into the container? Is the proper weight placed in the container? Is the container filled to the proper headspace? Is the headspace measured? At what point in the line is it measured? At what frequency is it measured? Is the measurement recorded? (See Fig. 6.24.)

Sealing: Are the containers checked for defects? What are the can seam inspection procedures? Is a visual inspection made? How often? Who makes the inspection? Does the seam inspector have proper supervision and/or training? Are records kept? Are they of only the defects? Are the remedial steps taken as required by the FDA's GMP's?

Retorting: Is the continuous processing retort properly equipped? Does it have a mercury thermometer? Is it properly installed, checked and readable? Does the retort have the proper recorder and recorder charts? Is the retort piping correct? Is the condensate drain operating? Are the bleeders properly installed, operating and of the right size? Is there an appropriate venting schedule available for the retort? Is the vent schedule posted? Does the retort operate under proper control? Is the reel speed correct? Is it checked? How often? Is the speed recorded? How is the speed checked?

Heat Process: Is the process correct? Who designed it? And when? Were all of the necessary critical factors considered when the process was originally designed? Are they still being adhered to? What are the process times and temperatures of operation? Are they recorded? How often? Are the process times and temperatures posted? Are they readily available? Do the retort operators have appropriate supervision? Are the records reviewed? When and how often? Is a production record kept? Is it appropriate and accurate? What is done with process deviations?

Cooling: Is the cooling water treated with a germicide? What is its concentration at the cooler outlet? Is this monitored regularly? Are the results recorded? Is the germicide effective? Is it approved for this usage by a regulatory agency?

Handling: Following the cooler, how are the cans handled? In what condition is the can handling equipment? What is the potential for container damage? Are records kept? What is the schedule for cleaning and care of can handling equipment? How are the dented cans handled?

Recording: Having made the observations, it is necessary to keep good, complete records. The records must reveal that observations have been made and at a given frequency. If a defect is observed, it should be recorded, and the remedy should also be recorded.

Responsibility of the Processor

The steps involved in the production of foods in hermetically sealed containers are regulated by law. The regulations, by their nature, can only be used as guidelines to assist in determining the critical points for a specific product line. The processor must determine what is critical and monitor it, even though it may not be specifically covered by regulations. Industry still has ultimate responsibility for the production of products free of public health significance.

SUMMARY

The high quality foods in great demand are also the highly perishable foods. Fortunately we now know how to preserve these foods, and canning remains one of the key building blocks of a sound, effective food supply. Further details concerning the process are presented in Chapter 14. New canning technology is permitting the introduction of new processing container types, forms and sizes. Major strides are already underway with easy opening containers and canning in flexible pouches. Weight reduction in containers and the increased use of aluminum and plastic units are aiding the ever increasing applications of the growing technology of food preservation by canning.

REFERENCES

AD HOC COMM. NATL. MEAT CANNERS ASSOC. 1975. Field performance of metal food containers with easy-open ends. Food Technol. 29, No. 2, 48-49.

ADAMS, H.W., and OWENS, W. 1972. Internal pressure of cans agitated during heating and cooling. Food Technol. 26, No. 7, 28-30.

AGNEW, B. 1973. Clean air systems for aseptic packaging. Food Technol. 27, No. 9, 58-62.

AMERICAN CAN CO. 1969. Canned Food Reference Bulletins. American Can Co., New York.

BALL, C.O. 1923. Thermal process time for canned foods. Bull. Natl. Res. Counc. 7.

BALL, C.O. 1928. Mathematical solution of problems on thermal processing of canned foods. Univ. Calif. Publ. Public Health 1.

BALL, C.O. 1943. Short time pasteurization of Milk. Ind. Eng. Chem. 35, 71-91.

BALL, C.O. 1947. New technic speeds sterilization of canned corn. Food Ind. 21, 307-311.

BALL, C.O., and OLSON, F.C.W. 1957. Sterilization in Food Technology. McGraw-Hill Book Co., New York.

BAUMAN, H.E. 1974. The HACCP concept and microbiological hazard categories. Food Technol. 28, No. 9, 30-34, 70.

BIGELOW, W.D. 1921. The logarithmic nature of thermal death time curves. J. Infect. Dis. 29, 528-532.

BIGELOW, W.D., BOHART, G.S., RICHARDSON, A.C., and BALL, C.O. 1920. Heat penetration in processing canned foods. Natl. Canners Assoc. Res. Lab. Bull. 16-L.

BITTING, A.W. 1924. A history of food preservation. Canning Age 5, 10-13.

BITTING, A.W. 1937. Appertizing or the Art of Canning. Its History and Development. The Trade Pressroom, New York.

BORGES, J.M., and DESROSIER, N.W. 1953. New high temperature short time multiaction pressure retort. Food Eng. 26, 54, 153-155.

BRENNER S., WODICKA, V.O., and DUNLOP, S.G. 1948. Effect of high temperature storage on the retention of nutrients in canned foods. Food Technol. 2, 207-221.

BRODY, A.L. 1972. Aseptic packaging of foods. Food Technol. 26, No. 8, 70-74.

BURTON, L.V. 1940. Heat sterilization in bags creates technological problems. Food Ind. 12, 31-32.

BYRAN, F.L. 1974. Microbiological food hazards today—based on epidemiological information. Food Technol. 28, No. 9, 52-66, 84.

CAMERON, E.J., and ESTY, J.R. 1940. Comments on the microbiology of spoilage in canned foods. Food Res. 5, 399-405.

CAMERON, E.J., PLICHER, R.W., and CLIFCORN, L.E. 1949. Nutrient retention during canned food production. Am. J. Public Health 39, 756-763.

CECIL, S.R., and WOODROOF, J.G. 1962. Long term storage of military rations. Ga. Agric. Exp. Stn. Tech. Bull. 25.

CHARM, S.E. 1958. The kinetics of bacterial inactivation by heat. Food Technol. 12, 4-3.

CURRAN, H.R. 1931. Influence of osmotic pressure on spore germination. J. Bacteriol. 21, 197-209.

CURRAN, H.R. 1935. Influence of some environmental factors upon the thermal resistance of bacterial spores. J. Infect. Dis. 56, 373-380.

CURRAN, H.R., and EVANS, F.R. 1952. Symposium on the biology of bacterial spores V. Resistance of bacterial spores. Bacteriol. Rev. 16, 111-117.

DASTUR, K., WECKEL, K.G., and VON ELBE, J. 1968. Thermal process for canned cherries. Food Technol. 22, No. 9, 126-132.

DAVIS, E.G., and ELLIOTT, A.G.L. 1958. Estimation of vacuum in unopened containers. Food Technol. 12, 473-478.

DAVIS, R.B., LONG, F.E., and ROBERTSON, W.F. 1972. Engineering considerations in retort processing of flexible packages. Food Technol. 26, No. 8, 65-68.

DEMONT, J. I., and BURNS, E.E. 1968. Effects of certain variables on canned rice quality. Food Technol. 22, No. 9, 136-138.

DESROSIER, N.W., and ESSELEN, W.B. 1950. Heat penetration and processing studies on home canned corn, hominy, pork and beans, and potatoes. Mass. Agric. Exp. Stn. Bull 456.

DESROSIER, N.W., ESSELEN, W.B., and ANDERSON, E.E. 1954. Behavior of heat resistant food spoilage spore forming anaerobes inoculated into soil. J. Milk Food Technol. 17, 206-210.

DESROSIER, N.W., and HEILIGMAN, F. 1956. Heat activation of bacterial spores. Food Res. 21, 54-62.

DOYEN, L. 1973. Aseptic system sterilizes pouches with alcohol and UV. Food Technol. 27, No. 9, 49-50.

DUFFY, P.J. 1973. Aseptic filling in flexible bags. Food Technol. 27, No. 9, 52-54.

ELVEHJEM, C.A. 1945. Vitamins and food processing. Agric. Eng. 26, 12-15.

ERICKSON, E.J., and FABIAN, F.W. 1942. Preserving and germicidal actions of various sugars and organic acids on yeast and bacteria. Food Res. 7, 68-79.

ESSELEN, W.B., JR. 1945. Botulism and home canning. Mass. Agric. Exp. Stn. Bull. 426.

ESSELEN, W.B., JR. 1949. Personal communication. Amherst, Mass.

EVANS, H.L. 1958. Studies in canning processes. II. Effects of the variation with temperature on the thermal properties of foods. Food Technol. 12, 276-280.

FAGERSON, I.S., and ESSELEN, W.B., JR. 1950. Heat transfer in commercial glass containers during thermal processing. Food Technol. 4, 411-414.

FELLERS, C.R., ESSELEN, W.B., JR., MACLINN, W.A., and DUNKER, C.F. 1939. Nutritive studies with fresh and processed fruits and vegetables. Mass. Agric. Exp. Stn. Bull. 355.

FOSTER, J.W., and WYNNE, E.S. 1948. The problem of dormance in bacterial spores. J. Bacteriol. 55, 623-625.

GRIFFIN, R.C., JR., HERNDON, D.H., and BALL, C.O. 1971. Use of computer derived tables to calculate sterilizing processes for package foods—part 3. Application to cooling curves. Food Technol. 25, No. 2, 36-45.

GOLDBLITH, S.A. 1971. The science and technology of thermal processing—part I. Food Technol. 25, No. 12, 44-50.

GOLDBLITH, S.A. 1972A. The science and technology of thermal processing—part II. Food Technol. 26, No. 1, 64-69.

GOLDBLITH, S.A. 1972B. Controversy over the autoclave. Food Technol. 26, No. 12, 62-65.

HASHIDA, W., MOURI, T., and SHIGA, I. 1968. Application of ribonucleotides to canned sea foods. Food Technol. 22, No. 11, 102-108.

HAYAKAWA, K. 1969. New parameters for calculating mean average sterilizing value to estimate nutrients in thermally conductive foods. J. Can. Inst. Food Technol. 2, 165-171.

HEILIGMAN, F., and DESROSIER, N.W. 1956. Spore Germination II. Inhibitors. Food Res. 21, 70-74.

HEIN, R.E., and HUTCHINGS, I.J. 1971. Influence of processing on vitamin-mineral content and biological availability in processed foods. In Symposium on Vitamins and Minerals in Processed Foods. Am. Med. Assn. Council on Foods and Nutrition and Food Industry Liaison Committee, New Orleans.

ITO, K. 1974. Microbiological critical control points in canned foods. Food Technol. 28, No. 9, 46-50.

JEN, Y., MANSON, J.E., STUMBO, C.R., and ZAHRADNIK, J.W. 1971. A procedure for estimating sterilization of and quality factor degradation in thermally processed foods. J. Food Sci. 36, 692.

KRAMER, A. 1971. A systems approach to quality assurance. Food Technol. 25, No. 10, 28-29.

KRAMER, A. 1973. Storage retention of nutrients. Food Technol. 28, No. 1, 50-60.

LABAW, G.D., and DESROSIER, N.W. 1953. Antibacterial activity of edible plant extracts. Food Res. 18, 186-190.

LACHANCE, P.A., RANADIVE, A.S., and MATAS, J. 1973. Effects of reheating convenience foods. Food Technol. 27, No. 1, 36-38.

LAMPI, R.A., SCHULZ, G.L., CIAVARINI, T., and BURKE, P.T. 1976. Performance and integrity of retort pouch seals. Food Technol. 30, No. 2, 38-48.

LANGE, L. 1973. Aseptic bag-in-box packaging. Food Technol. 27, No. 10, 80-82.

LAZAR, M.E., LUND, D.B., and DIETRICH, W.C. 1971. A new concept in balancing IQB reduces pollution while improving nutritive value and texture of processed foods. Food Technol. 25, No. 7, 684-686.

LUND, D.B. 1973. Effects of heat processing. Food Technol. 27, No. 1, 16-18.

MARTIN, W.M. 1948. Flash process, aseptic fill are used in new canning unit. Food Ind. 20, 832-834.

MERMELSTEIN, N.H. 1976. The retort pouch in the U.S. Food Technol. 30, No. 2, 28-37.

MIDDLEKAUF, R.D. 1976. 200 years of U.S. food laws: a gordian knot. Food Technol. 30, No. 6, 48-54.

NATL. CANNERS ASSOC. 1955. Processes for low acid canned foods in glass containers. Natl. Canners Assoc. Res. Lab. Bull. 30-L.

NATL. CANNERS ASSOC. 1968. Laboratory Manual for Food Canners and Processors, 3rd Edition, Vols. 1 and 2. AVI Publishing Co., Westport, Conn.

PERKINS, W.E., ASHTON, D.H., and EVANCHO, G.M. 1975. Influence of the z value of Clostridium botulinum on the accuracy of process calculations. J. Food Sci. 40, 1189-1192.

SCHROEDER, H.A. 1971. Losses of vitamins and trace minerals resulting from processing and preservation of foods. Am. J. Clin. Nutr. 24, 562.

SILLIKER, J.H., GREENBERG, R.A., and SCHACK, W.R. 1958. Effect of individual curing ingredients on the shelf stability of canned comminuted meats. Food Technol. 12, 551-554.

SOGNEFEST, P., and BENJAMIN, H.A. 1944. Heating lag in thermal death time cans and tubes. Food Res. 9, 234-236.

STEVENSON, C.A., and WILSON, C.H. 1968. Nitrogen closure of canned applesauce. Food Technol. 22, No. 9, 93-95.

STUMBO, C.W. 1948. Bacteriological considerations relating to process evaluation. Food Technol. 2, 116-122.

STUMBO, C.W. 1949. Further consideration relating to evaluation of thermal processes for foods. Food Technol. 3, 126-130.

STUMBO, C.R. 1973. Thermobacteriology in Food Processing, 2nd Edition. Academic Press, New York.

STUMBO, C.R., PUROHIT, K.S., and RAMAKRISHNAN, T.V. 1975. Thermal process lethality guide for low acid foods in metal containers. J. Food Sci. 40, 1316-1323.

TEIXEIRA, A.A., DIXON, J.R., ZAHRADNIK, J.W., and ZINSMEISTER, G.E. 1969. Computer optimization of nutrient retention in the thermal processing of conduction-heated foods. Food Technol. 23, 845-851.

TOEPFER, C.T., REYNOLDS, H., GILPIN, G.L., and TAUBE, K. 1946. Home canning processes for low acid foods. U.S. Dep. Agric. Tech. Bull. 930.

TOLEDO, R.T., and CHAPMAN, J.R. 1973. Aseptic packaging in rigid plastic containers. Food Technol. 27, No.11, 68-76.

TOWNSEND, C.T., ESTY, J.R., and BASELT, J.C. 1938. Heat resistance studies on spores of putrefactive anaerobes in relation to determination of safe processes for canned foods. Fod Res. 3, 323-330.

TOWNSEND, C.T. et al. 1949. Comparative heat penetration studies on jars and cans. Food Technol. 3, 213-218.

TRESSLER, D.K., and JOSLYN, M.A. 1971. Fruit and Vegetable Juice Processing Technology, 2nd Edition. AVI Publishing, Co., Westport, Conn.

VINTERS, J.E., PATEL, R.H., and HALABY, G.A. 1975. Thermal process evaluation by programable computer calculators. Food Technol. 29, No. 3, 42-48.

WECKEL, K.G., MARTIN, J.A., LAKAMA, R., and LYLE, M. 1962. Effect of added sugar on consumer acceptance and physiochemical properties of canned cream-style corn. Food Technol. 16, 131-133.

WILBUR, P.C. 1949. Factors influencing process determination in agitating pressure cookers. Convention Issue, Inf. Letter, Natl. Canners Assoc. 1219.

WILLIAMS, O.B. 1929. The heat resistance of bacterial spores. J. Infect. Dis. 44, 421-465.

Chapter 7

Principles of Food Preservation by Drying

DRYING—A NATURAL PROCESS

Drying is one of man's oldest methods of food preservation. It is a process copied from nature; we have improved certain features of the operation. Drying is the most widely used method of food preservation.

All the cereal grains are preserved by drying, and the natural process is so efficient it hardly requires added effort by man. However, there have been periods in history when climatic factors were such that grains failed to dry properly in the fields. In these instances, man attempted to assist the natural action by supplying heat to the grains which otherwise would decompose. Grains, legumes, nuts and certain fruits mature on the plants and dry in the warm wind. More fruits are preserved by drying than by any other method of food preservation. The natural sun drying of foods yields highly concentrated materials of enduring quality, yet a highly complex civilization cannot be so dependent upon the elements—they are unpredictable. Sun drying remains the greatest food preservation action (Fig. 7.1).

Dehydration—Artificial Drying

The use of heat from a fire to dry foods was discovered independently by many men in the New and Old Worlds. Ancient man dried foods in his shelters; pre-Columbus American Indians used the heat from fire to dry foods. However, it was not until about 1795 that a hot air dehydration room was invented. The team of Masson and Challet in France developed a vegetable dehydrator which consisted of a hot air (40°C) flow over thin slices of vegetables. It is worth noting that both canning and dehydration came into being at approximately the same time, nearly a century and a half ago.

Evaporation and desiccation are terms which perhaps note the

Courtesy of U.S. Dept. Agr.

FIG. 7.1.   GRAPES ON TRAYS DRYING IN THE SUN

same action. The term dehydration has taken the meaning in the food industry as that process of artificial drying.

Dehydration vs. Sun Drying

Dehydration implies control over climatic conditions within a chamber, or microenvironment control. Sun drying is at the mercy of the elements. Dried foods from a dehydration unit can have better quality than sun-dried counterparts. Less land is required for the drying activity. Sun drying for fruit requires approximately one unit of drying surface per 20 units of crop land.

Sanitary conditions are controllable within a dehydration plant, whereas in open fields contamination from dust, insects, birds and rodents are major problems.

Dehydration obviously is a more expensive process than sun drying, yet the dried foods may have more monetary value from dehydration due to improved quality. The yield of dried fruit from a dehydrator is higher inasmuch as sugar is lost due to continued respiration of tissues during sun drying, and also due to fermentation.

The color of sun-dried fruit may be superior to dehydrated fruit under optimum conditions of operation of both. Color development in certain immature fruits continues slowly during sun drying. This does not occur during dehydration.

In cooking quality dehydrated foods are usually superior to sun-dried counterparts. However, sun-dried animal flesh and fish can be highly acceptable.

On the basis of cost sun drying has advantages, but on the basis of time to dry and quality, dehydration has merits. Furthermore sun drying can not be practiced widely due to unfavorable weather conditions in many areas where man lives and agriculture is rewarding.

Why Dried Foods?

Dried and dehydrated foods are more concentrated (Fig. 7.2) than any other preserved form of foodstuffs. They are less costly to produce; there is a minimum of labor required, processing equipment is limited, dried food storage requirements are at a minimum, and distribution costs are reduced (one carload of dried, compressed food may equal ten carloads of the fresh commodity; see Tables 7.1 and 7.2).

TABLE 7.1

RELATIVE SPACE REQUIREMENTS PER UNIT
(FRESH BASIS) OF FOOD

Product	Fresh	Dehydrated	Canned or Frozen
Fruit	50-55	3-7	50-60
Vegetables	50-85	5-25	50-85
Meats	50-85	15-20	50-60
Eggs	85-90	10-15	35-40
Fish	50-75	20-40	30-75

TABLE 7.2

APPROXIMATE COMPOSITION OF DRIED FRUITS
AND VEGETABLES PER 100 G

Products	Water	Proteins	Carbohydrates	Fat	Ash
Apples	23	1.4	73.2	1.0	1.4
Apricots	24	5.2	66.9	0.4	3.5
Pigs	24	4.0	68.4	1.2	2.4
Prunes	24	2.3	71.0	0.6	2.1
Raisins	24	2.3	71.2	0.5	2.0
Cabbage	4	14.4	72.5	1.9	7.2
Carrots	4	4.1	84.5	1.4	6.0
Potatoes, white	7	7.1	82.2	0.7	3.0
Tomato flakes	3	10.8	76.7	1.0	6.2

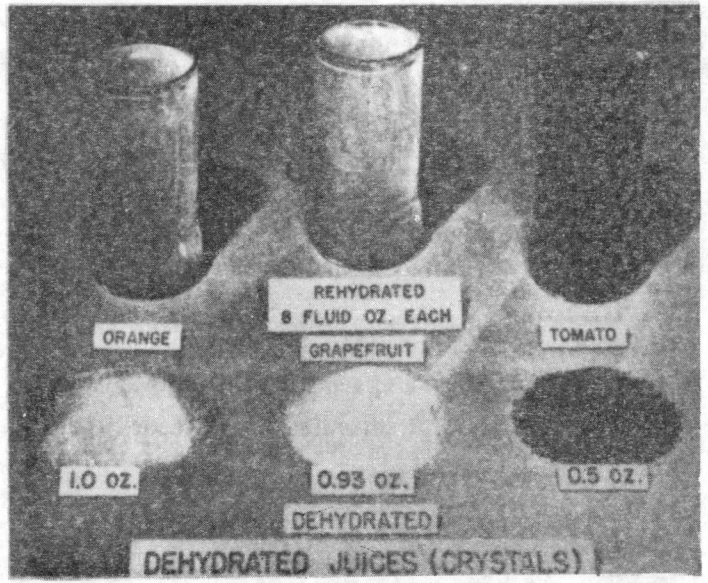

Courtesy of QM. Food and Container Institute

FIG. 7.2. DEHYDRATED FOODS HAVE ADVANTAGES IN PACKAGING,
STORAGE AND DISTRIBUTION
(Note: 8 fl oz = 236.6 ml; 1 oz = 28.35 g; 0.93 oz = 26.36 g; 0.50 oz = 14.175 g.

Dehydration Permits Food Preservation

There are chemical and biological forces acting upon the food supply man desires. Man controls the chemical forces in dehydrated food by packaging and certain chemical additives. The biological forces are controlled by reducing the free water content and by heating. To be a suitable substrate to support growth of microorganisms, a food must have free water available for the microorganisms. By reducing the free water content, thereby increasing osmotic pressures, microbial growth can be controlled.

Humidity—Water Vapor Content of Air

The weight of water vapor in air may be determined from the equation:

$$W = \frac{18.016}{28.967} \frac{(p)}{(P - p)}$$

where W is the grams of water vaper per gram of air, p is the partial

pressure of water vapor, and P is the total pressure.
The percent saturation of air with moisture vapor is obtained
from the equation :

$$\text{Percent saturation} = \frac{W}{W_s}(100)$$

where W_s is the value for saturated air.
The percent relative humidity of air is obtained from the equation:

$$\text{Percent RH} = \frac{p}{p_s}(100)$$

where p_s is the pressure of saturated water vapor at the existing
temperature.

Air — The Drying Medium

Foodstuffs may be dried in air, superheated steam, in vacuum, in
inert gas, and by the direct application of heat. Air is generally used
as the drying medium because it is plentiful, convenient, and overheating of the food can be controlled. Air is used to conduct heat to the
food being dried, and to carry liberated moisture vapor from the food.
No elaborate moisture recovering system is required with air, as is
needed with other gases. Drying can be accomplished gradually, and
tendencies to scorch and discolor are within control.

Function of Air in Drying.—Air conveys heat to the food; causing
water to vaporize, and is the vehicle to transport the liberated moisture vapor from the dehydrating food.

Volume of Air Required in Drying.—More air is required to conduct heat to the food to evaporate the water present than is needed
to transport the vapor from the chamber. If the air entering is not
dry, or if air leaving the dehydration chamber is not saturated with
moisture vapor, the volume of air required is altered. As a rule, 5 to 7
times as much air is required to heat food as is needed to carry the
moisture vapor from the food. The moisture capacity of air is dependent upon the temperature.

The volume of a gas at standard pressure increases 1/273 in volume
for each 1°C rise in temperature. Each 15°C increase in temperature
doubles the moisture holding capacity of air.

Heat Required to Evaporate 454 g of Water from Food.—As a
working figure, 4400 kgc are required to change 454 g of water

to vapor at common dehydration temperatures. The heat of vaporization is actually temperature dependent (1 lb = 454 g; 1 kgc = kilogram calorie = 4 Btu).

Rate of Evaporation from Free Surfaces.—The greater the surface area, the more porous the surface, and the higher will be the drying rate of food. The drying rate increases as the velocity of air flowing over food increases. The higher the temperature of air and the greater the temperature drop, the faster the rate of drying will be, providing case hardening does not develop. Almost as much time may be consumed in reducing the final 6% moisture as is required to bring the moisture content of 80% down to 6%. The drying time increases rapidly as the final moisture content approaches its equilibrium value.

The rate of evaporation from free surfaces may be estimated from the following equation:

$$W = 0.093\left(1 + \frac{V}{230}\right)(e' - e)$$

where W is the grams of water evaporated from a surface per meter^2 per hour, V the lineal velocity of air over the surface in meters per minute, e' is the vapor pressure of water at the temperature being investigated, and e is the vapor pressure of water in the atmosphere.

At 70 m per min, drying times are twice as rapid as in still air. At 140 m per min drying occurs three times more rapidly than in still air.

Case Hardening.—If the temperature of the air is high and the relative humidity of the air is low, there is danger that moisture will be removed from the surface of foods being dried more rapidly than water can diffuse from the moist interior of the food particle, and a hardening or casing will be formed. This impervious layer or boundary will retard the free diffusion of moisture. This condition is referred to as case hardening. It is prevented by controlling the relative humidity of the circulating air and the temperature of the air.

Types of Driers.—There are many types of driers used in the dehydration of foods, the particular type chosen being governed by the nature of the commodity to be dried, the desired form of the finished product, economics, and operating conditions.

The types of driers and the products upon which they are used are generally as follows:

Drier	Product
Drum drier	Milk, vegetable juices, cranberries, bananas
Vacuum shelf drier	Limited production of certain foods
Continuous vacuum drier	Fruits and vegetables
Continuous belt (atmospheric) drier	Vegetables
Fluidized-bed driers	Vegetables

Drier	Product
Foam-mat driers	Juices
Freeze driers	Meats
Spray driers	Whole eggs, egg yolk, blood albumin and milk
Rotary driers	Some meat products, usually not used for food
Cabinet or compartment driers	Fruits and vegetables
Kiln driers	Apples, some vegetables
Tunnel driers	Fruits and vegetables

Dehydration is an operation in which both heat transfer and mass transfer take place. Heat is transferred to the water in the product and the water is vaporized. Then the water vapor is removed.

Driers can be divided into two classes: (1) Adiabatic driers in which the heat is carried into the drier by a hot gas. The gas gives up heat to the water in the food and carries out the water vapor produced. The hot gas may be the product of combustion or heated air. (2) Heat transfer through a solid surface where the heat is transferred to the product through a metal plate which also carries the product. The product is usually held under a vacuum and the water vapor is removed by a vacuum pump. In some cases the product is exposed to the air and the vapor is removed by circulating the air.

It is possible to supply the heat by infrared, dielectric and microwave heating methods.

ADIABATIC DRIERS

Cabinet Driers.—The drier consists of a chamber in which trays of product can be placed. In large driers the trays are placed on a truck for ease of handling. In small units trays may be placed on permanent supports in the drier. Air is blown by a fan past a heater (usually finned steam coils) and then across the trays of material being dried.

The cabinet drier is usually the least expensive drier to build, is easy to maintain, and is quite flexible. It is commonly used for laboratory studies in the dehydration of vegetables and fruits, and in small scale and seasonal commercial operations.

Tunnel Driers.—These driers are the most common in use for dehydrating fruits and vegetables (Fig. 7.3). They consist of tunnels 10 to 20 m long into which trucks containing the trays of food are placed. Hot air is blown across the trays. Production is scheduled so that when a truck of finished product is removed from one end of the tunnel, a truck of fresh produce is put in the other end.

Air movement may be in the same direction as the movement of the product (parallel flow). This has the advantage that the hottest air contacts the wettest product, therefore hotter air can be used. On the

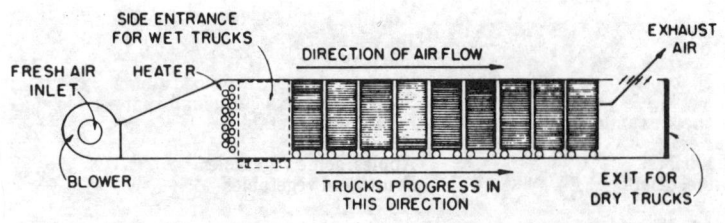

FIG. 7.3.   SIMPLE CONCURRENT TUNNEL (ELEVATION)

Courtesy of Van Arsdel

other hand, the air at the outlet end becomes cool and moisture laden and the final product may not be sufficiently dry.

The air movement may be in the opposite direction of the material flow. In this case the hot dry air contacts the driest product first so that a very dry product can be obtained. Care must be taken not to overload the drier as the moist charge may stand in the warm, moist air too long without being dried to any extent. This would allow time for product spoilage. On the other hand, the dry product should not be left in the drier too long since it is in contact with the hottest air and could become overheated. In general, the counter flow tunnel uses less heat and produces a drier product than a parallel flow tunnel.

In some cases the two types of tunnels are combined into one unit. The product is first placed in a parallel tunnel to take advantage of the high initial rate of drying. It can then be placed in a counter current tunnel to get a very dry end product.

In operation of these tunnels, the drying conditions are not constant. When a fresh tray of material is put into the tunnel, the air which reaches the air-exit-end of the tunnel may be cooler and wetter at the beginning of the cycle than at the end of the cycle. There will be a rise in the air temperature and a drop in the moisture content as the product at the air-inlet-end is dried.

In some tunnels a moving conveyor is used instead of trucks and trays. This has the advantage of reducing labor cost and of having more uniform drying conditions. However, a larger installation and investment are required.

Kiln Driers.—These are commonly two-story buildings. The floor of the upper story is composed of narrow slats on which the food product is spread. Hot gas is produced by a furnace or stove on the first floor and passes through the product by natural convection or with the aid of a fan. The material is turned and stirred frequently and a relatively long time is required for drying. Kiln driers are used for

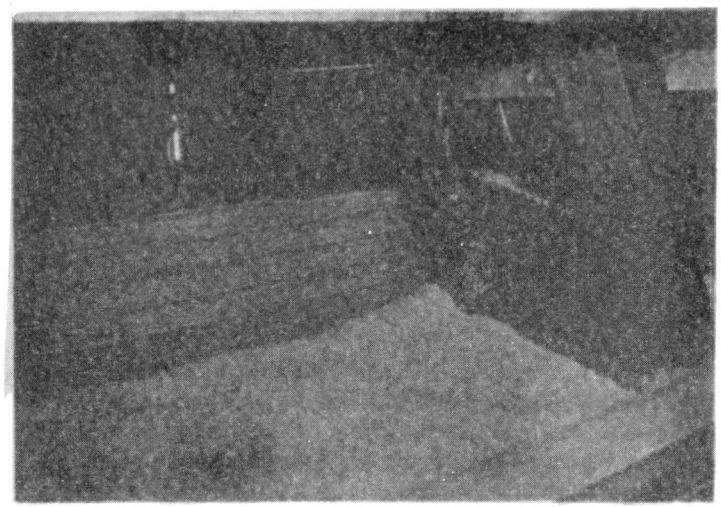

Courtesy of Food Processing

FIG. 7.4.   MODIFIED KILN DRIER

Trough dehydration in which warm air enters from beneath, then passes up through tumbling food particles.

drying such products as apple slices, hops, and occasionally for potatoes (Fig. 7.4).

The above driers are generally used to dehydrate relatively large pieces of material. The rate of drying is affected by the properties of the drying air and the properties of the solid. The important properties of the air are temperature, humidity and velocity. The properties of the solid to consider are: the type and variety of vegetable or fruit; the free moisture content; the method of preparation prior to drying; and the shape and size of the piece.

It has been found that the drying process can be divided into two parts: the constant drying rate period and the falling rate period. During the former the rate of drying is governed by how rapidly the air can supply heat to the water in the food particle and remove the water vapor produced. During this period the water is diffusing to the surface of the particle as fast as it can be evaporated. The temperature of the particle is generally the wet bulb temperature of the air in contact with the piece. However, a point is reached where the water can no longer diffuse to the surface as rapidly as it is evaporated. Then the rate of drying is controlled by the rate of diffusion. As the moisture content decreases, the rate of diffusion drops and the rate of drying slows. The solid material in the particle begins to absorb heat from the air, and the temperature of the piece begins to approach the dry bulb temperature of the air.

The rate of drying during the constant rate period is primarily governed by the properties of the drying air. As the water in the solid absorbs heat from the air, the air is cooled. Since the water in the solid is at the wet bulb temperature, the heat available is determined by the difference between the wet bulb and dry bulb temperature of the air rather than the absolute temperature of the air. The vapor pressure of the water in the solid is that for water at the wet bulb temperature of the air, while the vapor pressure of the water in the air is lower. The difference in the two vapor pressures determines the rate at which water vapor can be absorbed by the air. Therefore the air cannot be cooled to the point that it cannot absorb the water vapor produced. The air velocity is important because the more air available per unit of time, the more heat there is available, and the more water that can be carried away in a given time. Also, the heat and mass transfer coefficients are a function of air velocity.

The difference between the wet bulb and dry bulb temperatures (wet bulb depression) governs that rate of drying for a given air velocity. The greater the wet bulb depression, the greater is the rate of drying. For a given wet bulb depression, the greater the velocity, the more rapid is the rate of drying. The rate of drying is also influenced by the method of loading the tray, since this affects the contact between the air and solid food particle. The shape of the solid has some effect since it determines the ratio of surface area to weight of the solid.

During the falling moisture loss rate period, the rate of drying is determined by the rate at which the water at the center of the food particle diffuses to the surface. The nature of the solid and the thickness of the food product are important. It is assumed that the surface of the product is at a moisture content which is in equilibrium with the drying air. This equilibrium moisture is called the critical moisture. The center of the piece has a moisture content higher than the critical moisture. The difference between these moisture levels is the driving force for diffusion. As this difference decreases, the rate of diffusion drops and therefore the rate of drying falls.

As water leaves the solid, it leaves voids in the solid which shrinks. At low drying temperatures, the outer surfaces of a particle shrink inward, producing a wrinkled appearance. This reduces the surface area. At higher temperatures, outer surfaces dry fast enough to form a tough outer shell which resists the forces pulling inwardly. In this case, the relatively soft inner portions of the particle are drawn to the outer surface, leaving a hollow center.

When the outer portion of the particle becomes drier than the center, the soluble sugars are in a more concentrated solution than the sugars in the wet center. This may establish a concentration gradient which causes the sugars to diffuse to the center of the piece. This may lead to darkening because of browning of the sugars.

Spray Driers.—These are adiabatic driers and many of the considerations for adiabatic drying of solids can be applied to spray driers. They differ in that they are used to dry solutions, pastes or slurries. The food product is not carried on a tray or support, but is dispersed as small droplets which are suspended in the drying air. These have the advantage of very short drying times. If properly operated, a large portion of the flavor, color and nutritive value of the food is retained.

Horizontal Con-current.—The drier consists of a long chamber. The product and the drying air are injected at one end of the chamber. The dried powder settles to the floor, where it is removed by a conveyor. This type is easy to operate and uses relatively low air velocities. The drying time is limited by the trajectory of the injected particles. The size of the finished particles is therefore limited **(Fig. 7.5).**

Courtesy of Buflovak

FIG. 7.5.   HORIZONTAL SPRAY DRIER

Food particles are projected into chamber, and recovered from bottom.

Vertical Downflow Con-current.—In this drier the hot gas and food product are introduced at the top of a tower and travel downwards. The powder collects at the bottom of the tower. This drier is very flexible but usually is a large installation.

Vertical Upflow Con-current.—The hot gas and food product can be introduced into the bottom of a chamber and travel upward. The dry product falls back to the bottom of the drier; the wet gas goes out the top. This type of drier is used where fine, quick drying materials are being handled. The cost is low and the unit is small.

Vertical Upflow Counter-current.—The hot gas is introduced at the bottom of the drier and the product at the top. This type of drier is not commonly used for food because the hottest air hits the driest product, which becomes overheated.

Mixed Flow.—The food product is introduced at the top of the drier. Hot gas is introduced at the top of the drier such that it follows a spiral path to the bottom of the drier. This allows a longer path for the particles to travel for a given tower height. A drier can be designed to have the gas spiral back up the drier and eject the powder from the top of the drier.

The hot gas used for drying can be the direct product of combustion or heated air. The food products may be sprayed in with nozzles or atomizers. One-fluid nozzles are the most commonly used, being cheap and highly flexible. Centrifugal atomizers are used for pastes and materials which plug nozzles. The dried product collects on the floor of the drier, or may be collected outside the drier in bag filters or cyclones.

Air Lift Driers.—Special driers have been used in producing such foods as potato granules. The fresh, moist, granular potato particles are mixed with hot air and carried up a narrow column. The air velocity is such that the granules are suspended in the air stream as they dry. At the top of the drying column, the product and air enter a section where the air velocity drops enough to permit the dried particles to settle into a collector. The wet air goes out the top of the drier.

Foam-mat Driers.—These are used primarily for liquids which are prefoamed by whipping, adding a low level of an edible whipping agent to liquids that do not whip readily. Foaming a liquid exposes enormous surface areas for quick moisture removal which also permits use of lower drying temperatures. Foam is deposited in a uniform layer on a perforated tray or belt through which hot air is blown. Foam layers of many foods can be dried to about 2 to 3% moisture in approximately 12 min.

HEAT TRANSFER THROUGH A SOLID SURFACE

Drum Driers.—Steam heated rotating drums 1 to 2 m in diameter are used for dehydration of fluid products. The slurry is deposited on the drum in a thin film. Heat is transferred through the drum wall to

Courtesy of Buflovak

FIG. 7.6.   VACUUM DOUBLE-DRUM DRIER

Food slurry is deposited between counter-rotating drums on which food solids
dehydrate, then are scraped off, collected, ground, and packaged.

the product film. The drum may be exposed to the atmosphere or it may be held under a vacuum. The dried product is removed from the drum by a scraper blade. The dried film then may be ground to a fine powder (Fig. 7.6 and 7.7).

Vacuum Shelf Driers.—These consist of a cabinet with hollow shelves. The product is placed in pans on the shelves or, if solid, it can be laid directly on the shelves. The unit is closed and a vacuum drawn. Steam, hot water, hot oil, Dowtherm or some other suitable heating medium is circulated through the hollow shelves, heating the product. These units are expensive and have been used mainly on such products as "puff-dried" citrus powders, tomato powder and other food products.

Continuous Vacuum Driers.—These driers consist of a stainless steel belt on which the product is deposited. The film on the belt passes over a heating source, a heated drum or a grid of steam coils, and heat passes through the belt to the product film. In some cases, additional heat may be supplied from the top by infrared lamps. The entire unit is enclosed and held under a vacuum (Fig. 7.8).

CRITERIA OF SUCCESS IN DEHYDRATED FOODS

An acceptable dehydrated food should be competitive pricewise

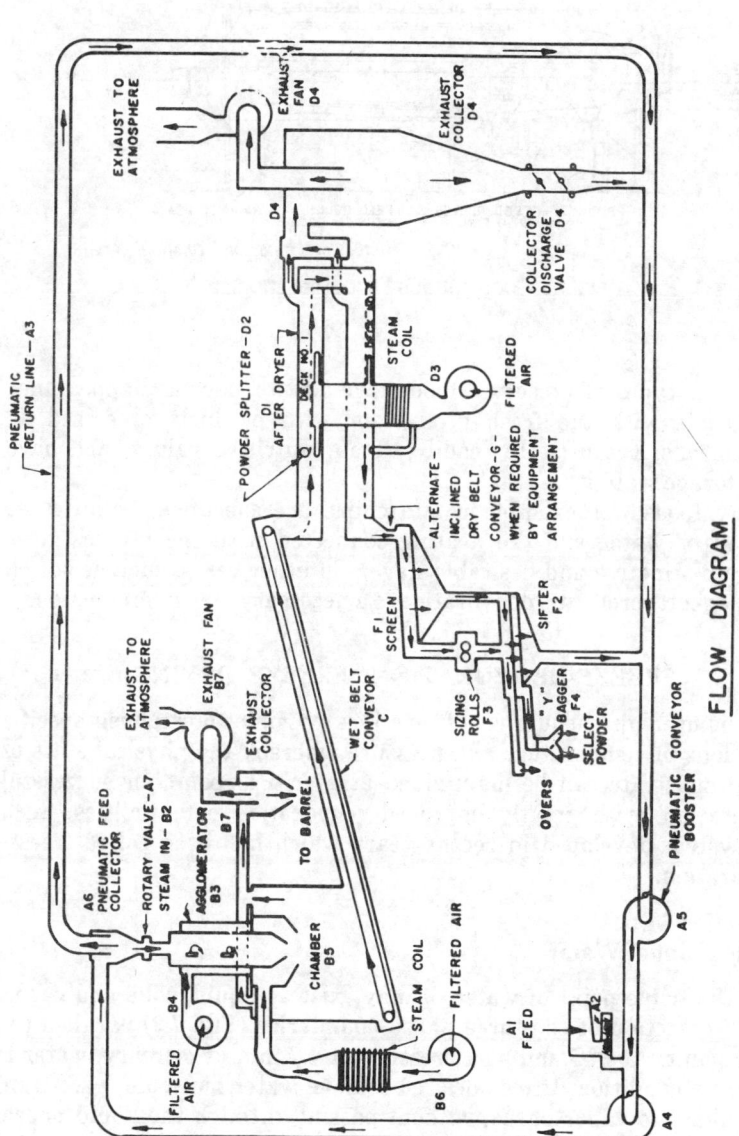

FLOW DIAGRAM

FIG. 7.7.   FLOW DIAGRAM OF BLAW-KNOX INSTANTIZING PROCESS

Courtesy of Hall and Hedrick

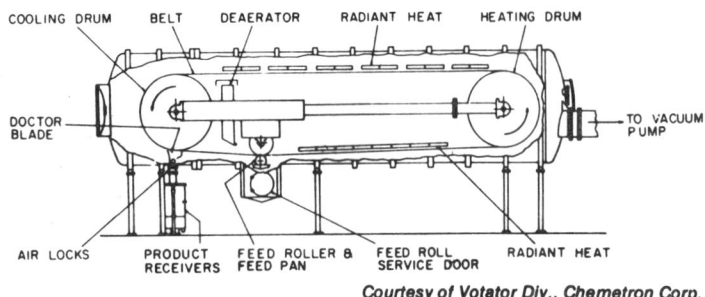

Courtesy of Votator Div., Chemetron Corp.

FIG. 7.8. CONTINUOUS VACUUM BELT DRIER

with other types of preserved food; have a taste, odor and appearance comparable with the fresh product or with products processed by other means; reconstitute readily; retain nutritive values; and have good storage stability.

Many foods, for example potato chips, are consumed because the removal of water and the heating connected with the process gives them a distinctive and desirable flavor. In other cases, such as starch and spaghetti products, dehydration is a necessary step in production.

FREEZE-DEHYDRATION (FREEZE DRYING)

By using high vacuum conditions it is possible to establish specific conditions of temperature and pressure whereby the physical state of a food substrate can be maintained at a critical point for successful dehydration, with greatly improved rehydration potentialities. Such is a system developed in recent years which has been called freeze-dehydration.

Triple Point of Water

At the triple point of water, it may exist as liquid, solid and vapor. The intersection of the three phase boundaries (Fig. 7.9) is called the triple point. At 0°C and at a pressure of 4.7 mm of mercury, water is in such a condition. If it is desired to have water molecules pass from solid phase to vapor phase, without passing through the liquid phase, it is seen from the diagram that 4.7 mm is the maximum pressure for the condition to occur, and a range of temperatures will be suitable.

At pressures above 4.7 mm the liquid phase can occur. In bringing a food substrate down to 5 mm boiling will occur. Blair has found that at 4 mm puffing occurs in some liquid substrates, and this puffing is controllable. Tomato juice subjected to such conditions boils and splatters at 5 mm but forms puffs and a spongy structure of great value at 3 mm. With meat, similar structure may be anticipated, although the flesh itself has sufficient inherent structure. A pressure of 1.5 mm has been found desirable. At 4 mm a food is usually below the triple point. It is at this pressure or lower that freeze-dehydration processes are designed.

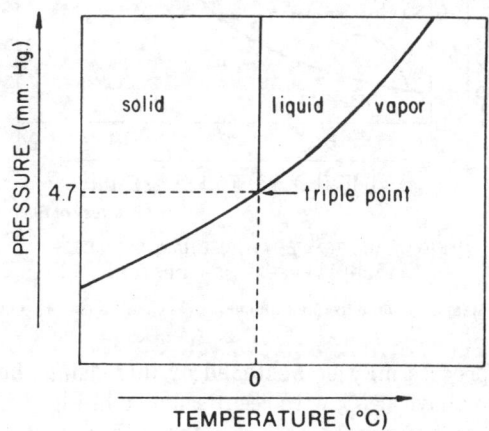

FIG. 7.9.   TRIPLE POINT OF WATER

Considerations in freeze-dehydration.

Temperature Changes in Meat Freeze-dehydration

Figure 7.9 indicates some temperature changes which occur during the freeze-dehydration process. The pressure is maintained below 4.7 mm and preferably nearer to 1 mm. Curve A illustrates conventional meat freeze-dehydration where the frozen meat is placed on a heated shelf maintained at 43°C throughout the process. So long as the meat contains subliming ice, the temperature remains low, rising as the remaining ice and its rate of disappearance decreases. When 0°C is reached the icy core in the piece of meat has disappeared, and the moisture content of the flesh is now near 6%. At this point the meat does not feel wet to the touch. There is not sufficient moisture present to "wet" the muscle fibers and cause their collapse. The structure is maintained. However, it will be noted that approximately one-half of the drying time has elapsed. The remaining 6% moisture requires as much time to be removed as the beginning 94%. The high insulation quality of the dried flesh resists the entrance of heat, required to sublime the remaining moisture.

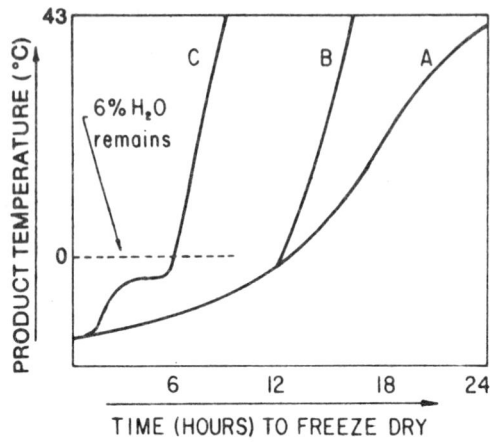

Courtesy of Blair

FIG. 7.10. TEMPERATURE CHANGES IN BEEF
DURING FREEZE-DEHYDRATION

Heat is required to supply the energy to vaporize ice present.

The drying process may be hastened by increasing the temperature of the system to above 43°C, with results shown in Fig. 7.10, curve B.

Curve C on this figure represents the addition of heat by other than conventional means, for example, microwave heating. Inasmuch as the water selects and absorbs the energy selectively, care must be taken in not permitting more energy to be absorbed than is needed, maintaining temperatures below freezing. Using such techniques, drastic reductions in the freeze-dehydration time can be obtained (**Fig. 7.11**).

Courtesy of Vickers Armstrongs, Ltd.

FIG. 7.11. MULTIPLE INSTALLATION OF ACCELERATED FREEZE-DRIERS

A critical relationship exists between pressure and temperature. At higher temperatures of heating, and low pressures, efficient exhaustion of the chamber must be effected. Moisture vapor obtained by sublimation must be rapidly evacuated to keep the system below the critical pressure level.

High temperatures are detrimental to product quality. Maintenance of temperatures above −3°C for prolonged periods, when much moisture remains, decreases quality.

There is difficulty in preventing the salting out of proteins from muscle fibers. The ice-water ratio is important. The following indicates the percent ice and watery solution in beef at several temperatures.

Temperature (°C)	Ice (%)	Watery Solution (%)
−1	35	65
−3	70	30
−10	94	6
−20	98	2

In the watery solution there will be varying concentrations of salts. At the above temperatures the following solids concentration levels are encountered. The drip from frozen beef has approximately a 1.5% soluble solids content.

Temperature (°C)	Watery Solution (%)	Ice (g)	Water (g)	Soluble Solids (g)	Soluble Solids (%)
−1	65	35	63.5	1.5	2.3
−3	30	70	28.5	1.5	5.0
−10	6	94	4.5	1.5	25.0
−20	2	98	0.5	1.5	75.0

Some deterioration of muscle protein occurs in a salt concentration above 5%. At temperatures during freeze-dehydration near −3°C there is such a concentration of salts in the watery solution. In freeze-dehydration, temperature control during the processing of flesh appears critical. Fewer difficulties are experienced with fruit and vegetable tissues.

With suitable temperature and pressure control excellent dried foods can be prepared, including meats. Rehydration, nutrient retention, color, flavor and texture characteristics of properly dried products indicate widespread application of the process in the future. The major differences between freeze-dehydration and conventional drying methods are given in Table 7.3 and shown in Fig. 7.12.

TABLE 7.3

CONVENTIONAL VS. FREEZE-DEHYDRATION

Conventional Dehydration	Freeze-Dehydration
successful for easily dried foods such as fruits, seeds and vegetables	successful for most foods but usually limited to those not successfully dried by other methods
meat generally not satisfactory	successful on cooked and raw animal products
continous processing	batch processing
temperatures between 37° and 93°C generally used	temperatures sufficiently low to prevent thawing used
usually at atmospheric pressure	pressures below 4 mm Hg used
drying time may be short, usually less than 12 hr	drying time generally between 12 and 24 hr
evaporation of water from food surface	moisture loss by sublimation from boundary of ever receding ice crystal zone
solid dried particle	porous dried particle
higher density than original food	lower density than original food
odor frequently abnormal	odor usually natural
color usually darker	color usually natural
slow rehydration, usually incomplete	rapid, complete rehydration possible
flavor may be abnormal	flavor generally natural
storage stability good, tendency to darken and become rancid	storage stability excellent
costs generally low, in the order of 2 to 7 cents a pound, water removed	costs generally high, in the order of four times more than conventional dehydration

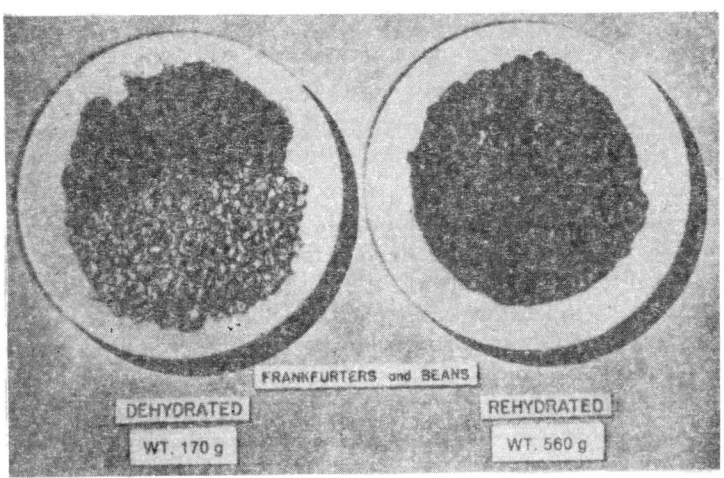

FRANKFURTERS and BEANS

DEHYDRATED   REHYDRATED

WT. 170 g   WT. 560 g

Courtesy of QM. Food and Container Institute

FIG. 7.12.   DEHYDRATED MEAL COMPONENTS
Meat freeze-dried, other ingredients conventionally treated.

The development of continuous freeze-drying equipment will do much to accelerate widespread application of this new technique. Freeze-dried pre-prepared meals having all the qualities of freshly prepared meals have been produced successfully for the Armed Forces.

INFLUENCE OF DEHYDRATION ON NUTRITIVE VALUE OF FOOD

In drying, a food loses its moisture content, which results in increasing the concentration of nutrients in the remaining mass. Proteins, fats and carbohydrates are present in larger amounts per unit weight in dried foods than in their fresh counterpart.

Beef (Chuck)	% Composition Fresh	Dried
Proteins	20	55
Fats	10	30
Carbohydrates	1	1
Moisture	68	10
Ash	1	4

In this major category dried foods yield reconstituted or rehydrated

Peas	% Composition Fresh	Dried
Proteins	7	25
Fats	1	3
Carbohydrates	17	65
Moisture	74	5
Ash	1	2

items comparable with fresh counterparts. However, as with any method of preservation, the food which is preserved cannot be of higher quality than the original foodstuff itself. With dried foods, there is a loss in vitamin content. The water-soluble vitamins can be expected to be partially oxidized. The water-soluble vitamins are diminished during blanching and enzyme inactivation. The extent of vitamin destruction will be dependent upon the caution exercised during the preparation of the foodstuff for dehydration, the dehydration process selected, the care in its execution, and the conditions of storage of the dried foods.

Ascorbic acid and carotene are subject to damage by oxidative processes. Riboflavin is light sensitive. Thiamin is heat sensitive and destroyed by sulfuring.

Fruits can be sun-dried, dehydrated or processed by a combination of the two. Sun drying causes large losses in carotene content. Dehydration, especially spray drying, can be accomplished with little loss in this nutrient. Vitamin C is lost in great proportions in sun-dried fruits. Freeze drying of fruits retains greater portions of vitamin C and other nutrients. The retention of vitamins in dehydrated foods is generally superior in all counts than in sun-dried foods.

Vegetable tissues dried artificially or in the sun tend to have losses in nutrients in the same order of magnitude as the fruits. The carotene content of vegetables is decreased as much as 80% if processing is accomplished without enzyme inactivation. The best commercial methods will permit drying with losses in the order of 5% of carotene. Thiamin content reduction can be anticipated to be in the order of 15% in blanched tissues, while unblanched may lose three-fourths of this nutrient. With ascorbic acid, rapid drying retains greater amounts than slow drying. Generally the vitamin C content of vegetable tissue will be lost in slow, sun drying processes. In all events the vitamin potency will decrease on storage of the dry food (Table 7.4).

TABLE 7.4

CHANGES IN VITAMIN C CONTENT IN DRIED PEACHES AS
INFLUENCED BY PROCESSING CONDITIONS

Treatment	Vitamin C Content (Dry Weight Basis) (mg/100 g)	
	Fresh	Dried
Sun dried, sulfured	52	34
Sun dried, blanched, sulfured	68	39
Dehydrated, sulfured	52	64
Dehydrated, blanched, sulfured	52	60

With milk products, the nutrient level of the raw milk and the method of processing will dictate the level of vitamins retained. Vitamin A is retained in good proportions in drum-dried and spray-dried milk. Vacuum packaged dry milk can be stored with good retention of vitamin A. Thiamin losses occur during both spray and drum drying, but losses are of a lower order of magnitude than with fruit and vegetable drying. Similar results are obtained with riboflavin. Ascorbic acid losses occur during the drying of milk. Being sensitive to heat and oxidation, vitamin C may be totally lost in a drying process. With careful processing, vacuum drying and freeze drying, ascorbic acid values can be retained in the same order of magnitude as fresh raw milk. The vitamin D content of milk is generally greatly decreased by drying. Fluid milk should be enriched with vitamin D prior to drying. Other vitamins such as pyridoxine and niacin are not materially lost.

Usually dried meat contains slightly less vitamins than fresh meat. Thiamin losses occur during processing, greater losses occurring at high drying temperature (Table 7.5). Vitamin C is in most part lost in dried meat. Small losses of riboflavin and niacin occur.

TABLE 7.5

THIAMIN RETENTION IN DEHYDRATED PORK AT VARIOUS
TEMPERATURES AND STORAGE TIMES

Time of Storage (Weeks)	Temperature of Storage (°C)	Thiamin Content[1] (mg)	Thiamin Retention (%)
8	-17	1.690	–
	21	1.130	67.8
	43	0.007	0.4
24	-17	1.920	113.5
	21	0.880	51.9
	43	0.010	0.6
	43-65	0.005	0.3
36	-17	1.220	72.5
	21	0.440	26.1
	43	0.001	0.1
	43-65	0.002	0.1

[1]Fat-free, moisture-free basis (100 g)

Influence of Drying on Protein.—The biological value of dried protein is dependent on the method of drying. Prolonged exposures to high temperatures can render the protein less useful in the dietary. Low temperature treatments of protein may increase the digestibility of protein over native material (Tables 7.6, 7.7 and 7.8).

TABLE 7.6

THE PEPSIN DIGEST-RESIDUE (PDR) AMINO ACID INDEX VALUES
OF FRESH, DEHYDRATED, COOKED, AND STORED BEEF

Treatment	PDR Index
None (frozen fresh)	76
Fresh cooked	77
Dehydrated (2% moisture)	77
Rehydrated	78
Rehydrated cooked	76
Frozen fresh, stored 12 mo. -28°C	68
Dehydrated (2% moisture), stored 12 mo. -28°C, flexible package	81
Dehydrated (8% moisture), stored 12 mo. 21°C, flexible package	76
Dehydrated (2% moisture), stored 12 mo. 37°C, canned	80
Dehydrated (8% moisture), stored 12 mo. 37°C, canned	76

TABLE 7.7

THE PDR INDEX VALUES OF FRESH, DEHYDRATED, COOKED, AND STORED HADDOCK

Treatment	PDR Index
None (frozen fresh)	78
Fresh cooked	86
Dehydrated	78
Rehydrated	76
Rehydrated, cooked	83
Frozen fresh, stored 12 mo., -28 °C	78
Dehydrated, stored 12 mo., -28 °C	80
Dehydrated, stored 12 mo., 4 °C	77
Dehydrated, stored 12 mo., 37 °C	78

TABLE 7.8

EFFECT OF COOKING AND DEHYDRATION UPON THE PDR INDEX VALUES OF KIDNEY, LIMA, AND NAVY BEANS

Product	Treatment	PDR Index
Kidney bean	None	69
Kidney bean	Precooked	68
Kidney bean	Precooked, dehydrated	70
Lima bean	None	74
Lima bean	Precooked	75
Lima bean	Precooked, dehydrated	74
Navy bean	None	71
Navy bean	Precooked	70
Navy bean	Precooked, dehydrated	70
Chili con carne[1]	Precooked, dehydrated	77

[1] Composition: 58.4% precooked, dehydrated red kidney beans; 25.3% precooked, freeze-dried hamburger; 6.68% tomato solids; 3.9% beef soup and gravy base; 2.97% chili powder; 2.23% sodium chloride; and 0.03% garlic powder.

Influence of Drying on Fats.—Rancidity is an important problem in dried foods. The oxidation of fats in foods is greater at higher temperatures than at low temperatures of dehydration. Protection of fats with antioxidants is an effective control.

Influence of Drying on Carbohydrates.—Fruits are generally rich sources of carbohydrates and poor sources of proteins and fats. The principal deterioration in fruits is in carbohydrates. Discoloration may be due to enzymatic browning or to caramelization types of reactions. In the latter instances, the reaction of organic acids and reducing sugars causes discolorations noticed as browning. The addition of sulfur dioxide to tissues is a means of controlling browning. The action is one of enzyme poisoning and antioxidant power. The effectiveness of this treatment is dependent upon low moisture contents. Carbohydrate deterioration is most important in fruit and vegetable tissues being dried. Slow sun drying permits extensive deterioration unless the tissues are protected with sulfites or other suitable agents. Burning sulfur is the least expensive method of obtaining such protection, and is done prior to drying. Animal tissues do not contain large amounts of carbohydrates, therefore, their carbohydrate deteriorations are of minor importance. Exceptions are found in milk and egg products. The critical moisture levels in browning appear to be between 1 and 30% moisture. Below 1%, browning occurs, but at greatly reduced levels. Above 30%, browning occurs apparently at equally low rates.

INFLUENCE OF DRYING ON MICROORGANISMS

Inasmuch as microorganisms are widely distributed throughout nature, and foodstuffs at one time or another are in contact with soil and dust, it is anticipated that microorganisms will be active whenever conditions permit. One obvious method of control is in the restriction of moisture for growth. Living tissues require moisture. The amount of moisture in food establishes which microorganisms will have an opportunity to grow. Certain parameters to microbial growth are established. Molds can grow on food substrates with as little as 12% moisture, and some are known to grow in foods with less than 5% moisture. Bacteria and yeasts require higher moisture levels, usually over 30%. Grains are dried to about 12% moisture, and therefore have the added protection of high solids content. Grain at 16% moisture may mold on storage. Fruits are dried to 16 to 25% moisture. These will mold if given high humidity conditions and exposed to air. Molds are aerobes.

Above 2% moisture, mold growth can be anticipated if environmental conditions are favorable.

From 30% moisture and higher, bacteria and yeasts can be anticipated to grow in foods if environmental conditions permit.

Sodium chloride is commonly employed in conjunction with drying. This salt itself establishes a control over the organisms which will grow. In general, putrefactive growth will be controlled with salt concentrations over 5%. The specific action of salt will be discussed under the heading of pickling.

Salt is useful in controlling microbial growth during sun drying and dehydration processes, i.e., meat and fish drying.

Pathogenic bateria are occasionally able to withstand the unfavorable environment for them in dried foods, then create a public health hazard when eaten. Notable are infections by enteric organisms and food poisoning organisms in general. It is a common technique to

TABLE 7.9

SURVIVAL OF BACTERIA IN DRIED WHOLE MILK POWDERS STORED AT
VARIOUS RELATIVE HUMIDITIES FOR EXTENDED PERIODS AT 38°C

RH (%)	Viable Bacteria After	
	72 Weeks (%)	103 Weeks (%)
70	51	—
60	55	—
50	57	—
40	59	—
30	69	60
20	70	—
10	68	69
5	67	30
0	0.1	0.02

dry cultures (by lyophilization techniques) for storage. Under these conditions there is a slow, steady decrease in the number of surviving populations. Similar experiences are found with bacterial contaminants of dried food (Table 7.9).

The most positive control would be to start with high quality foods having low contamination, pasteurize the material prior to drying, process in clean factories, and store under conditions where the dried foods are protected from infection by dust, insects, rodents and other animals.

The parasites which plague man through his food supply survive the drying process. It is therefore important, for example, to destroy organisms such as *Trichinella spiralis* by heat treatment prior to drying pork containing products.

INFLUENCE OF DRYING ON ENZYME ACTIVITY

Enzymes are sensitive generally to moist heat conditions, especially where temperatures range above the maximum for enzyme activity. Moist heat temperatures near the boiling point of water finds enzymes nearly instantaneously inactivated. There are exceptions, but as a rule, a minute at 100°C renders enzymes inactive.

When exposed to dry heat at the same temperature, such as used in drying, enzymes are notably insensitive to the effect of the energy. Short exposure to temperatures near 204°C have little effect on enzymes if the heating medium and the enzyme preparation is dry.

It is therefore important to control enzyme activity either by subjecting the food material to moist heat conditions or to chemically inactivate the enzymes. In any event, enzymes should be inactivated.

Two enzymes are used as indicators generally of residual enzyme

activity. These are catalase and peroxidase. Catalase is less resistant
to heat treatments than is peroxidase. Testing methods are simple
and quickly performed.

Enzymes require moisture to be active. Enzyme activity is reduced
with decreasing moisture, but a concentration of enzyme and substrate
simultaneously occur. Enzyme reaction rates are dependent upon the
concentration of both. Enzyme activity is nil at moisture levels below 1%.

INFLUENCE OF DRYING ON PIGMENTS IN FOODS

The color of foods is dependent upon the circumstances under
which food is viewed, and the ability of the food to reflect, scatter,
absorb or transmit visible light. As indicated previously, foods in their
native form are usually brightly colored.

Drying foods changes their physical and chemical properties, and
can be expected to alter their abilities to reflect, scatter, absorb and
transmit light, hence modifying the color of the foods.

The carotenoids have been found to be altered during the drying
process. The higher the temperature and the longer the treatment,
the more the pigments are altered. The anthocyanins too are injured
by drying treatments. Sulfur treatments tend to bleach anthocyanin
pigments, while at the same time exerting a strong inhibitory action
on oxidative browning.

The browning of ruptured plant tissues is induced by the oxidase
enzyme systems in the tissues. Oxidative changes are detrimental to
the quality of the food being dried. This discoloration may be controlled by heat inactivation of enzymes.

During heating, caramelization is likely to occur in substrates with
high carbohydrate concentrations.

The natural green pigments of all higher plants is a mixture of
chlorophyll a and chlorophyll b. The retention of the natural green-
ness of chlorophyll is directly related to the retention of magnesium
in the pigment molecules. In moist heating conditions the chlorophyll
is converted to pheophytin by losing some of its magnesium. The
color then becomes an olive green rather than a grass green. The establishment of a slightly alkaline environment is a positive control over
this magnesium transfer. However, this treatment does little to improve the other qualities of foodstuffs.

The interaction of amino acids and reducing sugars (Maillard
reaction) occurs during conventional dehydration of fruits. If the
fruits are sulfured, enzymatic browning can be inhibited, and the
Maillard reaction retarded. Browning can be retarded drastically by
having moisture contents below 1% in the dried products, although

such levels of dehydration may result in overheating and product deterioration from it. Browning can be slowed by storing dried products at cool temperatures.

Continuous vacuum dehydration units have been developed which greatly improve the quality of dried fruits, especially comminuted products such as purees and juices. This equipment has been used successfully in producing fruit powder crystals.

In-package desiccation has become useful in removing moisture from dehydrated foods during storage. Packets of calcium oxide placed within hermetically sealed containers of dried products is an effective means of lowering moisture levels. Orange juice crystals placed into such containers at 3% moisture levels can be brought down to less than 1% moisture content with in-package desiccants. This is an effective method of achieving low moisture contents without the danger of degrading the food with the additional heating otherwise required.

DEHYDRATION OF FRUITS

A great volume of fruit is sold as dry fruit with a moisture content of 15 to 25%. This can be achieved by placing fruit in trays for either sun drying, kiln drying or tunnel drying. Some fruit juice powders have been produced by adding a corn syrup to the juice and vacuum or spray drying.

Dehydration affords a means of producing dried fruits of new forms and of better quality than is possible by sun drying. Some of the fruits which are commercially dehydrated are:

Apples are usually sorted, washed, peeled and trimmed, sulfite treated and dried in kilns.

Apricots, peaches, and nectarines are commonly dried in the sun. Apricots should be sulfured for 3 to 4 hr, and peaches and nectarines for 4 to 6 hr. These fruits are usually steam blanched before drying.

Pears to be dehydrated are blanched prior to sulfuring. Drying usually requires 24 to 30 hr.

Prunes are thoroughly washed then dried in a tunnel drier for 18 to 24 hr. Prunes can be tree dried.

Grape dehydration depends in part on the particular variety of grape. Seedless grapes may be lye-dipped and sulfured before drying in the sun. Other raisins are handled similarly except that after sulfuring, they may be dehydrated in a tunnel drier.

Figs undergo considerable drying on the tree. Those harvested for drying may be spread on trays and sun-dried, or they may be washed and dehydrated.

Other fruits that are commercially dehydrated are cherries and loganberries. Some banana puree is drum-dried. Freeze-dried fruits are of exceptionally fine quality. Such dried pineapple rivals fresh fruit in acceptance ratings. The cost is high, however.

DEHYDRATION OF VEGETABLES

In dehydration of vegetables, enzyme systems must be inactivated. This is accomplished usually by heating in boiling water or steam. Many vegetables are more stable if given a treatment with sulfur dioxide or a sulfite. The moisture content of vegetables should be less than 4% if satisfactory storage life and quality retention are to occur. Residual moisture contents can be reduced to these levels in practice by in-package desiccants.

Vegetables are usually dried in tunnel, cabinet or belt driers. For some powdered vegetable products, drum driers and spray driers have been used. The amount of dried vegetables on the market is relatively small and limited in variety. Potatoes are the largest single item. Most of the other products are items such as onion, celery, parsley, and their powders which can be used as flavoring ingredients. Some dehydrated vegetables are sold in soup mixes and some are used in manufacturing canned products. Much developmental work needs to be done.

Before blanching, vegetables must be prepared. Cabbage is shredded, carrots and potatoes may be sliced or diced, green beans are sliced (Fig. 7.13) and dry beans may be cooked. Potatoes, carrots and beets are usually lye peeled. Blanching times vary. In general, 1 to 3 min is adequate for leafy vegetables, 2 to 8 min for peas, beans, and corn, and 3 to 6 min for potatoes, carrots and similar vegetables. Safe drying temperatures for most vegetables are between 60° and 62°C, but 73°C is permissible for cooked, dried beans, 68°C for carrots, 71°C for corn, and 57°C for onions and squash. The final moisture content desired for most vegetables is 4%, while 2 to 3% is possible for most powders.

The combination of sulfite and low moisture greatly retards the changes occurring in dried vegetables when held at a temperature of 37°C. For dried white potatoes, the time for a just visible degree of browning is increased 20-fold by using 30 ppm SO_2 and reducing the moisture content from 9 to 4%.

A calcium chloride treatment of potatoes for dehydration offers a means for helping to control heat damage during drying and to control non-enzymatic browning during storage at warm temperature.

Courtesy of QM. Food and Container Institute

FIG. 7.13.   DEHYDRATED AND RECONSTITUTED GREEN BEANS

For sweet-corn dehydration, this sequence of steps is effective: (1) Cutting the corn from the cob. (2) Blanching kernels for a minute at 100°C. (3) Treating with sulfite to develop about 2000 ppm SO_2. (4) Dehydrating with a forced air drier to 4% moisture. (5) Hermetically sealing the product in a suitable container.

This procedure usually yields a dehydrated sweet corn product which will remain acceptable after six months storage at 37°C.

DEHYDRATION OF ANIMAL PRODUCTS

Meat is usually cooked before it is dehydrated. The moisture content of the meat at the time of entry into a drier is about 50% and when dried should be approximately 4% for beef or pork (Fig. 7.14 and 7.15). Mildly cured vacuum packed bacon spoils because of the growth of many species of microorganisms. Mildly cured, vacuum packed, sliced bacon has a shelf-life of at least six months if it is sufficiently dehydrated before canning to have a moisture content to salt ratio of 5:1.

The types of meat and poultry products which are dehydrated include: (1) chicken flakes, pieces and powder; (2) turkey flakes and pieces; (3) beef pieces, slices, powder and chunks; and (4) pork pieces and chops.

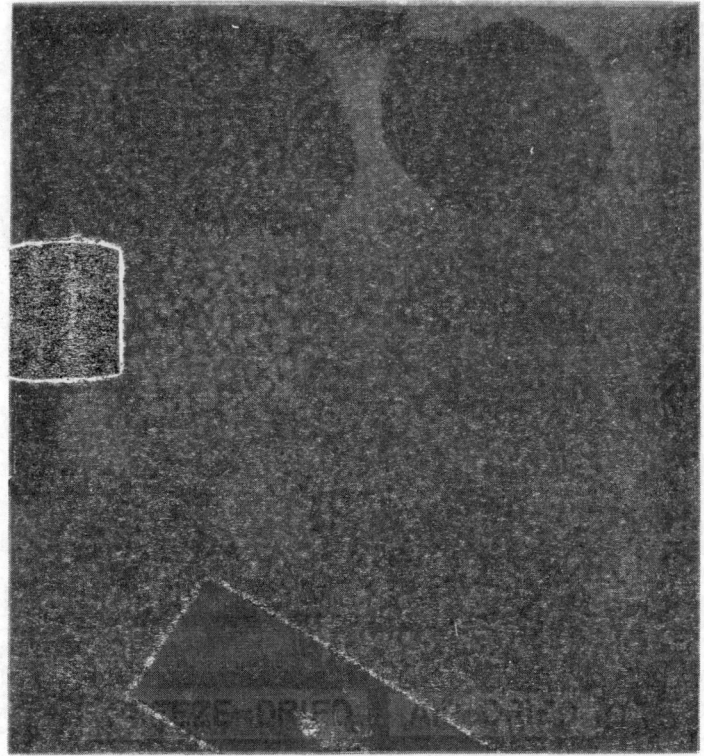

Courtesy of QM. Food and Container Institute

FIG. 7.14.   COMPARISON OF FREEZE-DRIED AN̠ CONVENTIONALLY DRIED MEAT AND VEGETABLES

Appearance of freeze-dried foods at left demonstrates the retention of structure in dried products, important in quality.

About three-fourths of the dehydrated meats and poultry are used in convenience foods such as dry soup mixes and meat-flavored dishes.

Dehydration of Fish

In general, the larger fish are usually cleaned and split down the back, salted and dried. Smaller fish may be salted whole. Fish may either be hot or cold smoked before drying.

Dehydration of Milk

Dried milk powders are of two types, either whole milk or nonfat

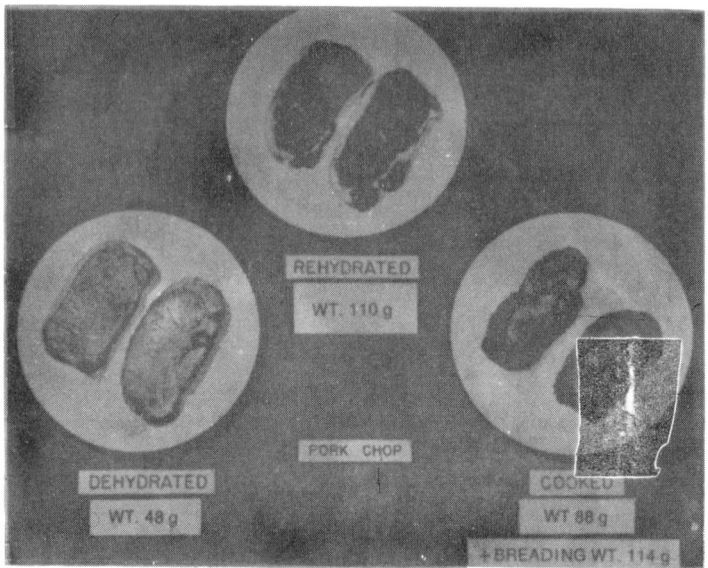

Courtesy of QM. Food and Container Institute

FIG. 7.15.   FREEZE-DRIED PORK CHOPS

skim milk. Suitable drying may be accomplished by the use of drum or spray driers. The procedure for drying milk follows the sequence: preheating, clarification, condensing, standardizing and drying.

Milk powder is used in large volumes by bakers and other food manufacturers as a raw ingredient. Usually low heat spray drying is preferred because the dried milk has a better flavor and color since it does not become hot during processing. However, for some uses drum-dried milk is preferred, and 10 to 20% of the powder currently is produced by this method. See *Food Dehydration* (Van Arsdel *et al.* 1973) for a complete presentation.

Dehydration of Eggs

Eggs may be dried as whole egg powder, yolks or egg white powder. Before drying, the glucose content should be reduced by enzyme treatment or fermentation to produce a stable product. Egg whites are usually spray-dried; cabinet drying or tunnel drying are occasionally used. Whole eggs are usually spray-dried. Dried egg white should have the following characteristics: high solubility, mild odor, low bacterial count, good beating properties and stability in the dry state.

Dried, yeast fermented egg white produces good angel food cakes.

Courtesy of QM. Food and Container Institute

FIG. 7.16.   VARIOUS PACKAGING MATERIALS ARE FUNCTIONAL FOR
DEHYDRATED FOODS
Benefits of container research in yielding improved and varied packaging materials
for dried foods.

Yeast fermented egg white is soluble in water, rehydrates readily and can be used to make high quality products.

PACKAGING OF DEHYDRATED FOODS

Eggs, meat, milk and vegetables are ordinarily packaged in tin containers. Occasionally fiberboard or flexible film material may be employed, although they are not as satisfactory as tin. Tin offers protection against insects, moisture loss or gain, and permits packaging with an inert gas. If packaged dehydrated foods are to be stored for a considerable period of time, it is advisable to use low temperature storage.

In-package desiccants improve the storage stability of dehydrated white potato, sweet potato, cabbage, carrots, beets and onions. The results show that a very substantial protection against browning can be obtained. The retention of ascorbic acid is markedly improved at temperatures up to 48°C and in either nitrogen or air. In treated foods the disappearance of sulfite is also greatly retarded.

For long-term storage of dehydrated foods, functional containers which are hermetically sealed and resistant to penetration by insects are required (Fig. 7.16).

The packaging requirements of dried foods are product specific.

INFLUENCE OF DRYING ON FOOD ACCEPTANCE

Dried foods find their greatest use in times of disaster, either natural or man-made. While it is true that drying is one of the principal methods of food preservation, and a natural force used by plants in preservation of seeds and fruits, dried vegetables have not been popular when wartime or national disaster does not enforce their use. Current day drying techniques yield highly acceptable food products, noting such success with dried skim milk, potato flakes, soup mixes, soluble coffee, prepared baby foods and the new freeze-dried prepared meals. The obvious economic benefits in the distribution of highly acceptable dried foods should find increasing emphasis being placed on this method of food preservation. Freeze-dried, prepared, pre-cooked meals of equal acceptance with freshly prepared counterparts have reached the pilot plant stage of development. However, the process still remains a batch production and is expensive. Time and the development of a continuous process should find its early exploitation.

The caramelization, discoloration, loss in texture and physical form, loss of volatile flavoring characteristics, and poor rehydration ability of many dried foods have left an imprint on the minds of consumers which is not quickly dispelled. Improved technology (Fig. 7.17) combined with consumer re-education will go far in giving high quality dried foods their due place on supermarket shelves.

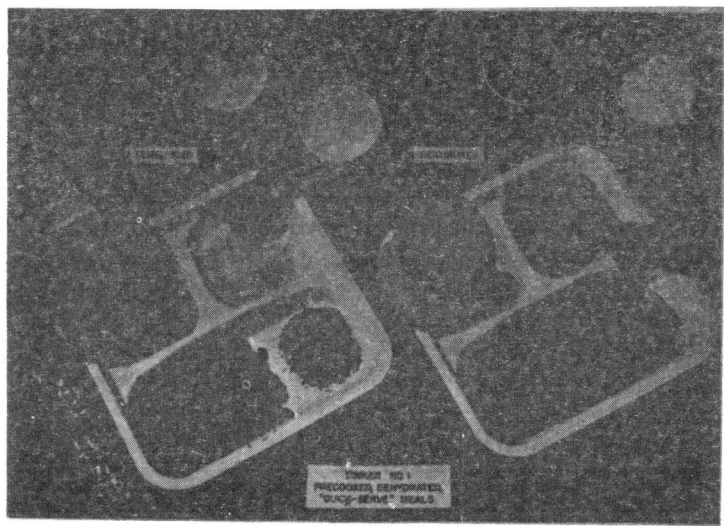

Courtesy of QM. Food and Container Institute

FIG. 7.17.   PRECOOKED, DEHYDRATED, QUICK-SERVE MEALS

An example of the advances being made in food dehydration.

TRENDS IN DRYING FOODS

Dehydration of foods involves many options. In practice, the decision on method is based on quality and economic factors. Inherent in the process is the prevention of microbiological deterioration or food poisoning. The moisture content of most foods must be lowered before protection is afforded.

When water is removed, the dissolved materials are concentrated in the remaining water unless they crystallize. As the temperature is raised to supply the energy to transform the water into vapor, chemical reactions accelerate during drying and during storage of the dehydrated material.

Of importance in understanding the deteriorative reactions during dehydration and in the dry state is a knowledge of the state of water at reduced water activity, such as has been detailed by Labuza (1973, 1976), including the kinetics of various reactions in dry systems.

Labuza summarized trends in dried foods as follows:

Vegetables

Major growth has occurred in dried vegetables, especially potatoes, onions, garlic and chili peppers.

Potato: Drum drying, once used as a step in granulation of potatoes prior to air drying, is now used for the whole process. Instant potato production has grown steadily.

Onions: The three-stage dryer has led to the successful development of dried onions. Production has doubled since 1970. Dry onion production now uses 20% of the onion crop. Garlic production accounts for 80% of the garlic crop. Paprika, carrots, bell peppers and tomatoes account for the remaining major dry vegetables.

Tomato: Vacuum foam drying with an "in-package dessicant" is a major achievement.

Drum drying continues to be used basically for potatoes, yeast and other vegetable slurries. The availability of fresh vegetables has reduced the use of dry counterpart products in homes. The major consumer of dry vegetables is the food industry.

Fruit

Dry fruit production has declined to less than 60% of the 1945 producton. Nevertheless, prune and raisin production account for 87% of the dried fruits, with prunes doubling and raisins decreasing. Yet, one-fourth of the grape crop is dried to raisins. There are currently 18 drying plants for raisins, 307 for prunes and 11 dried apple operations compared to 33 potato operations and 12 for other vegetables (Labuza 1976).

Meat, Fish and Eggs

Commercial production of dried meat, poultry and fish accounts for less than 1% of all meat, excluding dry sausage products in about six plants. Less than 0.5% of the fish catch is dried in the United States, excluding fish meal. The low consumption is probably because fresh and frozen fish are of higher quality than the dried.

Dried egg solids have become a major industrial ingredient (three-fourths of all dried eggs). The breakthrough came once glucose could be removed from eggs, either by a yeast fermentation or directly through the use of glucose oxidase/catalase enzyme system. A palatable product could be produced, which led to a whole new industry. About 20 egg-drying plants are in operation in the United States at present.

Milk

The dry milk industry with more than 600 plants involved in drying dairy products is the largest segment. The modern spray drier used in the United States has led the way to the development of many of our instantly soluble products, such as milk, tea and coffee. The use of instantizers did not occur until recently.

Coffee and Tea

Other specialty products promoted the development of different drying methods. The coffee-drying industry has grown to capture about 40% of the market.

Freeze drying, introduced on a large scale to produce antibiotics, has the least detrimental effect of any drying procedure on heat-sensitive materials. Because of its expense it has only captured a major use in coffee. Of the 20 plants in the United States drying coffee or tea, 10 are freeze drying.

GRAIN DRYING

It is clear that the length of time grain can be stored varies with the moisture content and the crop. Experts have a rule of thumb: lower moisture content 2% below that considered safe for 1 year if you plan to store it for 5 years.

TABLE 7.10

MOISTURE CONTENT DURING HARVEST AND FOR SAFE STORAGE (PERCENT, W.B.)

Cereal	Maximum During Harvest	Optimum at Harvest for Minimum Loss	Usual When Harvested	Required for Safe Storage (1 yr)	(5 yr)
Barley	30	18-20	10-18	13	11
Corn	35	28-32	14-30	13	10-11
Oats	32	15-20	10-18	14	11
Rice	30	25-27	16-25	12-14	10-12
Rye	25	16-20	12-18	13	11
Sorghum	35	30-35	10-20	12-13	10-11
Wheat	38	18-20	9-17	13-14	11-12

Source: Brooker *et al.* (1974).
Note: Above applies to climates where cereal grain is usually grown. For example, oats in a warm climate may need to be at 10-11% for safe-keeping; rye in the Soviet Union is stored at 13-15%.

The grain moisture contents in Table 7.10 are those reported by Brooker *et al.* (1974) for the safe storage of grain for one year. The highest moisture content in the bin should be used as the point to judge the grain for safe storage, other factors being equal (Table 7.10).

Grain driers are classified by method of drying, type of drier container to hold the grain during drying, batch versus continuous flow, and whether the drier is movable.

Nearly all driers use air to transfer heat to the grain and to carry away the moisture evaporated. These are typical convection or forced air driers. They have either direct or indirect heating units and may be classified accordingly. In a direct heat drier the products of combustion are mixed with the drying air. This is a common type found (Van Arsdel *et al.* 1973).

In larger driers at grain elevators and processing plants, either oil or natural gas may be used as fuel. The oil fired driers are constructed with refractory surfaces (Brooker *et al.* 1974) to assure complete combustion.

Grain driers are usually of the vertical, horizontal or bin types. The latter is used mainly on farms. Examples of grain driers are shown in Figures 7.18, 7.19 and 7.20.

For a full development of the subject the reader is directed to *Drying Cereal Grains* (Brooker *et al.* 1974).

SUMMARY

Drying continues to be the most widely used method of food preservation. Recent advances in manufacturing instantly rehydrating

Courtesy of Hart-Carter Co.

FIG. 7.18.   TYPICAL TOWER DRIER WITH
SCREENED COLUMN

food products indicate a continuing bright future.

Agglomerated, instantly rehydrating powders have become widely used in formulating important new and tasty foods, ranging from instant breakfasts to milk shakes, salad dressings and free-flowing flours. Mixes of many types and forms are virtually flooding the markets. Such is the nature of technologically based developments in food preservation.

Courtesy of Campbell Industries, Inc.

FIG. 7.19.   TYPICAL TOWER DRIER OF THE RACK
OR BAFFLE TYPE

Courtesy of AFE Industries, Inc.

FIG. 7.20.   COLUMN-TYPE HORIZONTAL DRIER

REFERENCES

ADACHI, R.R., SHEFFNER, L., and SPECTOR, H. 1958. The in vitro digestibility and nutrient quality of dehydrated beef, fish, and beans. Food Res. 23, 401-406.

ALGUIRE, D.E. 1973. Ethylene oxide gas sterilization of packaging materials. Food Technol. 27, No. 9, 64-67.

ANGELINE, J.F., and LEONARDS, G.P. 1973. Food additives, some economic considerations. Food Technol. 27, No. 4, 40-50.

ANGELOTTI, R. 1973. FDA regulations promote quality assurance. Food Technol. 29, No. 11, 60-62.

BAKEN, J.A. 1973. Microencapsulation of foods and related products. Food Technol. 27, No. 11, 34-44.

BALLS, A.K. 1942. The fate of enzymes in processed foods. Fruit Prod. J. 22, 36-39.

BEAZLEY, C.C., and INSALATE, N.F. 1971. The roles of fluorescent-antibody methodology and statistical techniques for salmonellae protection in the food industry. Food Technol. 25, No. 3, 24-26.

BLAIR, J. 1958. Personal communication. QM. Food and Container Institute, Chicago, Ill.

BOMBEN, J.L., GUADAGNI, D.G., and HARRIS, J.G. 1969. Aroma concentration for dehydrated foods. Food Technol. 23, No. 1, 83-87.

BONE, D. 1973. Water activity in intermediate moisture foods. Food Technol. 27, No. 4, 71-76.

BROOKER, D.B., BAKKER-ARKEMA, F.W., and HALL, C.W. 1974. Drying Cereal Grains. AVI Publishing Co., Westport, Conn.

CALDWELL, E.F., and SHMIGELSKY, S. 1958. Antioxidant treatment of paperboard for increased shelf life of packaged dry cereals. Food Technol. 12, 589-591

CHAIN BELT CO. 1950. The Chain Belt vacuum dehydrator. Chain Belt Co., Milwaukee, Wis.

COLE, L.J.N. 1962. The effect of storage at elevated temperature on some proteins in freeze-dried beef. J. Food Sci. 27, 139-144.

HALL, H.E. 1971. The significance of Escherichia coli associated with nut meats. Food Technol. 25, No. 3, 34-36.

HENIZ, Y., and MANHEIM, C.H. 1971. Drum drying of tomato concentrate. Food Technol. 25, No. 2, 59-62.

IFT EXPERT PANEL AND CPI. 1975. Sulfites as food additives. Food Technol. 29, No. 10, 117-120.

JAMES, D.E. 1971. Managing raw materials and good manufacturing. Food Technol. 25, No. 10, 30-31.

KAN, B., and DEWINTER, F. 1968. Accelerating freeze drying through improved heat transfer. Food Technol. 22, No. 10, 67-75.

KAREL, M. 1974. Packaging protection for oxygen-sensitive products. Food Technol. 28, No. 8, 50-60, 65.

KAUFFMAN, F.L. 1974. How FDA uses HACCP. Food Technol. 28, No. 9, 51,84.

KING, C.J., LAM, W.K., and SANDALL, O.C. 1968. Physical properties important for freeze-dried poultry meat. Food Technol. 22, No. 10, 100-106.

KOHL, W.F. 1971. A new process for pasteurizing egg whites. Food Technol. 25, No. 11, 102-110.

LABUZA, T.P. 1973. Effects of dehydration and storage. Food Technol. 27, No. 1, 20-26.

LABUZA, T.P. 1976. Drying food, technology improves on the sun. Food Technol. 30, No. 6, 37-46.

LAZAR, M.E., and MIERS, J.C. 1971. Improved drum-dried tomato flakes are produced by a modified drum drier. Food Technol. 25, No. 8, 72-74.

LEININGER, H.V., SHETON, L.R., and LEWIS, K.H. ·1971. Microbiology of frozen cream-type pies, prozen cooked-peeled shrimp, and dry food-grade gelatin. Food Technol. 25, No. 3, 28-30, 33.

MAJORACK, F.C. 1971. FDA's quality assurance programs: tools for compliance. Food Technol. 30, No. 10, 38-42.

MCFARREN, E.F. 1971. Assay and control of marine biotoxins. Food Technol. 25, No. 3, 38-48.

MEADE, R.E. 1973. Combination process dries crystallizable materials. Food Technol. 27, No. 12, 18-26.

MEER, G., MEER, W.A., and TANKER, J. 1975. Water-soluble gums, their past, present, and future. Food Technol. 29, No. 11, 22-30.

MITSUDA, H., KAWAI, F., and YAMAMOTO, A. 1972. Underwater and underground storage of cereal grains. Food Technol. 26, No. 3, 50-56.

NAGEL, A.H., PETERSON, E. et al. 1972. Forum: voluntary food standards. Food Technol. 26, No. 11, 57-64.

NAKAMURA, M., and KELLY, K.D. 1968. Clostridium perfringens in dehydrated soups and sauces. J.Food Sci. 33, No. 4, 424-426.

NEUBERT, A.M., WILSON, C.W., and MILLER, W.H. 1968. Studies on celery rehydration. Food Technol. 22, No. 10, 94-99.

NOTTER, G.K., TAYLOR, D.H., and BREKKE, J.E. 1958. Pineapple juice powder. Food Technol. 12, 363-366.

ORENT-KEILES, E., HEWSTON, E.M., and BUTLER, L. 1946. Effects of different methods of dehydration on vitamin and mineral values of meats. Food Res. 11, 486-493.

PORTER, W.L. 1958. The use of dehydrated and dehydrofrozen potato slices in the production of potato chips. M.S. Dissertation, Purdue Univ., Lafayette, Ind.

ROCKWELL, C.R., LOWE, E., SMITH, G.S., and POWERS, M.J. 1954. New through flow drier for the partial drying of apple slices. Food Technol. 8, 500-502.

ROSS, K.D. 1975. Estimation of water activity in intermediate moisture foods. Food Technol. 29, No. 3, 26-34.

SCARLETT, W.J. 1957, Control of moisture in granular products. Food Technol. 11, 532-535.

SCHWARZ, H.W. 1948. Dehydration of heat sensitive materials. Ind. Eng. Chem. 40, 2028-2033.

SELTZER, E., and SETTELMEYER, J.T. 1949. Spray drying of foods. In Advances in Food Research, Vol. 2, E.M. Mrak, and G.F. Stewart (Editors). Academic Press, New York.

SIMON, M., WAGNER, J.R., SILVERA, V.S., and HENDEL, C.E. 1955. Calcium chloride as a non-enzymic browning retardant for dehydrated white potatoes. Food Technol. 9, 271-274.

SINHA, R.S., and MUIR, W.E. 1973. Grain Storage: Part of a System. AVI Publishing Co., Westport, Conn.

SINNAMON, H.I., ACETO, N.C., and SCHOPPET, E.F. 1971. The development of vacuum foam-dried whole milk. Food Technol. 25, No. 12, 52-64.

SMITH, C.A., JR., and SMITH, J.D. 1975. Quality assurance system meets FDA regulations. Food Technol. 29, No. 11, 64-68.

STRASHUN, S.I., and TALBURT, W.F. 1953. WRRL develops techniques for making puffed powder from juices. Food Eng. 25, 59-60.

TALBURT, W.F., and LEGAULT, R.R. 1950. Dehydrofrozen peas. Food Technol. 4, 286-291.

TALBURT, W.F., WALTER, L.H., and POWERS, M.J. 1950. Dehydrofrozen peas. Food Technol. 4, 496-498.

TAPPEL, A.L., MARTIN, R., and PLOCHER, E. 1957. Freeze dried meat. V. Preparation properties, and storage stability of precooked freeze dried meats, poultry, and seafoods. Food Technol. 11, 599-604.

TISCHER, R.G., and BROCKMAN, M.C. 1958. Freeze drying ups quality of QM quick serve rations. Food Eng. 30, 110-112.

TRESSLER, D.K. 1949. New instant mixes contain dehydrated fruits. Food Ind. 21, 46-47.

TRESSLER, D.K. 1956, New developments in the dehydration of fruits and vegetables. Food Technol. 10, 119-124.

TRESSLER, D.K., and LEMON, J.M. 1949. Marine Products of Commerce. Reinhold Publishing Corp., New York.

VAN ARSDEL, W.B., COPLEY, M.J., and MORGAN, A.I. 1973. Food Dehydration, 2nd Edition, Vols. 1 and 2. AVI Publishing Co., Westport, Conn.

ZABIK, M.E., and FIGA, J.E. 1968. Comparison of frozen, freeze dried and spray dried eggs. Food Technol. 22, No. 9, 119-125.

ZIMMERMAN, P.L., ERNST, L.J., and OSSIAN, W.F. 1974. Scavenger pouch protects oxygen-sensitive foods. Food Technol. 28, No. 8, 63-65.

Chapter 8

Principles of
Food Concentrates

CONCENTRATED BUT MOIST

A food substrate concentrated to 65% or more soluble solids which contains substantial acid may be preserved with mild heat treatments provided it is protected from air. With more than 70% solids, high acid content is not required.

The manufacture of fruit jellies and preserves is one of the important fruit by-products industries, and is based upon the high solids-high acid principle. Not only are such concentrates very important products of fruit preservation, but equally, in modern day commerce, it is an important utilization of fruits which, though otherwise of excellent qualities, do not possess attractiveness to the sight. Such fruits do not enter usual fresh market channels. In addition to the pleasing taste of such preserved fruits, they possess substantial nutritive values.

Jelly making, once strictly a household process, has taken its place as an important food manufacturing activity. In contrast with most other food industries, the preserving plants are more frequently located near the population centers rather than in the fruit production areas.

Other food products are concentrated as a step in preservation, including sweetened condensed milk. This product contains the high solids content required, but not the high acid. Nevertheless, it is feasible to so preserve milk.

HIGH SOLIDS—HIGH ACID FOODS

Jelly, jam, preserves, marmalades and fruit butters are products prepared from fruit and/or plants with added sugar after concentrating by evaporation to a point where microbial spoilage cannot occur. The prepared product can be stored without hermetic sealing, although such protection is useful. Mold growth on the surface of fruit preserves is controlled by exclusion of oxygen, i.e., covering with paraffin. Modern practice replaces the paraffin with vacuum sealed containers; moisture losses, mold growth and oxidation are thus brought under control.

The preserving industry requires consistency controls on gels. This developed only after information concerning pectin gel formation became available.

Jelly

A jelly is strictly defined in the United States as that semisolid food made from not less than 45 parts by weight of fruit juice ingredient to each 55 parts by weight of sugar. This substrate is concentrated to not less than 65% soluble solids. Flavoring and coloring agents may be added. Pectin and acid may be added to overcome the deficiencies that occur in the fruit itself.

The name of the fruit used in making the jelly must be stated with other ingredients, in order of importance by weight, on such products offered for sale in the United States.

Four substances are essential in obtaining a fruit gel. These components are pectin, acid, sugar and water (Fig. 8.1).

Jam.—A jam has similar definitions as a jelly with the exception that it is the fruit ingredient that is used rather than the fruit juice. Concentration is carried to at least 65% for all jam, some requiring 68% solids to achieve desired qualities. Not less than 45 kg of fruit are permitted to each 55 kg of sugar in the United States.

Fruit Butter

Fruit butter is the smooth, semisolid food prepared from a mixture containing not less than 5 parts by weight of fruit ingredients to each 2 parts of sugar.

Marmalade

Marmalade is a product made from citrus fruit (usually) and is the jelly-like product made from properly prepared juice and peel, with sugar. It is concentrated to achieve its gel structure, similar to jelly, with approximately the same standards, except for the use of sliced peel.

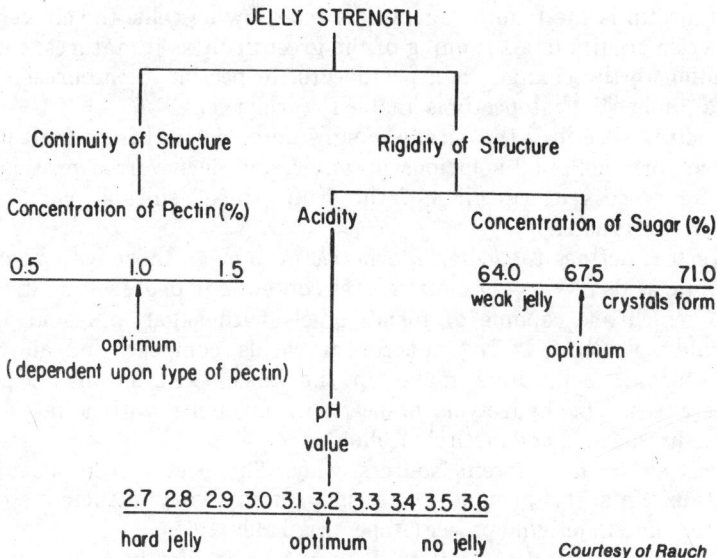

FIG. 8.1   JELLY FORMATION DEPENDENT UPON PECTIN, SUGAR AND ACID
COMBINATION
Narrow limits of operation for successful jelly manufacture.

PECTIN AND GEL FORMATION

Fruits and their extracts obtain their jelly-forming characteristics from a substance called pectin. It was first discovered in France by Braconnot in about 1825. A jelly forms when a suitable concentration of pectin-sugar-acid in water is reached.

Because pectin is important to fruit gel formation, it warrants attention. Plant tissues contain water-soluble pectin, insoluble pectic acid, protopectin, and a compound containing some pectic substance and cellulose.

Protopectin.—The middle lamellae of plant cells consists of protopectin, plus other constituents, which when boiled in acid (fruit) solution, such as in jelly making, is hydrolyzed to soluble pectin. Protopectin is the insoluble precursor of pectin. The change can be brought about either by enzymatic or acid hydrolysis. Protopectin is abundant in fruits, leaves, and fleshy roots.

When pectic substances are in a water-insoluble form, unhydrolyzed as in the plant tissues, the substances are termed protopectin. When the substances become soluble, they are termed pectin.

During the ripening of fruits, protopectin is converted to pectin enzymatically. During the further maturation, pectin may be decomposed to form methyl alcohol and insoluble pectic acid. Inasmuch as protopectin is the binding agent between growing cells, the conversion to pectin results in a softening of the green fruit as it matures, corresponding to the change from protopectin to pectin. In the presence of sugar and acid, protopectin is unable to form a gel.

Pectin.—Pectin is that group of substances derived from fruit juices which form colloidal solutions in water, and derive from protopectin in the process of ripening of the fruit. Under suitable conditions, pectin forms a gel.

Kertesz defines the general term pectin to mean those water-soluble pectinic acids of varying methyl ester content and degree of neutralization which are capable of forming gels with sugar and acid under suitable conditions. Polygalacturonic acids composed of anhydrogalacturonic acid units make up the basic skeleton of all pectic substances. The hydrolysis of pectin from apples with alkali yields galacturonic acid and methyl alcohol.

Pectins from different sources vary. The pectin from beet root contains an acetyl group. The pectins of fruits vary in their methoxyl content and in jellying power (Appendix, Table A.6).

Pectin is a reversible colloid. It may be dissolved in water, precipitated, recovered and dried, and redissolved in water without losing its gel forming capacities. Pectin is precipitated by alcohol, and it is used not only in identification but in preparation of commercial pectins. Recently, low methoxyl pectins have been found to have abilities to form gels with lower concentrations of sugar, and even without sugar, under special conditions.

Pectic Acid.—Pectin on hydrolysis yields pectic acid. There are several intermediates in the transformation, including the pectinic acid step. The units of pectin are reported to be pectic acid with carboxyl groups esterified by methyl alcohol.

Gel Formation.—The following explanation and sequence of events has been thought to outline the formation of a pectin-sugar-acid-water gel.

In an acid fruit substrate, pectin is a negatively charged colloid. The addition of sugar influences the pectin-water equilibrium established, and destabilizes the pectin. It conglomerates and establishes a network of fibers. This structure is able to support liquids. The continuity of the network formed by the pectin and the denseness of the formed fibers are established by the concentration of pectin. The higher the concentration, the more dense the fibers in the structure. The rigidity of the network is influenced by the sugar concentration and acidity. The higher the concentration of sugar, the less water supported by the structure. The toughness of the fibers in the structure is controlled by the acidity of the substrate. Very acid conditions result in a tough gel structure, or destroy the structure by action of hydrolysis of the pectin. Low acidity yields weak fibers, unable to support the liquid, and the gel slumps.

Gel formation occurs only within a narrow range of pH values. Optimum pH conditions are found near 3.2 for gel formation. Values below this point find gel strength decreasing slowly; values above 3.5 do not permit gel formation at the usual soluble solids range. The optimum solids range is slightly above 65%. It is possible to have gel formation at 60% solids, by increasing the pectin and acid levels. Too high a concentration of solids results in a gel with sticky characteristics.

The quantity of pectin required for gel formation is dependent upon the quality of the pectin. Ordinarily, slightly less than 1% is sufficient to produce a satisfactory structure.

Besides jellies made from sugar-acid-pectin systems, it is possible to form pectinic acid gels by treatment with metal ions, with which lower soluble solid concentrations are required. These may find their place in this calorie conscious age.

Syneresis is a term employed in describing jellies which have free liquid, i.e., liquid released from the gel. This is commonly called weeping of jelly.

INVERT SUGAR

During the process of boiling sucrose solutions in the presence of acid, a hydrolysis occurs, in which reducing sugars are formed (dextrose and levulose). Sucrose is converted into reducing sugars, and the product is known as invert sugar. The rate of inversion is influenced by the temperature, the time of heating, and the pH value of the solution.

Invert sugar is useful in jelly manufacturing, as crystallization of sucrose in the highly concentrated substrate is retarded or prevented. A balance is required between the sucrose and invert sugar content of the jelly. Low inversion of sucrose may result in crystallization; high inversion results in granulation of dextrose in the gel. The amount of invert sugar present should be less than the amount of sucrose. It appears that a 40:60 ratio is desirable.

Inasmuch as the acidity of fruits varies, and boiling conditions vary, the maintenance of a desired invert sugar-sucrose ratio is difficult. Control of acidity, pH and boiling requirements must be effected. With fruits of low acidity or pectin content where prolonged boiling is required, control of the ratio may be difficult.

In vacuum concentration little inversion of sucrose occurs. In this instance a portion of the sucrose should be replaced with pre-inverted sugar. Invert sugar is available commercially and is usually acid hydrolyzed, although there are invertase enzymes which can be used to accomplish the hydrolysis.

JELLY MAKING

Ideally fruits for jelly making should contain sufficient pectin and acid to yield good jelly. Such fruits are crabapples, sour apple varieties which are not overmature, sour berries, citrus fruits, grapes, sour cherries and cranberries. Sweet cherries, quinces and melons are rich in pectin but low in acid content. Strawberries and apricots contain sufficient acid but are low in pectin. Peaches, figs and pears are low generally in both acid and pectin. Because pectins are available commercially and edible acids are plentiful, it is possible to correct defects in acid content or pectin content of fruit in jelly making.

The process of jelly making involves boiling the fruit to extract pectin (converting protopectin), to obtain maximum yields of juice and extract flavoring substances characteristic of the fruit. Water may be added to the fruit during this extraction. The amount added depends upon the juiciness of the fruit. Excessive moisture must be boiled away during concentration. Therefore a minimum is added compatible with good juice yields, prevention of scorching and pectin extraction. Pectin hydrolyzing enzymes are destroyed during this boiling extraction.

The boiled fruit juice is next expressed from the fruit pulp by straining or pressing. The press cake remaining may be re-extracted with water and boiled a second time to obtain maximum pectin yields. The pressed juice contains suspended solids and usually these are removed by filtration.

The acidity, pH value, pectin content and soluble solids content of the juice is determined by analysis. Deficiencies in pectin may be remedied by its addition. Generally powdered pectin is mixed with ten times its volume in dry sugar, mixed thoroughly, and added to the juice. This ensures uniform distribution and controls lumping.

Sugar is added to the juice either as a solid or as a syrup. Care is exercised in adding the sugar to ensure its being dissolved. The juice is stirred and heated during the sugar adding stage.

Boiling is one of the important steps in jelly making. The juice must be rapidly concentrated (Fig. 8.2) to the critical point for gel formation of the pectin-sugar-acid system. Prolonged boiling not only

Courtesy of Food Machinery and Chemical Corp.

FIG. 8.2.   JELLY COOKING ASSEMBLY WITH CONTROL SYSTEM

causes hydrolysis of pectin and volatilization of acid, but losses in flavor and color. Vacuum concentration yields greatly improved jelly over products concentrated at atmospheric pressure.

The point at which evaporation is halted is determined by the level of soluble solids in the substrate. The usual means of identification is by the use of a refractometer. Tables are available relating refractive index to soluble solids content of sugar solutions. At standard pressure the end-point of concentration usually will be reached when the boiling temperature equals 104.4° to 105°C in atmospheric cookers. In vacuum concentration units the soluble solids content must be determined by refractive index or by the use of hydrometers (Appendix, Table A.7).

Continuous jelly making processes have been developed (Fig. 8.3). If acid must be added to compensate for deficiencies in fruit composition the acid is best added at the end of the evaporation cycle. The addition of acid late in the concentration ordinarily permits the jelly to be filled successfully into containers prior to gel formation. Setting of jelly can be controlled in part by the addition of buffer salts, such as sodium citrate, and certain phosphates. Their presence tends to delay the onset of gel formation. In jam and preserve manufacture,

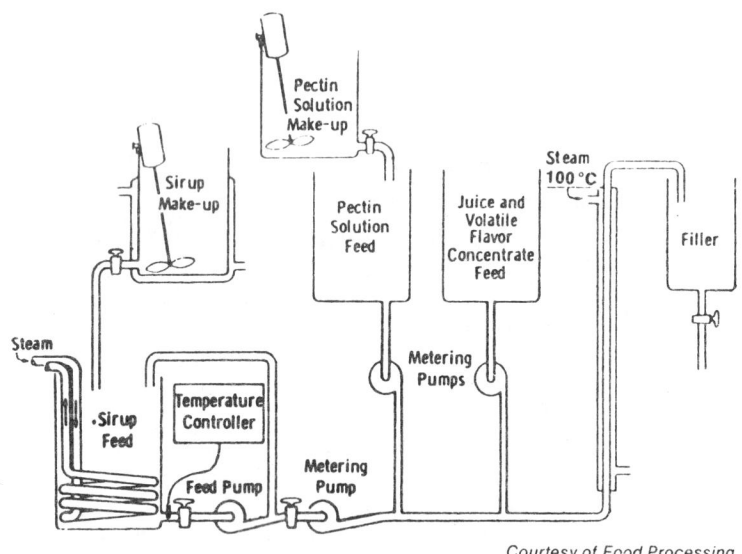

Courtesy of Food Processing

FIG. 8.3.   SCHEMATIC DIAGRAM OF CONTINUOUS PROCESS FOR JELLY
MANUFACTURE

when fruit is to be suspended in the formed gel, quick setting pectins
may be added in order to ensure that the fruits become entrapped in
the gel structure.

Jelly and preserved fruit products filled hot into containers (temperature near 87.8°C) and then sealed require no further sterilization
treatment.

Ordinarily 454 g of extracted fruit juice to which is added 454 g of
sugar, and heated to a temperature of 104.4° to 105°C at near standard
pressures, will form a jelly (Appendix, Table A.8).

Rhubarb, carrot juice and tomatoes may be preserved in the form of
jelly products, although vegetables are not generally preserved in such
a form.

OTHER FRUIT PRODUCTS

Fruit preserves are made much in the same manner as are jellies
except that the initial material for concentration contains fruit pulp
and fruit. Fruit pastes are the concentrated products prepared by
evaporation of water from fruit purees, producing a product with a
high solids content. Usually no gel structure is developed. Fruit
butters have been briefly noted previously.

Candied and Glaceed Fruits

The candying of fruits essentially involves their slow impregnation with syrup until the sugar concentration in the tissue is sufficiently high that it prevents the growth of spoilage microorganisms. The candying process is conducted in such a manner that the fruit does not soften and become merely jam, or becomes tough and leathery. By treating fruits with syrups with progressively increasing sugar concentrations, desired results may be obtained. The reader is referred to the publications of Cruess and coworkers for detailed information on candied fruits. Following the impregnation of the fruit with sugar, the fruit is washed and dried. The candied fruit (as it is now called) may be packaged and marketed in this condition, or the fruit may be coated with a thin glazing of sugar. In this case candied fruits are dipped into syrup and again dried. The sugar coated candied fruit is called glacéed fruit.

Fruits which have been preserved in sulfurous acid must be freshened in hot water. Cherries are stemmed and pitted before being freshened. Fresh fruits or even canned fruits are sometimes used successfully in these products.

Maraschino Cherries

With the development of canned fruit salads and fruit cocktails, large amounts of cherries of the Royal Anne and certain other varieties came into use for the preparation of Maraschino cherries. They are also used in the candy, ice cream and bakery industries. Generally the cherries are harvested after reaching full size, but prior to being fully matured in color and texture. Stems are generally not removed. In California fruit may be packed into paraffin-lined spruce barrels filled with a preservative brine containing approximately 1% of sulfur dioxide and 0.05% of unslaked lime. The barrels are sealed tightly and stored for about four weeks. The sulfur dioxide bleaches the color that may be present and the calcium hardens the tissue. After storage the cherries may be pitted and stemmed, followed by a freshening process during which the cherries are leached in water for 24 hr or more to remove most of the sulfur dioxide. Next the fruit is heated to boiling, with constant changing of water, to reduce the sulfur dioxide content to below 20 ppm. This leaching process also removes much of the natural fruit acids that were present, being useful later in enhancing the penetration of the dye into the flesh.

The leached cherries may be boiled for several minutes in dilute suitable dye solution at about 0.05%, allowed to stand in this solution overnight, then given a second boiling in dilute (0.25%) citric

Courtesy of S. and W. Fine Foods

FIG. 8.4.   MARASCHINO CHERRIES IN PREPARATION
Cherries being impregnated with sugar and flavoring.

acid solution. This greatly reduces the tendency of the cherries to bleed their color and stain other fruits.

In preparing the cherries for use in cocktails, the leached fruit is boiled for a few minutes in a 30° Brix sugar solution in which sodium benzoate, citric acid and dye have been added. The cherries are allowed to come into equilibrium with this solution overnight, then the sugar concentration is increased to 40° Brix. Artificial cherry flavor is added, the fruit and syrup are heated to boiling, and packed hot into containers (Fig. 8.4).

The benzoates are useful in preserving the cherries after the container (usually glass) has been opened. If cherries are packed at or above 88°C in small containers the chemical preservative may be omitted, but the fruit will be subject to spoilage after being opened.

SWEETENED CONDENSED MILK

John A. Jaynes

Sweetened-condensed milk was the earliest commercially successful milk product developed by Borden nearly a century ago. Approximately 18 kg of sugar are added to 100 kg of milk, and concentrated to 70% solids. The product is given but a slight heat treatment and is not sterilized; the high sugar content acts as the preservative.

Definition

Sweetened condensed milk is a mixture of pure, whole cow's milk and sugar with 60% of the water removed before it is packed in hermetically sealed cans. An approximate analysis of the product is as follows:

	(%)
Milk fat	8.5
Milk-solids-not-fat (MSNF)	20.5
Sugar	44.0
Moisture	27.0
Total	100.0

Sweetened condensed milk is actually very similar to homemade jelly, except that sugar is being used to preserve milk solids rather than the juice of fruits or berries, as is the case with jelly.

Manufacturing Process

Raw milk arrives fresh from the farm at the condensory's receiving platform where it is checked for quality and temperature. The milk is tested for fat and total solids and then standardized. This standardized milk is blended with sugar and pumped into large stainless steel tanks called "hot wells" to pasteurize the milk and destroy any pathogenic bacteria that might be present. The heat treatment in the hot well is varied with the seasons of the year to control physical instability which otherwise might occur in the product.

The blend of heated, standardized whole milk and sugar is then drawn into a vacuum condensing pan to reduce it to the required consistency and total solids content. To preserve the original flavor and color, the milk is boiled at a low temperature of approximately 54°C under vacuum. When the correct concentration is reached, the milk is drawn quickly from the vacuum pan to the vacuum cooling tank where the temperature is rapidly lowered to approximately 21°C. During this cooling procedure, a predetermined amount of finely ground milk sugar (lactose) is added to the product to control the crystallization of the sugar present. At this point, the preparation of the product is completed. It is now thick, smooth in texture, and creamy in color. Until canned, the milk is held in insulated storage tanks that are sealed and sterile.

Before filling, the consumer size cans are washed and sterilized inside and out by live steam and then fed automatically into a filling machine that is protected to prevent bacterial, yeast, and mold contamination to the product. The milk flows from the storage tank to a continuous filling machine where the filled cans are immediately sealed. Unlike most canned foods, the container is completely filled to prevent the presence of air which could result in the growth of yeast and/or mold in the product. After the cans are coded with the factory date of manufacture, they are then labeled and finally packed for shipping, 24 cans to a case. When final approvals for bacteriological, chemical, and physical analyses are given, the milk is shipped out to supermarkets across the United States and around the world.

Today, most of the sweetened condensed milk in the United States is still produced by Borden, Inc., and the product is still essentially the same as that produced by Gail Borden back in 1856. The product is packed in 14-oz cans and is used in the preparation of fancy desserts. Also, a considerable amount of bulk product is produced for use in the baking and confectionery industries.

CONCENTRATED — NOT MOIST

A different set of problems is confronted in regards to food concentrates which are not moist. An example is found in peanut butter.

The history of peanut butter is obscure but dates from about 1890 when it was discovered that a very palatable paste or butter could be obtained by grinding peanuts. It soon became known as peanut butter and presumably was made from raw peanuts.

Although roasted (parched) peanuts have been ground into a paste and mixed with honey and cocoa in South America for centuries, peanut butter as a North American food was apparently invented independently. The first product was shelled roasted peanuts chopped, ground, or beaten into a pulp in a cloth bag with salt added.

About 1900 it was learned that peanuts could be ground into a paste butter in the home. Soon it became a staple food and commercial production was stimulated. Its first general use was for sandwiches served at outdoor parties and picnics; soon it was served in combination dishes, then in candies, cookies, ice cream, and in other ways in the home, at school, and in public eating places.

Commercial manufacture and consumption of peanut butter is largely an American art. The production has increased more rapidly than the population; thus the per capita consumption has steadily increased. More than 500 million pounds are produced yearly in the United States. Peanut butter contains about 1.8% moisture, 27% protein, 49% fat, 17% carbohydrates, 2% fiber and 3.8% ash.

Peanut butter is a cohesive, comminuted food product prepared from dry roasted, clean, sound, mature peanuts from which the seed coat and "hearts" are removed, and to which salt, hydrogenated fat and (optional) sugars, antioxidants and flavorings are added. Due to the low moisture content and package protection, it is stable in general against microorganisms and rancidity.

MANUFACTURE OF PEANUT BUTTER

J. G. Woodroof

The manufacture of peanut butter consists of shelling, dry-roasting, and blanching the peanuts, followed by fine-grinding. Salt is used to improve the flavor, small quantities of other materials such as hydrogenated fat and dextrose are usually added; and sometimes corn syrup solids, or glycerin to prevent oil separation, lecithin and/or antioxidants to control rancidity are added.

Peanut butters may be made from any variety of peanuts. However, a blend of two parts Spanish or Runner peanuts with one part Virginia peanuts is considered best for the most desirable consistency. The peanuts are blended in the desired proportions as they are fed into the grinder.

Manipulation of temperature and grinding to produce a uniform, high-quality, sanitary product requires a definite procedure. This is indicated by the wide range of colors, flavors, and consistencies of peanut butters on the market, all made from almost identical varieties of peanuts.

Some peanut butters have a heavy-roast color and flavor, others have light-roast characteristics, and there are many gradations between. In an attempt to make peanut butters that are different in color, flavor, and texture, some manufacturers include large pieces of peanuts.

Beginning with shelled peanuts, steps required in the manufacture of peanut butter (Fig. 8.5) are as follows:

(1) Roasting.—Peanuts used in the manufacture of peanut butter are dry roasted by one of two methods—batch or continuous. In the opinion of some operators, batch roasters have many advantages that cannot be met by one big continuous roaster. In the first place, peanut butter may call for a blending of peanuts, and the different varieties must be roasted separately.

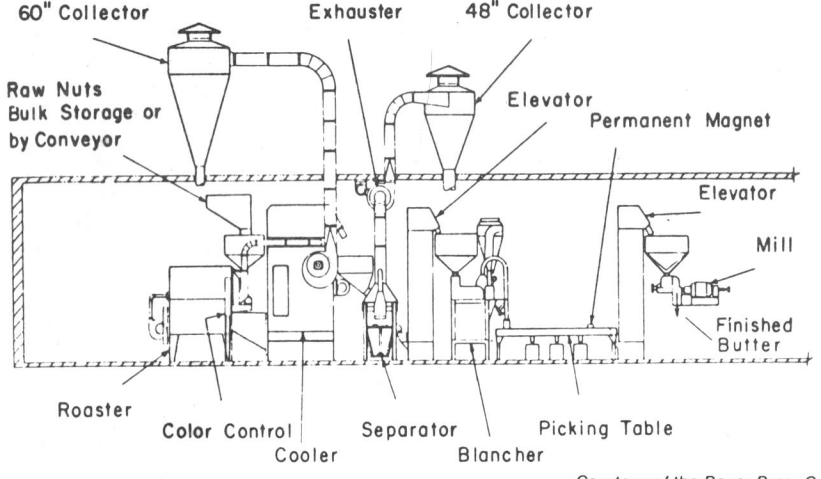

Courtesy of the Bauer Bros. Co.

FIG. 8.5. TYPICAL 2000 LB PER HOUR PEANUT BUTTER PLANT

Today the trend among large manufacturers is clearly toward continuous roasters. Advantages claimed are: saving in labor, more uniform roasting, smoother operations, and decreased spillage.

The first effect of roasting is rapid drying, in which the moisture content is reduced from a normal of about 5 to 0.5%. This is followed by the development of oily translucent spots on the surface of cotyledons, called "steam blisters," caused by oozing of oil from the cytoplasm as free oil.

Change in color is due to the cell walls becoming wet with oil. This stage is referred to as "white roast." The skins, too, become wet with oil, and darker in color. The final stage of roasting is the development of a brown color, at which time the peanuts are "brown roasted." Color and flavor of peanut butter is dependent on the extent to which brown roasting is allowed to proceed.

Roasted peanut kernels have a higher oil content (47.9%) than the corresponding raw kernels (47.4%). The difference results from loss of volatile components and moisture during roasting. Sorted cotyledons remaining after elimination of germs, skins, and objectionable material during blanching, screening, and sorting contain a higher percentage of oil (49.5%) than the roasted kernels. Separation of components of lower oil content is responsible for this difference. The skins of raw peanuts contain very little oil, but when separated from roasted nuts they contain 27% oil. The skins absorbed oil from the cotyledons during roasting. The free-fatty acids extracted from raw, roasted peanuts, sorted cotyledons, and peanut butters are uniformly low.

(2) Cooling.—Heat should be removed from roasted peanuts as quickly as possible in order to stop cooking at a definite point and to produce a uniform product.

(3) Blanching.—This removes germs and remaining skin and foreign matter.

(4) Picking and Inspecting.—The blanched nuts should be screened and inspected to remove scorched and rotten nuts, and stones or other foreign or undesirable matter. This is preferably done on a conveyor belt. Light nuts are removed by blowers, discolored nuts by electric eye, and metal parts by magnets. If all previous handling steps have been properly carried out, hand picking and inspecting can be done rapidly.

(5) Grinding, Cooling and Packaging.—Grinding is one of the simplest, yet one of the most delicate operations in the plant. Various devices used for grinding peanuts into butter are referred to as comminuters, attrition mills, homogenizers, disintegrators, hammer mills, or colloidal mills. Most of these devices are constructed so they can be adjusted over a wide range. This permits considerable variation in the quantity of peanuts ground per hour, the fineness of the product, and the amount of oil freed from the peanuts. Peanut butter is usually made by two grinding operations. The first reduces the nuts to a medium grind and the second to a fine, smooth texture. For fine grinding, clearance between plates is 0.032 to 0.003 in. "Chunky" peanut butter is made by mixing peanut pieces approximately the size of 1/8 kernel with regular peanut butter or by removing a rib from the grinder to cause incomplete grinding. A specially designed plate is substituted for the regular grinding plate. About 2% salt, and other ingredients if any, are fed into the grinder with the peanuts. The heat generated during grinding may melt hydrogenated fat, or the fat may need to be pre-melted. To prevent overheating, mills are cooled by a water jacket. The grinding mill should be adjustable, have automatic feed for peanuts and salt, and be easy to clean.

Peanuts should be kept under constant pressure from the start to the finish of the grinding process in order to assure uniform grinding, and protect the product from air bubbles. A heavy screw feeds the peanuts into the grinders and may also be used to deliver the deaerated peanut butter into containers in a continuous stream under even pressure.

For liquid additives such as stabilizers, there are metering units complete with proportioning pumps, heating coils, and other features which require particular ingredients. Stabilizers for peanut butter may be partially hydrogenated vegetable oils, monoglycerides, diglycerides of vegetable oils, or combinations of these. For dry additives, such as salt or sugar, there is a vibratory tray feeder, or a screw-type feeder. Either of these is synchronized with the peanut feed rate. To ensure complete and uniform assimilation of all additives into the peanut butter, as well as adding others, it may be discharged into mixing pump. From here it goes to the filling machine. This is homogenized peanut butter. Mixing should be with propeller-type blades with both forward and backward pitch.

The upper level of stabilizer is 5½%, with 3¼% most common; thorough mixing is extremely difficult. This is especially true when stabilizer is fed into single pass mill from the top. Recommended temperature for blending stabilizer is 60° to 74°C. Some mills run 10 to 15 degrees higher. A second mill for texturizing or homogenizing is recommended.

Mixing must be complete before the product reaches the filler, since little can be done there. Recommended temperature for filling jars is 29° to 44°C.

Filled containers should be disturbed as little as possible during the first 48 hr after filling. Storage temperatures should be about 10°C.

Recent studies have shown that the heat generated by grinding and mixing should be removed immediately to ensure proper crystallization of the fats. Votators, a type of heat exchanger, are used to cool peanut butter from about 49° to 76°C or less before it is packaged. After it is put in the final containers it should be allowed to remain undisturbed until crystallization throughout the mass is complete. Ill effects of improper cooling are cracking or shrinking in the center or pulling away from the containers.

Any procedure or temperature which disturbs the setting, and allows a reset, seems to increase firmness and separation of oil on the surface. This is the case when freshly filled containers of peanut butter are shipped within 48 hr. Vacuum packing peanut butter results in reduced firmness, more uniform texture, and less tendency for oil separation.

The main factor in preventing oxidation is proper packaging. If there is no oxygen available, there can be no oxidation, so the real problem is to exclude air from the container as nearly as possible. Vacuum packing is commonly used as an aid in minimizing residual air in the headspace and in the butter. However, even without vacuum packing, a completely filled and sealed jar will contain insufficient oxygen to cause rancidity other than in the layer in direct contact with the headspace.

The stability of oils in peanut butters toward antioxidative rancidity is quite high at the time of manufacture and remains high, even after storage at 27°C in the absence of light for two years. Neither the extent of roasting of the peanuts, nor incorporation of salt or salt and hydrogenated peanut oils have much effect on the stability of peanut butters. Stability is reduced by oxygen in the headspace, especially after the container is opened.

IMPROVEMENT IN PEANUT BUTTER

Scores of patents have been issued on improvements in formulation of peanut butter. No one of these has greatly influenced the industry, since all brands of peanut butter consist of at least 90% ground roasted peanuts and about 2% salt.

Many attempts have been made to change or improve the flavor, texture, consistency, or nutritional qualities of peanut butter. Such attempts have been to: (1) prevent the separation of oil on the surface with hydrogenated fat, dextrose, and powdered sugar; (2) improve the smoothness and ease of spreading; (3) improve consistency and eliminate stickiness by the addition of honey, lecithin, yeast, hydrogenated fat, and coarse grinding; (4) develop a type that can be blocked and sliced; and (5) accentuate the flavor with sugar, honey, malt, vanilla, essential oils, spices, dried fruits, synthetic and other flavorings.

The flavor of roasted peanuts is developed largely during the heating process and it is quite volatile. A great deal of the best flavor and aroma are unavoidably lost during roasting and cooling. Upon standing, especially when exposed to air at high humidity, the flavors continue to be lost, until within a few hours the aroma and flavor of freshly roasted peanuts gradually give way to a less desirable stale flavor. This is to be avoided by rapid and continuous processing, and by promptly and properly packaging the product.

SUMMARY

The preservation of foods as sugar concentrates has yielded many flavorful and nutritious foods. In what was previously a somewhat dormant area of study, we now find a good deal of activity. As will be seen in Chapter 11, the exploitation of this technology, beyond what Borden found, is underway. Meanwhile, the preservation of many foods as concentrates, ranging from fruits to milk to butters continues to contribute to the well being of man.

REFERENCES

ANON. 1976. Preservers Handbook. Sunkist Growers, Ontario, Calif.

BAKER, G.L. 1948. High polymer pectins and their deesterification. *In* Advances in Food Research, Vol. 1, E.M. Mrak, and G.F.Stewart (Editors). Academic Press, New York.

BEAVEN, G.H., and NORRIS, F.W. 1939. Molecular structure of pectic acid. Chem. Ind. *58*, 363-365.

BECK, K.M. 1957. Properties of the synthetic sweetening agent, cyclamate. Food Technol. *11*, 156-158.

BEVERAGE MANUF. IND. 1973. A survey of beverage drinking patterns in the U.S. and Canada. Report to Calorie Content Council. Washington, D.C.

BROWN, B.I. 1969. Processing and preservation of ginger by syruping under atmospheric conditions. Food Technol. *23*, No. 1, 87-92.

BUTTKUS, H. 1974. On the nature of the chemical and physical bonds which contribute to some structural properties of protein foods: A hypothesis. J. Food Sci. *39*, 484.

CAMPBELL, L.B., and KEENEY, P.G. 1968. Temper level effects on fat bloom formation on dark chocolate coatings. Food Technol. *22*, No. 9, 100-101.

CRUESS, W.V. 1958. Commercial Fruit and Vegetable Products. McGraw-Hill Book Co., New York.

CUMMING, D.B., STANLEY, D.W. and DEMAN, J.M. 1972. Texture-structure relationships in texturized soy protein. 2. Textural properties and ultrastructure of an extruded soybean product. J. Inst. Can. Sci. Technol. *5*, 124.

DAHLBERG, A.C., and PENCZEK, E.S. 1941. The relative sweetness of sugars as affected by concentration. N.Y. State Agric. Exp. Stn. Tech. Bull. *258*.

FELLERS, C.R., and GRIFFITHS, F.P. 1928. Jelly strength measurement of fruit jellies by the Bloom gelometer. Ind. Eng. Chem. *20*, 857-863.

GORDON, J. 1968. Intensity-concentration relationships for sugar and salt solutions. J. Food Sci. *33*, No. 5, 483-488.

HALL, R.L. 1975. GRAS—concept and application. Food Technol. *29*, No. 1, 48-53.

HOOVER, W.J., NELSON, A.I., and STEINBERG, M.P. 1955. Development of a new concept for processing fruit jelly. Food Technol. *9*, 377-379.

HU, K.H. 1972. Time-temperature indicating system "writes" status of product shelf life. Food Technol. *26*, No. 8, 56-62.

INGLETT, G.E. 1974. Symposium: Sweeteners. AVI Publishing Co., Westport, Conn.

JOHNSON, A.D., and PETERSON, M.S. 1974. Encyclopedia of Food Technology. AVI Publishing Co., Westport, Conn.

KERTESZ, Z.I. 1951. The Pectic Substances. Interscience Publishers, New York.

LITWILLER, E.M. 1950. Manufacture of jellies and preserves. Oreg. Agric. Exp. Stn. Bull. *490*.

LOPEZ, A., and LI-HSIENG, L. 1968. Low methoxyl pectin apple gels. Food Technol. *22*, No. 8, 91-95.

LUH, N., KAREL, M., and FLINK, J.M. 1976. A simulated fruit gel suitable for freeze dehydration. J. Food Sci. *41*, 89-92.

MEER, G., MEER, W.A., and TANKER, J. 1975. Water-soluble gums, their past, present, and future. Food Technol. 29, No. 11, 22-30.

MERMELSTEIN, N.H. 1973. Nutrient labeling and the independent laboratory. Food Technol. 27, No. 6, 42-46.

MIDDLEKAUF, R.D. 1976. 200 Years of U.S. food laws: a gordian knot. Food Technol. 30, No. 6, 48-54.

OSER, B.L., and HALL, R.L. 1972. GRAS substitutes. Food Technol. 26, No. 5, 35-42.

OSER, B.L., and FORD, R.A. 1973. GRAS substances. Food Technol. 27, No. 11, 56-57.

OSER, B.L., and FORD, R.A. 1974. GRAS substances. Food Technol. 28, No. 9, 76-80.

OSER, B.L., and FORD, R.A. 1975. GRAS substitutes. Food Technol. 29, No. 8, 70-72.

OSMAN, E.M. 1975. Interaction of starch with other components of food systems. Food Technol. 29, No. 4, 30-35, 44.

PETROWSKI, G.F. 1975. Food-grade emulsifiers. Food Technol. 27, No. 7, 52-62.

POHLAND, A.E., and ALLEN, R. 1970. Analysis and chemical confirmation of patulin in cheese. J. Assoc. Off. Anal. Chem. 53, 636.

PONTING, J.D., SANSHUCH, D.W., and BREKKE, J.E. 1959. Continuous jelly manufacture. Food Technol. 12, 252-254.

RAUCH, G.H. 1950. Jam Manufacture. Leonard Hill, Ltd., London, England.

ROSS, K.D. 1975. Estimation of water activity in intermediate moisture foods. Food Technol. 29, No. 3, 26-34.

SMITH, C.A., and SMITH, J.D. 1975. Quality assurance system meets FDA regulations. Food Technol. 29, No. 11, 64-68.

SZCZESNIAK, A.S. 1968. Artifical fruits and vegetables. U.S. Pat. 3,363,831.

STANLEY, D.W., CUMMING, D.B., and DEMAN, J.M. 1972. Texture-structure relationships in texturized soy protein. 1. Textural properties and ultrastructure of rehydrated spun soy fibers. J. Inst. Can. Sci. Technol. 5, 118.

TIEMASTRA, P.J. 1968. Degradation of gelatin. Food Technol. 22, No. 9 101-103.

TOLSTOGUSOV, W.B. 1974. Physikochemische Aspekte der Herstellung Kunstlicher Nahrungsmittel. Die Nahrung 18, 523.

TREYBAL, R.E. 1955. Mass-Transfer Operations. McGraw-Hill, New York.

U.S. FOOD AND DRUG ADMIN. 1952. Fruit butters, fruit jellies, fruit preserves and related products: definition and standards of identity. Fed. Regist. 17, 2596-2598.

VAN BUREN, J.P., LABELLE, R.L., and SPLITTSTOESSER, D.F. 1967. The influence of SO_2, pH and salts on brined Windsor cherries. Food Technol. 21, No. 7, 90-92.

WOOD, F.W., WOODS, M.F., and YOUNG, R. 1974. Food products of nonuniform texture. Can. Pat. 944,604.

Chapter 9

Principles of
Semi-moist Foods

As we have seen in previous chapters, reducing the amount of water available to microorganisms in food substrates to the point of creating antimicrobial conditions is an effective means of preserving foods. Historically this technology has been successfully applied to the preservation of jams and jellies, dry sausage, country hams, prunes, figs, dates and other fruits, certain bakery products including fruit cakes, sweetened condensed milk and a long list of highly acceptable products. Even the American Indian learned to preserve pemmican, their storage stable meat product, with this method. Yet, a new series of food products is evolving, which will carry this technology even further.

Essentially what has happened is that the availability of wholesome, inexpensive antimycotics, effective at low levels of addition, has permitted the development of a new category of "semi-moist" products or "intermediate moisture" products. The purpose of this chapter is to discuss semi-moist technology as it is currently evolving. The potential public health problems of canned white bread and their resolution will be indicated; next the availability of sorbic acid (and its salts), the combination of the two in the most successful application to date—storage stable, inexpensively packaged semi-moist pet foods—will be discussed; then the preservation of semi-moist human foods will be considered.

CANNED WHITE BREAD

Storage Stability

The conditions of growth for *Clostridium botulinum*, as discussed

278

in Chapters 1 and 6, are: pH value ranging from about 4.5 to 8.0, moisture levels normally found in moist foods; an adequate source of microbial nutrients; and anaerobic conditions. A review of canned bread in the early 1940s revealed that it was a potential health hazard, since *Cl. botulinum* could grow in some of the processed product available at that time.

In a series of studies by Dack (1953) it was demonstrated that the toxin of *Cl. botulinum* could be generated by the organism in canned white bread. Further, he found that by careful attention to moisture, pH and solids levels in the finished product, the canned bread could be brought to a condition wherein anaerobic spore forming bacteria could not germinate and produce toxin. Essentially these conditions are reached when the moisture content of the bread falls below about 35%.

Holding moisture levels in canned bread to between 32 and 35% was subsequently shown to be compatible with a reasonable canned white bread quality, at the same time exerting a control over the growth of anaerobic spore forming bacteria. It has been indicated in Chapter 6 that this category of microorganisms is the one requiring drastic exposures at elevated temperatures to destroy their spores, and hermetically sealed containers to prevent reinoculations.

With the technology developed by Dack and many others, lowering the moisture content to below 35% and hot packing at atmospheric pressures resulted in: eliminating anaerobic spore-forming bacteria as a potential problem; destroying enzymes and non-heat-resistant bacteria, yeasts and molds present by exposing the substrates to temperatures near 100°C for relatively short times; and sealing the hot products in hermetic containers to prevent reinoculation. Since that time this technology has been widely applied to canned bread.

FUNGISTATIC AND FUNGICIDAL AGENTS

Sorbic Acid

As indicated in Chapter 12 the polyunsaturated fatty acids are effective fungistatic agents. Under some conditions they have also been found to be fungicidal. Some of their applications in preventing molding of meat, cheese, and many other refrigerated, packaged foods will be shown later. At this point it would be useful to review in more detail some of the characteristics of sorbic acid and its potassium salt. However, sodium and calcium salts should not be overlooked.

Solubility.—Sorbic acid is soluble in anhydrous propylene glycol to an extent of 5.5% at room temperature. Dilution with water reduces the solubility sharply, to about 0.5% for a 50:50 mixture of glycol and water.

Potassium sorbate solutions in propylene glycol present no precipitation problems at 20% or lower concentrations at 20°C. For example, a 15% solution was chilled below −0°C and seeded, but no precipitation could be induced. Addition of water increases the solubility of salt, as would be expected.

Various mixtures lof a id and salt may be dissolved in propylene glycol. For example, 4.0% sorbic acid and 4.6% potassium sorbate makes a permanent solution equivalent to 10% potassium sorbate in available sorbate radical (Doherty 1965).

Stability.—Both potassium sorbate and sorbic acid are known to discolor when held at elevated temperatures for extended periods. Storage temperatures below 37°C are recommended for both these solids. However, 45°C was selected for a test program arbitrarily to simulate adverse summer storage conditions and to allow observation of corrosion tendency under accelerated conditions. Four representative solutions were stored in contact with each of four representative materials of construction and followed with chemica and color determination for a period of 16 weeks. Test results reported by Doherty (1965) for UV absorption and acidity are shown in Table 9.1.

The ultraviolet (UV) absorption data show some experimental scatter but no consistent trend which could be interpreted as appreciable loss of active ingredient. Acidity titrations on the two solutions containing sorbic acid confirmed the inherent stability of these materials, even under unfavorable conditions.

As would have been predicted on the basis of experience with storing the solids, colors gradually increased while the solutions aged at 40°C. However, loss of active ingredient accompanying color generation was not measurable by the analytical methods employed by Doherty.

She confirmed the accelerating effect of the elevated temperature by storing 5% and 10% solutions of potassium sorbate in propylene glycol at 23°C for 12 weeks. While the original colors were unchanged after aging, a 15% solution discolored slightly under these conditions.

Long-term storage of concentrated solutions of potassium sorbate in water is not recommended because of color buildup. One part of a 10% solution of potassium sorbate in propylene glycol was diluted with 19 parts water to bring the salt concentration to 0.5%. Refluxing this solution 6 hr at 100°C showed no loss of sorbate as measured by UV absorption. A 5% solution of sorbic acid in propylene glycol was similarly diluted with water to bring the final acid concentra-

TABLE 9.1

STORAGE STABILITY OF PROPYLENE GLYCOL SOLUTIONS OF POTASSIUM SORBATE AND/OR SORBIC ACID AT 45°C

Sample Identification [1]	UV Absorption at 253 mμ (weeks)					Acidity as % Sorbic Acid (weeks)				
	0	2	4	8	16	0	2	4	8	16
2% Potassium sorbate										
Control, amber glass	0.649	0.650	0.685	0.683	0.705					
UCC No. 2 coating	—	0.660	0.687	0.665	0.710					
Aluminum 3003	—	0.660	0.680	0.674	0.704					
Polyethylene	—	0.670	0.680	0.670	0.706					
10% Potassium sorbate										
Control, amber glass	0.690	0.675	0.680	0.693	0.690					
UCC No. 2 coating	—	0.680	0.685	0.692	0.691					
Aluminum 3003	—	0.670	0.688	0.693	0.695					
Polyethylene	—	0.680	0.690	0.688	0.693					
2% Sorbic acid										
Control, amber glass	0.905	0.853	0.750	0.850	0.870	1.98	2.00	1.95	2.00	1.95
UCC No. 2 coating	—	0.875	0.800	0.640	0.745	—	1.99	1.96	1.95	1.92
Aluminum 3003	—	0.870	0.805	0.832	0.827	—	1.99	1.97	2.04	1.95
Polyethylene	—	0.850	0.854	0.820	0.800	—	2.00	1.99	1.97	1.95
4.6% Potassium sorbate—4% sorbic acid										
Control, amber glass	0.669	0.658	0.682	0.680	0.681	3.88	3.90	3.88	3.85	3.80
UCC No. 2 coating	—	0.690	0.677	0.680	0.680	—	3.90	3.85	3.86	3.84
Aluminum 3003	—	0.669	0.669	0.669	0.684	—	3.90	3.88	3.88	3.85
Polyethylene	—	0.658	0.669	0.670	0.675	—	3.89	3.87	3.92	3.74

Source: Doherty (1965).
[1] Described in text

tion to 0.25%. A similar 6 hr reflux period resulted in no appreciable loss. In both cases, both UV absorption and total acidity indicate little or no loss of preservative effects during short periods at relatively high temperatures.

Container Corrosion Tests.—There was no evidence of corrosion in the 16-week accelerated test period at 43°C. No significant differences in color behavior could be detected and associated with the different materials. Doherty concluded that aluminum, glass, polyethylene, and No. 2 plastic-lined steel were all satisfactory materials and having no adverse affect on the test solutions.

Application of Semi-moist Technology to Pet Foods.—The combination of intermediate moisture technology with antimycotic technologies was inevitable. The first major breakthrough was made by Burgess et al. (1965) in the pet food area.

SEMI-MOIST PET FOODS

Animal foods, particularly dog and cat foods, are commonly prepared for the consumer in two forms: (1) the meal-type ration which has a dry, cereal-like texture and a low moisture content (typically about 10%) and (2) the canned-type ration which has a meat-like texture and a high moisture content in the neighborhood of 75%. Due in large measure to the difference in moisture content, these two forms of animal foods have widely divergent product characteristics, some desirable and some undesirable. Such foods are generally formulated from: (1) meat and/or meat by-products, or (2) one or more vegetable protein sources as well as combinations of these together with (3) other nutritional supplements (Burgess et al. 1965).

Meal-type animal foods generally have a very high nutritional and caloric value, providing a complete and balanced diet for the animal, and excellent storage characteristics, thus permitting the use of relatively inexpensive packaging techniques. However, the palatability of many dry meal-type animal foods is poor. In many cases the animal will not eat them at all in dry form, necessitating the addition of liquids prior to their consumption. Liquid addition often fails to solve the palatability problem, since the products become mushy or doughy and are rejected by the animal if there are any other foods available. Moreover, such reconstitution fails to bring forth the inherent initial palatability factor possessed by meat and meat by-products. Therefore, the desirable nutritional characteristics of this form of animal food may be defeated by its relatively poor palatability. In general, product stabilization against microbiological spoilage is achieved in such a product by maintaining the moisture content below the critical level for vegetative growth of such organisms as yeasts, molds and bacteria.

Canned-type animal foods, on the other hand, are generally received very favorably by animals, apparently due in part to their meat-like texture, consistency and aroma. However, the elevated moisture content of such products necessitates thermal processing in sealed containers to obtain a commercially sterile product, thereby adding considerably to product cost. Furthermore, once such a can is opened it must be quickly consumed, since the product supports microbiological growth and hence will deteriorate very rapidly unless stored under refrigeration.

In general, the concept of an intermediate moisture product, that is, one having a moisture content in excess of 10% and substantially below 75% had been largely overlooked.

To be sure, an increase in the moisture level of many animal or pet food products will increase palatibility. However, any significant elevation of the moisture level of such foods above 10% leads to microbiological decomposition, unless such products are packaged in a hermetically sealed container and commercially sterilized or maintained in a frozen or refrigerated state throughout the period of distribution and storage by the consumer. Such packaging or preservation methods are expensive and not convenient to the consumer under all anticipated conditions of use (Burgess et al. 1965).

Hallinan et al. (1963) reported on a "condensed" product at a moisture level of about 30 to 40% as well as products equivalent in moisture level to conventional canned-type products, provided certain specified processing and packaging conditions were met. Their approach involved adjustment of product pH from 2.0 to 5.0, and the optional employment of minor levels of dissolved solids to effect an initial destruction of microorganisms in processing and to help inhibit growth of surviving organisms.

Essential to the requisite inhibition of microbiological growth by the Hallinan et al. process were: a thorough pasteurization of the product; maintenance of these pasteurization conditions while filling; sealing containers with avoidance of occluded air pockets; and shaking container and contents to assure hot product contact with the package. Further, they stated that such a package must be sealed sufficiently tight to prevent the entry of microorganisms. The packaging material, besides performing this critical function, must also meet other requirements; i.e., retain water vapor and resist attack by the water, acid and fat present in the product itself.

Hallinan *et al.* (1963) noted that certain packaging materials, such as polyvinylidine chloride, were satisfactory to package such formulations. However, they apparently found that pet food compositions per se support mold growth despite this hot packaging precaution due to loss of package integrity, giving rise to an unsightly appearance and in some cases obnoxious gas pockets within the package and contamination by mold growth. Further, we now recognize that certain molds can be demonstrated to produce substances toxic to animals and man.

In accordance with the Burgess process, a variety of intermediate moisture animal foods of high palatability and nutrition as well as practical shelf-life, even when packaged in inexpensive water-impermeable packaging material, was provided by formulating a pathogen-free matrix of normally biological growth-supporting constituents and dispersing them in an aqueous phase of water-soluble solids, principally sugar, present at a level at least high enough to exert a bacteriostatic effect sufficient to stabilize the animal food. The animal food had a moisture content less than 30% and greater than 15% and a level of water-soluble solids between 15 and 35% by weight of the total composition. As is explained below, the water-soluble solids, while predominantly sugar, could contain other low average molecular weight materials (such as sorbitol, propylene glycol, and sodium chloride) capable of endowing the animal food with microbiological protection, due to the osmotic pressure effect of these water-soluble solids per se. Since the animal food has its primary and advantageous application in the marketing of a product adapted to be aerobically packaged under ambient temperatures, the composition must contain that level of water-soluble solids which will exert a bacteriostatic effect on the mesophilic organisms but uniquely does not call for overt employment or creation of any acidulant to control bacterial growth.

By virtue of the ability of process and package the intermediate moisture animal food without resorting to commercial sterilization or other means to arrest bacterial spoilage, such processing and packaging could give rise to mold and yeast development. The animal food must therefore include an antimycotic, serving to prevent the growth of yeasts and molds which are adaptable to high soluble solids concentrations at the intermediate moisture range employed. Indeed, an antimycotic was found to be essential. The manner of incorporating such an agent was not critical. Some types of antimycotics could be incorporated with the ingredients being processed to form the animal food, whereas others could simply be sprayed or otherwise surface coated on the product. Still others could be applied to the packaging which would be in contact with the product surface. An antimycotic must be added to prevent yeast and mold growth. Sorbate salts, such as potassium sorbate as well as sorbic acid, can be used either separately or in combination. Propylene glycol, which may be used alone or with other humectants like sorbitol to impart a degree of product softness or tenderness, has also been found to serve as an antimycotic. Other antimycotic agents are apparent.

The amount of antimycotic agent added is selected so as to produce the desired results and still constitute a minor proportion of the product, from about 0.1 to 2.5% of the total weight, depending on the particular antimycotic and the particular product composition. Potassium sorbate in a water solution can be sprayed onto the surface of the animal food or the food can be dipped in this solution. Other antimycotics lend themselves to such surface application as esters of the parabens (para-hydroxy benzoate) such as propyl and methyl parabens (methyl parahydroxy benzoate), as described by Burgess.

Sugar is employed as the principal source of water-soluble solids and may range upwardly in weight percentage of the composition from 15 to 35% depending upon the particular sugar or sugar mixture relied upon to offer the desired bacteriostatic protection. As the moisture content of the product increases in the intermediate moisture range, the level of a given sugar will correspondingly increase in order to maintain a sufficient bacteriostatic effect. The level of sugar chosen will also vary depending upon the presence and level of auxiliary water-soluble solids, also offering a similar increase in osmotic pressure to the aqueous phase of the composition. Thus, a variety of low average molecular weight materials may be included as part of the water-soluble solids in the aqueous phase and will augment the sugars in their role of providing sufficient osmotic pressure to prevent bacterial decomposition.

The term water-soluble solids is understood, therefore, to apply to any animal feed or feed additive material which is substantially soluble in water at room temperature or at temperatures comparable to those practiced in processing the ingredients of the dog food composition. Included in the class of water-soluble nonsugar solids that can be employed are certain inorganic salts used at a level compatible with palatability requirements, e.g., sodium chloride and potassium chloride. Burgess also found that certain compounds like the diols and polyols, propylene glycol, sorbitol, glycerol and the like which have another function, i.e., as antimycotic and/or texturizers, may also be relied upon to afford the soluble solids employed in the aqueous phase for bacteriostatic protection. The propylene glycol is useful in this respect since it is capable of serving a multiple role as mold inhibitor anu plasticizing humectant for texture, as well as contributing to the water-soluble solids of the aqueous phase. For these reasons propylene glycol is most functional as an additive for use in combination with the matrix materials.

Process for Semi-moist Pet Foods

In practice the meat is pasteurized in combination with an emulsifier and all liquid ingredients at a temperature of about 100°C for about 10 min. Upon completion of pasteurization, colors, antimycotic agent, crystallization retardant, flavors and nutrients are added. Immediately following, all the remaining dry ingredients are added (such as soy flakes, soy hulls, bone meal, dried skimmed milk and sugar) in one charge. The temperature of the total mass, due to the addition of these dry ingredients, drops to approximately 60°C. The total product is then cooked at a temperature of about 85°C for 7 min. Upon completion of the final cook, the product is cooled to a temperature of about 21°C.

The cooked mixture, upon achieving uniformity of texture, can be packaged according to conventional wrapping procedures. It was an additional feature of this process that such mixtures could be packed in ordinary moisture-impermeable wrapping material without any need for sterilization.

Further, the product may be formed into cylindrical patties. The procedure for forming such patties includes the extrusion of the finished mixture at low pressures and at a temperature sufficiently low to prevent stickiness. The extruded product may be molded under low pressure into the desired hamburger-like patties (Burgess et al. 1965) (Fig. 9.1).

The prepared product is characterized by a completely meat-like appearance, color, consistency, texture and general handle-ability. On being chewed, it has a meat-like chewy texture. It is particularly characterized by its extended storage life. Under normal ambient conditions it may be stored for long periods (six months or longer) without damage in a loose moisture-impermeable wrapping, which is sufficient to protect the product under normal conditions of handling. However, the composition can be stored in an open unpackaged condition without undergoing bacterial deterioration. Wrapped in this inexpensive manner, the product requires no refrigeration to retain its advantageous characteristics, even for extended storage periods, without undergoing an undesirable darkening in color.

The density of the product will depend on the technique of fabrication including the degree of pressure used to form the patties.

TABLE 9.2

INTERMEDIATE MOISTURE LOOSELY WRAPPED PRODUCT

Ingredient	Parts by Weight
Chopped meat by-products (tripe, udders, cheek trimmings, tongue trimmings, gullets, etc.)	32.0
Defatted soy flakes	31.0
Sucrose	21.7
Flaked soy bean hulls	3.0
Dicalcium phosphate	3.0
Dried nonfat milk solids	2.5
Propylene glycol	2.0
Bleachable fancy tallow	1.0
Mono- and di-glycerides	1.0
Sodium chloride	1.0
Potassium sorbate	0.3
FD and C red dye	0.006
Garlic	0.2
Vitamin and mineral premix	0.06

Typically the density will be 0.7 to 1.9 kg per cubic decimeter and most commonly about 0.85 kg per cubic decimeter. It is apparent that the bulk density of the product may be controlled as desired.

It is particularly significant that the product of this technology is fully as palatable (i.e., as well liked by animals) as is the best equivalent canned animal food—this being far in excess of the generally low palatability or receptiveness of conventional low moisture content animal foods. It is also significant that two 85-g patties of

FIG. 9.1.   THE SEMI-MOIST PRODUCT IS EXTRUDED AT LOW PRESSURE AT WHICH POINT IT APPEARS SOMEWHAT LIKE FRESHLY GROUND MEAT

It is then molded into hamburger-like patties and packaged.

FIG. 9.2.   ANOTHER EXAMPLE OF EXTRUDED SEMI-MOIST
PET FOOD; THIS ONE RESEMBLES GROUND MEAT

the product of this process are equal in feeding value to a 454-g can of a ration-type animal food.

Thus, it may be observed that the product of the Burgess process possesses the desirable features of both canned and dry animal foods without including the undesirable features. A typical formula is shown in Table 9.2. Examples of some of the many semi-moist pet foods currently available are shown in Fig. 9.1 and 9.2.

Marbled, Textured Product

Bone (1968) discussed a technology for producing marbled semi-moist products (Fig. 9.3). This product is a meaty semi-moist animal food resembling marbled meat, comprising a red portion and a white portion. The red portion contains a caseinate salt binder in an amount sufficient to cause tackiness at temperatures above 48°C. The white portion randomly distributed in the red portion, and the red and white portions are joined to form an integral mass having

FIG. 9.3.   A MEATY SEMI-MOIST ANIMAL FOOD
RESEMBLING MARBLED MEAT

substantially distinct interfaces between layers.

The method for the manufacture of a semi-moist animal food resembling marbled meat comprises the following steps:

(1) Extruding by a cooking-extrusion process a first sheet of unexpanded storage-stable semi-moist animal food having the appearance of lean red meat and containing a caseinate salt adhesive in an amount sufficient to provide stickiness at temperatures above 48°C.

(2) Extruding by a cooking-extrusion process a second sheet of unexpanded storage-stable semi-moist animal food having the appearance of fat, with the cooking-extrusion process as follows: (a) admixing comminuted animal food ingredients which include a total moisture content between about 20 and 50%, the ingredients having been selected to provide a substantially solid, storage-stable nutritious pet food after cooking the resulting admixture; (b) subjecting the resulting admixture to cooking conditions at a superatmospheric pressure sufficient to prevent substantial expansion of the resulting plastic mass in the extruder; and (c) extruding the resulting plastic mass through a die into the atmosphere.

(3) Superimposing one of the sheets on the other.

(4) Forming superimposed sheets into a loaf-like mass, the forming being done in such a manner as to distribute randomly the fat-like portion in the resulting mass; the forming taking place while average temperature of the first and second sheets is about 76°C (Bone 1968).

Steak-like cuts can be prepared by slicing rolls of appropriate diameter. Cuts resembling chops are prepared by slicing smaller diameter rolls or loaves which may have a relatively heavy fat-like overlay. While a loaf can be formed around a prefabricated bone-like structure, if a structure resembling bone is used, it must be either inert with respect to water pick-up or it must have approximately the same water activity as the surrounding material.

Water Activity

Generally the formulation of a pet food to provide storage stability is one in which the water activity of the product is less than 0.90 and also contains an effective amount of mold inhibitor, such as sorbic acid or its salts.

Water activity (a_w) as used herein is defined as:

$$a_w = f/f_0$$

TABLE 9.3

COMPOSITION OF MARBLED SEMI-MOIST PRODUCT

Ingredient	Base (%)	Marbling (%)	Total Formula (%)
Beef by-products	29.0000	42.50	32.3750
Sugar	26.5778	26.50	26.5584
Sodium caseinate	15.0000	15.00	15.0000
Beef trimmings	13.5000	—	10.1250
Cornstarch	7.5000	7.50	7.5000
Propylene glycol	4.5000	4.50	4.5000
Dicalcium phosphate, dihydrate	2.5000	2.50	2.5000
Salt, iodized	1.2000	1.20	1.2000
Potassium sorbate	0.1000	0.10	0.1000
Vitamin E supplement	0.0670	—	0.0503
Titanium dioxide	—	0.20	0.0500
Riboflavin supplement (4 g/lb)	0.0270	—	0.0202
Vitamin A supplement (30,000 IU/lb)	0.0670	—	0.0503
Irradiated dried yeast	0.0040	—	0.0030
FD and C Red No. 2	0.0027	—	0.0020
FD and C Yellow No. 6	0.0027	—	0.0020
Thiamin mononitrate	0.0010	—	0.0007
Total	100.0000	100.0000	100.0000

Source: Bone (1968).

where f = fugacity of water vapor in the specified system

f_0 = fugacity of pure water at the specified system temperature and 1 atm total pressure.

Water activity in the desired range is achieved by including appropriate quantities of soluble materials (such as sugars and glycols) in the formulation, as described earlier.

Formulations such as those listed in Table 9.3, using sugar in an amount between about 15 and 30%, propylene glycol in an amount between about 2 and 10% and sorbic acid or its salts in amounts from 0.06 to 0.3% provide adequate storage stability for the products of this process which contain moisture in an amount between about 20 and 40% by weight based on the weight of the product.

WATER ACTIVITY

The term semi-moist foods has entered our vocabulary during the past decade to identify a heterogeneous group of foods which resemble dry foods in their resistance to microbial deterioration, but which contain too much moisture to be regarded as dry. *Although these products normally contain 20 to 30% water, they resist microbiological spoilage without refrigeration.*

These foods are plastic, easily masticated and do not produce an oral sensation of dryness. However, these foods are subject to the same types of adverse chemical changes as observed with fully dehydrated foods. As a generalization, Brockman (1970) suggests that foods in this intermediate moisture range are more susceptible to the Maillard reaction than dry foods but less susceptible to fat oxidation. Unless precautions are taken to inactivate enzymes, some moisture foods are susceptible to a variety of enzymic reactions, as one might expect.

Activity of Water

The binding, immobilization and other types of restraint imposed on the behavior of the water present in food is mediated by a multiplicity of systemic factors such as: the nature and concentration of dissolved components; the number and binding capacity of polar residues including those with negative coefficients; the configuration of hydrophobic and hydrophilic areas; and presumably, the mechanisms which alter the structure of water itself (Brockman 1970).

The availability of water for spore germination and microbial growth is closely related to the relative vapor pressure, now called water activity. Water activity (a_w) is defined as the ratio of the vapor pressure (P) of water in the food to the vapor pressure of pure water (P_0) and at that temperature, $a_w = P/P_0$. Within the range favorable to the growth of mesophilic microorganisms, a_w is practically independent of temperature (Brockman 1970). The limits of water activity for microbial growth are given in Table 9.4.

By incorporating an effective antimysotic agent, heating to destroy vegetative organisms, and adjusting to a_w of approximately 0.85, foods sealed in inexpensive plastic pouches have an excellent record for stability under experimental conditions.

TABLE 9.4

APPROXIMATE LOWER LIMITS OF WATER ACTIVITY
FOR MICROORGANISM GROWTH

Organism	Minimum a_w
Bacteria	0.91
Yeasts	0.88
Molds	0.80
Halophilic bacteria	0.75
Xerophilic fungi	0.65
Osmophilic yeasts	0.60

Source: Labuza et al. (1972A).

Potential Advantages

Semi-moist foods offer potential advantages for various situations. They are relatively low in moisture and hence concentrated from the standpoint of weight, bulk and caloric content, plus ease in packaging and storage.

In contrast to freeze-dried products, semi-moist foods could be acceptable for direct consumption without giving rise to the sensation of dryness. Their texture is much closer to normal food than fully dehydrated foods.

In contrast to the heat sterilized canned products, the wholesomeness or safety of intermediate moisture foods is less dependent upon the integrity of containers.

The availability of a wide variety of semi-moist foods of demonstrated acceptability is regarded as prerequisite to expanded use of such foods.

Current Problems

A major obstacle in converting familiar foods into semi-moist items is seen from the moisture sorption data summarized by Brockman (1970) (Table 9.5). Foods are usually too dry to be acceptable to eat even at a_w 0.85. Table 9.6 extends the data of Table 9.5 to include a_w values for aqueous solutions containing 100 g of anhydrous sodium chloride, glycerol and sucrose. Sodium chloride and to a lesser extent glycerol have the capacity to depress the vapor pressure of large amounts of water.

Comparing the values for food in Table 9.5 to those in Table 9.6, it should also be noted that 22 g of water raises a solution containing 100 g of sucrose from a_w 0.85 to 0.90. On the other hand, 3.8 g of additional water raises the a_w of wheat from 0.85 to 0.90.

Raoult's law approximates the relationship between the moles of water and the moles of solute in a solution and its a_w. At present the addition of additives for depressing water activity and antimycotic to suppress growth of mold and yeast are required. However, the technology continues to expand.

Ross (1975) described a basis for making reasonably accurate estimations of water activity in complex food substrates: *a_w is simply the product of a_w values for simple solutions of each solute measured at the same concentration as in the complex solution; or $a_w = (a_w°)_1 \times (a_w°)_2 \times (a_w°)_3$* ... where $(a_w°)_X$ is a component of the food system. This equation carries great insight concerning semi-moist foods.

As previously indicated, when equilibrium exists between the water content of a food and the percent relative humidity of its environ-

TABLE 9.5

MOISTURE SORPTION DATA FOR COMMON FOODS AND FOOD COMPONENTS

Product	°C	Water g/100 g Dry Material [1]				
		for a_w =				
		0.70	0.75	0.80	0.85	0.90
Rice	30	13.2	14.5	16.2	18.4	
Milk, nonfat	30	11.9	13.5	15.8	19.8	
Egg albumin[2]	25	14.5	15.8	17.8	21.0	25.5
Wheat	25	16.3	17.5	19.3	22.0	25.8
Chicken[2,3]	20	18.2	20.3	22.7	25.9	
Beef[2]	25	16.7	19.8	23.4	27.5	34.6
Peas[2]	20	16.4	19.6	23.0	29.0	
Shrimp[2]	35	16.9	19.2	23.2	29.4	
Gelatin	25	21.6	23.5	26.7	31.6	40.2
Currants[2]	20	25.2	31.5	39.4	51.0	
Apple[2]	30	30.2	35.8	43.7	60.6	

Source: Brockman (1970).
[1] Some values calculated or interpolated from cited data
[2] Freeze-dried
[3] Fat-free basis

ment, a_w is directly relatable to RH, a_w = RH/100. The relationship between a_w (or RH at equilibrium) and the moisture content of a food can be precisely described by its moisture sorption isotherm.

The physical-chemical phenomena underlying sorption isotherms have been the subject of extensive analyses by Labuza (1968). Without reference to theories which provide an insight into the mechanisms which lower the vapor pressure of water in food, it is apparent that as the moisture content of a food decreases, a progressive restraint is exercised on the vapor pressure, evidenced by a_w values. Thus, 40 g of water remaining in 100 g of dry egg albumin would have a vapor pressure equal to only one-tenth that of pure water.

TABLE 9.6

MOISTURE SORPTION DATA FOR PURE CHEMICAL COMPOUNDS

Compound at 25°C	Water g/100 g Anhydrous Compound				
	For a_w =				
	0.70	0.75	0.80	0.85	0.90
Sodium chloride	1	277	332	423	605
Glycerol	56	72	96	132	203
Sucrose	1	1	1	49	71

Source: Brockman (1970).
[1] Beyond solubility limit

Based on Structure

The interaction between water and solutes has been reported by Ross (1975) to be based on the three-dimensional structure of water. Water is recognized not as a simple homogeneous mixture of separate molecules, but rather molecules linked together into a three-dimensional network and possibly into several different three-dimensional structures in constant rearrangement and interaction. When a solute is added to water, first, the concentration of water is reduced, and second, the interaction of the solute with the water may break or increase water structures. LiOH breaks the water and increases the effective concentration of water (water activity) over its actual concentration (Brockman 1970). Sucrose, on the other hand, creates greater structures in water and reduces water activity possibilities below what they should be due to concentration alone.

Food Structure Important

The food structure itself is important in preservation systems. Bone (1973) predicted that the water activity of a water-in-oil emulsion would be higher than if the oil and water phases were separate. The vapor pressure of a liquid is recognized to be greater when it is in the form of small droplets than it is in the bulk phase; hence, so too, water activity. An emulsion with fine droplets of aqueous solution would have a higher water activity than an emulsion with coarse droplets. According to Bone (1973) the effect of droplet size on water activity is predicted by Kelvin's equation:

$$P/P_0 = \frac{2\ G\ M}{d\ RT\ r}$$

where G = surface tension of water
M = molecular weight of water
d = density of water
RT = gas constant $\times °K$
r = radius of curvature.

Simply reducing droplet size increases the water activity of pure water as a result of surface phenomena.

SEMI-MOIST HUMAN FOODS

The broadening of the preceding technologies to include many kinds of human food is already underway. While the field has barely been explored, in light of recent developments in closely allied fields it appears to hold considerable promise.

Hollis *et al.* (1968) reported their findings on the stabilization of semi-moist or intermediate moisture food products. They define intermediate moisture foods as foods that are partially dehydrated and have a suitable concentration of dissolved solids to bind the remaining water sufficiently to inhibit growth of bacteria, mold and yeast.

Hollis *et al.* (1968) developed two methods whereby food pieces could be infused yet retain the appearance, texture and flavor of the natural food. The two methods differ primarily in the initial moisture of the food piece being used in the process. It was found that both dehydrated and wet foods could be effectively infused.

One method treats foods that were vacuum-dried or freeze-dried by soaking in an infusing solution to a point where, after draining, the food pieces have the proper moisture content and water activity to make them edible without further hydration and to maintain them microbiologically stable at room temperature. It was found that if a dehydrated, porous, absorptive food is soaked in a solution whose viscosity is not too high to penetrate, the food will absorb an amount of solution approximately equal to the amount of water it would normally absorb.

The other method consists of soaking a *normal moisture* food in an equilibrium solution so that, after draining, the moisture content and the water activity have been reduced to the desired levels to make the food edible without rehydration and to maintain it microbiologically stable at room temperature. The food can be precooked or raw.

It was found that raw food can be cooked in the equilibrium solution. The solution's composition is based on the "target" formula of the finished product and the amount of water in the food to be treated. The processing formula is based on the assumption that the additives will diffuse into the food to an extent that the final concentration in the food and in the solution will be similar. The amount of solution should be enough to immerse the food.

They found that the processing formula could be expressed by the mathematical relationship:

$$C = \frac{W_S \, C_S}{W_S + W_f}$$

where C = desired concentration of non-aqueous solutes in equilibrated solution

C_S = initial concentration of non-aqueous solutes in external (infusion) solution

TABLE 9.7

SEMI-MOIST COARSE GROUND BEEF

Item	Percentage	Grams
Beef, ground chuck		
Solids	39.0	2655.9
Moisture	61.0	4154.1
Total	100.0	6810.0
Infusion solution for coarse ground beef		
Glycerol	65.61	5916.0
Water	15.61	1407.6
Beef soup spice base	11.92	1074.4
Propylene glycol	5.92	537.2
Potassium sorbate	0.90	81.6
Total	100.00	9016.8
Moisture content (vacuum—oven method)	37.00	
a_w	0.78	

Source: Hollis et al. (1968).

W_s = initial weight of above

W_f = initial weight of water in food.

To help visualize what is occurring in these products, Hollis et al. (1968) think of the food structure as a sponge-type having a liquid holding capacity. In the dry state it absorbs the infusing solution, and in the wet state part of the water is "washed out" with another liquid by diffusion.

Coarse Ground Beef and Beef Cubes

Shelf-stable intermediate moisture coarse ground beef was prepared for storage evaluation by applying the cook-soak equilibration method to fresh raw chuck hamburger. Hamburger was immersed, heated to cooking temperature, and cooked at approximately 99°C for 15 min in a solution calculated to infuse the proper amount of additives. The mixture was then cooled, soaked overnight under refrigeration and reheated. The cooked intermediate moisture coarse ground beef was drained and dried. The final product was palatable and had acceptable moisture content and water activity. Ingredients used are shown in Tables 9.7 and 9.8.

OTHER PRODUCTS BEING DEVELOPED

Five other human food products were researched by Hollis et al. (1968): diced white chicken meat, diced carrots, beef stew, barbecued pork and apple pie filling. All consisted of distinct food pieces which did not lend themselves to the soft-moist pet food technology of

TABLE 9.8

SEMI-MOIST RAW BEEF PIECES (½ TO 1 IN.)

Item	Percentage	Grams
Beef, cubed chuck		
Solids	35.0	1617.0
Moisture	65.0	3003.0
Total	100.0	4620.0
Infusion solution		
Glycerol	56.6	2112.0
Beef soup spice base	19.8	739.2
Water	8.0	300.0
Sodium chloride	7.4	277.2
Propylene glycol	7.1	264.0
Potassium sorbate	1.1	39.6
Total	100.0	3732.0
Moisture content (vacuum-oven method)	31.0	
a_w	0.77	

Source: Hollis et al. (1968).

TABLE 9.9

MOISTURE, a_w AND LIPID CONTENT OF FOOD ITEMS AFTER STORAGE

Item	H_2O (%)	a_w	Lipids (%)
Chicken #5850-56	27.0	0.73	0.92
Coarse ground beef #5850-60	36.9	0.81	7.06
Diced carrots #5850-58	37.5	0.74	0.13
Beef stew with gravy #5492-73	27.2	0.75	17.45
Beef stew (no gravy) #5492-74	35.1	0.76	3.31
peas	37.5	0.80	0.56
carrots	43.3	0.75	0.33
potatoes	35.3	0.75	0.25
beef	29.5	0.75	10.25
Barbecued pork #5850-64	29.0	[1]	6.97
Apple pie filling #5850-54	25.7	0.71	0.50

Source: Hollis et al. (1968).
[1] Not measured due to volatile acetic acid in sauce

mixing and grinding together wet and dry materials. The pieces had to be infused with additives to adjust water activity to the desired point (0.71 to 0.81) (Table 9.9).

The intermediate moisture products were microbiologically stable under nonsterile conditions (Table 9.10) and required no sterilization. They also were acceptable to the taste panel (Table 9.11).

STORAGE STABILITY

No microbiological development or significant chemical, physical or sensory changes have been noted during four months storage

TABLE 9.10

MICROORGANISM COUNTS BEFORE AND AFTER STORAGE FOR FOOD ITEMS

Item	Standard Plate Initial	4 Mo.	Molds Initial	4 Mo.	Yeast Initial	4 Mo.
Chicken #5850-56	10	< 10	< 10	< 10	< 10	<10
Chicken #5856-53	140		< 10		< 10	
Coarse ground beef #5850-60	< 10	< 10	< 10	< 10	< 10	< 10
Diced carrots #5850-58	< 10	< 10	< 10	< 10	< 10	< 10
Beef stew with gravy #5492-73	40	60	< 10	< 10	< 10	10
Beef stew, no gravy #5492-74	< 10	40	< 10	< 10	< 10	400
Barbecued pork #5850-64	30	10	< 10	< 10	< 10	< 10
Apple pie filling #5850-54	20	20	10	< 10	< 10	< 10

Source: Hollis et al. (1968).

TABLE 9.11

SENSORY EVALUATION RATINGS

Item	Storage Time (Months)	Control	0°C As Is	38°C As Is	0°C Rehydrated	38°C Rehydrated
Chicken pieces						
study 1	4	10	5.1	5.1	7.4	6.6
study 2	2	10	6.6		7.7	
Coarse ground beef	4	10	8.1	7.1	9.6	9.1
Diced carrots	4	10	6.6	6.1	8.0	7.9
Beef stew with gravy	4	10	6.3	5.8	—	—
Beef stew (no gravy)						
rehydration I	4	10	—	—	7.1	6.9
rehydration II	4	10	—	—	7.7	7.3
Barbecued pork						
rehydration I	4	10	7.9	7.2	7.9	7.6
rehydration II	4	10	7.9	7.2	8.3	7.6
Apple pie filling	4	10	5.9	5.1	7.6	6.2

Source: Hollis et al. (1968).

in sealed containers at 38°C (Brockman 1970).

Some difficulty was encountered with a phase separation of certain sauces under the above storage conditions. Standard plate counts performed at the time of preparation and after four months storage revealed a well defined trend toward a reduction in viable bacteria during storage. Counts for yeasts and molds less than 10 per g were found (Brockman 1970).

Two intermediate moisture casserole items, chicken a-la-king at $a_w = 0.85$ and ham in cream sauce at the same a_w were inoculated with S. aureus (12,000 to 14,000 organisms per g). After 1 month at 38°C the viable count had fallen to less than 2000; after 4 months it was listed for both products as 0.4 organisms per g. Comparable observations were made on the same two items after inoculation with *E. coli*, Salmonella, and vegetative cells of *Cl. perfringens*.

Brockman (1970) has also reported that *food which has been subjected to heat processing sufficient to destroy vegetative microorganisms, and which is protected from subsequent contamination by reliable packaging, presents an acceptable risk at a $_w$ as high as 0.95.*

SUMMARY

As a working generalization, an a_w scale can be an "availability" index for water in the vital processes incident to microbial growth. Under favorable conditions *S. aureus* has been observed to grow at a_w as low as 0.86. Halophilic bacteria may grow at a_w as low as 0.75. The common species of yeast and mold are suppressed at a_w 0.88 and 0.80, respectively, while the limits for xerophilic molds and osmophilic yeast are stated to be 0.65 and 0.60, respectively. Those organisms are not heat resistent.

Based solely on the potential limits of microbial growth, semimoist food requires an a_w = 0.60 or less. At this a_w most common foods are difficult to distinguish from dry products.

The literature summarized indicates that spores cannot germinate and relatively few species of bacteria, including only one food pathogen, can multiply at a_w = 0.90 to 0.95. Substantial support for such products can be drawn from the experience with pasteurized bacon, prefried bacon and several varieties of canned bread. Additional evidence for the microbiological safety of a variety of canned products exists, some of which also qualify as semi-moist foods.

Further developments are urgently needed, not only in the developed areas of the world, but especially in the developing areas. It is apparent that this technology is in its infancy and its future growth potential is barely visible.

REFERENCES

ACOTT, K., and LABUZA, T.P. 1975. Inhibition of *Aspergillus niger* in an intermediate moisture food system. J. Food Sci. *40*, 137.

ASSOC. OFF. AGRIC. CHEM. 1965. Official Methods of Analysis, 10th Edition. Association of Official Agricultural Chemists, Washington, D.C.

AYERST, G. 1965. Determination of the water activity of some hygroscopic food materials by a dew-point method. J. Sci. Food Agric. *16*, 71.

BONE, D.P. 1968. Method of preparing a solid, semi-moist marbled meat pet food. U.S. Pat. 3,380,832. Apr. 30.

BONE, D.P. 1969. Water activity—Its chemistry and applications. Food Prod. Dev. *3*, No. 5, 81.

BONE, D.P. 1973. Water activity in intermediate moisture foods. Food Technol. *27*, No. 4, 71-76.

BOYLAN, S., ACOTT, K., and LABUZA, T.P. 1976. *Staphylococcus aureus* F265 challenge studies in an intermediate moisture food. J. Food Sci. (In press)

BROCKMAN, M.C. 1970. Development of intermediate moisture foods for military use. Food Technol. *24*, 896.

BRODY, A.L. 1972. Aseptic packaging of foods. Food Technol. *26*, No. 8, 70-74.

BULL, H.B. 1944. Adsorption of water vapor by proteins. J. Am. Chem. Soc. *66*, 1499.

BULL, H.B., and BREESE, K. 1968. Protein hydration. 1. Binding sites. Arch. Biochem. Biophys. *128*, 488.

BURGESS, H.M., and MELLENTIN, R.W. 1965. Animal food and method of making same. U.S. Pat. 3,202,514.

BURGESS, H.M., and MELLENTIN, R.W. 1969. Method of making animal food. U.S. Pat. 3,482,985.

BYRAN, F.L. 1974. Microbiological food hazards today—based on epidemiological information. Food Technol. *28*, No. 9, 52-66, 84.

CHRISTIAN, J.H.B. 1963. Water activity and the growth of microorganisms. Recent Adv. Food Sci. *3*, 248.

DACK, G.M. 1953. Food Poisoning. Univ. of Chicago Press, Chicago.

DOHERTY, C.J. 1965. Storage, Handling and Assay of Sentry Preservatives and Humectant in Pet Food Applications. Union Carbide Corp., Tarrytown, New York.

EICHNER, K., and KAREL, M. 1972. The influence of water content and water activity on the sugar-amino browning reaction in model systems under various conditions. J. Agric. Food Chem. *29*, 218.

FETT, H.M. 1974. Water activity determination in foods on the range 0.80 to 0.99. J. Food Sci. *38*, 1097.

FRAZIER, W.C. 1967. Food Microbiology, 2nd Edition. McGraw-Hill Book Co., New York.

GENERAL FOODS CORP. 1968. Parameters for moisture content for stabilization of food products. Rep. *68-22-FL.* Contract DAAG 19-129-AMC-860. AD-675473.

GENERAL FOODS CORP. 1970. Parameters for moisture content for stabilization of food products. Report in preparation. Contract DAAG 19-129-AMC-860, Phase II.

HALLINAN, F.J., CZARNETZKY, E.J., and COOMBES, A.I. 1963. Pet food and method of packaging same. U.S. Pat. 3,115,409. Dec. 24.

HOLLIS, F., KAPLOW, M., KLOSE, R., and HALIK, J. 1968. Parameters of moisture content for stabilization of food products. U.S. Army Natick Lab. Tech. Rep. *69-26-FL.*

LABUZA, T.P. 1968. Sorption phenomena in foods. Food Technol. *22,* 15.

LABUZA, T.P. 1970. Properties of water as related to the keeping quality of foods. In SOS/70 Proceedings. Institute of Food Technologists, Chicago, Ill.

LABUZA, T.P. 1971. Kinetics of lipid oxidation in foods. CRC Crit. Rev. Food Technol. *2,* 355.

LABUZA, T.P., CASSIL, S., and SINSKEY, A.J. 1927A. Stability of intermediate moisture foods. 2. Microbiology. J. Food Sci. *37,* 160-162.

LABUZA, T.P. *et al.* 1972B. Stability of intermediate moisture foods. 1. Lipid oxidation. J. Food Sci. *37,* 154-158.

LABUZA, T.P. *et al.* 1976. Determination of water activity: A comparative study. J. Food Sci. (In press).

LABUZA, T.P., TANNENBAUM, S.R., and KAREL, M. 1970. Water content and stability of low-moisture and intermediate-moisture foods. Food Technol. *24,* 35.

LABUZA, T.P., TSUYUKI, H., and KAREL, M. 1969. Kinetics of linoleate oxidation in model systems. J. Am. Oil Chem. Soc. *46,* 409.

LAJOLLO, F., TANNENBAUM, S.R., and LABUZA, T.P. 1971. Reaction at limited water concentration. 2. Chlorophyll degradation. J. Food Sci. *36,* 850.

LAMPI, R.A. 1963. Infiltration of Porous Foods with High Caloric Non-Aqueous, Edible Materials. Rep. Nos. 1-3, Quartermaster Contract DA19-129-AMC-84(N) (OI 9056), Natick, Mass.

LEAGUE, D.N. 1953. In-can baking perfected. Food Eng. *25,* No. 5, 86.

LEUNG, H., MORRIS, H.A., SLOAN, E.A., and LABUZA, T.P. 1976. Development of an intermediate moisture processed cheese food product. Food Technol. *30,* No. 7, 42-44.

LIVINGSTON, A.L., ALLIS, M.E., and KOBLER, G.O. 1971. Amino acid stability during alfalfa dehydration. J. Agric. Food Chem. *19,* 947.

LONCIN, M., BIMBENET, J.J., and LENGES, J. 1968. Influence of the activity of water on the spoilage of foodstuffs. J. Food Technol. *3,* 131.

MELPAR, INC. 1965. Interrelationships between storage stability and moisture sorption properties of dehydrated foods. Contract DA 19-129-AMC-252. Final Rep. Natick, Mass.

MOSSEL, D.A.A., and INGRAM, M. 1955. The physiology of the microbial spoilage of food. J. Appl. Bact. *18,* 232.

NATL. BUR. STAND. 1951. Methods of Measuring Humidity and Testing Hygrometers. Natl. Bur. Stand. Circ. *512.*

OLYNYK, P., and GORDON, A.R. 1943. The vapor pressure of aqueous solutions of sodium chloride at 20, 25 and 30° for concentrations from 2 Molal to saturation. J. Am. Chem. Soc. *65,* 224.

QUAST, D., and KAREL, M. 1972. Computer simulation of storage life of foods undergoing spoilage by two interacting mechanisms. J. Food Sci. *37,* 679.

RANK, R.G. 1967. Plastic food composition for animal and process for preparing said compositions. Fr. Pat. 1,483,971. May 2.

RICE, E.E. *et al.* 1944. Preliminary studies on stabilization of thiamin in dehydrated foods. Food Res. *9,* 491.

ROCKLAND, L.B. 1969. Water activity and storage stability. Food Technol. *23,* 1241.

ROSS, K.D. 1975. Estimation of water activity in intermediate moisture foods. Food Technol. *29,* No. 3, 26-34.

ROUBAL, W.T. 1970. Trapped radicals in dry lipid-protein systems undergoing oxidation. J. Am. Oil Chem. Soc. 47, 141.

SCOTT, W.J. 1957. Water Relations of Food Spoilage Microorganisms. In Advances in Food Research, Vol. 7. E.M. Mrak, and G.F. Stewart (Editors). Academic Press, New York.

SCOTT, W.J. 1962. Available water and microbial growth. Proc. Low Temp. Microbial. Symp. 89. Campbell Soup Co., Camden, N.J.

SLOAN, A.E., and LABUZA, T.P. 1975. Investigating alternative humectants for foods. Food Prod. Dev., Sept., p. 25.

THOMAS, M.A., and HYDE, K.A. 1972. The manufacture of processed cheese—a working manual. N.S.W. Dep. Agric. Bull. D40.

TROLLER, J.A. 1973. The water relations of food-borne pathogens. J. Milk Food Technol. 36, 276.

VOS, P.T., and LABUZA, T.P. 1974. Technique for measurement of water activity in the high aw range. J. Agric. Food Chem. 22, 328.

WARMBIER, H.C., SCHNICKELS, R., and LABUZA, T.P. 1976. Effect of glycerol on non-enzymatic browning in a solid intermediate moisture model food system. J. Food Sci. 41, 528.

Chapter 10

Principles of Food
Preservation by Fermentation

LIFE WITH MICROORGANISMS

Microorganisms no doubt outnumber other living entities on this planet and can be found existing actively or passively wherever living organisms occur. While the energy for life on this planet is captured by green plants in the photosynthetic process, microorganisms are generally responsible for the final decomposition of the photosynthetic products. Animals play a minor role in the cycle.

Inasmuch as bacteria, yeasts and molds are to be found throughout the environment of man, it is to be anticipated that these microorganisms are in direct competition with other living entities for the energy for life. Whenever the conditions of nutrients and environment are favorable for microbial activity, it will be found (Table 10.1).

Man must compete with all other living entities on earth. In order to retain food supplies for himself, he must interfere with natural processes. Through his study, and as a fruit of his curiosity, man has evolved a number of control systems. One is the preservation of food by controlling, yet encouraging, the growth of microorganisms. Under such a condition, man may employ microorganisms to create unfavorable conditions for other microbes, yet retain in the foodstuffs the nutrients desired.

While microorganisms were not identified as the important agents in food spoilage until a century ago, wine making, bread baking, cheese making and salting of foods have been practiced for more than four thousand years. For all those years mankind practiced food preservation using unknown, invisible, active, living organisms.

While food preservation systems in general inhibit the growth of

TABLE 10.1

EXPLOSIVE INCREASE IN BACTERIAL POPULATIONS UNDER FAVORABLE
CONDITIONS FOR GROWTH: MILK AT ROOM TEMPERATURE

Storage (hr)	Bacterial Count (per ml)
0	137,000
24	24,674,000
48	639,885,000
72	2,407,083,000
96	5,346,667,000

microorganisms, all such organisms are not detrimental. In fact some are commonly utilized in food preservation. The production of substantial amounts of acid by certain organisms creates unfavorable conditions for others.

To review terms for a moment, respiration is that process whereby carbohydrates are converted aerobically into carbon dioxide and water with the release of large amounts of energy. Fermentation is a process of anaerobic, or partially anaerobic, oxidation of carbohydrates. Putrefaction is the anaerobic degradation of proteinaceous materials.

Sodium chloride is useful in a fermentation process of foods by limiting the growth of putrefactive organisms and by inhibiting the growth of large numbers of other organisms. Yet some bacteria tolerate and grow in substantial amounts of salt in solution.

FERMENTATION OF CARBOHYDRATES

The word fermentation has undergone evolution itself. The term was employed to describe the bubbling or boiling condition seen in the production of wine, prior to the time that yeasts were discovered. However, after Pasteur's discovery, the word became used with microbial activity, and later with enzyme activity. Currently the term is used even to describe the evolution of carbon dioxide gas during the action of living cells. Neither gas evolution nor the presence of living cells is essential to fermentative action, however, as seen in lactic acid fermentations where no gas is liberated, and in fermentations accomplished solely with enzymes.

There is a clear difference between fermentation and putrefaction. Fermentation is a decomposition action on carbohydrate materials; putrefaction relates to the general action of microorganisms on proteinaceous materials. Fermentation processes usually do not evolve putrid odors, and carbon dioxide is usually produced. In putrefaction the evolved materials may contain carbon dioxide, but the characteristic odors are hydrogen sulfide and sulfur-containing protein decomposition products. A putrid fermentation is usually a contaminated fermentation. Putrid kraut or pickles result from microbial growths decomposing protein, rather than the normal fermentation of carbohydrates to acid.

Industrially Important Organisms in Food Preservation

There are three important characteristics microorganisms should have if they are to be useful in fermentation and pickling. (1) The microorganisms must be able to grow rapidly in a suitable substrate and environment, and be easily cultivated in large quantity. (2) The organism must have the ability to maintain physiological constancy under the above conditions, and yield the essential enzymes easily and abundantly in order that the desired chemical changes can occur. (3) The environmental conditions required for maximum growth and production should be comparatively simple.

The application of microorganisms to food preservation practices must be such that a positive protection is available to control contamination.

The microorganisms used in fermentations are notable in that they produce large amounts of enzymes. Bacteria, yeasts and molds, being single cells, contain the functional capacities for growth, reproduction, digestion, assimilation and repairs in a cell, that higher forms of life have distributed to tissues. Therefore, it is to be anticipated that single celled complete living entities (such as yeasts) have a higher enzyme productivity and fermentative capacity than found with other living creatures (Table 10.2).

Enzymes are the reactive substances which control chemical reactions in fermentation. The microorganisms of each genus and species are actually a warehouse of enzymes, with its own special capacity to produce and secrete enzymes. Man has yet to learn to synthesize them.

A dry gram of an organism endowed with high activity lactose fermenting enzymes is capable of breaking down 10,000 g of lactose per hour. This great chemical activity is associated with the simple life-process requirements of the organisms, the ease with which they obtain energy for life, their great growth capacity and reproduction rate, and their great capacity for maintenance of the living entity. One generation may occur in a matter of minutes.

But there is a balance in effort. In living, the organisms consume energy. The product of their actions is a substrate of lower energy

TABLE 10.2

SOME ENZYMES OF YEASTS, SUBSTRATES ACTED UPON,
AND PRODUCTS FORMED

Enzyme	Substrate	Products Formed
Hydrolases		
Carbohydrases		
Sucrase (Saccharese, invertase, invertin)	$C_{12}H_{22}O_{11}$ Sucrose	$C_6H_{12}O_6 + C_6H_{12}O_6$ Glucose Fructose
Maltase	$C_{12}H_{22}O_{11}$ Maltose	$2C_6H_{12}O_6$ Glucose
Lactase	$C_{12}H_{22}O_{11}$ Lactose	$C_6H_{12}O_6 + C_6H_{12}O_6$ Galactose Glucose
Melibiase	$C_{12}H_{22}O_{11}$ Melibiose	$C_6H_{12}O_6 + C_6H_{12}O_6$ Galactose Glucose
Trehalase	$C_{12}H_{22}O_{11}$ Trehalose	$2C_6H_{12}O_6$ Glucose
Glycogenase	$(C_6H_{10}O_5)x$ Glycogen or $C_6H_{12}O_6$ Glucose	$C_6H_{12}O_6$ Glucose $(C_6H_{10}O_5)x + xH_2O$ Glycogen
Proteolytic enzymes		
Proteases	Yeast proteins	Proteoses, peptones, and polypeptides
Peptidases	Peptides	Amino acids
Esterases		
Phosphatases		
Polynucleotidase	Nucleic acid	Mononucleotides
Phosphatase	Hexose + H_3PO_4	Hexosephosphate
Amidases		
Asparaginase	$H_2N \cdot CO \cdot CH_2 \cdot CHNH_2 \cdot COOH$ Asparagine	$HOOC \cdot CH_2 \cdot CHNH_2 \cdot COOH + NH_3$ Aspartic acid
Desmolases		
Zymase group		
Oxydoreductase (Mutase, dehydrase)	RCHO Aldehyde	$RCH_2OH + RCOOH$ Alcohol Acid
Glycerolphosphoric dehydrogenase	Glycerolphosphoric acid	Glyceraldehyde phosphoric acid
Carboxylase	$CH_3 \cdot CO \cdot COOH$ Pyruvic acid	$CH_3 \cdot CHO + CO_2$ Acetaldehyde Carbon dioxide
Methylglyoxalase	$CH_3CO \cdot CHO$ Methylglyoxal	$CH_3CHOHCOOH$ Lactic acid
Hexokinase	Hexoses	Active hexoses

than that native material upon which they were planted. However, the product of the activity in the instance of wine is one which man generally enjoys more than the native juice from which the wine was produced.

Order of Fermentation

Microorganisms have available carbohydrates, proteins, fats, minerals and minor nutrients in native food materials. It appears that microorganisms first attack carbohydrates, then proteins, then fats. There is an order of attack even with carbohydrates; first the sugars, then alcohols, then acids. Since the first requirement for microbial activity is energy, it appears that the most available forms, in order of preference, are the CH_2, CH, CHOH, and COOH carbon linkages. Some linkages such as CN radicals are useless to microorganisms.

Types of Fermentations of Sugar

Microorganisms are used to ferment sugar by complete oxidation, partial oxidation, alcoholic fermentation, lactic acid fermentation, butyric fermentation and other minor fermentative actions.

(1) Bacteria and molds are able to break down sugar (glucose) to carbon dioxide and water. Few yeasts can accomplish this action.

(2) The most common fermentation is one in which a partial oxidation of sugar occurs. In this case, sugar may be converted to an acid. The acid finally may be oxidized to yield carbon dioxide and water, if permitted to occur. For example, some molds are used in the production of citric acid from sugar solutions.

(3) Yeasts are the most efficient converters of a dehydes to alcohols. Many species of bacteria, yeasts and molds are able to yield alcohol. The yeast, *Saccharomyces ellipsoideus*, is of great industrial importance in alcoholic fermentations. The industrial yeasts yield alcohol in recoverable quantities. While other organisms are able to produce alcohol, it occurs in such mixtures of aldehydes, acids and esters that recovery is difficult. The reaction from sugar to alcohol is many stepped.

(4) Lactic acid fermentations are of great importance in food preservation. The sugar in foodstuff may be converted to lactic acid and other end products, and in such amounts that the

environment is controlling over other organisms. Lactic acid fermentation is efficient, and the fermenting organisms rapid in growth. Natural inoculations are such that in a suitable environment the lactic acid bacteria will dominate, i.e., souring of milk.

(5) Butyric fermentations are less useful in food preservation than those noted previously. The organisms are anaerobic, and impart undesirable flavors and odors to foods. The anaerobic organisms capable of infecting man causing disease are commonly butyric fermenters. Carbon dioxide, hydrogen, acetic acid and alcohols are some of the other fermentation products.

(6) In addition to the above there is a fermentation which involves much gas production. It is useful in food preservation, although gas production has disadvantages. Energy-wise it is less efficient to produce gases (carbon dioxide and hydrogen) which have little or no preserving power in concentrations found in comparison with lactic acid. Also, the important food spolage organisms are capable of growing in such environments. In gassy fermentations sugar molecules are altered to form acids, alcohols and carbon dioxide. It is usually necessary to include some other controlling influence, such as adding sodium chloride to a substrate, with this form of fermentation.

(7) There are many fermentative actions possible in foods which are detrimental to the acceptability of treated foods. Generally the organisms capable of attacking higher carbohydrates such as cellulose, hemicelluloses, pectin, and starch will injure the texture, flavor and quality of treated foods.

Fermentation Controls

Foods are contaminated naturally with microorganisms and will spoil if untended. The type of action which will develop is dependent upon the conditions which are imposed. The most favorable to a given type of fermentation under one condition will be altered by slight changes in a controlling factor. Untended meat will naturally mold and putrefy. If brine or salt is added, entirely different organisms will take over.

The pH Value of Food is a Controlling Factor.—Most foods in native, fresh form which man consumes as food are acid. Vegetables range in pH value from 6.5 down to 4.6. Fruits range from 4.5 down to 3.0. Animal flesh when freshly killed is approximately neutral (7.2) but within two days the pH value will be approximately 6.0. Milk has a pH value near 6.4.

Inasmuch as the two important fermentations in such foods are oxidative and alcoholic, the growth of organisms will be controlled by the acidity of the medium. In fruits and fruit juices, yeasts and molds will quickly establish themselves. In meats, yeasts are less active than bacteria. In milk, an acid fermentation is established in the matter of a few hours.

Source of Energy.—Inasmuch as the immediate need of microorganisms is a source of energy, the soluble, readily available carbohydrates influence the microbial population that will dominate. In milk the sugar is lactose; those organisms which quickly mount in numbers are the lactose fermenting organisms. Because suitable energy sources are generally available to microorganisms in man's foods, energy sources are not usually a limiting factor, with certain exceptions (such as milk).

Availability of Oxygen.—The degree of anaerobiosis is a principal factor controlling fermentations. With yeasts, when large amounts of oxygen are present, yeast cell production is promoted. If alcohol production is desired, a very limited oxygen supply is required (Fig. 10.1).

Molds are aerobes, and are controlled by the absence of oxygen. Bacterial populations which will dominate a substrate may be manipulated by their oxygen requirements and its availability.

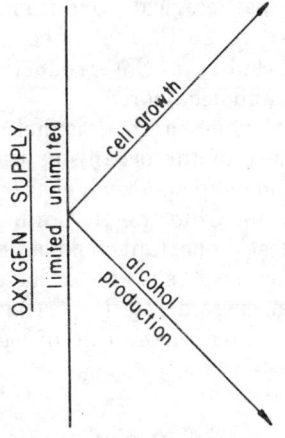

FIG. 10.1.   YEAST ACTIVITY CONTROLLED BY OXYGEN SUPPLY

The end product of a fermentation can be controlled in part by the oxygen tension of the substrate, other factors being optimum.
Temperature Requirements.—Each group of microorganisms has an optimum temperature for growth; the temperature of a substrate therefore exerts a positive control on their growth. To obtain the maximum performance during fermentation, the optimum temperature for the organisms must be created. Examples are available in milk fermentation and also in acetic acid production.
Milk held at 0°C has little microbial activity, and retarded expansions in bacterial numbers. At 4°C there is slight growth of organisms and more rapid development of off-flavors.
At 21°C *Streptococcus lactis* growth is usually dominant. At 37°C *Lactobacillus bulgaricus* populations commonly domina e in milk. At 65°C most organisms will be killed, but *Lactobacilus thermophilus* can grow. At 71°C souring of milk is generally demonstrated to be due to the presence of *Bacillus calidolactis*. The temperature of milk is a controlling factor in the growth of the above organisms, other factors being equal.
The acetic acid bacteria are temperature sensitive, having definite and peculiar temperature relations important in vinegar making. At 7°C the organism grows slowly; cells are short and usually broad. At 21°C the cells appear normal, growing and developing into chains of cells of varying number. Cell walls become swollen. At 37°C long thread-like transparent filaments with no visible cross walls and with irregular bulging, sometimes branching, have been observed. This appears to be a temperature induced, physiologically malfunctioning, pathological condition. Lowering the temperature of the substrate to 26°C, that used in the United States in vinegar production, results in the production of some cells with normal characteristics and behavior.
The temperature at which a food is held will determine within certain limits the nature of the organisms capable of either yielding the desired fermentation or spoilage, whichever the case may be.
The Action of Sodium Chloride in Controlling Fermentations.—Salt is one of the most important food adjuncts in food preservation. In drying it has been shown to have beneficial effects. In fermentations salt can exert a role in sorting the organisms permitted to grow. We will return to the role of salt and its multiple uses in the next chapter.

WINE

Wine and beer or similar fermented products originated in

antiquity. Alcoholic beverages were discovered by man in many areas on earth, with the exception of the early American Indian. There was a fermented cactus juice known to certain tribes in the Southwest, but evidently alcohol was not a commodity in America as it was in the ancient Egyptian, Greek, Roman and Oriental civilizations. The ability to produce pleasant, palatable effervescent beverages by fermentation of natural fruit juices is a demonstration of man's inherent ingenuity. From earliest recorded history wine and beer have been important items of trade.

In the United States the term wine means the product of an alcoholic fermentation of the juice of sound, ripe grapes. Any other fruit juice wine must carry the notation, for example, that it is cherry or blackberry wine, indicating the source of the fruit juice.

The formation of alcohol from sugar is accomplished by yeast enzymes which are contributed by the growing yeasts. *Saccharomyces ellipsoideus* is the true wine yeast. There is normally a population of yeasts on grape skins. When a viable yeast cell has access to sugar from the juice of a crushed grape, fermentation begins. Usually the true wine yeast is present in sufficient quantities to dominate the fermentation. However, the juice may be heated to kill contaminating organisms, and reinoculated with pure culture wine yeasts. In certain special instances (i.e., effervescent sparkling wine) it is essential that a secondary fermentation occurs in the bottle to develop a carbon dioxide content.

In a suitable environment and substrate, the amount of alcohol produced depends upon the amount of sugar present and the efficiency of the yeast in converting the sugar to alcohol. Grapes for wine making vary in sugar content from 12 to 25%. The ability of different yeast strains to survive in increasing alcohol concentrations varies. Eventually the yeast is unable to tolerate the fermentation products, usually when 12 to 16% alcohol has been developed. Some yeasts can tolerate 18% alcohol. Fortified wines are natural fermentation products to which alcohol has been added. For each 100 parts of sugar present in the juice, approximately 51% alcohol and 49% carbon dioxide are produced.

Dry wines are wines containing little unfermented sugar (Table 10.3).

Aging of wines improves the flavor and bouquet due to oxidation and formation of esters. These esters of higher acids formed during aging give the ultimate pleasing bouquet to well aged wine. Aged wine may be polished by filtration to give a clear, bright appearance prior to bottling.

Sulfur dioxide may be added to crushed grapes to retard the

TABLE 10.3

AVERAGE CHEMICAL ANALYSIS OF PRIZE-WINNING HIGH QUALITY WINES

Component	Dry White	Dry Red	(g per 100 ml) Sweet White	Sweet Red	Spark- ling
Alcohol by volume, (%)	2.45	12.61	18.38	19.30	13.22
Alcohol	9.88	10.00	14.58	15.31	10.48
Glycerol	0.7019	0.6355	0.3025	0.5089	0.4177
Ash	0.196	0.247	0.203	0.311	0.153
Total acids	0.586	0.649	0.412	0.502	0.658
Volatile acids	0.101	0.128	0.092	0.122	0.082
Reducing sugars	0.134	0.146	11.30	10.26	3.409
Protein	0.162	0.150	0.162	0.232	0.214
Tannins	0.039	0.236	0.036	0.096	0.035
Specific gravity	0.9917	0.9943	1.0298	1.0276	1.0045

Source: Amerine et al. (1972).

growth of contaminating organisms. Mustard seeds have been used for centuries in Europe and the Near East as additives to crushed grapes for wine making. Recently it has been found that mustard seed contains an antibacterial agent, in the essential oil of mustard.

White wines do not have extracted red pigments from grape skins and are lower in tannins and extracts.

Preservation

Pasteurization is applied in one of three ways: (1) by flash pasteurizing and returning to the storage tank; (2) flash pasteurizing into the final bottle; and (3) pasteurization by heating the filled and sealed bottle. There is also some bulk pasteurization and hot holding at a temperature of 49°C for several days to stabilize wines, usually dessert types.

Bulk pasteurization is used for three purposes: first, to stabilize the wine chemically and physically by coagulating certain heat-coagulable colloids; second, to stabilize it microbiologically by destroying bacteria and yeasts; and third, to hasten aging, particularly of ordinary dessert wines.

Pasteurizers should be constructed of stainless steel and be so designed that they can be easily disassembled and cleaned. Two designs are in use. The first consists of a tubular heat exchanger made up of several horizontal metal tubes encased in a metal jacket, in which incoming wine is heated nearly to pasteurizing temperature by the outgoing hot pasteurized wine, and of a similar set of tubes heated by steam. All parts of the pasteurizer coming in contact with the wine must be of metal not attacked by the wine. Wines containing appreciable sulfur dioxide are particularly corrosive.

The second pasteurizer is known as the A.P.V. design. The heating section consists of a series of hollow plates clamped together in a metal frame. Steam or hot water on one side of each plate acts as a heating medium. The liquid to be heated is on the other side. A similar set of plates forms the heat interchanger. Heating and cooling are reported to be rapid and efficient, and the plates are easily removed for cleaning. For further information refer to Amerine et al. (1972).

In flash pasteurizing in bulk, wine is heated for about 1 min to 85°C and is then cooled continuously against the incoming wine and water or refrigerant cooled pipes.

The time-temperature relationship for pasteurization of wines is: vegetative yeast cells are killed at 40°C while yeast spores are only killed at 57°C.

To stabilize ordinary white dry table and dessert wines, a common method consists of adding about 100 ppm of sulfur dioxide, heating to about 60°C, adding bentonite, holding at 52°C for 1 to 3 days, racking, cooling and filtering. It is often very effective for removing heat-sensitive proteins.

Sterilization Filtration

Wine and fruit juices can be rendered sterile, germ- or yeast-free, by filtration through sterile, very tight filter pads. The Seitz process of sterile filtration includes three stages: sterilizing, filling and corking (Fig. 10.2 and 10.3). Sterilizing of bottles is accomplished by use of sulfur dioxide gas, by removal of the gas with sterile water, filling under pressure, and corking with sterile corks. When sulfurous acid solutions are employed to sterilize the bottles, the bottles must be well drained in order to prevent excessive pickup of sulfur dioxide (Amerine et al. 1972). The filter plates, pipes, hose lines, etc., beyond the filter to the bottling machine, and the latter, and the bottles and corks, must be sterilized and kept sterile during filtration and bottling.

In the best installations the entire filtration, bottling and corking is done in a small room which can be sterilized (Fig. 10.2 and 10.3). The secret of success is to operate at a low, even filtration pressure. The wine should be as brilliant as possible before sterile filtration.

Most experts report sterile filtration of fine Bordeaux wines "seems to fatigue the wines."

Membrane (Millipore) filters are the latest innovation in sterile filtration. The filters are porous membranes composed of inert cellulose esters. Because of the uniformity of pore size, absolute

Courtesy of Seitz-Werke, Kreuznacn

FIG. 10.2.   FULLY AUTOMATIC STERILE BOTTLING ROOM

From left to right: sterilizing, filling and corking. The sterilizing is done with sulfur dioxide gas.

Courtesy of Seitz-Werke, Kreuznach

FIG. 10.3.   STERILE BOTTLING ROOM FOR HAND OPERATION

Note rotary table in wall for introduction of bottles. The machine in the background is a semi-automatic rotary filter and on the right is a semi-automatic corking machine, capacity 1400 bottles per hour.

sterility can be achieved by choosing a filter of the correct pore size. Since pore volume occupies approximately 80% of the total filter volume, extremely large flow rates are obtained provided the wine is prefiltered (Fig. 10.4).

Wine making remains an art and a science, dependent on nature to yield environmental conditions for grape culturing. Certain localized areas have the proper combination of soil type, sunlight, temperature and rainfall required for grape growing. These regions have become known throughout the world. Examples are found in the Rhine and Rhone valleys of Europe, and in areas of New York and California. Expert wine tasters are able to locate generally the vineyard area from which the grapes for a wine have come.

BEER

Beer and ale are fermentation products with characteristic flavor and aroma from malt and hops. The alcoholic content ranges from 3 to 7%. (Table 10.4). The strain of yeast used, the composition of the wort, and the temperature of fermentation are controlling factors. Hops has at least two antibacterial components in addition to having a role in flavoring. Boiled wort must be innoculated with yeast at the start of the fermentation. Packaged beer is usually pasteurized, while barrelled beer is not. Pasteurization temperatures of 64°C are commonly used.

Cold Pasteurization

For many years attempts were made to capture the unheated flavor of casked draft beer in bottles and cans for home use. Recently

TABLE 10.4

APPROXIMATE COMPOSITION OF ALCOHOLIC BEVERAGES [1]

Type	Sugar (%)	Gas (vol)	pH	Acidulant	Alcohol (%)
Malt beverages					
Beer (lager)	1.48	2.4-3.0	4.2-4.5	Lactic	4.3-4.6
Ale	1.56	2.4-3.0	4.1-4.3	Lactic	5.1-6.0
Sparkling wines					
Champagne	1.0-4.0	4.5-7.0	3.0-3.4	Tartaric	11.5-13.0
Sparkling wine	1.0-5.0	4.5-5.6	2.5-3.8	Tartaric	10.5-14.0

Source: Johnson and Peterson (1974).
[1] The Wine Institute (1966) classifies wines as: Group 1—less than 1.5% reducing sugar; Group 2—1.5-2.5%; and Group 3—over 2.5%, respectively. Sugar data for both malt beverages and sparkling wines indicate reducing sugar.

success was achieved by using a cold pasteurization technique. This employs newly developed microporous membrane filters which physically remove the majority of bacteria and larger yeasts. Using filtration of this kind rather than heat, microbial numbers are effectively decreased and the flavor of draft beer is largely preserved. This is the method that has given rise to draft beer in cans. Figure 10.4 illustrates a filtration assembly housing two sets of microporous membrane discs connected in series. From here the beer is aseptically filled into the commercially sterile cans or bottles to the right. Cold pasteurization with microporous filters also is finding application in the processing of other heat sensitive liquids such as pulp-free fruit juices and wines.

Courtesy of Millipore Corporation

FIG. 10.4.   FILTRATION ASSEMBLY HOUSING MICROPOROUS MEMBRANE DISCS

VINEGAR FERMENTATION

Vinegar is a condiment prepared from various sugary or starchy materials by alcoholic and subsequent acetic fermentation. It consists principally of a dilute solution of acetic acid in water, but also contains flavoring, coloring and extracted substances, fruit acids, esters, and inorganic salts, varying according to its origin. Although acetic acid is the active ingredient of all vinegars, these additional substances give distinctive and pleasing quality to the product. Vinegar may be produced from the juices of most fruits such as apples, grapes, cherries and pears, but cider vinegar has always been the most popular in the American home. The principles involved and method of preparing cider vinegar will be discussed herein.

The Food and Drug Administration of the United States defines cider vinegar or apple vinegar as the product of alcoholic and subsequent acetic acid fermentation of the juice of apples. According to regulations, the word vinegar, unqualified, can be applied only to vinegar made from apples.

Vinegar contains at least 4 g of acetic acid per 100 ml (1.0 fl oz per 24.1 fl oz). Dealers generally use the term grain strength rather than percent acetic acid. Percent acid times ten gives the strength of the vinegar in grains. Thus a vinegar containing 4 g of acetic acid per 100 cu cm (approximately 4% acid) is referred to as 40 grain. Vinegar offered for sale nust contain at least this much acid, must be wholesome, made from sound edible fruit, and must be properly labeled.

Cider vinegar which during the course of manufacture has developed in excess of 4% of acetic acid, may be reduced to a strength of not less than 4%. Cider vinegar so reduced is not regarded as adulterated, but must be labeled to this effect as diluted cider vinegar.

Principles of Vinegar Fermentation

The manufacture of vinegar requires two fermentation processes. The first transforms the sugar into alcohol, by yeast. The second changes the alcohol into acetic acid and is brought about by vinegar bacteria. One of the chief causes of failure in preparing vinegar, and a factor not often considered, is that vinegar-making involves two very distinct and different fermentations. The first must be completed before the second begins.

Alcohol Fermentation.—The formation of alcohol from sugar is accomplished by alcohol-producing yeasts of which *Saccharomy-*

ces ellipsoideus is the best example. The change which occurs is usually described in the following equation:

$$C_6H_{12}O_6 + Saccharomyces\ ellipsoideus = 2\ C_2H_5OH + 2\ CO_2$$

Simple Sugar + Yeast=Alcohol + Carbon Dioxide

Besides sugar, of which the solids of cider largely consist, the substance represented by the acidity and ash are quite necessary for the yeast to carry on their fermentation as well as the subsequent acetic fermentation. The acidity of cider is mainly malic acid, which serves to protect the cider from the development of undesirable bacteria. The minerals in the ash are as essential to the microbial growth as they are to human development.

Yeasts normally are present in cider. A favorable temperature of 23° to 26°C should be maintained during fermentation. At a temperature of near 37°C the fermentation becomes abnormal, and ceases at about 40°C. The alcoholic fermentation will occur naturally with the growth and activity of the yeasts present in the cider. Such a fermentation, however, is unreliable, usually wasteful of sugar, and the resulting vinegar may be of varying and uncertain quality. It is best to add a "starter" yeast. The flavor of the resulting vinegar can be improved by using a culture of pure wine yeast which has been especially cultivated for vinegar production.

With the exception of aeration at the start, air is not necessary during alcoholic fermentation. It is objectionable in the later stages because it may result in growth of undesirable bacteria with a subsequent loss in alcohol. The alcoholic fermentation should be conducted in containers in which the juice is not unduly exposed to air. A barrel lying horizontally with the bunghole plugged with cotton, or an air trap, is satisfactory. For small quantities a large bottle, the mouth of which is plugged with cotton, may be used. The container must not be sealed airtight as it may burst, due to the pressure from the gas produced. Room for frothing also must be allowed.

The juice is allowed to ferment until practically all the sugar is converted into alcohol and carbon dioxide (Table 10.5).

After the alcoholic fermentation, the juice should be freed from yeast, uplp and sediment by settling and racking (drawing off carefully) or by filtering before beginning the acetic fermentation. If left in the hard cider, the sediment may give a bad flavor and interfere with the acetic fermentation.

Acetic Fermentation.—The formation of acetic acid results from oxidation of alcohol by vinegar bacteria in the presence of oxygen from the air. These bacteria, unlike the alcohol-producing

Product	Alcohol Yield (U.S. gal.)
Wheat	85
Corn	84
Raisins	81
Rice	79
Barley	79
Rye	79
Prunes	72
Molasses	70
Sweet potatoes	34
Potatoes	23
Sugar cane	15
Grapes	15
Apples	14
Pears	11
Peaches	11
Carrots	10

[1] For metric conversion see Appendix.

yeasts, require a generous supply of oxygen for their growth and activity. The change which occurs is generally described by the following equation:

$$C_2H_5OH + O_2 + Acetobacter\ aceti. = CH_3COOH + H_2O$$

Alcohol + Oxygen + Vinegar Bacteria = Acetic Acid + Water

The number of acetic bacteria usually present in fermented juice is small and they are often of an undesirable or inactive type. Therefore, a suitable starter should be added to supply the proper kind of bacteria and to make conditions favorable to their growth and activity. The best means of preventing the growth of undesirable organisms is to add strong, unpasteurized vinegar to the fermented juice *after the alcoholic fermentation is complete*. The addition of such vinegar innoculates the "hard" cider heavily with vinegar bacteria.

Vinegar bacteria grow in the liquid and on the surface exposed to the air. They may form a smooth, grayish, glistening, gelatinous film. The film does not always form, for some strains of the organisms grow only in the liquid and not on the surface. If the film is not disturbed, the liquid remains rather clear until converted into vinegar, The rate of conversion of alcohol to acetic acid depends on the activity of the organism, the amount of alcohol present, the temperature, and the amount of surface exposed per unit of volume. At a favorable temperature of 26°C the limiting factor may be

FIG. 10.5.   COMMERCIAL VINEGAR GENERATOR

Alcoholic substrate percolates down through packed generator. Packings vary, but develop extremely large surface areas.

the surface area exposed. The time required for the slow process, barrel type, is about three months or more. In large-scale generator processes (Fig. 10.5), where the surface exposed to air is large, the time for fermentation may be shortened to a matter of hours.

After the acetic fermentation has been completed, the vinegar should not be exposed to the air because the vinegar may be further oxidized to carbon dioxide and water, reducing rather rapidly to a worthless condition. To overcome this the vinegar should be placed in completely filled, tightly sealed containers. The vinegar may also be pasteurized in a manner to be described later.

Acetic acid fermentation occurs most rapidly when the "hard" cider contains from 6 to 8% alcohol, but 12% alcohol may be tolerated. The action is slow when 1 to 2% alcohol is present. During active fermentation, heat is produced and this may be sufficient to raise the temperature of the generator. Fermentation activity will continue at temperatures between 20° and 35°C.

Yields of Alcohol and Acetic Acid.—Theoretically for every 100 parts of sugar present in the cider, 51 parts of alcohol and 49 parts of carbon dioxide are produced. In practice, between 45 to 47 parts of alcohol are obtained, since some of the sugar is utilized by the yeast or is lost in the production of other substances. Starting with a cider containing 10% sugar, approximately 4.6% alcohol should be obtained by complete yeast fermentation.

In acetic acid fermentation, 100 parts of alcohol should yield 130 parts of acetic acid. Actually, because of losses due to evaporation and other causes, less than 120 parts are obtained.

Hence, if one starts with 100 parts of sugar, it is possible to obtain 50 to 55 parts of acetic acid, under very favorable conditions. Therefore, to produce a vinegar of the legal minimum acid content of 4 g per 100 ml (4% or 40 grain strength), it is necessary to use a juice that contains at least 8% sugar.

A bushel of apples should yield about 3 gal. of such vinegar.

VINEGAR MAKING

The juice of ripe apples has been found to vary in the sugar content between 7 to 15% with an average for a large number of varieties in several states being near 11% Generally, the juice from summer apples has the lowest sugar content, winter apple juice the highest, and fall apple juice somewhere between the two. Mature apples in the ripe stage contain the largest amount of sugar, green apples a much smaller amount, and over-ripe fruit less than ripe fruit. The juice should be expressed and collected.

Preparation of Yeast Starter

The yeast is inoculated into 3.8 liters of cider which has previously been brought to boiling and cooled overnight before the yeast is added. Aerate by pouring the juice back and forth, and set aside in a warm place. Within 24 hr the juice will be fermenting vigorously and can be used to inoculate a 95- or 190-liter barrel of fresh juice. If desired, the contents of this barrel can be used later to inoculate the contents of other containers of juice.

Alcoholic Fermentation

The freshly cleaned and steam-sterilized barrel or cask should be placed on its side, bunghole up and open. The fresh cider should be added until the barrel is nearly full. To the juice approximately 10% by volume of active starter, prepared as above, should be added as the barrels are filled. Thus, to start a 190-liter batch, add about 19 liters of starter. When this barrel is actively fermenting, the juice from it may be used to inoculate other barrels of juice. These, in turn, may be used to inoculate still other barrels as the crushing season progresses.

The fermentation room should be at or above 21°C but not above 32°C.

If a yeast starter is employed in the manner previously described, the change from sugar to alcohol should be completed in two weeks or less at a temperature near 23°C. When fermentation is complete, gas is no longer evolved and the juice tastes "dry" or free of sugar.

The fermented juice should be allowed to settle after the alcoholic fermentation is complete. This removes from suspension in the liquid, the pulp, yeast and other sediment. During this process, the barrels should be kept full to prevent the growth of wine flowers (film yeast) on the exposed surface. This may be accomplished by combining the liquid from one barrel with another. All storage tanks should be closed.

When the sediment has settled, the fermented juice should be carefully siphoned from the barrel and added to the vinegar generator. After siphoning, the remaining liquid may be filtered through several thicknesses of cloth and combined with the other clear hard cider.

Acetic Fermentation

A simple, practical vinegar generator consists of a barrel of sufficient size to produce the desired amount of vinegar. The barrel should be placed on its side, bunghole up. Several fittings are desirable. A funnel of glass, aluminum, or new enamel with an extension which goes to the bottom of the barrel facilitates introducing hard cider to the generator without unduly distributing the surface growth. Holes 3.8 to 5.0 cm in diameter should also be bored to allow free circulation of air during the acetic fermentation. These openings should be covered with cheesecloth to prevent entrance of fruit flies and insects. A glass tube level indicator may be inserted at one end of the barrel at the bottom, or an outlet faucet may be inserted in its place. The level indicator would serve not only in this capacity but could also be used to remove finished vinegar from the barrel.

The barrel, properly prepared, is now filled to an inch from the lower hole with alcoholic cider.

The transformation of the alcohol after fermentation of the juice is brought about by vinegar bacteria. Unless the juice after alcoholic fermentation is acidified and inoculated with these bacteria, there will hardly be enough present to start quickly or work efficiently. Both the above objectives may be accomplished by the addition of new vinegar to the fermented juice after it has been racked. Approximately 1 liter of strong, new, nonpasteurized vinegar should be added to 5 liters of the fermented juice. The barrel generator should be stored at a temperature between 23° and 29°C until the vinegar is produced. At temperatures lower than this, vinegar formation is slow. After sufficient acid has been formed, part of the vinegar is withdrawn and replaced with more fermented juice. Approximately one-half the vinegar present in the generator should be withdrawn. The record of a typical generator is given in Table 10.6.

From this record, it has been found that very little change in acid occurs in 1 week, but that in 3 weeks, the vinegar reaches 4.94% acid. At that time, 11.4 liters were removed and the barrel was refilled as above. Two weeks later, an acidity of 5.35% was attained and 18.9 liters were removed from the generator. By repeating this process, they produced 68 liters of vinegar in 7 weeks in a 57-liter barrel. This demonstrates a far more rapid rate of vinegar generation in a small barrel-type generator than ordinarily found in barrels of cider stock where circulation of air is not adequate. They found that numerous samples of vinegar which had been fermented in ordinary barrels were below standard in acidity even after a 2- or 3-year fermentation.

TABLE 10.6
RATE OF GENERATION OF VINEGAR IN BARREL

Time (days)	Cider Added		Acidity of Vinegar (%)	Amt of Vinegar Removed (liters)
	Amt (liters)	Acidity (%)		
0	30.2	1.04	1.04	—
7			1.23	
14			2.56	
21			4.94	11.4
21	15.1	1.00	3.30	
28			4.30	
35			5.35	18.9
35	19.5	1.20	3.00	
42			4.10	
49			5.20	18.9
49	18.9	1.30	3.30	

The strength of acid in vinegar determines its value for pickling purposes. The method for measuring the acidity of vinegar is by titration with a standard alkali solution. After all the alcohol present has been converted to acetic acid, the vinegar bacteria, as well as other bacteria, then attack the acid itself and cause a rapid decrease in the total acidity, in the presence of air. This situation may be prevented by filling barrels completely after the acetification, by combining vinegar from previous generations, and sealing them tightly to exclude the air. The flavor and quality of vinegar are improved by aging, but it may be used immediately after generation. Aging should take place in barrels that are full and tightly sealed.

After the vinegar has been produced, it should be bottled. If the vinegar is cloudy, the cloud may be removed by filtration. It is desirable to heat the vinegar to pasteurize it and destroy bacteria which may be present. Pasteurization may be accomplished by heating the vinegar to 65°C, filling into hot bottles, and sealing immediately. Such vinegar is stable on storage.

CHEESE

As noted previously, commercial cheese production is a modification of ancient practice, with careful control over fermentation. By controlling the microorganisms, temperature and substrate, hundreds of differently flavored cheeses can be prepared. Some have become world famous. The final yield of cheese amounts to approximately 10% of the weight of the milk. The compositions of several common cheeses are listed in Table 10.7. No new important cheese

TABLE 10.7
APPROXIMATE PERCENTAGE COMPOSITION OF SOME VARIETIES OF CHEESE

Variety	Moisture	Fat	Protein	Ash (Salt-free)	Salt	Calcium	Phosphorus
Brick	41.3	31.0	22.1	1.2	1.8	—	—
Brie	51.3	26.1	19.6	1.5	1.5	—	—
Camembert	50.3	26.0	19.8	1.2	2.5	0.68	0.50
Cheddar	37.5	32.8	24.2	1.9	1.5	0.86	0.6
Cottage							
uncreamed	79.5	0.3	15.0	0.8	1.0	0.10	0.15
creamed	79.2	4.3	13.2	0.8	1.0	0.12	0.15
Cream	54.0	35.0	7.6	0.5	1.0	0.3	0.2
Edam	39.5	23.8	30.6	2.3	2.8	0.85	0.55
Gorgonzola	35.8	32.0	26.0	2.6	2.4	—	—
Limburger	45.5	28.0	22.0	2.0	2.1	0.5	0.4
Neufchatel	55.0	25.0	16.0	1.3	1.0	—	—
Parmesan	31.0	27.5	37.5	3.0	1.8	1.2	1.0
Roquefort	39.5	33.0	22.0	2.3	4.2	0.65	0.45
Swiss	39.0	28.0	27.0	2.0	1.2	0.9	0.75

has ever been developed in the United States. North Central Europe has developed most of the commercially valuable cheeses.

Kosikowski (1976) gives an excellent presentation on cheeses and their manufacture. The following is abstracted from *Cheese and Fermented Milk Products*.

Courtesy of FAO

FIG. 10.6.   NEPALESE CHEESEMAKER FOLLOWS HERDS OF
YAK IN HIMALAYAS

Kinds of Cheese

Cheese is the product made from the curd of the milk of cows and other animals. The curd is obtained by the coagulation of the milk casein with an enzyme (usually rennin), an acid (usually lactic acid), and with or without further treatment of the curd by heat, pressure, salt and ripening (usually with selected microorganisms). All cheese types begin with curdmaking and then involve various treatments of the curd or whey.

There are over 800 kinds of cheeses. However, there are only

TABLE 10.8

ANALYSIS OF COMMERCIAL CHEESES AND PROCESSED CHEESES

Cheese	Manufacturer	a_w		pH	%H_2O	% Fat	g H_2O/g Dry Matter (fat-free basis)
		Fett-Vos	VPM				
Cracker Barrel (Cheddar)	Kraft	0.97	0.98	5.3	36.3	32.7	1.156
Canadian Cheddar	Purity	0.95	0.95	5.2	34.5	32.4	1.042
Longhorn Colby	Byerly's	0.99	0.98	5.3	38.1	34.0	1.365
Bongard's Colby	Bongard's	0.99	1.00	5.0	44.4	23.4	1.380
Monterey Jack	Purity	0.99	1.00	5.8	43.5	29.0	1.685
Mozzarella	Kraft	1.00	1.00	5.8	45.2	20.4	1.314
Parmesan (grated)	Kraft	0.76	0.75	5.4	17.3	27.1	0.311
Romano	Kraft	0.97	1.00	5.3	38.7	22.3	0.992
Provolone	Kraft	0.98	1.00	5.4	43.4	27.5	1.468
Swiss	Kraft	1.00	0.97	5.7	39.0	27.3	1.211
Gouda	Purity	0.99	1.00	4.7	—	27.3	1.499
Edam	Purity	1.00	¹	5.3	—	26.1	1.175
Muenster	Purity	1.00	0.99	5.4	—	27.1	1.387
Farmer Cheese	Purity	0.99	0.96	5.8	—	30.3	1.858
May Bella	Purity	0.98	¹	5.0	47.9	26.5	1.638
Jackie	Denmark	0.99	¹	5.4	38.3	38.3	
Whey	Sweden	0.91	0.88	5.8	—	8.8	0.359
American (processed cheese)	Kraft	0.98	0.97	5.8	38.9	36.1	1.298
Hoffman's processed cheese	Anderson Clayton	0.97	0.97	5.8	40.0	31.1	1.386
Pot (processed cottage cheese)	Milwaukee	1.00	1.00	5.1	72.9	0.4	3.151
Fondue Swiss Knight	Switzerland	1.00	¹	5.5	—	15.4	2.521
Kaukauna Klub (cold-pack Cheddar)	International Multifoods	0.96	1.00	4.8	—	32.0	1.073
Ye Old Tavern (Cheddar)	Cheese food	0.96	0.95	5.1	—	22.9	1.570
Swiss-American spread	Kraft	0.96	0.93	5.6	—	21.3	1.499
Velveeta (spread)	Kraft	0.96	0.99	5.8	—	2.10	2.029
Brie	Fromagerie Bongrain Inc.	1.00	1.00	7.4	—	28.9	2.436
Camembert	France	1.00	¹	6.1	—	21.6	2.611
Camembert	Denmark	0.99	¹	7.0	—	23.0	2.206
Blue	Treasure Cave	0.94	¹	5.1	44.2	27.2	1.551
Blue Stilton	England	0.94	¹	5.8	39.5	30.0	1.294

Source: Acott and Labuza (1975).
¹ Not adaptable to a_w measurment by VPM.

about 18 distinct types of natural cheeses. These include brick, Camembert, Cheddar, cottage, cream, Edam, Gouda, hand, Limburger, Neufchatel, Parmesan, Provolone, Romano, Roquefort, sapsago, Swiss, Trappist, and whey cheeses. All of the major types of cheese generally fit into such classification. The compositions of several of these cheeses are given in Tables 10.7 and 10.8.

Cottage Cheese

Cottage cheese is a soft cheese, generally coagulated with lactic acid rather than rennin. The curd is not pressed, aged or ripened.

Starting with pasteurized skim milk rather than whole milk, the curd-forming operations have many similarities to the early stages of all cheese making (Fig. 10.7). The steps are:

(1) Pasteurized skim milk is warmed in a vat to 20°C.
(2) A lactic starter (1% level) is added to produce acid. The *S. lactis* starter usually contains *Leuconostoc citrovorum*, a flavor-producing bacterium.
(3) The vat is fermented for 14 hr.
(4) The coagulated milk is cut into 0.6-cm cubes (Fig. 10.7, upper left).
(5) The curd cubes are cooked for about 90 min with stirring, gradually increasing the temperature to 49°C.
(6) After cooking, the whey is drained and the curd is washed with cold water (Fig. 10.7, upper right).
(7) The curd is trenched to drain (Fig. 10.7, lower left and right).
(8) The curd may now be mildly salted.
(9) The curd may be blended with sweet or sour cream to give 4% fat, and called creamed cottage cheese.

Cottage cheese is packaged as loose curd particles and undergoes no further processing. It is highly perishable and must be kept refrigerated.

Variations in cottage cheese making have to do with the length of fermentation time in the vat. The 14 hr holding time at 20°C is known as the long set method. By using a larger amount of lactic starter (about 6%) and a higher temperature, the (32°C) time needed for coagulation and curd cutting can be reduced to 5 hr. This is the short set method. Another variation employs rennet plus starter.

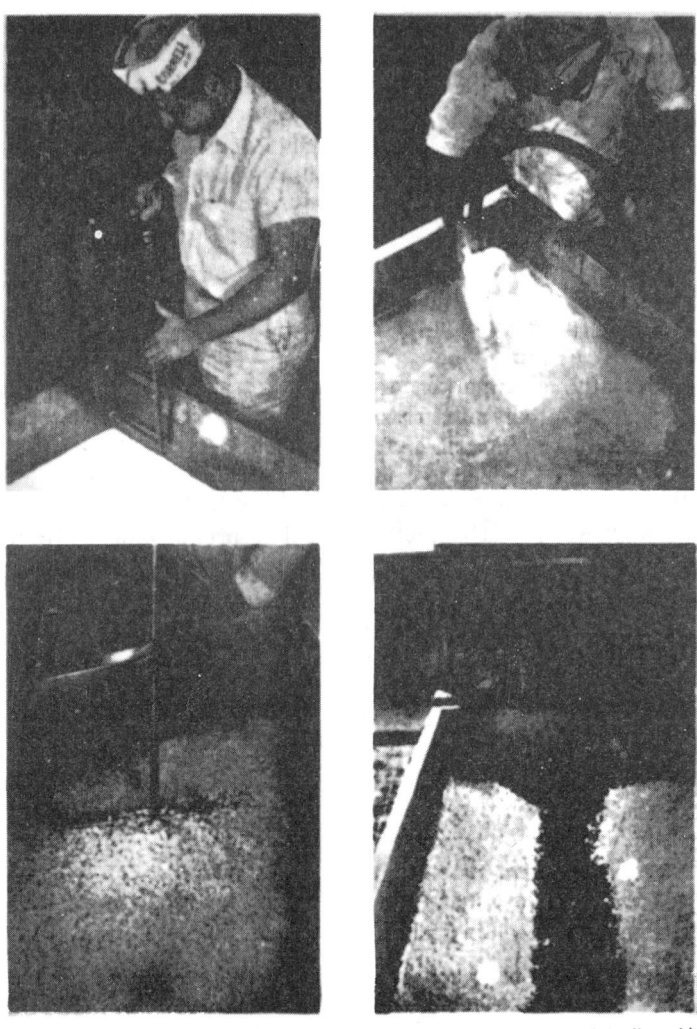

FIG. 10.7. BASIC STEPS IN COTTAGE CHEESE MAKING

Swiss Cheese

This is a hard-type cheese, characterized by large holes or eyes and a sweet nutty flavor, obtained through the activities of *Propionibacterium shermanii*. This organism follows the growth of lactic acid organisms and ferments lactic acid to propionic acid and carbon dioxide. The propionic acid contributes to the nutty flavor

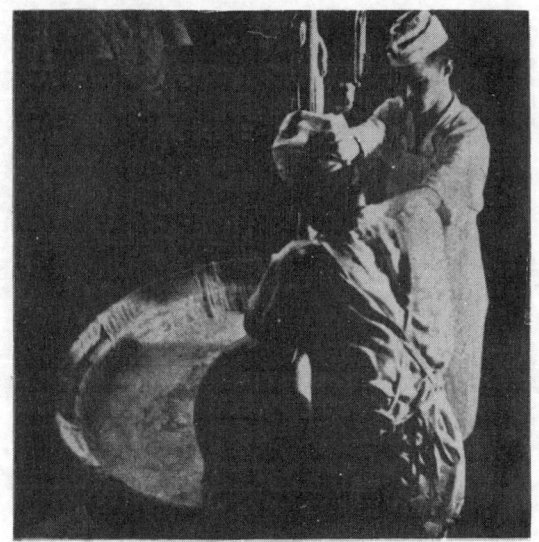

Courtesy of Valio Finnish Coop.
Dairies Assoc.

FIG. 10.8.   DRAINING SWISS CHEESE CURD PRIOR TO
PLACING IT IN THE HOOP

and the carbon dioxide gas collects in pockets within the ripening curd and forms the eyes.

Swiss cheese is generally made from raw milk. To the milk in large kettles is added a multiple organism starter. Present are lactic acid organisms including *Lactobacillus bulgaricus,* and a heat tolerant streptococcus, *Streptococcus thermophilus,* which produces lactic acid through the rather high cooking temperatures of about 54°C which the curd undergoes during processing. The starter may contain the eye-forming *Propionibacterium* or this may come in with the raw milk. Following an initial period of lactic acid fermentation, rennet is added to the kettle to coagulate the milk. The curd is cut with a harp-like wire knife into rice-size particles. The curds and whey are heated and cooked at about 54°C for about 1 hr, stirred and allowed to settle. A cloth is slid under the curd and the mass is drained (Fig. 10.8).

The curd is placed in a single large hoop and pressed for one day to form a beginning rind. The cheese wheel is removed from the hoop and placed in a brine tank at about 10°C (Fig. 10.9). It floats in this brine for about three days and its top is periodically salted (Fig. 10.9). The salt removes water from the cheese surfaces and produces the heavy protective rind.

Courtesy of Swiss Cheese Union

FIG. 10.9.   SALTING WHEELS OF SWISS CHEESE

The cheese is removed from the brine to a ripening room at about 20°C and 85% RH. The cheese remains there for about 5 weeks during which time the eyes are formed by the growth of the *Propionibacterium* at the relatively warm temperature. As the eyes are formed the cheese becomes somewhat round (Fig. 10.10).

After about 5 weeks, the cheese is moved to a colder curing room at about 7°C. It remains there 4 to 12 months to develop the characteristic full sweet nutty flavor (Fig. 10.10 and 10.11).

Blue Cheeses

These cheeses are typified by a semisoft texture and blue mold growing throughout the curd. There are four well-known varieties. Three are made from cow's milk: Blue cheese, Stilton, and Gorgonzola. The fourth is Roquefort, which is made from sheep's milk.

All these cheeses acquire the characteristic blue marbling by having their curd inoculated with the blue-green mold *Penicillium roqueforti* prior to being hooped and pressed. Mold growth is encouraged during the ripening period which can be from 3 to 10 months at a cool, moist 8°C and 90% RH.

Courtesy of Swiss Cheese Union

FIG. 10.10.   SWISS CHEESE IN CURING ROOM

Courtesy of Swiss Cheese Union

FIG. 10.11.   TYPICAL EYE FORMATION IN QUALITY SWISS CHEESE

Courtesy of Danish Dairy Assoc.

FIG. 10.12.   BLUE CHEESE SHOWING HEAVY MOLD GROWTH
ALONG AIR CHANNELS

Molds are aerobic. To encourage their growth throughout the
cheese mass it is common practice to pierce the pressed cheese
when it is placed in the curing room, to allow air to penetrate the
cheese and support growth throughout the mass. The color is from
mold spores. The darkened lines in Fig. 10.12 indicate where mold
growth is heaviest, along the pierced air channels. *Penicillium
roqueforti* produces the mottled blue color and also gives rise to
fatty acids and ketones which contribute to the characteristic
sharp, peppery flavor of these cheeses.

Camembert

Another mold-ripened cheese is Camembert. It is characterized
by a soft cream-colored curd and white mold growth, which covers
its entire surface (Fig. 10.13). The mold is *Penicillium camemberti.*
It is inoculated onto pressed curd by spraying a mist of mold spores
onto the cheese surfaces. Ripening is under damp conditions of
about 8°C and 95% RH. Ripening time is about three weeks.

HAZARD ANALYSIS IN CHEESES

Thirty assorted cheeses were purchased in 1975 by Labuza and
coworkers from a supermarket. The rinds were removed and an internal sample was taken. The a_w of duplicate samples was determined at room temperature (23°C) by the Fett-Vas method and the
vapor-pressure manometer (VPM) technique (Leung *et al.* 1976).

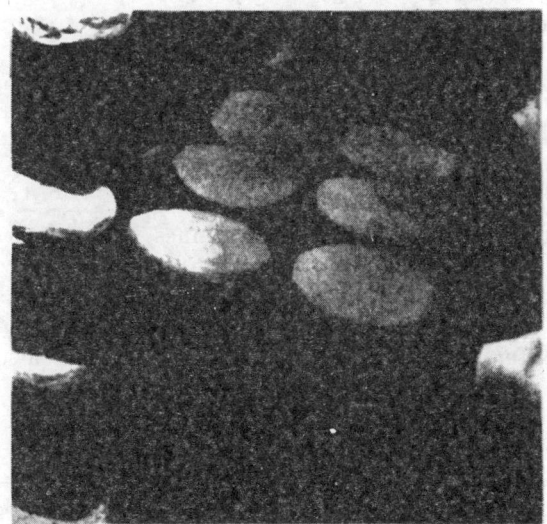

FIG. 10.13.   CAMEMBERT CHEESE SHOWING WHITE
SURFACE MOLD DURING CURING

These methods measure a_w to 0.01 units. The a_w and pH values of
the 30 commercial cheeses tested by them are listed in Table 10.8.
With the exception of Parmesan and whey cheese, all the samples
tested showed an a_w greater than 0.94. Many cheeses had an a_w
very close to 1.0. The a_w values determined by the two different
techniques were within 0.02 a_w units, they report.

As shown in Table 10.8 most cheeses had a very high a_w and medium acid pH. The pH of most cheeses ranged from 4.7 to 5.8 in most
cases. However, Camembert and Brie cheeses exhibited a high pH
(6.1, 7.0, 7.4), due probably to the greater enzymatic degradation
of proteins during ripening. They believe these cheeses could be
subject to microbial spoilage and growth of *Staphylococcus aureus*, especially if not held at refrigerated temperature. The inhibitory effect of the acids in cheese is only partially responsible for
retarding the growth of any microorganism present if the cheese is
held at refrigerated temperature.

The cheese foods have a lower a_w than do the natural cheeses,
even at higher moisture levels, because of added low molecular
weight ingredients which control the a_w through water binding.
Muenster cheese has a moisture content of about 1.2 g water per g
nonfat solids, and shows an a_w of approximately 1.0. Ye Old Tavern cheese food and Swiss-American spread, both with higher

water contents (1.6 and 1.5 g water per g nonfat solids, respectively), have an a_w of 0.95 (Table 10.8). Acott and Labuza (1975) also studied the effectiveness of various microbial inhibitors in a food base having an a_w of 0.85 at pH 4.2 and 5.7. The results are given in Table 10.9.

Mycotoxins and Cheese

Consumption of natural cheeses continues to increase steadily in the United States. Mold growth on these products is a common occurrence during aging and storage at low temperatures. Since certain molds are capable of producing toxic and carcinogenic metabolites, proliferation of these organisms on foods also must be regarded as a potential public health hazard.

TABLE 10.9

MICROBIAL INHIBITORS IN FOOD WITH a_w 0.85

Inhibitor	W/W (%)	Time for First Appearance of A. niger[1] (weeks) pH 5.7	pH 4.2
Potassium sorbate	0.15	2	ng
	0.30	ng	ng
Calcium propionate	0.1	2	ng
	0.2	19	ng
	0.3	ng	ng
Benzoic acid	0.2	7	ng
	0.3	ng	ng
Methyl paraben	0.03	ng	ng
	0.05	ng	ng
	0.10	ng	ng
Propyl paraben	0.01	ng	ng
	0.03	ng	ng
	0.04	ng	ng
Parabens Me/Pro	0.05	ng	ng
(2:1)	0.10	ng	ng
Pimaricin	0.001	1	4.5
	0.002	ng	ng
	0.005	ng	ng
1,3 Butanediol	1.0	1	22
	2.0	ng	ng
	4.0	ng	ng
Propylene glycol	1.0	ng	ng
	2.0	ng	ng
	4.0	ng	ng
Mannitol	1.0	ng	ng
	2.0	ng	ng
Sorbitol	1.0	ng	ng
	2.0	ng	ng
Glycerol	1.0	ng	ng
	1.0	ng	ng
Control		1	4.5

Source: Acott and Labuza (1975).
[1] 9 months storage at 23°C; ng = no mold growth.

Bullerman (1976) reported that 82% of the molds found on refrigerated Cheddar cheese belonged to the genus Penicillium, 7% were Aspergillus species and 1% were of Fusarium species. A number of species of these genera are indeed known to be capable of mycotoxin production. Toxicological screening of isolates from Cheddar cheese indicated that approximately 20% of the isolates were toxic to chicken embryos, causing 50% mortality or more.

Thin-layer chromatographic examination of the extracts of mold cultures from the cheese showed the presence of known mycotoxins in 7% of the culture extracts. The mycotoxins detected were patulin, penicillic acid, ochratoxin A and aflatoxins. However, the presence of these toxins in the cheeses themselves could not be demonstrated.

TABLE 10.10

SOURCE AND TYPE OF MOLD ISOLATES FROM SWISS CHEESE
CLASSIFIED BY GENUS AND TEMPERATURE AT WHICH ISOLATED

Genus	Total Isolated	5 °C	Total Isolated 21 °C
All genera	183	102	81
Penicillium	159	95	64
Aspergillus	1	—	1
Other genera	23	7	16

Source: Bullerman (1976).

TABLE 10.11

KNOWN MYCOTOXINS AND CHICK EMBRYO TOXICITY FOUND
IN CULTURE EXTRACTS OF MOLDS ISOLATED FROM SWISS CHEESE
AND GROWN ON YEAST EXTRACT SUCROSE (YES) BROTH OR RICE
POWDER CORN STEEP AGAR (RPCS)·

Toxin	Number of Isolates + by TLC YES	RCPS	Toxicity to Chicken Embryos (% mortality)
Aflatoxins	1	1	100
Ochratoxin A	ND[1]	ND	—
Sterigmatocystin	ND	ND	—
Patulin	4	ND	90-100
Penicillic Acid	5	3	40-60
Citrinin	ND	ND	—
Luteoskyrin	ND	ND	—
Zearalenone	ND	ND	—
Undetermined	3		100
Undetermined	7		90
Undetermined	17		80
Undetermined	13		70
Undetermined	12		60

Source: Adapted from Bullerman (1976).
[1] ND = None detected.

While the majority of the molds isolated from Cheddar cheese were not toxic, further studies on the molds on additional products are warranted to clarify with no uncertainty the extent of the potential hazard that may exist with mold growth on cheeses. Bullerman (1976) states that additional information about the numbers and kinds of toxin-producing molds found on cheeses is needed to assess fully the potential danger to people (Tables 10.10 and 10.11).

SUMMARY

Fermented foods are available in a wide variety of forms. They add pleasure to eating in many ways. Some of our most appetizing and nutritious foods have developed through the agency of microorganisms. Perhaps the most unexploited area in food technology is also the most ancient. There is little doubt that the future will find increasing attention being devoted to this important area.

REFERENCES

ACOTT, K., and LABUZA, T.P. 1975. Inhibition of *Aspergillus niger* in an intermediate moisture food system. J. Food Sci. 40, 137.

AMERINE, M.A., BERG, H.W., and CRUESS, W.V. 1972. The Technology of Wine Making, 3rd Edition. AVI Publishing Co., Westport, Conn.

ANGELOTTI, R. 1973. FDA regulations promote quality assurance. Food Technol. 29, No. 11, 60-62.

BAUMAN, H.E. 1974. The HACCP concept and microbiological hazard categories. Food Technol. 28, No. 9, 30-34, 70.

BEAZLEY, C.C., and INSALATA, N.F. 1971. The roles of fluorescent antibody methodology and statistical techniques for salmonellae protection in the food industry. Food Technol. 25, No. 3, 24-26.

BELL, T.A., ETCHELLS, J.L., and JONES, I.D. 1966. A method for testing cucumber salt-stock brine for softening activity. U.S. Dep. Agric., Agric. Res. Serv. Rep. 72-5.

BONE, D. 1973. Water activity in intermediate moisture foods. Food Technol. 27, No. 4, 71-76.

BULLERMAN, L.B. 1974. A screening medium and method of detection of mycotoxins in mold cultures. J. Milk Food Technol. 37, No. 1, 46.

BULLERMAN, L.B. 1976. Examination of Swiss Cheese for incidence of mycotoxin producing molds. J. Food Sci. 41, No. 1, 26-28.

BULLERMAN, L.B., and OLIVIGNI, F.J. 1974. Mycotoxin producing potential of molds isolated from Cheddar cheese. J. Food Sci. 30, 1166.

BURKHOLDER, L. et al. 1968. Fish fermentation. Food Technol. 22, No. 10, 76-82.

BYRAN, F.L. 1974. Microbiological food hazards today—based on epidemiological information. Food Technol. 28, No. 9, 52-66, 84.

EICHNER, K., and KAREL, M. 1972. The influence of water content and water activity on the sugar-amino browning reaction in model systems under various conditions. J. Agric. Food Chem. 29, 218.

FETT, H.M. 1974. Water activity determination in foods on the range 0.80 to 0.99. J. Food Sci. *38*, 1097.

HALL, C.W., and HEDRICK, T. I. 1971. Drying of Milk and Milk Products, 2nd Edition. AVI Publishing Co., Westport, Conn.

HALL, C.W., and TROUT, G.M. 1968. Milk Pasteurization. AVI Publishing Co., Westport, Conn.

HENDERSON, J.L. 1971. The Fluid-Milk Industry, 3rd Edition. AVI Publishing Co., Westport, Conn.

HERRINGTON, B.L. 1963. Dairy Products. In Food Processing Operations, Vol. 1. M.A. Joslyn, and J.L. Heid (Editors). AVI Publishing Co., Westport, Conn.

IFT EXPERT PANEL AND CPI. 1975. Sulfites as food additives. Food Technol. *28*, No. 10, 117-120.

JACOBSON, R.E., and BARTLETT, R.W. 1963. The ice cream and frozen dessert industry changes and challenges. Ill. Agric. Exp. Stn. Bull. *694*.

JOHNSON, A.H., and PETERSON, M.S. 1974. Encyclopedia of Food Technology. AVI Publishing Co., Westport, Conn.

JUDKINS, H.F., and KEENER, H.A. 1960. Milk Production and Processing. John Wiley & Sons, New York.

KOSIKOWSKI, F. 1976. Cheese and Fermented Milk Foods, 2nd Edition. Published by the author, Ithaca, New York

LABUZA, T.P. 1968. Sorption phenomena in foods. Food Technol. *22*, 15.

LABUZA, T.P. 1970. Properties of water as related to the keeping quality of foods. In SOS/70 Proceedings. Inst. of Food Technologists, Chicago, Ill.

LABUZA, T.P. 1971. Kinetics of lipid oxidation in foods. CRC Crit. Rev. Food Technol. *2*, 355.

LABUZA, T.P., CASSIL, S., and SINSKEY, A.J. 1972. Stability of intermediate moisture foods. 2. Microbiology. J. Food Sci. *37*, 160-162.

LABUZA, T.P., MCNALLY, L., GALLAGHER, R., HAWKES, J., and HURTADO, F. 1972. Stability of intermediate moisture foods. 1. Lipid oxidation. J. Food Sci. *37*, 154-158.

LABUZA, T.P. et al. 1977. Determination of water activity: A comparative study. J. Food Sci. (In press).

LABUZA, T.P., TANNENBAUM, S.R., and KAREL, M. 1970. Water content and stability of low-moisture and intermediate-moisture foods. Food Technol. *24*, 35.

LABUZA, T.P., TSUYUKI, H., and KAREL, M. 1969. Kinetics of linoleate oxidation in model systems. J. Am. Oil Chem. Soc. *46*, 409.

LAJOLLO, F., TANNENBAUM, S.R., and LABUZA, T.P. 1971. Reaction at limited water concentration. 2. Chlorophyll degradation. J. Food Sci. *36*, 850.

LAMPERT, L.M. 1965. Modern Dairy Products. Chemical Publishing Co., New York.

LAMPI, R.A. 1963. Infiltration of Porous Foods with High Caloric Non-Aqueous, Edible Materials. Rep. Nos. 1, 2, 3. Quartermaster Contract DA19-129-AMC-84 (N) (019056). Natick, Mass.

LEAGUE, D.N. 1953. In-can baking perfected. Food Eng. *25*, No. 5, 86.

LEUNG, H., MORRIS, H.A., SLOAN, E.A., and LABUZA, T.P. 1976. Development of an intermediate moisture processed cheese food product. Food Technol. *30*, No. 7, 42-44.

LONCIN, M., BIMBENET, J.J., and LENGES, J. 1968. Influence of the activity of water on the spoilage of foodstuffs. J. Food Technol. *3*, 131.

MELPAR, INC. 1965. Interrelationships between storage stability and moisture sorption properties of dehydrated foods. Contract DA19-129-AMC-252. Final Rep. Natick, Mass.

MILK IND. FOUND. 1960. Milk Facts 29. Milk Industry Foundation, Washington, D.C.

MOSSEL, D.A.A., and INGRAM, M. 1955. The physiology of the microbial spoilage of food. J. Appl. Bacteriol. 18, 232.

NATL. BUR. STAND. 1951. Methods of Measuring Humidity and Testing Hygrometers. Natl. Bur. Stand. Circ. 512.

NICKERSON, R.A. 1960. Chemical composition of milk. J. Dairy Sci. 43, 598-606.

OHLMEYER, D.W. 1957. Use of glucose oxidase to stabilize beer. Food Technol. 11, 503-508.

OLSON, T.M. 1950. Elements of Dairying. Macmillan Co., New York.

PEDERSON, C.S. 1930A. Floral changes in fermentation of sauerkraut. N.Y. Agric. Exp. Stn. Tech. Bull. 168.

PEDERSON, C.S. 1930B. The effect of pure culture inoculation on the quality and chemical composition of sauerkraut. N.Y. Agric. Exp. Stn. Tech. Bull. 169.

PEDERSON, C.S., and ALBURY, M.N. 1954. The influence of salt and temperature on the microflora of sauerkraut fermentation. Food Technol. 8, 1-5.

PEDERSON, C.S., and BEATTIE, H.C. 1943. Vinegar making. N.Y. Agric. Exp. Stn. Circ. 148.

PEDERSON, C.S., and WARD, L. 1949. The effect of salt upon the bacteriological and chemical changes in fermenting cucumbers. N.Y. Agric. Exp. Stn. Bull. 288.

POTTER, N.N. 1973. Food Science, 2nd Edition. AVI Publishing Co., Westport, Conn.

RAPER, K.B., and FANNELL, D.I. 1965. The Genus Aspergillus. Williams and Wilkins, Baltimore, Md.

RAPER, K.B., and THOM, C.A. 1968. A Manual of the Penicillin. John Wiley & Sons, New York.

REINBOLD, G.W., ANDERSON, R.F., and HERMAN, L.G. 1973. Microbiological methods for cheese and other cultured products. In Standard Methods for the Examination of Dairy Products, 13th Edition, W. Hausler (Editor). American Public Health Assoc., Washington, D.C.

SAMISH, S. 1955. Studies on pickling Spanish-type green olives. Food Technol. 9, 173-176.

SAMISH, S., COHEN, S., and LUDIN, A. 1968. Progress of lactic acid fermentation of green olives as affected by peel. Food Technol. 22, No. 8, 77-80.

SANDERS, G.P. 1953. Cheese Varieties and Descriptions. U.S. Dep. Agric., Agric. Handb. 54.

SCOTT, P.M., LAWRENCE, J.W., and VAN WALBEEK, W. 1970. Detection of mycotoxins by thin-layer chromatography: Application of screening to fungal extracts. Appl. Microbiol. 20, 839.

SMALL, E., and FENTON, E.F. 1964. A summary of laws and regulations affecting the cheese industry. U.S. Dep. Agric., Agric. Handb. 265.

SOMMER, H.H. 1951. The Theory and Practice of Ice Cream Making, 6th Edition. Published by the author, Madison, Wisc.

U.S. DEP. HEALTH, EDUC. AND WELFARE. 1965. Grade A Pasteurized Milk Ordinance. U.S. Public Health Serv. Publ. 229.

VAN SLYKE, L.L., and PRICE, W.V. 1952. Cheese. Orange Judd Co., New York.

VAN WALBEEK, W.   1973.   Fungal toxins in foods.   Can. Inst. Food Technol.
   J. 6, 96.
VERRETT, M.J., MARTIAC, J.P., and MCLAUGHLIN, J., JR.   1964.   Use of
   embryo in the assay of aflatoxin toxicity.   J. Assoc. Off. Agric. Chem. 74, 1003.
VOS, P.T., and LABUZA, T.P.   1974.   Technique for measurement of water
   activity in the high a_w range.   J. Agric. Food Chem. 22, 326.
WALLERSTEIN LAB.   1975.   Bottle Beer Quality.   Wallerstein Laboratories,
   New York.
WEBB, B.H., JOHNSON, A.H., and ALFORD, J.A.   1974.   Fundamentals of
   Dairy Chemistry, 2nd Edition.   AVI Publishing Co., Westport, Conn.
WEBB, B.H., and WHITTIER, E.O.   1971.   Byproducts from Milk, 2nd Edition.
   AVI Publishing Co., Westport, Conn.
WHO.   1962.   Milk Hygiene.   FAO/WHO Monogr. Ser. 48.

Chapter 11

Principles of Food Pickling and Curing

PRINCIPLES OF FOOD PICKLING AND CURING

An important food preservation system combines salting to selectively control microorganisms and fermentation to stabilize the treated materials. The process when applied to fruits and vegetable matter is called pickling. When applied to meats it is called curing.

SOURCES OF SALT

Salt is one of man's highly valued accessory materials. It has been used even to control societies, i.e., the salt taxes in India. Nations attempt to become self-sufficient in salt production today, as they did five thousand years ago. Man's need for salt does not diminish.

There are three major sources of salt. Solar salt is obtained by evaporation of salt water, either from the oceans or from inland salt lakes. Mined salt, commonly referred to as rock salt, is obtained from mines, operating a thousand feet and more below the surface of the earth. Some salt is pumped from deeper subterranean salt deposits, using water as the transporting medium, and is called welled salt.

Salt obtained from solar distillation contains chemical impurities, and salt tolerant, halophilic microorganisms. Mined and welled salts are generally free from these contaminating organisms.

The amount of salt added determines whether or not any organisms can grow and what types will grow, as in the control of the fermentation activity, other factors being equal (Table 11.1).

Salt in solution in food substrates exerts a growth repressing action on certain microorganisms, can be made to limit moisture availability

TABLE 11.1

EFFECT OF SALT ON DEVELOPMENT OF BACTERIA AND ACID
IN SAUERKRAUT FERMENTATION AT 22°C

Salt Concentrations (%)	Fermentation Time (Days)	Total Plate Count (100,000/ml)	Total Acid (%)
1	1	4740	0.38
	5	625	1.21
	10	1095	1.55
	21	130	1.91
2.25	1	1320	0.15
	5	1270	1.19
	10	2300	1.57
	21	251	1.78
3.5	1	431	0.08
	5	1180	1.25
	10	1200	1.55
	21	57	1.91

and can dehydrate protoplasm, causing plasmolysis.

In foods containing salt as a preservative, the salt has been ionized, collecting water molecules about each ion. This process is called ion hydration. The greater the concentration of salt, the more water employed to hydrate ions. A saturated salt solution at a temperature is one that has reached a point where no further energy is available to dissolve the salt. At this point (26.5% sodium chloride solution at room temperature) bacteria, yeasts and molds are unable to grow. It has been postulated that there is no free water available for microbial growth.

In describing the preservative action of salt, one must consider the dehydration effect, the direct effect the Cl^- ion, reduced oxygen tension, and interference with the action of enzymes.

Organisms can be selected or sorted on the basis of salt tolerance. This is a well employed procedure in identification of bacteria. It is also functional in controlling fermentations.

Certain lactic acid bacteria, yeasts and molds either tolerate or adapt to moderate salt solutions. Spore-forming aerobes and anaerobes are not tolerant of salt solutions, or they are sufficiently inhibited that subsequent production of acid by lactic acid bacteria so supplements the inhibition influence of salt that spore-forming bacteria are of little concern, providing both conditions of salt and acid are operative.

Proteolytic bacteria and pectolytic organisms are also inhibited by salt and acid solutions. However, these organisms are more sensitive to acid than salt. In conditions where mold (tolerant to salt and a utilizer of acid) is permitted to grow, decreasing the acidity of the substrate, putrefactive and pectolytic organisms can be anticipated to increase in numbers, causing food spoilage.

Salt, vinegar and spices are commonly used in complementary action in many fermented food products. Spices vary in their antibacterial activity, some (mustard oil) being very active, others (black pepper) having little activity.

Pickled Fruits and Vegetables

Fresh fruits and vegetables placed into a watery solution will soften in 24 hr and begin a slow, mixed fermentation-putrefaction. It is necessary to suppress undesirable microbial activity and create a favorable environment for the desired fermentation. The addition of salt permits the naturally present lactic acid bacteria to grow, thereby rapidly producing sufficient acid to supplement the action of the salt. One of the important changes that occurs in the pickling process is that the fermentable carbohydrate reserve of the cucumber, for example, is changed to acid. The level of acid developed ranges from 0.8 to 1.5%, expressed as lactic. The color of the cucumber changes from bright green to an olive or yellow-green. The tissue changes to translucent from the normal chalky white and opaque. During this time the cucumber absorbs salt. The commercial production of salt stock, or fermented salted cucumbers, is such a process.

The salt concentration is maintained at 8 to 10% during the first week, and increased 1% a week thereafter until 16% salt is obtained in solution. It is to be noted that the cucumber contains nearly 90% moisture, and as the equilibrium is established between tissue and salt solution, the concentration of the latter slowly decreases. Constant vigilance is required in maintaining the salt concentration and protective effect of the salt in solution in controlling the fermentation. At the end of 4 to 6 weeks, after the fermentation is completed, noted by the change in tissue characteristics within the cucumber, the salt concentration is raised to about 16%. If properly controlled, the salted fermented cucumber, now called "salt stock," may be held for several years. During storage, salt stock is subject to spoilage by yeasts and molds.

The salt concentration is usually referred to in terms of degree Salometer. Degree Salometer is based upon the saturation of water with 25% sodium chloride at room temperature. This point is referred to as 100 Salometer. As a working rule, percent salt in solution multiplied by four equals degree Salometer.

TABLE 11.2

CHANGES IN COMPOSITION OF FRESH CUCUMBERS BY
FERMENTATION TO SALT STOCK

Component	Fresh Cucumbers (%)	Salt Stock (%)
Moisture	95.0	90.2
Total solids	5.0	9.8
Sugars	1.20	0.01
Protein	1.43	1.04
Fat	0.06	0.15
Ash	0.52	6.48

TABLE 11.3

LOSS OF BRIGHT GREEN COLOR IN CUCUMBER PICKLES: CHLOROPHYLLS CONVERTED
TO PHEOPHORBIDES AND PHEOPHYTINS

Days in Brine	Micromoles per 100 Grams, Fresh Weight Basis					
	Chlorophyll a	Chlorophyll b	Pheophorbide a	Pheophorbide b	Pheophytin a	Pheophytin b
0	6.32	2.90	—	—	—	—
2	2.09	1.83	2.60	0.97	0.84	0.21
4	0.85	1.08	3.53	1.39	0.93	0.32
7	0.04	0.02	5.15	2.44	1.32	0.75

Use of Salt Stock

Salt stock is not considered a consumer commodity. When desired, salt stock is freshened and prepared into consumer times (Tables 11.2 and 11.3). This is accomplished by leaching the salt from the cured cucumber with warm water (43° to 54°C) for 10 to 14 hr. This is repeated at least twice. In the final washing water alum may be added to firm the tissue, and tumeric may be added to improve color. Most vegetables and fruit (green) may be preserved in this manner. It is necessary to exercise due caution in maintaining product quality, for example, with cauliflower. Care in handling the latter is required to prevent physical damage to the tissue.

Sour Pickles, Sweet Pickles, Processed Dill Pickles

The freshened salt stock is used in the preparation of many types of pickles and relishes. Sour pickles are prepared by reprocessing freshened salt stock with weak vinegar, and packing into consumer units. A final acidity is maintained not lower than 2.5%. Sweet pickles are prepared in much the same manner except that a sweet, spiced vinegar solution is added to freshened salt stock. Processed dill pickles are prepared much as sour pickles, but with the addition of dill herb and spices to the acidified brine.

Genuine dill pickles are prepared from fresh cucumbers, and are not prepared from salt stock. Fresh cucumbers are packed into barrels with dill herb and salt solution, with added spices. The acid developed is maintained. The marketed genuine dill pickles are not usually packaged in vinegar. Genuine dill pickles are perishable, unless repackaged and pasteurized. Care must be exercised in destroying enzymes in pickles. Heating to approximately 82°C is effective in enzyme control.

Sauerkraut

Cabbage as a commodity may be preserved in its natural state for a short period of time (3 or 4 months) or it can be subjected to bacterial fermentation, controlled with salt. During the fermentation acid is developed and acts as a preservative in addition to developing a desired flavor. Sauerkraut is the German word describing this fermented, salted, shredded cabbage. It has been popular for centuries in Western Europe (Table 11.4).

Cabbage for kraut should contain upwards of 3.5% sugar. Shredded cabbage is mixed with salt in the proportion of 2.25% salt by weight. Salt draws juice from the shredded cabbage and immediately establishes a control over putrefactive growths. On the average, 1.5 to 2.0% acid expressed as lactic acid is produced in the final fermentation. There is an interesting series of fermentations which occurs. Laboratory experiments indicate the sequence (Fig. 11.1).

Initially, *Leuconostoc mesenteroides*, a coccus and an acid and gas former, establishes growth. If the temperature is near 25°C and the salt concentration near 2.5%, after 2 days, from 0.7 to 1.0% acid is developed. The kraut at this stage has a pleasant odor, but still is raw and uncured.

The former organism subsides, and *Lactobacillus plantarum* and *Lactobacillus brevis* mount in numbers. After the fifth day, the acidity eventually reaches 1.5 to 2.0% and these organisms subside.

As a general rule, approximately one-half as much acid is produced as there is sugar present in the cabbage. The fermentation may be complete in a little over a week. Actually the commercial process requires from two weeks to several months. This may be explained on the basis that the temperatures are not controllable within such

TABLE 11.4

EFFECT OF FERMENTATION TEMPERATURE ON BACTERIAL GROWTH AND
ACID PRODUCTION IN SAUERKRAUT WITH 2.25% SALT ADDED

Temperature (°C)	Days	Total Acid (%)	Total Bacterial Count (100,000/ml)
7	1	0.04	40
	10	0.48	2640
	20	0.70	2105
17	1	0.16	2150
	10	1.23	2330
	20	1.71	560
31	1	0.71	6400
	10	2.02	725
	20	—	—
36	1	72	15,600
	10	1.76	48
	20	—	—

narrow tolerance, and the growth of the organisms is not this well defined.

The controls which operate in sauerkraut fermentation are the concentration of salt, the temperature of fermentation, the kind

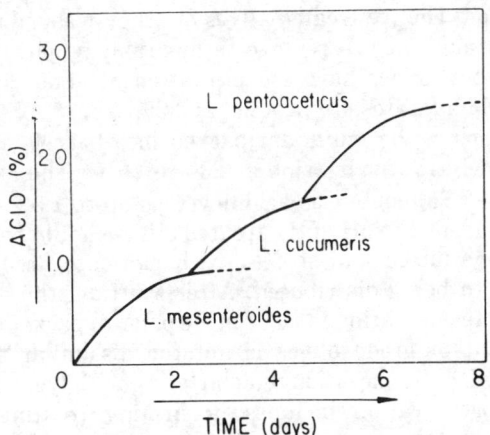

FIG. 11.1.   ACID PRODUCTION IN SAUERKRAUT FERMENTATION

and quality of cabbage, and the sanitary conditions under which the fermentation is permitted to occur (Table 11.5).

TABLE 11.5

EFFECT OF SALT ADDED ON TEXTURE OF SAUERKRAUT
(SOFT TEXTURE OF 1% SALT SAUERKRAUT UNACCEPTABLE COMMERCIALLY)

Salt Added (%)	Tenderometer Reading
0.5	29
1.0	66
1.5	117
2.0	130
2.5	151
3.0	169
3.5	150

Olives

Green olives are harvested when fully sized, but before fully ripe. The fruit at this point contains a bitter alkaloid, oleuropein, which must be controlled. The bitterness is removed by treating olives with a 2% sodium hydroxide solution at room temperature. The lye is permitted to penetrate nearly two-thirds through the fruit, but not completely to the pit. By stopping the alkaline penetration at this point, a small amount of bitterness remains in the flesh and imparts a pleasing flavor. The lye treated olives are then washed until lye-free. Care must be taken not to remove fermentable sugars. As grown in the United States, olives have a composition of about 70% moisture, 20% oil, 5% carbohydrates, 3% protein, and 2% ash.

After debittering, the fruits are packed into barrels and permitted to ferment slowly in a salt brine at 23° to 26°C. The salt content is built to 25° - 30° Salometer, as acid develops up to 1.5%. Fermentable sugar may be added. The pH of the treated olives should be 3.8 or less.

Olives may be pitted and stuffed with pimiento or other suitable material prior to being distributed. After stuffing, the olives may be given a further fermentation in a 30° brine prior to packaging.

The treatment of green olives is not too dissimilar to the curing of cucumbers, with the exception of debittering with lye.

Ripe olive preservation includes: a firming treatment in brine; treatment with lye to promote oxidation of the color of ripe olives; a further lye treatment to debitter the fruit; washing to remove lye; curing in 8° to 10° brine for up to a week; and preserving. The pH value of ripe olives is near 7.0 and they require heat sterilization in hermetically sealed containers for preservation (Table 11.6).

TABLE 11.6

LACTIC ACID CONTENT OF FERMENTED OLIVES AFTER VARYING PRETREATMENTS

| Variety | Pretreatment | | Lactic Acid In Fermented Fruit (%) |
	Blanching (hr)	Lactic Acid Added to Brine (%)	
Sevillano	36	0.0	0.41
	5	0.2	1.12
	5	0.3	1.14
	5	0.4	1.20
Merhavia	36	0.0	0.51
	5	0.2	1.31
	5	0.3	1.20
	5	0.4	1.28

FERMENTED AND PICKLED PRODUCTS

Deterioration

If untended, fermented foods and pickled products will deteriorate quickly. In general, protection is necessary against the action of molds which metabolize the acid developed, paving the way for the advance of other microorganisms. In cool storage fermented and pickled food products can be expected to be stable for several months. Longer storage periods demand more complete protection from spoilage agents; for example, the canning of fermented fruits, vegetables and meats, thereby inactivating enzymes and protecting losses in quality due to microbial and chemical agents.

The common method of preserving pickled products is to pasteurize them at atmospheric pressure. Hot water sprays of varying temperatures are used to increase temperatures gradually, and cool gradually, to prevent thermal shock to containers. In Fig. 11.2 the pickles are packed hot into jars, sealed, and passed through hot water sprays on a continuous belt. The method is simple and effective.

Nutritional Value

Nutrient retention in fermented and pickled products is about equal to that in other methods of food preservation. In the instances of carbohydrates, there is usually a conversion to acid or to alcohol, but these are of value nutritionally. The stabilized foods contain other nutrients in adequate amounts, when compared with the original perishable tissues (Table 11.7). Nutrient levels are sometimes increased due to the presence of yeasts.

FIG. 11.2.   LOADING END OF CONTINUOUS FLOW PASTEURIZER

TABLE 11.7

PERCENT RETENTION OF NUTRIENTS IN FRESHENED SALT STOCK
AS COMPARED TO FRESH CUCUMBERS

Nutrient	Retained (%)
Ascorbic acid	14
Carotene	72
Thiamin	18
Calcium	100
Iron	100

Source: Fellers (1960).

BLOATER DAMAGE CONTROL

Bloater damage in brined cucumbers usually increases with concentration of dissolved CO_2 in the brine. The damage is accentuated with larger cucumbers and at higher incubation temperatures (Fleming et al. 1975). Microorganisms which produce large amounts of CO_2 (such as coliform bacteria, yeasts and heterofermentative lactic acid bacteria) have been implicated as causes of bloater damage in natural fermentations commercially.

They also found that CO_2 which originates from the cucumber is sufficient to cause bloater damage. Factors affecting CO_2 buildup and retention in the brine, such as brining depth, also influence bloater development. Removal of dissolved CO_2 from brines by purging with nitrogen gas reduced or eliminated bloater damage. Brine circulation may also reduce buildup of CO_2.

The controlled fermentation of cucumbers brined in bulk involves: (1) directing the fermentation by thorough washing of the cucumbers, chlorination and acidification of the cover brine, addition of sodium acetate as a buffer, and inoculation of the brine with a starter culture of *L. plantarum*; and (2) purging of dissolved CO_2 from the brine with nitrogen.

Present objectives were to: study factors affecting CO_2 removal from brines by purging with nitrogen; compare effects of nitrogen and air, as purging gases, on the fermentation and quality of the brine-stock cucumbers; and, compare the effectiveness of purging in reducing bloater damage in natural as compared to controlled fermentations. Larger sizes of cucumbers have greater susceptibility to bloater damage. The trend in recent years has been toward the harvest of the larger sizes.

No important differences in flavor were noted between the nitrogen-purged and nonpurged cucumbers.

Controlled Fermentations in Commercial Brining Tanks

The suggested procedure for nitrogen purging (Etchells *et al.* 1973) is as follows: the tanks are purged intermittently with either 55 cu ft per hr of nitrogen for 1 hr at 8-hr intervals during the first 4 days after brining; for 1 hr at 12-hr intervals during the first 4 days after brining; for 1 hr at 12-hr intervals on the 5th through 7th days; and for 1 hr on the 8th and 9th days after brining. The cured brine-stock that showed no evidence of mechanical damage of the cucumbers prior to brining was evaluated by experienced judges and contained 3% bloaters. Nonpurged natural fermentation averaged 25 to 35% serious damage due to bloaters, typical for this size stock.

(1) Removal of CO_2 from controlled fermentations by nitrogen purging of the brine greatly reduced or eliminated bloater formation. Purging of CO_2 from natural fermentations did not consistently reduce bloater damage.

(2) Primarily because of its inertness, nitrogen is preferred over air as the purging gas. Air removed CO_2 from brine. However, effects on quality factors, such as loss of firmness and discoloration of the cured cucumbers, strongly suggests that air not be used for purging.

(3) Film yeasts were not a problem in closed, tight containers

either not purged or purged with nitrogen, due apparently to the inert atmosphere created over the brine surface. Air purging of cucumbers in closed containers resulted in brine surface growth of film yeast and mold.

(4) Neither nitrogen nor air purging, at the levels used, appreciably affected the rate of acid production or final brine acidities in controlled fermentations.

(5) Removal of CO_2 from brine by purging reduces buoyancy pressures created by bloated cucumbers and, therefore, could serve to relieve excessive pressures on containers in which cucumbers are brined.

(6) Commercial application of the controlled fermentation process resulted in greatly improved brine-stock quality.

BRINE RECOVERY

An estimated 300,000 tons of cucumbers are brined annually in the United States. This generates approximately 60 million gallons of spent brine. The brine contains 14 to 18% NaCl and 10,000 to 15,000 ppm BOD. Reuse of this brine would at least partially solve the serious waste disposal problems in the cucumber pickling industry. Studies by Palnitkar and McFeeters (1975) were undertaken to assess the chemical changes produced by NaOH treatment of the brine and to determine the feasibility of multiple recycling by this procedure. The present data show that heat-treated brines give a normal fermentation.

Data are available for heat inactivation of pectinase from *Aspergillus niger*, 11 sec at 95°C in spent brine, as a guide for heat treatment. A reasonable degree of caution must be exercised, however. First, polygalacturonase may be found in commercial brines with even greater heat stability. Secondly, brines will vary in pH, organic matter composition, salt concentration and mineral composition. These variations may significantly affect the heat stability of softening enzymes.

Table 11.8 shows data from brine samples when pickles were removed at the end of each cycle. On the third, fourth and fifth cycles, BOD was determined before and after NaOH treatment. The final pH of the brine increases with each cycle. This indicates the increasing buffer capacity of the brine. There is a limited buildup of BOD and COD. A small increase in brine titratable acidity occurred after each fermentation cycle. A rather large decrease in suspended solids occurred by the fourth cycle.

Subjective evaluations showed no discernible differences in color,

TABLE 11.8

CHARACTERISTICS OF SPENT BRINE FROM REPETITIVE CYCLING

	Cycle 1	Cycle 2	Cycle 3	Cycle 4	Cycle 5
Days of fermentation	—	23	20	21	41
Initial pH	—	6.18	6.18	6.55	5.98
Final pH	3.30	3.95	3.97	4.10	4.21
Lactic acid (%)	0.31	0.69	0.72	0.73	0.76
Reducing sugar (mg/ml as glucose)	—	0.79	0.54	0.54	0.40
Suspended solids (mg/l)	274	2,316	1567	550	—
Volatile suspended solids (mg/l)	149	1,200	260	390	—
Total solids (%)	9.18	9.37	9.55	8.96	9.14
BOD (untreated) (mg/l)	12,300	17,360	20,100	18,400	21,320
BOD after NaOH (mg/l) treatment	—	—	19,400	17,600	19,400
COD (untreated) (mg/l)	11,306	24,437	31,700	43,812	42,000

Source: Palnitkar and McFeeters (1975).

odor, texture or flavor between pickles fermented in recycled spent brine and those fermented in fresh brine.

Defect Reduction

The only consistent difference between the salt stock quality of cucumbers fermented in recycled spent brine compared to fresh brine was that less severe bloating occurred in recycled brine. These experiments show that spent brines treated with NaOH and neutralized with HCl result in salt stock equivalent in quality to that produced by fermentation in fresh brine. Repetitive brine cycling through five cycles of use does not cause a discernible reduction in salt stock quality.

THE PRINCIPLES OF FISH SALTING

The preservation of fish by salting is an ancient art. It remains a most important means of preserving fresh water and marine products.

The Influence of the Composition of Salt

The composition of the salt used has been found by Tressler and Lemon (1951) to be of great importance, not only in affecting the rate of its penetration into the tissues of the fish, but also in determining the physical qualities of the product. The chief impurities in commercial salt are calcium salts, magnesium salts, sulfates, and organic matter. Sea salts are almost universally used for the salting of fish.

Calcium and magnesium salts and all sulfates effect a retardation of the rate of penetration of sodium chloride (salt) into fish during the salting process. This retardation of the rate of salting permits more decomposition of the protein of the fish tissues during the process of salting. Therefore, it is important to use nearly pure salt when stale fish are to be salted or when the salting process takes place under adverse conditions, such as in warm climates. Calcium salts retard penetration of salt into fish to a greater extent than either magnesium salts or sulfates and are, therefore, objectionable if present in appreciable amounts. Sulfates are seldom present in sufficient quantities to be objectionable.

Calcium and magnesium salts, present as impurities in salt used for the salting of fish, affect the color and firmness of the product to a remarkable extent. Fish salted with pure salt are soft, flabby, and of a very light yellow or cream color. Such fish are easily freshened and, when cooked, closely resemble fresh fish. They possess few of the qualities commonly associated with salted fish. The presence of as small an amount as 1% of calcium or magnesium in salt causes a remarkable whitening and stiffening of the flesh. Salts of both of these metals give a strong, bitter taste, characteristic of commercial salted fish. By varying the proportion of calcium chloride in common salt from 0 to 5%, salted fish can be prepared of any desired shade, from pale yellow to snow, chalky whiteness. This should be of considerable value to the fish salter, for it enables him to control the color and character of his product. Salt containing as much as 2% calcium as calcium chloride produces very stiff, brittle fish, quite unlike fresh fish. A like amount of magnesium chloride or sulfate produces a similar though slightly stiffer result.

Commercial Methods of Salting Fish

There are two classes of commercial methods of salting fish in common use today: brine-salting and dry-salting. The term dry-salted fish refers to the method of salting and not to the procedure followed in packing or storing fish. It should not be confused with dried, salted fish.

Brine-salting.—Brine-salting is of relatively little importance commercially as compared with dry-salting. The chief fish that is salted by brine-salting is the alewife or river herring; only the principle of the method will be considered at this point. The cleaned fish are placed in large vats, partially filled with concentrated salt solution. A small amount of salt is put on top of the fish floating in the brine. The fish are stirred daily to prevent the brine from becoming too dilute at any one point. The details of practice vary considerably in various localities, but the general procedure is as outlined above.

Dry-salting.—The exact procedure followed in dry-salting fish depends upon the kind of fish and the custom in the particular locality. But, for a general consideration of the subject the following description is sufficiently detailed. The round, gibbed, beheaded, or split fish are washed and then packed in watertight containers with an excess of dry salt. The proportion of salt to fish varies from 10 to 35% of the weight of the fish, depending upon the kind of fish, the weather, and the custom of the fish salter. The fish are usually rubbed in salt just before packing and each layer is sprinkled with salt. After a few hours sufficient pickle has formed to cover the fish. They are not disturbed until completely salted, when they are either repacked in fresh pickle or removed and dried.

Comparative Efficiency of Brine-salting and Dry-salting.—There has been much discussion of the relative value of the two methods. Tressler conducted extensive investigations concerning their merit. Using the squeteague and alewife the dry-salt method was found to obtain more rapid penetration of salt into the fish and to inhibit decomposition more quickly. Evidently in the dry-salting process the brine remains more nearly saturated. This is probably because of the greater surplus and better distribution of the dry salt (Fig. 11.3).

Smoke-curing Processes

It must be emphasized that only fish or shellfish of good quality should be used for smoke-curing. Smoking will not mask or conceal the poor flavor and quality of inferior raw material. The strictest sanitation should be maintained at all points in the procedure. Most smoked and kippered fish are perishable and should be handled accordingly. Careless handling will result in a product of poor quality and short keeping time. Clean utensils, as well as clean, pure salt and water, should be used in all brining and washing processes.

The smoke cure may be, as previously stated, either cold or hot. Cold smoking, which is accomplished at temperatures below 32.2°C, may in turn be divided into two general types, depending upon the preliminary brining or salting (i.e., whether the fish receive a heavy or a light salt curing previous to smoking). For further details refer to Tressler and Lemon (1951).

Cold-smoking (Heavy Salt Cure).—Fish which are cured and smoked in this manner have fair to excellent keeping qualities, depending upon the length of the salting and smoking operations. After

Courtesy of U.S. Fish and
Wildlife Service

FIG. 11.3. SIDE OF SALMON AS IT IS BEING SALTED

this heavy preliminary salting the texture and taste of the final product little resemble that of the fresh raw material. Preparation of a number of typical examples will be described in the following. In all cases reference is made to procedures used in old-style smokehouses, unless otherwise stated. In a controlled oven the time is always less, and other details may have to be varied slightly.

Smoked Salmon.—Cold-smoked salmon, or just "smoked salmon" as Tressler and Lemon have found in the trade, is prepared almost exclusively from king salmon sides that have been subjected to a preliminary mild-curing process, as described previously. Mild-cured salmon has been used extensively for smoking in Germany, England and the Scandinavian countries. However, the largest market is in the urban areas of the eastern part of the United States, especially in the large Jewish centers, where smoked salmon or lox is part of the daily diet. Lesser amounts are smoked in the Middle West and on the Pacific Coast.

Before the mild-cured sides can be given the smoke cure, it is necessary to remove the excess salt. This is done by soaking or freshening the sides in a tank of running cold water for a period of 12 to 24 hr, depending on how long the salmon has been in brine and on the demands of the trade. After proper freshening the sides are trimmed of any rough edges and then given a final wash with a soft cloth to remove any attached foreign particles. It is customary to pile the sides on a platform for a short while before hanging in the smoke oven. This pressing or "waterhorsing" serves a dual purpose; pressure not only removes considerable moisture, but also smooths the cut surfaces of the fish, both of which make for better appearance and quicker smoking. Some smokers eliminate this step by hand-pressing and smoothing each side as it is being hung in the oven.

The salmon sides are hung from rods placed across the smoke oven at a sufficient height above the fires to avoid danger of overheating. Hanging is done by means of a special type of steel wire hook, having one end curved to fit over a rod and the other equipped with six sharp prongs which penetrate the skin and flesh near the nape. Care should be exercised that the sides do not touch one another. Crowding or overloading of the oven should be avoided as it might interfere with proper air circulation.

The first fires that are kindled under the salmon should be "drying" fires, in which the combustion is quite complete (i.e., those that produce "blown" smoke). All drafts and ventilators should be opened and the drying continued until all surface moisture is removed and a glossy pellicle has been formed. The time required will be 24 to 48 hr, depending on weather conditions. This is followed by building fires which smoke heavily, in which a considerable amount of sawdust is generally used. The drafts and ventilators are almost closed so that combustion is incomplete and "distilled" smoke is produced. During this stage, which takes from 1 to 3 days, depending on the local market requirements, the sides will assume a distinct smoky flavor and the color will darken. The temperature of the smokehouse should be held below 29.4 C. If the sides are exceptionally fat and oily, it may be necessary to keep the temperature below 26.7°C to prevent excessive drip. The weight loss or shrinkage from the mild-cure weight during smoking will run from 0 to 30%, depending on the length of cure and the size and quality of the sides.

After smoking the sides are wrapped in oilproof paper and packed for shipment in boxes to a net weight of approximately 15 kilos. Smoked salmon is perishable and should be stored at 1° to 2°C.

Storage below freezing at −17.8°C is not detrimental if the period of storage is not extensive. In an endeavor to broaden the market and as an aid to the retailer in merchandising smoked salmon, several fish smokers now package sliced smoked salmon in 1- and 5-lb tin cans. Although the cans are hermetically sealed, they are not heat processed and should, therefore, be stored and displayed under refrigeration.

Hard-smoked Salmon.—This produce is sometimes referred to as "Indian smoked" salmon of the west coast. Hard-salted salmon may be used, but a better produce can be prepared by starting with fresh or frozen salmon. By so doing the smoker can control the preliminary salting himself and thereby secure the maximum extraction of moisture from the flesh. After the salmon have been split into two sides with the backbones removed, and thoroughly cleaned and washed, they are salted in barrels or tierces (some prefer dry-salting) for 3 to 5 days. This is followed by a soaking in fresh water for a few hours to remove the slight excess salt near the surface of the fish. Hard-salted salmon must be soaked 24 to 48 hr, and difficulty is experienced in getting a uniform freshening.

After draining and "waterhorsing" (pressing) the sides are hung from rods in the smoke oven by means of hoods, similar to those used for mild-cured sides. The first fires to be kindled are of the drying type in order to remove the surface moisture and to form a pellicle. From 2 to 4 days are required, during which time the dampers and louvers are kept open so as to maintain good circulation. When the sides are sufficiently dried, smoking fires are built and the air circulation is reduced by partially closing the dampers and louvers. The length of this stage will depend on the type of product desired, and may be as long as 10 days for a really hard-dried smoked salmon with good keeping qualities. However, 4 to 6 days will suffice for the average markets. The temperature should not exceed 29° to 32° C. Shrinkage in the smoking process may run as high as 50% (Fig. 11.4).

These hard-smoked sides may be packed for shipment and sold without further preparation. But more often they are cut into strips, chunks or slices and packed in jars or cellophane bags for distribution to the tavern trade. Since this produce is fairly dry and somewhat salty, it keeps quite well without refrigeration (Tressler and Lemon 1951).

MEAT CURING AND SMOKING

The effect of curing and smoking is relatively simple to define.

Courtesy of R.K. Pedersen

FIG. 11.4.   HARD-SMOKED SALMON SMOKED IN CONTROLLED SMOKEHOUSE

Both methods have a bacteriostatic effect by creating an environment hostile to bacteria. The drying effect of smoke and the smoke components (creosote, acetic acid and pyroligneous acids) inhibit bacteria and retard fat oxidization. Color and flavor changes that occur are very acceptable.

Although salt was used for preserving fish as far back as 3500 B.C., the origin of its use in curing meat is lost. It is known that sausage was made by adding salt to meat. By the 5th century B.C. production of salted meat products had become commonplace. Most people were well aware of the preservative action of salt. It is also believed that the color-preserving properties of saltpeter (nitrate) were discovered as a result of its presence as an impurity in salt. Present rapid curing methods, such as brine injection procedures, are more recent origin.

Many innovations have occurred in meat curing during the last 25 years, which have been made possible by advances in equipment design and greater mechanization. Meat curing was used originally as a means of preserving meat during times of plenty to carry over to times of scarcity. Today, cured meat products are generally mild-cured and must be stored under refrigeration.

Pickled Meats

An important method of preserving animal flesh is by curing, or pickling. Meat preservation may be accomplished by dry curing or with a pickling solution. The ingredients in curing and pickling are sodium nitrate, sodium nitrite, sodium chloride, sugar, and citric acid (or vinegar). For hams the ingredients may be used in the following proportions:

	(%)
Sodium chloride	24.0
Sodium nitrite	0.1
Sodium nitrate	0.1
Sugar	2.5

This dry mix dissolved in water yields a nearly saturated salt solution. Citric acid may be added in small amounts and speeds the conversion of nitrite to nitrous acid, which in turn combines with myoglobin, forming the stable red color of cured meats (for example, nitrosomyoglobin and pink nitrosohemochrome). Sodium chloride is used for its preservative properties. Sugar is added for flavoring. Nitrite and nitrate, in addition to their forming stable pigments, exert some control over anaerobic microorganisms.

Pickling and salting of meats to be cured is accomplished at temperatures between 1.6° and 3.3°C. Curing occurs at a rate of about an inch per week through treated flesh, depending on the thickness of the meat and the concentration of curing agents. When the meat cut is large and penetration of the cure is slow, the cure may be pumped into the meat via an artery (as is common with large hams) or the cure may be injected with multiple needles into slabs of bacon bellies (Fig. 11.5).

An excellent presentation on cured meats has been made by Kramlich et al. (1973). The following summary of cured meat technology has been abstracted from Processed Meats.

Salt

Salt is basic to all curing mixtures and is the only ingredient necessary for curing. Salt acts by dehydration, altering the osmotic pressure so that is inhibits bacterial growth and subsequent spoilage. Use of salt alone, however, gives a harsh, dry, salty product that is not very palatable. In addition, salt alone results in a dark, undesirable colored lean that is unattractive and objectionable to consumers.

Courtesy of Swift and Co.

FIG. 11.5.   AUTOMATIC MULTIPLE-NEEDLE INJECTION OF
CURE INTO BACON

As a consequence of undesirable effects salt is generally used in combination with both sugar and nitrite and/or nitrate. A limited number of products still are sometimes cured with salt alone. Cuts such as clear plates, fat backs, jowls, or heavy bellies that are intended for seasoning of other food products, including pork and beans, may occasionally be cured with salt alone. Even in the case of such fatty cuts, nitrite and/or nitrate is sometimes used. When salt is used alone, it is added in excess since the extreme saltiness is commonly modified by cooking with other food products.

Sugar and Corn Syrup Solids

The addition of sugar to cures is primarily for flavor. Sugar softens the products by counteracting the harsh hardening effects of salt by preventing some of the moisture removal and by a direct moderating action on flavor. Sugar also interacts with amino groups of the proteins and, upon cooking, forms browning products which enhance the flavor of cured meats. Corn syrup, molasses, and other natural sugar substitutes are sometimes used in place of sugar. The extent of substitution is largely a matter of cost after determining the relative effects upon flavor and color.

Nitrite and/or Nitrate

The function of nitrite in meat curing is: (1) to stabilize the color of the lean tissues; (2) to contribute to the characteristic flavor of cured meat; (3) to inhibit growth of a number of food poisoning and spoilage microorganisms; and (4) to retard development of rancidity. Although color stabilization was originally the primary purpose of adding nitrite to curing mixtures, its effects upon flavor and inhibition of bacterial growth are even more important. Nevertheless, the attractive pink-red color of cured meat adds to its desirability. The effect of nitrite on the flavor of cured meat has only recently been demonstrated. However, the most important reason for adding nitrite to meat cures appears to be its effect on microbial growth. It has been clearly demonstrated that nitrite is effective in preventing the growth of the *Clostridium botulinum* organism. Evidence also suggests the levels of nitrite found in cured meat may also aid in preventing the growth of other spoilage and food poisoning organisms. Nitrate serves principally as a source of nitrite.

Nitrosamines.—The reaction of nitrous acid (which is formed by the breakdown of nitrite) with secondary amines to produce nitrosamines is a well-known reaction in organic chemistry. The reaction of nitrous acid with dimethylamine is:

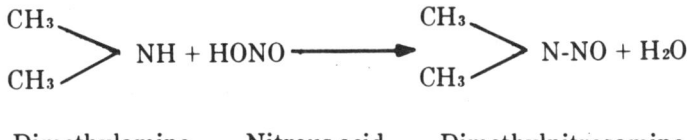

$$\underset{\text{Dimethylamine}}{\overset{CH_3}{\underset{CH_3}{>}} NH} + \underset{\text{Nitrous acid}}{HONO} \longrightarrow \underset{\text{Dimethylnitrosamine}}{\overset{CH_3}{\underset{CH_3}{>}} N\text{-}NO} + H_2O$$

Until recently nitrosamines were not of concern in meat curing. However, demonstration that nitrosamines are carcinogenic compounds has raised the question as to their occurrence in cured meats. Since nitrite is present in the cure and meat contains an abundance of secondary amines, it was assumed that nitrosamines may be present in cured meat. To date, results suggest that nitrosamines are not normally formed during curing, but they have been isolated from cured meats in a few instances. Work is now under way to determine the factors that control their formation The legal level permitted for nitrite in finished products is 156 ppm at the time of formulation.

Phosphates

Phosphates are added to the cure to increase the water-binding capacity and thereby the yield of finished product. The action of phosphates in improving water retention appears to be twofold: (1) raising the pH and (2) causing an unfolding of the muscle proteins, thereby making more sites available for water binding. Only alkaline phosphates are effective for improving water binding, since acid phosphates may lower the pH and cause greater shrinkage. Phosphates will improve the retention of brine and improvements in yields have been noted. With addition of phosphates to pumping brines, it is not difficult to obtain finished yields for intact hams and shoulders of over 100%.

Sodium tripolyphosphate, sodium hexametaphosphate, sodium acid pyrophosphate, sodium pyrophosphate, and disodium phosphate have all been approved for production of cured primal cuts, but only sodium acid pyrophosphate is permitted in sausages. Legal limits for added residual phosphates are set at 0.5% in the finished product. Since meat contains 0.01% of natural phosphate, this must be subtracted in calculating the level added during curing.

Sodium Erythorbate

The salts of ascorbic acid and erythorbic acid are commonly used to hasten development and stabilize the color of cured meat. Ascorbates serve three main functions. (1) Ascorbates take part in the reduction of metmyoglobin to myoglobin, thereby accelerating the rate of curing. (2) Ascorbates react chemically with nitrite to increase the yield of nitric oxide from nitrous acid. (3) Excess ascorbate acts as an antioxidant, thereby stabilizing both color and flavor.

CURED MEAT COLOR

There are a number of muscle pigments in meat, including myoglobin, hemoglobin, the cytochromes, catalase, the flavins, and other colored substances. Quantitatively, myoglobin and hemoglobin are by far the most abundant.

Myoglobin and hemoglobin are both complex proteins that undergo similar reactions in meat. However, their roles in living tissue are quite different. Hemoglobin is the red pigment found in blood and acts as the carrier for oxygen to the tissues. Myoglobin is the predominant pigment in muscle and serves as the storage mechanism for oxygen at the cellular level. Because of the differences in function, myoglobin has a greater affinity for oxygen. The rapid uptake of oxygen by myoglobin is evident in meat upon exposure of a freshly cut surface to the air, as manifested by the rapid brightening in color as it takes up oxygen.

An essential difference in structure between myoglobin and hemoglobin is that myoglobin complexes only one heme group per molecule, whereas hemoglobin contains four hemes per molecule. Thus, myoglobin has a molecular weight of 16,000 to 17,000 as compared to approximately 64,000 for hemoglobin. The chemical structure of myoglobin is shown in Fig. 11.6. The central atom does not contribute any electrons but has accepted 6 pairs of electrons from other atoms—5 pairs from nitrogen and 1 pair from oxygen. Four of the nitrogen atoms contributing electrons are in the porphyrin ring, while the other nitrogen atom is from the imidazole group of the histidine molecule in the amino acid chain of globin. The nature of the group attached to the iron atom of the heme at the position shown to be occupied by the OH2 radical determines the color of the pigment, both of myoglobin and hemoglobin.

Myoglobin and hemoglobin can be both oxidized and oxygenated, both resulting from the presence of oxygen. The relative proportion of metmyoglobin and oxymyoglobin depends upon the partial oxygen

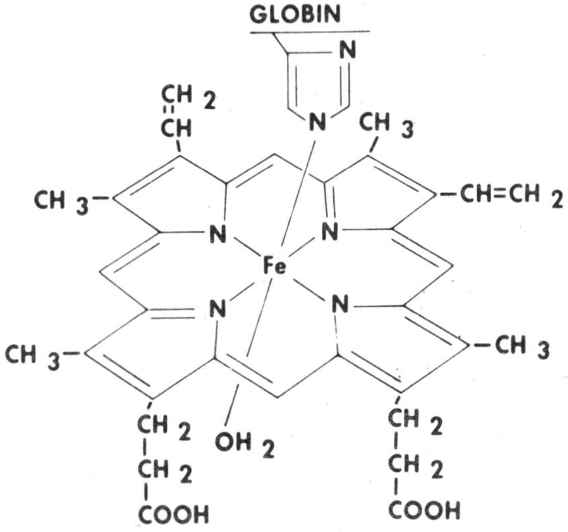

Courtesy of Kramlich et al.

FIG. 11.6.   DIAGRAM OF MYOGLOBIN MOLECULE SHOWING THE WAY THE HEME AND GLOBIN COMPLEX TOGETHER AND THE WAY IRON IS COMPLEXED WITH THE NITROGEN IN THE PYRROLE RINGS

pressure. At low oxygen pressures formation of metmyoglobin is favored, whereas high oxygen pressures favor oxymyoglobin formation. Thus, reduced myoglobin is constantly being both oxygenated and oxidized.

Role of Nitrite and/or Nitrate in Meat Color

Nitrite and/or nitrate are used in curing meat to counteract the undesirable effects of salt upon color. Not only is the color of fresh meat protected from degradation, but the pigments react with nitric oxide to produce the stable pigments characteristic of cured meat. These pink pigments are important to the acceptability of most processed meat products (Fig. 11.7).

Meats may be cured by mixing the product with a dry mix of curing agents, by soaking in a pickling solution, by pumping the pickling solution into the flesh through veins, by injection of the pickling solution with needles, or by a combination of these methods.

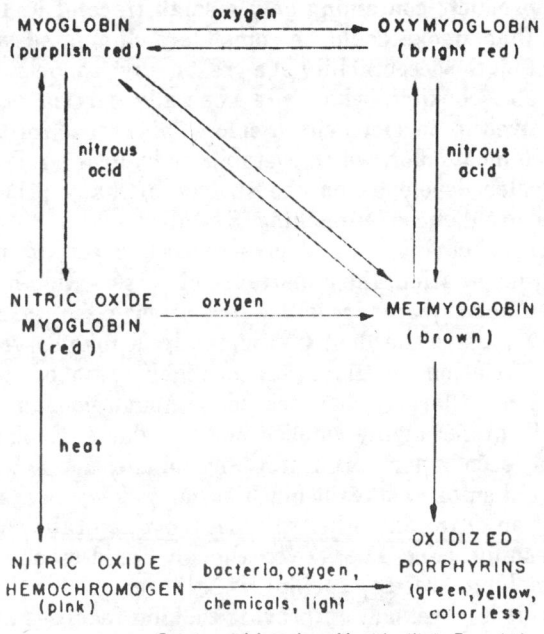

Courtesy of American Meat Institute Foundation

FIG. 11.7. COLOR CHANGES OBSERVED DURING THE CURING AND HANDLING OF MEATS

Curing may be combined with smoking in preserving animal products. Smoke acts not only as a dehydrating agent, but deposits a coating on the surface of the meat of materials obtained in the destructive distillation of wood. Small amounts of formaldehyde and other ingredients from the smoke are present. Smoking acts as a complementary preservative to the curing process. Pork and beef tissues are commonly preserved by these methods, especially in temperate climates in modern times. This is a very ancient process which still finds widespread use.

PURPOSES OF SMOKING

The primary purposes of smoking meat are: (1) development of flavor, (2) preservation, (3) creation of new products, (4) development of color, and (5) protection from oxidation. By smoking of meat and creation of new flavors, an entirely different group of meat products has been developed. Formerly, many of these products were heavily smoked. The trend is now toward less smoke flavor, many commercial products containing only a small trace of it. In fact, it is conceivable that smoke could be completely eliminated without any serious effect on the acceptability of a great many products.

Smoking and cooking, which are generally carried out together, are also involved in development of color. This is true for the development of cured meat color, which is stabilized by heating. Furthermore, the brown color developed on the surface of many processed meat products is also enhanced by smoking (Kramlich et al. 1973).

Smoking, like curing, has a preservative effect on meat. With mechanical refrigeration, the importance of preservation had declined. Today mild-smoked cured products are often eaten to add variety and attractiveness to the diet. Consequently, smoke serves primarily to provide variation in flavor. The highly smoked products of earlier times have largely disappeared, although consumers in some countries still prefer highly smoked meat products. Examples of this are commonly seen in northern European countries and in Iceland.

Use of liquid smoke makes it much easier to keep equipment clean, since deposition of tar and other residues from natural smoke requires frequent cleaning. (Fig. 11.8). Even though liquid smoke does away with the smoking process, cooking is still required for most meat items. Thus, it is necessary to provide cooking facilities even though the smoke generator is eliminated.

The continuous smoking system offers some major advantages in space saving, speed of processing and labor savings. It also allows more specific control of processing conditions (Fig. 11.9).

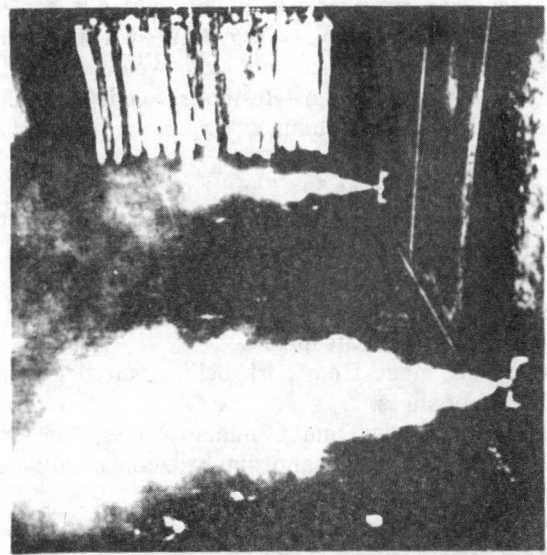

Courtesy of Red Arrow Corp.

FIG. 11.8. SMOKEHOUSE WITH LIQUID SMOKE INJECTION

Courtesy of The National Provisioner

FIG. 11.9. CONTINUOUS SMOKEHOUSE

Modern smokehouses are insulated, hermetically sealed, and thermostatically controlled. Smoke is generated with a device which burns sawdust or wood chips. Fruit tree woods as well as hickory, maple, oak and ash are popular. Resinous woods such as pine should not be used. Slight flavor changes can be achieved with different woods.

Trichinae are destroyed at 59°C and enzymatic activity is halted at about 60°C. Internal temperatures of 60°C must be sustained if the label states, under USDA authorization, "fully-cooked" or "ready to eat."

Many smoked products are cured before smoking. Members of the smoked/cured family of meats include:

Breakfast bacon, made from pork bellies, cured and smoked. It is available in slabs or slices.

Canadian bacon, made from Canadian backs, the sirloin portion of the pork loin. It is cured and smoked, and available in visking, canned, and sometimes ready to eat (Fig 11.10).

Courtesy of Swift and Co.

FIG. 11.10.   HAMS BEING REMOVED FROM A CABINET-TYPE SMOKEHOUSE

Hams, made from whole or boneless pork legs. They are generally cured and smoked. "Ready-to-eat" ham is a most popular product.

Virginia ham is not smoked but is barrel cured for about 7 weeks in salt, then rubbed with a mixture of molasses, brown sugar, black and cayenne pepper, and sodium nitrate, and cured for 2 more weeks. It is then hung for 1 to 12 months or more.

Smithfield ham is rubbed with salt and sodium nitrate and shelf cured for 5 days plus 1 day per pound. After curing, it is washed, rubbed with black pepper, shelved for 30 days, and then aged for 10 to 12 months.

Prosciutto ham (Italy) and Wesphalian ham (Germany) are hard, ready-to-eat hams of distinctive flavor. Both are outstanding for use as hors d'oeuvres.

Sausage denotes fresh ground pork items, plus a host of items sometimes but not always smoked, including wieners and salamis. Over 50 million kilos of sausage are produced annually in the United States.

Today there are several ways to generate smoke remotely and then circulate it into a smoke room or smoke tunnel (Fig. 11.11). Smoke can be generated in a special device without fire by high speed frictional contact with the wood. The smoke also can be given an electric charge and electrostatically deposited onto the meat surface. There also are synthetic solutions of the chemicals from smoke, but their use is legally restricted to only limited products.

Courtesy of The National Provisioner

FIG. 11.11. FORCED AIR SMOKEHOUSE

Sausages and Table-ready Meats

Cured meats especially, and uncured meats to a lesser extent, find their way into enormous quantities of sausage products. There are over 200 kinds of sausage products sold in the United States, the most popular of which are frankfurters. Most have their origins in countries outside the United States and are sold in the larger American cities where the population is of many origins.

DRY SAUSAGE MANUFACTURE

Pepperoni is a highly spiced (with or without paprika), fermented, fully-dry sausage prepared from pork or a mixture of pork and beef. Except for the drying step, its processing should be similar to that of other fermented sausages such as Lebanon bologna or summer sausage. However, neither details of the process nor the chemical and microbiological changes which occur during the various steps have been reported in the literature. Palumbo et al. (1976) studied sausages with characteristics similar to those of commercial pepperoni and followed the chemical and microbiological changes throughout processing.

Sugars-spices-cure used the following sugar/spice mixture for pepperoni:

Sugar/Spice	g/kg Meat Mix
sucrose	10
glucose	10
ground cayenne pepper	2
crushed red pepper	2
pimiento	5
whole anise seed	2½
garlic powder	0.1

The curing agent ($NaNO_3$) was added to 1.2 g per kg of meat. This level was a compromise with the approximately 1.7 g per kg permitted by Meat Inspection Regulations (Meat and Poultry Inspection Program, APHIS, USDA, 1973, Washington, D.C.).

Boneless pork shoulders or picnics were ground through a ¾-in. plate and frozen at −27°C and held until needed. Prior to use, meats were thawed at 10°C.

Pepperoni was fermented either by the natural lactic microflora of the meat, encouraged through the aging of salted (3%) meat at 5° C for 10 days, or by the addition of lactic acid starter culture.

Processing

The following general procedure is used for the experimental pepperoni. The frozen meat was thawed and mixed with 3% salt (NaCl). The salted meat was then aged for about 10 days (the aging step was omitted when starter culture was used). After aging, the sugar-spice mixture and $NaNO_3$ were added to the meat, which consisted of either pork or beef or a 1:1 mixture of the two, and mixed. This mixture was then ground through a 3/16-in. plate, stuffed into 55 mm clear fibrous casings, coated with paraffin (mp, 52°C), and hung at 35°C and 85% RH for 1 to 3 days to allow fermentation. The paraffin, which was used to prevent mold growth and excessive moisture loss during fermentation, was removed after fermentation and the sausage was dried for 6 weeks (40 to 42 days) at 12°C and 65% RH.

Commercial pepperoni were purchased and analyzed. All were prepared from a pork and beef mixture, most with a mixed (nitrate/ nitrite) cure. About one-half contained paprika and one-third were fermented with starter culture.

Compositional and chemical evaluations of commercial and experimental pepperoni are given in Table 11.9. The fat content of commercial products ranged from 38.1 to 52.8%. The protein varied from 17.9 to 24.8%, ash from 5.1 to 6.4%, and moisture from 17.0 to 31.5%. For all commercial and pilot plant pepperoni, moisture/ protein ratios (M/P) were less than 1.6/1.0, the maximum permitted for pepperoni.

Fermentation

Changes in the microflora occur during fermentation of pepperoni (Tables 11.10 and 11.11). Selective agars are extremely useful in defining the microbial sequence; yeasts die and are not detected after the first day of fermentation, and lactobacilli and micrococci then predominate. The number of lactobacilli which developed naturally during aging of salted beef and pork are sufficient to carry out the fermentation and the reduction of nitrate to nitrite.

FUTURE TRENDS

Some of man's most appetizing and nutritious foods are developed through the agencies of microorganisms and salt/sugar systems. Perhaps the most unexploited area in food technology is also the most ancient. There is little doubt that the future will find increasing attention being devoted to this important area.

TABLE 11.9

COMPOSITIONAL ANALYSES AND CHEMICAL MEASUREMENTS OF COMMERCIAL AND PILOT PLANT PEPPERONI

Company	Moisture (M), (%)	Ash (%)	Fat (%)	Protein (P), (%)	M/P Ratio	a_w	pH	Acid as Lactic (%)
Commercial Pepperoni								
A	30.9	5.4	39.2	20.2	1.48	0.87	5.5	0.49
B	17.0	5.6	50.0	21.2	0.80	0.80	5.2	0.71
C	17.0	5.3	52.8	20.1	0.85	0.81	6.1	0.29
D	31.5	6.4	38.1	20.5	1.54	0.83	4.8	0.84
E	22.6	6.1	44.1	21.5	1.05	0.84	5.8	0.40
F (loc 1)[1]	30.8	5.5	41.7	21.1	1.46	0.87	4.9	0.56
F (loc 2)[1]	27.1	5.8	43.7	24.8	1.09	0.84	4.9	0.61
F (loc 2)[1] sandwich style	26.3	5.4	47.7	20.4	1.29	0.85	4.8	0.65
G	25.4	5.6	42.2	20.8	1.22	0.86	5.1	0.55
H	18.4	5.4	52.5	21.9	0.84	0.81	4.7	0.73
Imported	21.1	5.1	51.0	17.9	1.18	0.83	5.3	0.37
Pilot Plant Pepperoni								
Expt. I all pork	24.1	6.3	37.5	28.3	0.85	0.81	4.7	not done
Expt. II all pork	20.8	6.7	40.6	28.6	0.73	0.80	4.8	1.14
Expt. III all beef	28.9	7.9	18.1	41.3	0.70	0.82	4.9	1.54
all pork	27.0	7.1	32.2	32.2	0.84	0.83	4.8	1.42
pork-beef	28.6	7.6	23.0	37.7	0.76	0.82	4.8	1.33

Source: Palumbo et al. (1975).

[1] The products of company F are manufactured at two plants, designated loc 1 and loc 2. Sandwich style designates a product produced in a wide (ca 45 mm finished size) casing; all other products were of the stick variety, finished diameter ca 25 mm.

TABLE 11.10

NUMBER OF VIABLE MICROORGANISMS PRESENT IN SELECTED COMMERCIAL PEPPERONI

Company	Viable Counts Per Gram of Pepperoni Plated On:[1]				
	APT	PD	Rog	EMB	MSA
A	1.3×10^8	3.3×10^3	3.7×10^7	1.0×10^5	1.2×10^7
B	1.0×10^4	$<1 \times 10^2$	1.0×10^2	3.0×10^2	$<1 \times 10^2$
C	2.5×10^7	9.6×10^3	1.4×10^5	3.0×10^5	1.1×10^7
D	2.3×10^7	1.0×10^5	4.0×10^4	$<1 \times 10^2$	5.0×10^4
E	6.0×10^6	1.0×10^2	1×10^2	3.5×10^2	1.5×10^5
F (loc 1)[2]	1.0×10^8	$<1 \times 10^2$	8.5×10^7	$<1 \times 10^2$	2.7×10^3
F (loc 2)[2]	3.3×10^6	1.0×10^2	1.6×10^6	$<1 \times 10^2$	2.0×10^3
F (loc 2)[2] sandwich style	1.0×10^4	$<1 \times 10^2$	2.7×10^3	3.9×10^3	1.1×10^4
G	7.0×10^6	1.1×10^5	6.9×10^5	3.0×10^4	1.3×10^6
H	1.2×10^7	1.6×10^4	6.0×10^6	1.8×10^3	7.0×10^3
Imported	2.5×10^6	$<1 \times 10^2$	1.1×10^6	1.0×10^4	1.7×10^6

Source: Palumbo *et al.* (1976).
[1] APT, APT agar; PD, acidified potato dextrose agar; Rog, rogosa SL agar; EMB, Eosin Methylene Blue agar; MSA, mannitol salt agar.
[2] The products of company F are manufactured at two plants, designated loc 1 and loc 2. Sandwich style designates a product produced in a wide (ca 45 mm finished size) casing; all other products were of the stick variety, finished diameter ca 25 mm.

TABLE 11.11

CELLULAR TYPES, AND GRAM AND CATALASE REACTIONS
OF THE MICROFLORA FOUND IN COMMERCIAL PEPPERONI

Company	Microbial Types[1] Found on the Following Media				
	APT	PD	Rog	EMB	MSA
A	a	e	a	f	c
B	b, d	—	f	c	—
C	c	e	a	c	c
D	a	e	e	—	e
E	c, b	e	—	d	c
F (loc 1)[2]	f	—	f	—	f
F (loc 2)[2]	c	e	c	—	b
F (loc 2)[2] sandwich style	b	—	a	b	b, d
G	a, c	e	a	f	c
H	f	e	f	b	b, e
Imported	a	—	a	f	c

Source: Palumbo *et al.* (1975).
[1] a = lactobacilli (catalase-negative, gram-positive rods).
b = bacilli (catalase-positive, gram-positive sporeforming rods).
c = micrococci (catalase-positive, gram-positive cocci).
d = catalase-positive, gram-negative rods (not typical coliforms).
e = yeast.
f = catalase-negative, gram-positive cocci.
[2] The products of company F are manufactured at two plants, designated loc 1 and loc 2. Sandwich style designates a product produced in a wide (ca 45 mm finished size) casing; all other products were of the stick variety, finished diameter ca 25 mm.

REFERENCES

ACTON, J.C., and SAFFLE, R.L. 1969. Preblended and prerigor meat in sausage emulsions. Food Technol. *23*, 93.

AM. MEAT INST. 1953. Sausage and Ready to Eat Meats. Univ. of Chicago Press, Chicago.

AMERINE, M.A., and JOSLYN, M.A. 1951. Table Wines. Univ. of California Press, Berkeley.

ANDERSON, E.E. *et al.* 1951. Pasteurized fresh whole pickles. II. Thermal resistance of micro-organisms and peroxidase. Food Technol. *5*, 364-368.

ANGELOTTI, R. 1973. FDA regulations promote quality assurance. Food Technol. *29*, No. 11, 60-62.

ANON. 1946. Meat Curing and Sausage Making. Morton Salt Co., Chicago.

ANON. 1965. Hydrocarbon residues in cooked and smoked meats. Nutr. Rev. *23*, 268.

BAILEY, E.J., and DUNGAL, N. 1958. Polycyclic hydrocarbons in Icelandic smoked fish. Brit. J. Cancer *12*, 348.

BAUMAN, H.E. 1974. The HACCP concept and microbiological hazard categories. Food Technol. *28*, No. 9, 30-34, 70.

BEAZLEY, C.C., and INSALATA, N.F. 1971. The roles of fluorescent-antibody methodology and statistical techniques for salmonellae protection in the food industry. Food Technol. *25*, No. 3, 24-26.

BELL, T.A., ETCHELLS, J.L., and JONES, I.D. 1955. A method for testing cucumber salt-stock brine for softening activity. U.S. Dep. Agric., Agric. Res. Serv. Rep. 72-5.

BODWELL, C.E., and McCLAIN, P.E. 1971. Proteins. *In* The Science of Meat and Meat Products, J. F. Price, and B.S. Schweigert (Editors). W.H. Freeman and Co., San Francisco.

BONE, D. 1973. Water activity in intermediate moisture foods. Food Technol. *27*, No. 4, 71-76.

BRADY, D.E., SMITH, F.H., TUCKER, L.N., and BLUMER, T.N. 1949. The characteristics of country style hams as related to sugar content of curing mixture. Food Res. *14*, 303.

BULLERMAN, L.B. 1974. A screening medium and method to detect mycotoxins in mold cultures. J. Milk Food Technol. *37*, 1.

BULLERMAN, L.B. 1976. Examination of Swiss Cheese for incidence of mycotoxin producing molds. J. Food Sci. *41*, No. 1, 26-28.

BULLERMAN, L.B., and OLIVIGNI, F.J. 1974. Mycotoxin producing potential of molds isolated from Cheddar cheese. J. Food Sci. *39*, 1166.

BURKHOLDER, L. *et al.* 1968. Fish fermentation. Food Technol. *22*, No. 10, 76-82.

BYRAN, F.L. 1974. Microbiological food hazards today—based on epidemiological information. Food Technol. *28*, No. 9, 52-66, 84.

CAMPBELL, C.H., ISKER, R.A., and MACLINN, W.A. 1954. Campbell's Book—A Manual on Canning, Preserving and Pickling. Vance Publishing Corp., Chicago.

CHAPLIN, M.H., and DIXON, A.R. 1974. A method for analysis of plant tissues by direct reading spark emission spectroscopy. Appl. Spectrosc. *28*, 5.

CHIPLEY, J.R., and SAFFLE, R.I. 1968. Various acid effects on the peeling performance, shelf life, color and flavor of frankfurters. Food Technol. *22*, No. 11, 128-131.

COSTILOW, R.R. 1957. Sorbic acid as a selective agent for cucumber fermentations. Food Technol. *11*, 591-594.

CRANFIELD, D. 1973. Cucumber brining and salt recovery. Pickle Packers International Seminar on Pickle Processing Wastes, Chicago.

CRUESS, W.V. 1958. Commercial Fruit and Vegetable Products. McGraw-Hill Book Co., New York.

DAVIS, N.D., DIENER, U.L., and ELDRIDGE, D.W. 1966. Production of aflotoxins B_1 and G_1 by Aspergillus flavus in a semi-synthetic medium. Appl. Microbiol. 14, 378.

DOERR, R.C., WASSERMAN, A.E., and FIDDLER, W. 1966. Composition of hickory sawdust smoke. Low-boiling constituents. J. Agric. Food Chem. 14, 662.

DUNKER, C.F., BERMAN, M., SNIDER, G.G., and TUBIASH, H.S. 1953. Quality and nutritive properties of different types of commercially cured hams. III. Vitamin content, biological value of protein and bacteriology. Food Technol. 7, 288.

DURKEE, E.L., LOWE, E., BAKER, K.A., and BURGESS, J.W. 1973. Field tests of salt recovery system for spent brine. J. Food Sci. 38, 507.

DURKEE, E.L., LOWE, E., and TOOCHECK, E.A. 1974. Use of recycled salt in fermentation of cucumber salt stock. J. Food Sci. 39, 1032.

ETCHELLS, J.L., BORG, A.F., and BELL, T.A. 1968. Bloater formation by gas-forming lactic acid bacteria in cucumber fermentations. Appl. Microbiol. 10, 1029.

ETCHELLS, J.L., FABIAN, F.W., and JONES, I.D. 1945. The Aerobacter fermentation of cucumbers during salting. Mich. State Univ. Agric. Exp. Stn. Tech. Bull. 200.

ETCHELLS, J.L. et al. 1973. Suggested procedure for the controlled fermentation of commercially brined pickling cucumbers—The use of starter cultures and reduction of carbon dioxide accumulation. Pickle Pak. Sci. 3, 4.

ETCHELLS, J.L. et al. 1975. Factors influencing bloater formation in brined cucumbers during controlled fermentation. J. Food Sci. 40, 569.

ETCHELLS, J.L., and NONTS, L.M. 1973. Advisory statement: Information on the nature and use of an improved system for recording quality control data during the brining of cucumbers. Pickle Packers International, Inc., St. Charles, Ill.

FAN, T.Y., and TANNENBAUM, S.R. 1972. Stability of N-nitroso compounds. J. Food Sci. 37, 274.

FELLERS, C.R. 1960. Effects of fermentations on food nutritients. In Nutritional Evaluation of Food Processing. R.S. Harris, and H. Von Loesecke (Editors). John Wiley & Sons, New York.

FELLERS, P.J. 1968. Loss of whiteness from fresh cucumber pickles. Food Technol. 22, No. 12, 105-107.

FLEMING, H.P., and ETCHELLS, J.L. 1967. Occurrence of an inhibitor of lactic acid bacteria in green olives. Appl. Microbiol. 15, 1178.

FLEMING, H.P., ETCHELLS, J.L., THOMPSON, R.L., and BELL, T.A. 1975. Purging of CO_2 from cucumber brines to reduce bloaters. J. Food Sci. 40, No. 6, 1304-1310.

FLEMING, H.P., THOMPSON, R.L., and ETCHELLS, J.L. 1974. Determination of carbon dioxide in cucumber brines. J. Assoc. Off. Agric. Chem. 57, 130.

FLEMING, H.P. et al. 1973A. Bloater formation in brined cucumbers fermented by Lactobacillus plantarum. J. Food Sci. 38, 499.

FLEMING, H.P. et al. 1973B. Carbon dioxide production in the fermentation of brined cucumbers. J. Food Sci. 38, 504.

GEISMAN, J.R., and HENNE, R.E. 1973A. Recycling food brine eliminates pollution. Food Eng. 45, No. 11, 119.

GEISMAN, J.R., and HENNE, R.E. 1973B. Recycling brine from pickling. Ohio Rep. 58, 76.

GIDDINGS, G.G., and MARKIKIS, P. 1972. Characterization of the red pigments produced from ferrimyoglobin by ionizing radiation. J. Food Sci. 37, 361.

GREENBERG, R.A. 1972. Nitrite in the control of Clostridium botulinum. In Proceedings of the Meat Industry Research Conference. American Meat Institute Foundation, Chicago.

HAQ, A., WEBB, N.B., WHITFIELD, J.K., and IVEY, F.J. 1973. Effect of composition on the stability of sausage type emulsions. J. Food Sci. 38, 271.

HAQ, A., WEBB, N.B., WHITFIELD, J.K., and MORRISON, G.S. 1972. Development of a prototype sausage emulsion preparation system. J. Food Sci. 37, 480.

HOAGLAND, R. et al. 1947. Composition and nutritive value of hams as affected by method of curing. Food Technol. 1, 540.

HOWARD, J.W., TEAGUE, R.T., JR., WHITE, R.H., and FRY, B.E., JR. 1966. Extraction and estimation of polycyclic aromatic hydrocarbons in smoked foods. I. General method. J. Assoc. Off. Agric. Chem. 49, 595.

HOWARD, J.W., WHITE, R.H., FRY, B.E., JR., and TURICCHI, B.W. 1966. Extraction and estimation of polycyclic aromatic hyocarbons in smoked foods. II. Benz(a)pyrene. J. Assoc. Off. Agric. Chem. 49, 611.

IFT EXPERT PANEL AND CPI. 1975. Sulfites as food additives. Food Technol. 29, No. 10, 117-120.

JENNESS, R., and PATTON, S. 1959. Principles of Dairy Chemistry. John Wiley & Sons, New York.

JOHNSON, A.H., and PETERSON, M.S. 1974. Encyclopedia of Food Technology. AVI Publishing Co., Westport, Conn.

KEMP, J.D., SMITH, R.H., and MOODY, W.G. 1968. Quality of aged hams as affected by alternating aging temperatures. Food Technol. 22, No. 10, 113-114.

KOLARI, O.E., and AUNAN, W.J. 1972. The residual level of nitrite in cured meat products. 18th Meeting European Meat Research Workers, American Meat Institute Foundation, Chicago.

KRAMLICH, W.E., PEARSON, A.M., and TAUBER, F.W. 1973. Processed Meats. AVI Publishing Co., Westport, Conn.

LOVE, S., and BRATZLER, L.J. 1966. Tentative identification of carbonyl compounds in wood smoke by gas chromatography. J. Food Sci. 31, 218.

MARQUARDT, R.A., PEARSON, A.M., LARZELERE, H.E., and GREIG, W.S. 1963. Use of the balanced lattice design in determining consumer preferences for ham containing 16 different combinations of salt and sugar. J. Food Sci. 28, 421.

MONROE, R.J. 1969. Influence of various acidities and pasteurization temperatures on the keeping quality of fresh pack dill pickles. Food Technol. 23, No. 1, 71-77.

MORRISON, G.S. et al. 1971. Relationship between composition and stability of sausage-type emulsions. J. Food Sci. 36, 426.

OHLMEYER, D.W. 1957. Use of glucose oxidase to stabilize beer. Food Technol. 11, 503-508.

PALNITKAR, M.P., and MCFEETERS, R.F. 1975. Recycling spent brine in cucumber fermentations. J. Food Sci. 40, No. 6, 1311-1315.

PALUMBO, S.A., ZAIKA, L.L., KISSINGER, J.D., and SMITH, J.L. 1976. Microbiology and technology of the pepperoni process. J. Food Sci. 41, 12-17.

PEARSON, A.M., BATEN, W.D., GOEMBEL, A.J., and SPOONER, M.E. 1962. Application of surface-response methodology to predicting optimum levels of salt and sugar in cured ham. Food Technol. 16, No. 5, 137.

PEDERSON, C.S. 1930A. Floral changes in fermentation of sauerkraut. N.Y. Agric. Exp. Stn. Tech. Bull. 168.

PEDERSON, C.S. 1930B. The effect of pure culture inoculation on the quality and chemical composition of sauerkraut. N.Y. Agric. Exp. Stn. Tech. Bull. 169.

PEDERSON, C.S., and ALBURY, M.N. 1954. The infulence of salt and temperature on the microflora of sauerkraut fermentation. Food Technol. 8, 1-5.

PEDERSON, C.S., and BEATTIE, H.G. 1943. Vinegar making. N.Y. Agric. Exp. Stn. Circ. 148.

PEDERSON, C.S., and WARD, L. 1949. The effect of salt upon the bacteriological and chemical changes in fermenting cucumbers. N.Y. Agric. Exp. Stn. Bull. 288.

PORTER, R.W., BRATZLER, L.H., and PEARSON, A.M. 1965. Fractionation and study of compounds in wood smoke. J. Food Sci. 30, 615.

POTTER, N.N. 1967. Salmonella contamination. Am. Dairy Rev. 29, No. 9, 70, 75-76, 124-127.

POTTER, N.N. 1973. Food Science, 2nd Edition. AVI Publishing Co., Westport, Conn.

RAPER, K.B., and FANNELL, D.I. 1965. The Genus Aspergillus. Williams and Wilkins, Baltimore, Md.

RAPER, K.B., and THOM, C.A. 1968. A Manual of the Penicillin. John Wiley & Sons, New York.

REINBOLD, G.W., ANDERSON, R.F., and HERMAN, L.G. 1973. Microbiological methods for cheese and other cultured products. In Standard Methods for the Examination of Dairy Products, 13th Edition, W. Hausler (Editor). American Public Health Assoc., Washington, D.C.

SAMISH, S. 1955. Studies on pickling Spanish-type green olives. Food Technol. 9, 173-176.

SAMISH, S., COHEN, S., and LUDIN, A. 1968. Progress of lactic acid fermentation of green olives as affected by peel. Food Technol. 22, No. 8, 77-80.

SANDERS, G.P. 1953. Cheese Varieties and Descriptions. U.S. Dep. Agric., Agric. Handb. 54.

SCOTT, P.M., LAWRENCE, J.W., and VAN WALBEEK, W. 1970. Detection of mycotoxins by thin-layer chromatography: Application of screening to fungal extracts. Appl. Microbiol. 20, 839.

SMALL, E., and FENTON, E.F. 1964. A summary of laws and regulations affecting the cheese industry. U.S. Dep. Agric., Agric. Handb. 265.

SOMMER, H.H. 1951. The Theory and Practice of Ice Cream Making, 6th Edition. Published by the author. Madison, Wisc.

TRESSLER, D.K., and LEMON, J.M. 1951. Marine Products of Commerce. Reinhold Publishing Co., New York.

U.S. DEP. HEALTH, EDUC. AND WELFARE. 1965. Grade A Pasteurized Milk Ordinance. U.S. Public Health Serv. Publ. 229.

VAN SLYKE, L.L., and PRICE, W.V. 1952. Cheese. Orange Judd Co., New York.

VAN WALBEEK, W. 1973. Fungal toxins in foods. Can. Inst. Food Technol. J. 6, 96.

VERRETT, M.J., MARTIAC, J.P., and MCLAUGHLIN, J., JR. 1964. Use of embryo in the assay of aflatoxin toxicity. J. Assoc. Off. Agric. Chem. 74, 1003.

WALLERSTEIN LAB. 1975. Bottle Beer Quality. Wallerstein Laboratories, New York.

WEBB, B.H., JOHNSON, A.H., and ALFORD, J.A. 1974. Fundamentals of Dairy Chemistry, 2nd Edition. AVI Publishing Co., Westport, Conn.

WEBB, B.H., and WHITTIER, E.O. 1971. Byproducts from Milk, 2nd Edition. AVI Publishing Co., Westport, Conn.

WHO. 1962. Milk Hygiene. FAO/WHO Monogr. Ser. 48.

Chapter 12

Principles of Chemical Preservation of Foods

The essential value of selected chemical compounds in preserving foods, the benefits derived by their judicious applications, and the general principles governing their uses have received attention in most governments and their laboratories. The matter is so important that it was and is the subject of many national and international conferences.

In the United States the National Academy of Sciences (NAS)—National Research Council (NRC), the President's Science Advisory Committee Panel on Chemicals and Health, and expert committees from governmental agenices, universities and industries have focused on the area. The Food and Agriculture Organization (FAO) and the World Health Organization (WHO) of the United Nations have also been active in this field, approaching the matter from an international point of view, generally embracing the important features of the thinking of most nations in the world.

As a result, a growing body of knowledge has become available which has coalesced and yielded insights into this important subject (Fig. 12.1).

WHAT ARE FOOD ADDITIVES?

In the United States the Food Protection Committee of the NAS/NRC defined a food additive as a substance or a mixture of substances, other than a basic foodstuff, which is present in food as a result of any aspect of production, processing, storage or packaging. The term does not include chance contamination.

Courtesy of FDA

FIG. 12.1.   FOOD ADDITIVE TEST ON MICE
AT FDA LABORATORY

A food additive in accord with the Food Additives Amendment of 1958 is any substance which is or may become a part of food. Additives can be classified on the basis of legal status, chemical effect, chemical classification, and natural or synthetic.

Middlekauff (1976) states that a food additive, according to the Federal Food, Drug and Cosmetic Act (U.S. Code, Section 321 (s) is a substance—the intended use of which results or may reasonably be expected to result, directly or indirectly, in its becoming a component or otherwise affecting the characteristics of any food—with certain exceptions. Specifically, the statute excludes from this classification   (Middlekauff 1976) a substance which is generally recognized among experts, qualified by reason of scientific training and experience to evaluate its safety, as having been adequately shown through scientific procedures to be safe under the conditions of its intended use. These are the GRAS (Generally Recognized As Safe) substances. Those GRAS substances in use prior to 1958 have as a basis of judgement either scientific procedures or experience based on common use in food. Those in use after 1958 have as a basis for judgement the need for scientific procedures to establish their safety in use.

It is clear that there are intentional additives which are added to perform specific functions. There are incidental additives which really have no function in a food product but become part of a food product through some phase of production, processing, storage or packaging. Food additives do not include those substance which may find their way into foods accidentally, inadvertently or unintentionally. The terms do not include pesticides, for example, nor color additives, new animal drugs or any substance used in accordance with a sanction or approval granted prior to the effective date of the Food Additives Amendment of 1958 (Middlekauff 1976).

DETERMINING USABILITY

If a company has a specific need for a particular additive and wants to know whether it may be used in food for the purpose intended and in the amount intended, it should first check to see if the substance is listed in one of the FDA regulations found in Title 21 of the *Code of Federal Regulations* (CFR). The various lists include:

Food Additives permitted in food for human consumption (*CFR*, Title 21, Section 121.1000).

Prior Sanction Substances and substances employed in the manufacture of food-packaging materials. Remember that the Act exempts from the definition of food additives those substances that had received a prior sanction (*CFR*, Title 21, Section 121.2001). FDA places the burden of coming forth with evidence of prior sanction on the person who asserts it (*Fed. Regist. 38*, 20048).

Indirect Food Additives (*CFR*, Title 21, Section 121.2500).

Interim Food Additives (*CFR*, Title 21, Section 121.4000). This grouping was created in December, 1972 to provide a classification for those substances which have a history of use in food for human consumption or food contact surfaces, but for which new information has raised a substantial question about their safety or functionality. Such substances are placed in this interim class when they are not harmful and when there is a reasonable certainty that no harm to the public health is expected to result from their continued use for a limited period of time, while the question raised is being resolved by further study (Middlekauf 1976).

The FDA list of substances generally recognized as safe is the so-called White List (*CFR*, Title 21, Section 121.4000).

If the substance is a flavor it may be on one of the GRAS lists of the Flavor and Extract Manufacturers' Association (Hall 1975).

If the substance is not found in any of the above lists, ask the FDA whether the substance has ever been determined as GRAS. It might be an unlisted substance being covered in the FDA GRAS review.

If a search discloses that no decision has ever been made on the substance by either the FDA or any other reliable group of experts, then request a formal determination as to whether the substance may be used in food for the purpose intended by writing to the FDA.

UN STUDY GROUPS

A report of a Joint FAO and WHO Expert Committee on Food Additives was published following meetings in Rome in 1956. This committee was requested to formulate general principles governing the use of food additives. They defined food additives as non-nutritive substances added intentionally to food, generally in small quantities, to improve its appearance, flavor, texture or storage properties. Substances added primarily for the nutritive value, such as vitamins and minerals, are not considered in this category. It is recognized, however, that in certain instances chemicals added to food to impart a desired quality or for some other functional purpose may also be of nutritional value.

A decade later, the Committee again reviewed the entire matter and felt it important to add that an increase in the number of permitted food additives does not imply an increase in the total amount of additives eaten per person. Different varieties of food additives are used largely as alternatives. Thus, from a toxicological point of view there is really less likelihood of high or cumulative dose levels being attained if a wide range of substances is available for use (WHO 1967).

Clearly, the availability of chemical substances for use in food is not to be confused with human consumption. Sugar, a chemical substance which is used in bread making, is largely fermented in the process. Sulfur dioxide used in the wine industry is volatilized. Many fats and oils may be used in cooking but drained away from the food prior to eating.

Importance of Chemical Additives

Additives can contribute substantially in the preservation of food; for example, they can help prevent the loss of seasonal surpluses. In economically underdeveloped countries, lack of functional storage facilities and the inadequacy of transportation and communications may increase the necessity of using certain food additives for purposes of food preservation. In tropical regions high temperatures and humidities favor microbial attack, and increase the rate of development of oxidative rancidity. A wider use of antimicrobial agents and antioxidants may be more justified in these countries than in those in more temperate climates. It is recognized that the increased risks associated with the increased use of food additives must be weighed against the benefits gained from preventing food losses and making more food available in areas in which it is needed. In such circumstances, food additives might be used to supplement the effectiveness of traditional methods of food preservation rather than to replace these methods.

On the other hand, the committee felt that in countries which are technically and economically highly developed, the availability of adequate facilities for refrigerated transportation and storage greatly reduces, even if it does not eliminate, the need for chemical preservatives. In these countries, however, there is an increasing demand for more attractive foods of uniform quality, and for a wide choice of foods at all seasons of the year. Moreover, large quantities of many of the foods consumed have to be transported from distant producing areas, a fact which may create special transportation and storage problems. For such purposes the variety of useful food additives is great and their employment improves the utilization of the available foods. The nature of the food additives and the extent to which food additives are likely to be needed vary considerably from region to region and even from country to country. In decisions concerning the use of an additive, attention should be given to its technological usefulness, protection of the consumer against deception, the use of inferior techniques in processing and to the evidence bearing on the safety of using the additive.

Legitimate Uses in Food Processing

Food additives have a legitimate use in the food processing and distribution systems of both technologically advanced and of less well developed countries, in promoting the utilization of available foods.

The use of food additives to the advantage of the consumer may be technologically justified when it serves the following purposes:

(1) The maintenance of the nutritional quality of a food.
(2) The enhancement of keeping quality or stability with resulting reduction in food losses.
(3) Making foods attractive to the consumer in a manner which does not lead to deception.
(4) Providing essential aids in food processing.

Undesirable Uses of Additives

The use of food additives is not in the best interest of the consumer in the following situations and should not be permitted:

(1) To disguise the use of faulty processing and handling techniques.
(2) To deceive the consumer.
(3) When the result is a substantial reduction of the nutritive value of the food.
(4) When the desired effect can be obtained by good manufacturing practices which are economically feasible.

Safety of a Food Additive

Safety in using an additive is an all important consideration. While it is impossible to establish absolute proof of the nontoxicity of a specified use of an additive for all human beings under all conditions, critically designed animal tests of the physiological, pharmacological, and biochemical behavior of a proposed additive can provide a reasonable basis for evaluating the safety of use of a food additive at a specified level of intake. Any decision to use an intentional additive must be based on the considered judgement of properly qualified scientists that the intake of the additive will be substantially below any level which could be harmful to consumers. The committee also recommended that permitted additives should be subjected to continuing observation for possible deleterious effects under changing conditions of use, and should be reappraised whenever indicated by advance in knowledge. Special recognition in such reappraisals should be given to improvements in toxicological methodology.

Clearly, other factors must be taken into account in food additives control. When a new food additive is proposed for use, substantial evidence must be available to show that benefits to the consumer will ensue. In those classes of foods which constitute a considerable proportion of the diet the FAO-WHO position is that the use of intentional additives should, in principle, be limited. It is generally felt that the presence of harmful impurities in food additives can be excluded most effectively by the establishment of specifications of purity. To be functional, food additives should be identifiable in chemical and physical terms.

It is agreed that the amount of an authorized additive used in a food should be the minimum necessary to produce the desired effect. The limit should be established with due attention to the following factors:

(1) The estimated level of consumption of the food or foods for which the additive is proposed.

(2) Minimal levels which in animal studies produce significant deviations from normal physiological behavior.

(3) An adequate margin of safety to reduce to a minimum any hazard to health in all groups of consumers.

For all peoples, legal control over the use of food additives is essential. This is best accomplished through the use of a permitted list, which effectively prevents the addition of any new substances to food until an adequate basis for judgment of their freedom from health hazard has been established. It is the consensus of opinion of experts that the alternative method of a "prohibited" list involves risk, since it may result in the use of a harmful additive before being adequately studied.

As a matter of principle, consumers should be informed of the presence of additives in their food. Label declaration has been found to be the most effective method of achieving this result. A simple declaration of the presence of a particular additive is thought to be sufficient.

Regulations governing the control of food additives are useless unless the laws can be enforced. Trained food inspectors, food control laboratories, and reliable analytical methods are of utmost importance.

While from an international viewpoint the above has great significance in promoting useful practices, national responsibilities demand specific attention.

FUNCTIONAL CHEMICAL ADDITIVE APPLICATIONS

Chemical additives in foodstuffs have various functions. The following is an analysis of the functions of chemical additives in food technology. They are used as:

1. Preservatives
   a. Microbiological spoilage
   b. Chemical deterioration
   c. Chemicals to control insects and rodents
2. Nutritional supplements
   a. Vitamins
   b. Amino acids
   c. Minerals
   d. Calories
3. Color modifiers
   a. Natural coloring matters
   b. Certified food dye
   c. Derived colors˙
4. Flavoring agents
   a. Synthetic
   b. Natural
   c. Flavor enhancers or extenders
5. Chemicals which affect functional properties of foods
   a. Control of colloidal properties
      (1) Gel
      (2) Emulsion
      (3) Foam
      (4) Suspensoid
   b. Firming agents
   c. Maturing agents

6. Chemicals used to process foods
   a. For sanitation, public health, or aesthetic purposes
   b. To facilitate the removal of unwanted coverings (skins, hides, feathers, hair, etc.)
   c. Antifoaming agents
   d. Chelating agents
   e. Yeast nutrients
7. Chemicals to control moisture
   a. Waxes
   b. Anticaking agents
8. Chemicals used to control pH
   a. Acids
   b. Bases
   c. Salts
9. Chemicals used to control physiological functions in relation to quality
   a. Ripening agents
10. Miscellaneous
    a. Gases—pressure dispensing (Fig 12.2)

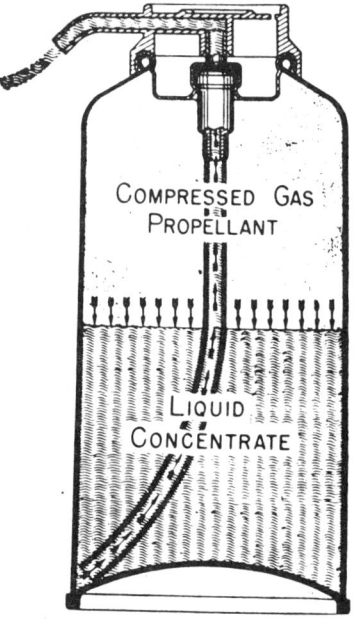

Courtesy of Food Processing

FIG. 12.2.   AEROSOL CONTAINER FOR FOOD
A development in dispensing food products

Historical Significance

The use of chemical additives in food preservation has a long history. Waterglass preservation of eggs (in sodium silicate solution) has been employed successfully in Europe for centuries. The solution seals the egg shell and maintains eggs in edible condition under cool storage over winter. Significantly, the tested methods of chemical preservation remain important to modern man. While there have been obvious misuses of chemicals in the past, fortunately the uses of chemical additives in foods have improved over the years with the expanding chemical technology.

Specific Uses of Chemical Additives

The varied uses for chemical additives commonly employed in food processing and manufacture have been compiled by the Food and Drug Administration in the United States, and include:

Acidifying
Alkalizing
Anticaking
Antidrying (humectants)
Antifoaming
Antihardening
Antispattering
Antisticking
Bleaching
Buffering
Chillproofing
Clarifying
Color-retaining
Coloring
Conditioning (dough)
Creaming
Curing
Dispersing
Dissolving
Drying
Emulsifying
Enhancing flavor
Enriching
Firming

Flavoring
Foam-producing
Glazing
Leavening
Lining food containers
Maturing agent (flour)
Neutralizing
Peeling
Plasticizing
Preserving (including antioxidants)
Pressure dispensing
Refining
Replacing air in food packages
Sequestering (removing unwanted
 metallic ions)
Stabilizing
Sterilizing
Supplementing nutrients (yeast food)
Sweetening
Texturizing
Thickening
Waterproofing (wrappers)
Water-retaining
Whipping

Additives Permitted and Prohibited in the United States

The Food and Drug Administration of the Department of Health, Education and Welfare administers the Federal Food, Drug and Cosmetic Act, as amended, in the United States. It is useful to consider the Food and Drug Administration position relative to chemical additives to foods. During hearings before the Congress of the United States, the Food and Drug Administration made available lists of chemical additives generally considered safe, those which are permitted for special purposes, those not satisfactory for use, and examples of additives that have no functional value when added to foods.

The following is a partial list of chemical food additives[1] generally recognized as safe (GRAS) by the Food and Drug Administration when used according to accepted cooking procedures or as used in good commercial practice.

Acacia gum
Acetic acid (dilute)
Agar-Agar
Aluminum and potassium sulfate (alum)
Aluminum sodium sulfate
Amino acids which are normal constituents in foods
Baking powder
Baking soda
Bay leaves
Brandy
Butter
Calcium carbonate
Calcium sulfate
Caramel
Carbon dioxide
Carragheen
Citric acid
Cloves
Coffee
Corn oil
Cornstarch
Corn syrup
Cottonseed oil

Cream
Cream of tartar
Dextrin
Dicalcium orthophosphate
Dried skim milk
Ethyl vanillin
Gelatin
Glycerin
Karaya gum
Lard
Lecithin
Lemon juice
Lemon extract
Mace
Magnesium carbonate
Margarine
Methyl- and propyl-para-hydroxy-benzoates
Molasses
Monocalcium phosphate
Mono- and di-glycerides of the fat-forming fatty acids (except lauric)
Mustard
Nitrogen gas
Oat gum

---

[1] These chemicals would have to be used so that the food complies with all sections of the Federal Food, Drug, and Cosmetic Act. It is understood that each substance should be of food grade.

The following is a partial list of chemicals that have been accepted for use in foods by the Food and Drug Administration, or for use in meats by the Meat Inspection Division, provided their use does not deceive or render the food adulterated or misbranded under sections of the act other than Sec. 402 (a) (2):

Antifoam A: Not more than 10 ppm

Benzoic acid: 0.1%

Benzoate of soda: 0.1%

Butylated hydroxyanisole[2]: In meat food products, not over 0.01% of the fat content of the meat; in other food, not over 0.02% of the fat content of the food

Butylated hydroxytoluene[2]: Not over 0.01% of the fat content of the food

Calcium propionate or sodium propionate or any mixture of the two: In bread, not more than 0.32 part for each 100 parts by weight of flour used

Cyclohexylamine: Up to 10 ppm of the amine (free or combined) in steam that may be in contact with food

Dilauryl thiodipropionate[2]: Not over 0.01% of the fat content of the food

Distearyl thiodipropionate[2]: Not over 0.01% of the fat content of the food

Monoisopropyl citrate: In margarine—in amount not to exceed 0.02% by weight of the finished margarine

Morpholine: Up to 10 ppm of the amine (free or combined) in steam that may be in contact with food

Propyl gallate[2]: Not more than 0.01% of the fat content of the food

Resin guaiac[2]: Not over 0.1% of the fat content of the food

Saccharin: In some special dietary foods

Sodium diacetate: In bread: permitted in not more than 0.32 part for each 100 parts by weight of flour used

Sodium silico-aluminate, precipitated hydrated: Satisfactory for use in salt, not more than 1%; baking powder, not more than 5%

Sulfur dioxide (or sodium sulfite): In molasses, dried fruits, and some other foods—200 to 300 ppm (not permitted in some foods, see separate list)

Thiodipropionic acid[2]: Not over 0.01% of the fat content of the food

Tocopherol: Not more than 0.03% of the fat content of the food

---

[2] Where two or more of these antioxidants are present in the same food, the total quantity of antioxidant should be limited.

The following is a partial list of chemical additive uses that would not have functional value:

Borates (boric acid or borax) in codfish and whole eggs as preservatives (to mask poor manufacturing or storage practices)

Chrome yellow on coffee beans for coloring (to make the beans appear to be of better value)

Copper in canned peas for coloring (to make the peas appear to be less mature)

Fluorine compounds in beer and wine to stop fermentation (as a substitute for the pasteurization used in food manufacture)

Formaldehyde in milk to kill bacteria, or in frozen eggs to conceal the odor of decomposition (to mask poor production or storage practices)

Hydrogen peroxide in cream and milk as a preservative (to mask poor production or storage practices)

Monochloracetic acid in various foods, such as carbonated and fruit-type beverages, pickles, and wine (to replace proper sanitation)

Salicylates in shrimp sauce (to retard decomposition from poor manufacturing or holding practices)

Sodium nitrite in fish fillets (as a preservative in lieu of good manufacturing and handling practices)

Sulfites (to redden stale meat)

The Federal Food, Drug and Cosmetic Act controls the addition of all additives to food. Chemicals may be incorporated legally in the fabrication of many foods, and many chemicals indeed serve to improve the quality of many foods. The Act does make a clear stipulation concerning the labeling of such treated foods.

If a food bears any chemical preservative, artificial flavoring, or artificial coloring, the food will be deemed misbranded unless its labeling states this fact. There are exceptions to the law, for example, which concern the labeling of butter, cheese and ice cream. This exception applies solely to artificial coloring agents for these products.

CHEMICAL PRESERVATIVES

The Federal Food, Drug and Cosmetic Act designates any chemical which when added to a food tends to prevent or retard its deterioration as a chemical preservative.

Every chemical added to a food during the process of its fabrication is not necessarily considered to be a preservative. For example, in canning, if the air in the headspace is displaced with nitrogen, the gas is not considered a chemical preservative added to the food.

The general regulations also exclude natural preservatives or condiments from the classification of chemical preservatives. These include common table salt, sugars, vinegars, spices and their oils, and those substances incorporated into food by direct exposure of the product to wood smoke. The presence of these materials in foods need not be disclosed as preservatives, although natural preservatives may have the same general action as chemical preservatives. Inasmuch as the natural preservatives do not appear to meet the same prejudices as artificial preservatives encounter, the natural ingredients are excluded from this consideration.

In the United States chemical preservatives added to foods must be stated on the label, stating the chemical as an ingredient. The name of the chemical must be declared in terms understandable to consumers. If two or more chemical preservatives are added to a food, each must be stated individually.

The amount of chemical preservation added to a food does not affect the necessity of stating that a chemical preservative has been added. It must be noted that where standards of identity have been established for food products, unless the addition of a chemical preservative is listed as an optional ingredient, the chemical cannot be added to the food, if the food is to enter commerce. There are permitted additions of chemical preservatives (for example, for fruit products) but these are specifically established, and in the amounts permitted.

The following is a list of preservatives permitted in foods by the Food and Drug Administration, where the addition is not in conflict with other sections of the Act.

Preservatives (Antimycotics)

Calcium propionate	Sodium propionate
Potassium sorbate	Sodium sorbate
Propionic acid	Sorbic acid

Specified Uses and Amounts

Product	Amount	Uses
Caprylic acid	—	Antimycotic in cheese wraps
Potassium bisulfite	—	Not in meats or in foods recognizable as a source of vitamin B-1
Potassium metabisulfite	—	ditto
Sodium benzoate	0.1%	No special use specified
Sodium bisulfite	—	Not in meats or in foods recognizable as a source of vitamin B-1
Sodium metabisulfite	—	ditto
Sodium sulfite	—	ditto

Preservatives (general)

Acetic acid
Citric acid
Phosporic acid
Sorbitol

Specified Use

Sulfur dioxide   Not in meats or foods recog-
                                                      nizable as a source of vi-
                                                      tamin B-1

Microbial Antagonists

There are two large categories of microbial antagonists useful in food preservation; those of inorganic nature and those of organic nature. Examples of each will be discussed.

Inorganic Agents.—*Sulfur Dioxide.*—Sulfur-containing compounds are extremely useful to mankind. Sulfur dioxide has been used in food preservation for centuries. It currently still finds widespread use throughout the world, principally in treating foods of plant origin. Being more effective against molds than yeasts, sulfur dioxide has found wide use in the fermentation industries, such as wine making. Sulfur dioxide is much more toxic to molds and bacteria than to yeasts.

Sulfur dioxide is used in concentrations upwards to 2000 ppm in the preservation of fruit concentrates. Sulfur dioxide is thought to be an enzyme poison, and finds use in controlling enzymatic browning during drying of foods. The burning of sulfur, and the occurrence of sulfur houses of one form or another in drying yards, has been observed in most countries (Table 12.1).

The ability of soluble sulfite salts to protect discoloration of foods has found many applications. Notable are the uses in treatment of prepared apple slices, peeled potatoes, and other fruits and vegetables to prevent browning.

Sodium sulfite has been used to a minor extent in preserving animal tissues, but generally in combination with other agents. It is *not* permitted in the United States for this purpose.

The advantage of sulfuring dried fruits, for example, is that not only does it protect certain nutrients and control discoloration, but it is equally effective in controlling microbial and insect activity. Furthermore, during the preparation of treated sulfured food for consumption, the sulfur dioxide is largely eliminated because it is evolved during boiling or heating when the food is reconstituted. Sulfur dioxide fumes are corrosive, and suitable caution must be exercised in handling it. After boiling, sulfur dioxide treated foods usually have about 1 ppm or less remaining in the foods (Table 12.1).

TABLE 12.1

EFFECTS OF TEMPERATURE AND TIME ON SULFUR DIOXIDE CONCENTRATION IN WINE

Storage Temperature (°C)	Total Sulfur Dioxide (mg/l) Time of Storage (months)			Free Sulfur Dioxide (mg/l) Time of Storage (months)		
	1	9	22	1	9	22
0	241	181	193	135	90	90
11	224	198	187	112	108	100
21	210	212	186	112	99	82
48	212	146	8	84	35	2

Further, the presence of small concentrations of sulfur dioxide in fruit juices and in vegetables with delicate flavors may protect these fresh flavors.

Hydrogen Peroxide.—Hydrogen peroxide occurs naturally in many living tissues, but is not permitted to become toxic or remain in the living tissues due to the protective action of enzymes. The breakdown products are oxygen and water. The germicidal properties of hydrogen peroxide have been widely employed.

There have been several patents which involve the use of this compound to preserve food. Concentration levels of 0.1% and lower are effective. Contrary to common belief, hydrogen peroxide is not decomposed instantly under usual conditions in acid solution when its decomposition is not catalyzed. The enzyme catalase has such ability and specific activity to decompose hydrogen peroxide.

A proposed process for sterilizing fluid whole milk involves the addition of 0.1% hydrogen peroxide to fluid milk, allowing a reaction time of several minutes for sterilization, adding sterilized catalase to decompose residual hydrogen peroxide, then heating to inactivate the enzyme. Combined with aseptic packaging, the process theoretically has potential. One objection is the damaging influence of the liberated oxygen.

Anaerobic spore forming bacteria can be killed with hydrogen peroxide. Surface sterilization of many commodities may be accomplished. In this regard, hydrogen peroxide finds wide usage in controlling surface infections of man.

Hydrogen peroxide is *not* a permitted additive to foods in the United States.

Chlorine.—Chlorine is a widely used chemical disinfectant, finding an important use in the treatment of water for drinking and processing purposes. Its action is most effective at low pH values (Table 12.2).

Carbon Dioxide.—Carbon dioxide has been found to have preservative properties at higher pressures than normally encountered in

TABLE 12.2

CHLORINE LOSSES AT VARIOUS PH VALUES OF
CLEANING SOLUTIONS HELD AT 65°C

pH	Avg ppm Free Chlorine After Exposure Time (min)				Chlorine Lost During Experiment (5 ppm)
	0	10	30	60	
4.5	4.9			3.3	1.6
6.8	5.1			3.5	1.6
7.8	4.8			3.2	1.6
4.5	10.0		8.6		1.4
6.8	9.9			7.4	2.5
7.8	9.0		7.6		2.6
4.5	21.3	19.0			2.3
6.8	20.1	17.8			2.3
7.8	19.0	15.0			4.0

the atmosphere. In addition to uses in carbonated beverages, there are applications in partially prepared foods, such as unbaked biscuits which have as their principal preservative increased atmospheres of carbon dioxide, developed in-package during storage. Storage at refrigerated temperatures permits successful distribution of such dough products for several months after preparation.

Carbon dioxide is currently being used in controlling the maturation and storage quality of fresh fruits. In this instance it is a physiological control.

Organic Antagonists.—*Benzoic Acid.*—Benzoic acid occurs naturally in cranberries. Cranberries are easily preserved in their native form, provided protection is given against moisture losses and respiration losses.

Benzoic acid and its salts and derivatives are a family of the widely used chemical preservatives. The use of benzoic acid in foods has been the subject of much discussion, since in sufficient concentrations it is objectionable and even poisonous. Such a statement naturally can be made of almost any chemical entity. The concentration employed is a deciding factor. Sodium chloride in large quantities is poisonous to man, as is carbon dioxide. Concentration is a dependent factor. In the concentrations used by man in food preservation no untoward physiological damage has ever been found. Sodium benzoate is widely used in preserving acid foods. The benzoates generally are more effective against yeasts and molds than against bacteria in concentrations of 0.1% or less, the amounts allowed. Benzoic acid is more effective against yeasts than molds, and when added to apple juice as a chemical preservative does not permanently preserve the fruit juice. If the production of the juice is such that the contamination of microorganisms is low, the action of benzoic acid is most effective,

TABLE 12.3

EFFECT OF VARYING FUNGISTAT CONCENTRATION AND PH ON SUGAR
UTILIZATION BY *S. ROUXII VAR. POLYMORPHUS*

Fungistat Concentration	Sugar Utilized (%)				
	pH 6.0	pH 5.5	pH 5.0	pH 4.5	pH 4.0
0.025% benzoate	21	29	26	25	12
0.050% benzoate	23	18	19	13	13
0.075% benzoate	15	14	7	0	0
0.100% benzoate	13	9	5	0	0

even at 0.05% levels. In combination with cold storage (0°C) benzo-ated cider may be maintained in acceptable condition for a month or six weeks. In highly contaiminated juice, 0.1% is only slightly useful.

While sodium and ammonium salts of benzoic acid are commonly used, it is the benzoic acid molecule itself that appears to be germicidal. The undissociated molecule is thought to be the active arrangement. The sodium salt is more soluble in water than the acid and the former finds preference in use.

The acidity of the substrate in which benzoates are added influence the effectiveness of the chemical preservative. In a food with a pH value of 7.0, benzoates are less effective than in an acid food, with a pH value near 3.0. In general it is felt that the germicidal activity of benzoic acid is increased ten-fold in the latter substrate over the former (Table 12.3). In a highly acid food, the germicidal action is in the order of 100 times more effective than in slightly alkaline foods.

Germicidal ices can be prepared containing benzoic acid. The freezing point of a 0.1% solution is just slightly more than 3°C and is an eutectic mixture. If permitted, such ices may have uses in food preservation and experimentally have been used in treatment of fresh fish. At 0.1%, benzoic acid may improve the keeping quality of fish. The formation of trimethylamine, the odor of spoiled fish, may be suppressed by this treatment without controlling bacterial populations. In this instance, the use of the chemical preservative is questionable.

Benzoates are commonly used in preserving apple cider, fountain syrups, margarine, pickle relishes and other acid foods. A combin-ation treatment with mild heating is complementary.

Benzoates in foods at concentrations of 0.1% may be noticeable, and can impart a disagreeable peppery or burning taste to the food. This may be especially noticeable in treated fruit juices.

Methyl and propyl parahydroxy benzoates are also permitted additives in the United States.

Fatty Acids.—The practice of adding vinegar to bread dough to prevent ropiness is more than a century old. In recent years fatty acids

TABLE 12.4

INFLUENCE OF TEMPERATURES ON THE EFFECTIVENESS OF
PROPIONATES TO INHIBIT MOLD GROWTH

Temperature (°C)	Propionates/100 lb Flour (oz)
under 21	1.50
21-26	2.25
26-32	3.00
32	3.50

have received increasing attention as preservatives.
Fatty acids containing from 1 to 14 carbon atoms are effective mold inhibitors. The presence of double bonds increases their preservative effects; branching of chains lowers it. Propionates can be extremely useful in controlling molding and ropiness in bread. Propionic acid may be added to salt in the fermentation of foods, to control molds, especially in pickle fermentations. Their effectiveness is temperature controlled, as with most preservatives. Lower temperatures, with low levels of microorganisms, permit efficient preservative action with low concentrations of these fatty acids and their derivatives (Table 12.4).

The use of propionates for mold inhibition, not only by incorporation of the sodium or calcium salts into food mixes but by application to packaging materials, is being explored. Their use in treatment of fruits and vegetables for controlling not only molding but bacterial attack has some promise. These treatments are not permanent preservation methods, but prolong the usable life of perishable commodities.

Many of the fatty acids have specific and sometimes objectionable odors of such intensity (for example, butyric acid) that their introduction into food for preservation purposes may lower the acceptability of the foodstuff. Careful selection of preservatives is indicated.

$$-\overset{\displaystyle |}{\underset{\displaystyle |}{C}}-\overset{\displaystyle |}{\underset{\displaystyle |}{C}}-\overset{\displaystyle O}{\overset{\displaystyle \|}{C}}-OH$$

Propionic Acid

Sorbic Acid.—The general class of polyunsaturated fatty acids are effective as fungistatic agents. In particular, sorbic acid has been found most useful in controlling mold growth in packaged cheese. In

$$-\overset{|}{\underset{|}{C}}-\overset{|}{C}=\overset{|}{C}-\overset{|}{C}=\overset{|}{C}-\overset{O}{\overset{\|}{C}}OH$$

Sorbic Acid

margarine made with cultured milk, it is reported that the addition of sorbic acid is functional in quantities one-third less than that required for preservation with sodium benzoate. It has subsequently been shown that sorbic acid is effective against many molds commonly found on meats, and applications to many food processing operations are appearing currently. The effectiveness of sorbic acid in protecting cheeses which are packaged and stored in humid conditions, hence ideally attacked by molds, has been widely recognized. Uses for the preservative are mounting, particularly in the fermentation industries.

It has been found that the inhibitory influence exerted on mold growth by sorbic acid is due to the inhibition of the dehydrogenase enzyme systems in molds. In the presence of low levels of mold growth, sorbic acid has been found to exhibit fungistatic, and may even exhibit fungicidal effects on the organisms. In the presence of high levels of mold growth, the sorbic acid is metabolized and does not appear to have any inhibitory influence.

Sorbic acid has been the subject of intensive study. The metabolism of this polyunsaturated fatty acid has been found to be normal in animals. The demonstration that the metabolism of sorbic and caporic acids is similar in animal organisms has been taken as unequivocal proof of the harmlessness of sorbic acid by many workers. Sorbic acid is permitted as a chemical preservative in the Unites States.

Glycols and Volatiles.—Propylene glycol and triethylene glycol have germicidal properties, the latter being more effective. Their use as mold inhibitors has been suggested. The application of glycols in decontaminating air in concentrations 1 to 2 ppm is useful. Insofar as preserving foodstuffs they have minor significance, except in disinfecting atmospheres in which foodstuffs may be stored or treated. Trichloroethylene, ethylene dichloride, and many chlorinated hydrocarbons have germicidal activity. The exposure of foods to

$$CH_3CHOHCH_2OH \qquad\qquad (CH_2OCH_2CH_2OH)$$

Propylene Glycol   Triethylene Glycol

such agents prolongs the storage life of perishable commodities. Applications have been found, for example, in controlling brown rot in peaches and molding of small fruits. The use of chlorinated hydrocarbons has not been encouraged although certain obvious application areas exist in handling highly perishable commodities, such as fruits in transit from growing area to consumption areas. Although the preservation action is temporary and the residues are below toxicity levels, further research and developement are needed before widespread application in the food industries can occur. Chlorinated hydrocarbons are *not* permitted as food additives in the United States.

Fumigants.—The use of fumigants such as methyl bromide and epoxides (i.e., ethylene and propylene oxides) is important in the treatment of large volumes of foodstuffs to destroy insects. In addition, ethylene oxide-carbon dioxide processes have been developed for greatly reducing the bacterial populations of heat sensitive, highly flavored substances; spices are given such treatments to reduce to tolerable amounts the incidence of thermophilic contaminants. Chocolate and cocoa powder may be partially sterilized with satisfactory results. The thermophilic contaminants in cocoa powders are such that the production of sterile canned chocolate milk drinks is most difficult, if the desired tastes and qualities are retained. Combination treatments are successful experimentally; killing thermophiles with ethylene oxide, sterilizing the milk by canning techniques. The use of epoxides is prohibited in the United States in foods with high moisture contents.

The disinfection of grains with space fumigants is worldwide in its application, as noted earlier.

Antibiotics

Living tissues produce metabolic products. Some of these products from microorganisms have been found to have germicidal properties, and are called antibiotics. Living tissues contain other materials which have antimicrobial activity (Table 12.5), but the term antibiotic is restricted to those materials yielded in metabolism by microorganisms. Spices may contain antimicrobial compounds, yet the active ingredients are not called antibiotics.

Since their discovery antibiotics have found wide use for mankind in controlling human pathogens, and the pathogens which attack his animals and plants. The application of antibiotics to the food preservation industry has been somewhat slower than progress in the field of medicine. There are several reasons for the delay. It has been thought that the unrestricted consumption of antibiotics by a man might develop an artificial selection of harmful pathogens. If

TABLE 12.5

EFFECT OF SEVERAL PLANT EXTRACTS ON SURVIVAL OF SPORES OF
FLAT SOUR NO. 787 HEATED AT 93°C

Time Heated (min)	Buffer Control	Number of Surviving Organisms/ml				
		Tomato Plant	Green Beans	Carrot	Lupulon	Indol-3-Acetic Acid
0	15,700	35,000	34,000	42,000	3,200	24,000
½	15,600					
1	14,500	500	9,000	1,000	100	—
2	14,100	—	100	—	—	—
3	—	65	0	200	0	5,000
4	14,000	—		—		—
5	—	0		0		—
6	11,000					300
8	9,000					—
9	—					65
10	7,500					—
15	4,000					0
18	—					0
20	980					0
29	225					
32	150					
35	0					

antibiotics were used promiscuously, and a time came when the person required the antibiotic for medicinal purposes, it might be of no benefit. The issue is complicated in any event, and the following appears reasonable for the future of antibiotics in food preservation.

(1) The antibiotic must have no harmful effect to man when taken by mouth, and preferably should be metabolizable, at least digestible to an inactive state and eliminated from the body.
(2) There should be good reason for the incorporation of the antibiotic in man's foods, i.e., no other feasible method.
(3) The economy of use, ease of use, and reliability of control and detection must exist.
(4) Preferably the antibiotic must be compatible with other methods of permanent preservation.
(5) The antibiotics should have a wide spectrum of attack against spoilage organisms.

The field of antibiotics for use in food preservation has yet to be exploited. The researchers active in the antibiotic field in the past have employed those criteria for a successful antibiotic which may or may not be compatible with being able to discover useful antibiotics specifically for food preservation. Sorting and isolating techniques used in the past dealt with the ability of an antibiotic to have activity against human pathogens. Those organisms infecting foods need not necessarily be sensitive to the action of an antibiotic which happens

TABLE 12.6

EFFECT OF ANTIBIOTICS USED IN CHILL WATER ON STORAGE LIFE OF CHICKENS

Experiment No.	Control	Storage Life (days) Chlortetracycline 20 ppm in Slush Water
1	4	6
2	4	5
3	3	4
4	3	4
5	6	9

to have application in controlling an infection in man. The future for a search for antimicrobial agents, commonly present in living tissues, metabolized by man, and useful in preserving food, appears promising.

Usually when an antibiotic has no value in treating humans, other uses are sought. Then someone might find by chance that there happens to be an application in food and agriculture. This was the instance of several applications of antibiotics, notably in controlling fire blight in fruit orchards, and in controlling the decomposition of dressed poultry. Certainly if a search were made specifically for antimicrobial agents capable of destroying or inhibiting spore-forming anaerobic bacteria, promising leads would result.

The experimental use of subtilin in controlling food spoilage organisms in canning is an example. The initial application by the researchers in the Western Regional Research Laboratories, U.S. Department of Agriculture, indicates that there are useful applications to be made of antibiotics to control such organisms as *Clostridium botulinum*, a public health hazard for mankind.

Combination treatments of antibiotics with other sterilizing agents, including heat, radiation, and combinations of all three, offer real promise.

The application of antibiotics including penicillin, streptomycin bacitracin, chlortetracycline and others is spreading to non-medicinal uses. Difficulties have been encountered in the past with antibiotic treated cows whose milk was used in cheese manufacturing. Other difficulties may be uncovered, however, by careful research it may be possible to extend the use of antibiotics. An example has been shown in Central America where beef spoils very quickly after slaughter. Refrigeration in homes is rare. Meat is purchased daily, and distribution of meats is difficult and inefficient. A pre-kill injection of antibiotic has resulted in the extension of storage life of fresh meat three- to four-fold. With care in controlling such applications, future exploitations may be anticipated in the field of antibiotics (Tables 12.6 and 12.7).

TABLE 12.7

EFFECT OF SIZE OF INOCULUM UPON THE INHIBITION OF
P. VULGARIS BY CHLORTETRACYCLINE IN BEEF

| Inoculum/g
x 10³ | Thousands of Bacteria in 24 hr/g | |
	Control	Chlortetracycline (5 ppm)
1.8	1.1	0.7
22	2,700	39
29	2,200	11
39	1,000	44
47	18,000	19 (10 ppm)
65	20,000	1,800
95	230	40
130	70,000	15,000

Quality Improving Agents

Many chemical compounds aid in food preservation by protecting the nutrients, flavor, texture, and storage stability of foodstuffs.

The major components of foods are the proteins, carbohydrates and fats. The spoilage of foods containing fats includes the deterioration of the fatty components; the principal difficulties found in their preservation are in the development of rancidity. The spoilage of fats and oils has received much attention, and substantial progress has been made in their preservation.

Antioxidants.—There are two major categories of rancidity; oxidative and hydrolytic. Other minor difficulties are found in light struck oils and ketonic degradation. Of the deteriorations of fats and oils, oxidative rancidity is most always present. Oxidative rancidity is influenced by air, light, heat, heavy metal ions, moisture and the presence of natural antioxidants.

As anticipated from their chemical structure, unsaturated fatty acids are subject to oxidation at their double bonds. In the presence of suitable catalysts, or in contact with lipoxidase enzymes, long chained unsaturated fatty acids may be fragmented to short chained fatty acids. The cleavage usually involves an intermediate peroxide formation step:

$$\underset{-C=C-}{\overset{H\ \ H}{}} \xrightarrow{O_2} \underset{\underset{O-O}{|\ \ |}}{\overset{H\ \ H}{C-C}} \longrightarrow \text{cleavage products}$$

The shorter chained fatty acids which are formed have characteristic odors, which are usually objectionable, and are primarily responsible for the rancid taste of fats exposed to oxygen. Among the many substances capable of acting as antioxidants are a number of phenols (hydroquinone, pyrogallol) and naturally occurring substances such as glutathione, ascorbic acid, and the tocopherols (vitamin E for example). Neither of these phenols is a permitted food additive in the United States.

Some of the biological activity of vitamin E has been attributed to its antioxidant properties. In cheese the disappearance of vitamin E occurs before the degradation of vitamin A and the onset of rancidity. The addition of vitamin E to solutions of vitamin A in oil inhibits the oxidation of vitamin A, so long as vitamin E remains in its antioxidant role.

It would appear that fat or oil molecules become activated, react with available oxygen and produce peroxides, which in turn transfer some energy to other fat molecules, which undergo similar activity. In the presence of antioxidants, the energy appears to be taken up by the antioxidant, which does not transmit the energy to other oil molecules, stopping the chain reaction, so long as there is an antioxidant operative. The oxidative action is not easily controlled if the oil has substantial reactive peroxides present prior to the addition of the antioxidant. The presence of metallic ions may destroy the efficiency of the antioxidant.

Some of the antioxidants used in increasing the storage life of fats are: ascorbic acid, ascorbyl palmitate, calcium ascorbate, erythorbic acid, sodium ascorbate and the tocopherols. Other products used as antioxidants for fats and foods containing fats include: butylated hydroxytoluene, butylated hydroxyanisole, dilauryl-thiodipropionate, propyl gallate and thiodipropionic acid which are commonly used at the 0.02 level. Gum guaiac is also used as an antioxidant, but 0.10% is required. Benzoic acid has some antioxidant properties.

There are two distinct types of antioxidant activities; the antioxidants themselves, and synergists (Table 12.8). The useful antioxidants in most part are tocopherols, phospholipids, gum guaiac, propyl gallate, butylated hydroxyanisole, ascorbic acid, etc.

Synergism has been demonstrated where the combined effect of two or more antioxidants produces an effect greater than a single one alone. Certain compounds have no antioxidant capacities alone, yet in combination with an antioxidant improved results are obtained. The actual mechanism of the synergism depends upon the nature of the synergist, the antioxidant, and the fat. An explanation has been forwarded using quinones and ascorbic acid. In the presence of ascorbic acid, quinone is reduced to a semiquinone, and stabilization of the

TABLE 12.8

CLASSIFICATION OF WATER-SOLUBLE ANTIOXIDANTS

Chemical Type	Antioxidant Activity
Aliphatic polyhydroxy	Weak
Aliphatic hydroxy acids	Moderate
Polybasic hydroxy acids	Strong
Aliphatic amino acids	Strong
Protein	Good
Ortho and pyrophosphoric acids	Very strong

fat molecule is then affected by the activated semiquinone.

It is possible to add a substance to fats and oils which selectively becomes oxidized, thereby protecting the unsaturated fatty acids. Or, the unsaturated fatty acids may be hydrogenated, reducing the number of double bonds, and converting unsaturated to corresponding saturated or partially saturated fatty acids.

Antibrowning Agents.—There is an enzymatic browning which occurs in fruit and vegetable tissues which have been ruptured by cutting, peeling, slicing and grinding. Immediately, when in direct contact with oxygen, the ascorbic acid in the exposed wet surface is oxidized. When this activity is completed, it appears that the mechanism of brown color development involves polyphenol oxidase, a suitable substrate, and oxygen. In the first stages, catechol is oxidized and o-benzoquinone is formed. Orthoquinone undergoes further oxidation to hydroxyquinone, which polymerizes to form the brown pigments commonly seen in such ruptured tissues.

The activity may be controlled by heating the substrate to inactivate the enzyme systems operative, or suitable chemical controls may be effected. Obviously ascorbic acid could be used and it has been found most suitable. Much use of ascorbic acid is made, taking advantage of its antibrowning capabilities. Treating susceptible tissue in a 0.1% ascorbic acid solution is an effective control. Citric acid has been found to be synergistic with ascorbic acid in this action. Salt solutions too have been used to temporarily control this discoloration and are commonly employed in home food preparation. Sulfur dioxide has been discussed previously as an effective controlling agent in browning.

Firming Agents.—Many plant tissues soften during processing. For example, tomatoes contain pectic compounds which form a firm structure between cells and tissues in the fruit. Commonly, as during canning, the structure is partially collapsed. The addition of calcium salts to the fruit establishes a calcium pectate gel which supports the tissues and maintains the structure, even after being heat processed.

Calcium salts are used in many fruit processing operations in concentrations below 0.1%. In processing delicate berries, in canning sliced apples, or in the freezing of such tissues, beneficial action can be obtained by the addition of the above quantities of firming agent. Alum is used in the firming of salt stock. As with all such additives, care must be exercised in their use. Excess quantities may reduce rather than improve the quality characteristics desired. A number of firming agents are permitted as food additives.

Waxing.—Plant tissues commonly have a coating of wax on their exterior surface. This protective wax is effective in reducing water losses due to evaporation, hence the wilting of food products. The application of wax coatings to foods is a time-tested process. The major advantage of waxing is in reducing shrinkage losses. Waxing has little preventative action in retarding the growth of spoilage organisms, except when germicidal agents are incorporated in the waxes. Oranges and cucumbers for example are often given a wax coating, as are turnips, eggplants, tomatoes, peppers and some melons. Waxes are applied either by dipping, spraying an emulsion, or dipping in an emulsion. These consist of paraffin or carnauba waxes plus emulsifying agents. Too thick a coating creates anaerobic respiration conditions. Too thin a coating offers little control over moisture losses.

OTHER CHEMICAL ADDITIVES

Artificial Flavoring

A number of ingredients are added to foods to contribute to their flavor. The Act controls the addition of flavoring matter produced synthetically. Any substance which contributes flavor or odor and is manufactured by a process of synthesis or similar artifice when added to food is considered an artificial flavoring.

Artificial flavoring refers only to those flavors artificially produced. This aspect differs substantially from the concept of added coloring agents. Any flavoring substance which has a natural source of origin is not considered to be an artificial flavor.

The amount of artificial flavoring agents added to food does not influence the statement that it has been added. If any artificial flavoring is added, regardless of quantity, it must be stated on the label.

It should be noted that the name of the artifical flavoring ingredient need not be specified in the labeling of the food. The simple declaration "artificial flavoring" in the ingredient list is sufficient.

If the word "flavoring" is not used, it may be substituted by the word "seasoning" provided that the name of the ingredient is stated. If the name of the ingredient is not stated, the label must designate that the food is artifically flavored.

Where standards of identity have been promulgated, and no specification is noted for artificial flavoring as an optional ingredient, the artificial flavoring may not be used.

Non-nutritive sweeteners are listed by FDA as permitted food additives when not in violation of other sections of the Act.

Artificial Coloring

The Act defines an artificial coloring broadly as any coloring matter which contains a dye or pigment. It is not significant whether the dye or pigment is manufactured synthetically or whether the dye or pigment is obtained by extracting a natural dye or pigment from plants or other sources. Therefore, caramel falls within this definition since it is a pigment made from sugar. The fact that the coloring may be either synthetic or naturally occurring requires at times different types of labeling on colored food products. All such foods could be thought of as artificially colored, but because the disclosure can be made in regard to coloring it becomes necessary to determine whether the coloring is artificial in character.

With caramel for example, the pigment is a natural one, and therefore to call it an artificial coloring is actually incorrect. The interpretation by federal authorities accepts such statements of added caramel as "burnt sugar coloring," "added caramel color," and "colored with caramel."

Where the coloring is truly artificial (synthetically produced) the statement required is "artificial coloring." The statement "food color added" is considered insufficient to indicate the presence of artificially produced color, because it may lead a consumer to believe that the color has been derived, for example, from a food.

The amount of coloring added to a food cannot affect the statutory requirements. The Act provides for the declaration of artificial coloring if any detectable quantity is used.

Where standards of identy have been issued for a food, if artificial coloring is not stated as an optional ingredient, artificial coloring cannot be added, even if label declaration is made. Standards for a number of foods permit the addition of artificial coloring to, for example, certain fruit products.

Certain colorants are listed by FDA as permitted food additives when not in violation with other sections of the Act.

UNTREATED CONTROL CIPC TREATED

303 DAYS

STORAGE TEMPERATURE 55° F.

Courtesy of QM. Food and Container Institute

FIG. 12.3. SPROUT INHIBITION OF POTATOES WITH CIPC

Treatment of potatoes by introduction of chemical into atmosphere, while in storage chamber. Not permitted for use as yet. (Note: 55 °F = 12 °C)

Other Agents

Many agents can be added to food products to improve their quality. Few have preservation roles. However, stabilizers are commonly used in the canning of evaporated milk, in manufacture of ice cream, in preparing sauces which will not curdle on freezing preservation, and in maintaining emulsions in salad dressings, sausage manufacture and other food products.

Monosodium glutamate is added to intensify the flavor of soups, fish products, sauces, meats, and preparations, and is not considered a chemical additive.

The technological uses of chemicals in the food industries are growing. The food industry has been progressive in attempting to improve the quality of foods offered to consumers. The addition of chemicals in the foodstuffs of man puts the burden of proof of harmlessness of the added agent on the food industry. This may actually retard progress in certain fields of endeavor (Fig. 12.3).

In modern food technology additives to man's foods are recognized for their functional values and/or nutritive value. The danger in using chemical preservatives is that they may be used as a substitute for cleanliness; food has aesthetic values placed upon it. Unfortunately most chemical entities which are capable of killing microorganisms in man's food are equally capable of killing man himself, if taken in sufficient quantity.

The following chemical additives are permitted in foods for specific purposes.

Buffers and Neutralizing Agents.—These are listed by the Food and Drug Administration as permitted food additives when not in violation of other sections of the Act. Included are: acetic acid, aluminum ammonium sulfate, aluminum sodium sulfate, aluminum potassium sulfate, ammonium bicarbonate, ammonium carbonate, ammonium hydroxide, ammonium phosphate (mono- and dibasic-), calcium carbonate, calcium chloride, calcium citrate, calcium oxide, calcium phosphate, citric acid, lactic acid, linoleic acid, magnesium carbonate, magnesium oxide, oleic acid, potassium acid tartrate, potassium bicarbonate, potassium carbonate, potassium citrate, potassium hydroxide, sodium acetate, sodium acid pyrophosphate, sodium aluminum phosphate, sodium bicarbonate, sodium carbonate, sodium citrate, sodium hydroxide, sodium phosphate (mono-, di-, tri-), sodium potassium tartrate, sodium sesquicarbonate, sulfuric acid, and tartaric acid.

Preservatives (sequestrants).—These are listed by the Food and Drug Administration as permitted additives when not in violation of other sections of the Act. Included are: calcium acetate, calcium chloride, calcium citrate, calcium diacetate, calcium gluconate, calcium hexametaphosphate, calcium phytate, citric acid, dipotassium phosphate, disodium phosphate, monocalcium acid phosphate, monoisopropyl citrate, potassium citrate, sodium acid phosphate, sodium citrate, sodium diacetate, sodium gluconate, sodium hexametaphosphate, sodium metaphosphate, sodium phosphate (mono-, di-, tribasic), sodium potassium tartrate, sodium pyrophosphate, sodium tartrate, sodium tetrapyrophosphate, sodium tripolyphosphate, and tartaric acid.

And for specified uses in amounts specified:

Product	%	Use
Isopropyl citrate	0.02	No special use specified
Sodium thiosulfate	0.1	Salt
Stearyl citrate	0.15	No special use specified

Nutrients.—Nutrients listed by the Food and Drug Administration as permitted additives when not in violation of other sections of the Act include: ascorbic acid, calcium carbonate, calcium oxide, calcium pantothenate, calcium phosphate (mono-, di-, tribasic), calcium sulfate, carotene, ferric phosphate, ferric pyrophosphate, ferric sodium pyrophosphate, ferrous sulfate, reduced iron, 1-lysine monohydrochloride, niacin, niacinamide, α-pantothenyl alcohol, potassium chloride, pyridoxine hydrochloride, riboflavin, riboflavin-5-phosphate, sodium pantothenate, sodium phosphate (mono-, di-, tribasic), thiamin hydrochloride, thiamin mononitrate, tocopherols, α-tocopherol acetate, vitamin A, vitamin A acetate, vitamin A palmitate, vitamin B-12, vitamin D-2, and vitamin D-3.

And for uses in amounts specified:

Product	%	Uses
Coper gluconate	0.005	In any food
Cuprous iodide	0.01	Used in table salt as source of dietary iodine
Potassium iodide	0.01	Ditto

Stabilizers.—Stabilizers listed by the Food and Drug Administration as permitted additives when not in violation of other sections of the Act include: agar-agar, carob bean, carragheen, and guar gum.

Emulsifying Agents.—These are listed by the Food and Drug Administration as permitted additives when not in violation of other sections of the Act. Included are: acetyl tartaric acid esters of mono- and diglycerides, except lauric; mono- and diglycerides, except lauric; mono-sodium phosphate derivatives of mono- and diglycerides, except lauric; propylene glycol; and cholic acid, desoxycholic acid and glycocholic acid (0.01%) in egg white.

Miscellaneous Additives.—Listed by FDA as permitted when not in violation of the Act, these include: aluminum sodium sulfate, aluminum sulfate, butane, tribasic calcium phosphate, carbon dioxide, carnauba wax, glycerin, glycerol monostearate, helium, magnesium hydroxide, monoammonium glutamate, nitrogen, papain, propane, propylene glycol, triacetin (glyceryl triacetic), tricalcium phosphate, sodium carbonate, sodium phosphate, and sodium polyphosphate.

And for specified uses and in amounts indicated:

Product	Amount	Use
Aluminum calcium silicate	2%	Table salt; anticaking agent
Calcium silicate	5%	As an anticaking agent in baking powder
Calcium silicate	2%	As an anticaking agent in table salt

Caffeine	⅓ to ½ grain in 177 ml bottles of cola drinks	In cola drinks
Ethyl formate	15 ppm	When used as fumigant for cashew nuts
Magnesium silicate	2%	Table salt; anticaking agent
Ox bile extract USP (solids)	0.01%	Egg whites stabilization
Taurocholic acid (or its sodium salt)	0.01%	Egg whites stabilization
Tricalcium silicate	2%	As an anticaking agent in table salt
Triethyl citrate	0.25%	Egg whites stabilization

CHEMICAL ADDITIVES AND THE FUTURE

A major problem area facing food scientists and technologists exists in the field of chemical additives. The problem is one in which we would prohibit the use of deleterious compounds and at the same time encourage the development of new and useful additives.

The food supply available to man today is in short supply. With the increasing world populations, concentrating in larger and larger urban units, the problem of providing them with an adequate food supply year-round becomes immensely complicated. The foods eaten become more and more composed of preserved foods in one form or another and less and less of fresh commodities.

Since the preservation of foods does inflict changes in the characteristics of foods, it would be most desirable to have preserved foods which would equal fresh commodities in nutritional values and in organoleptic qualities. However, this cannot occur without experimentation and the discovery of suitable additives to foods to protect the qualities desired. In this sense, the field of chemical additives has yet to be fully explored. The exploration appears desirable and in the public interest.

Legislation which protects the public health is absolutely necessary, as past history has clearly revealed. Yet, such laws should not stifle the development of materials which are needed and which would have legitimate uses. Legislation may have such an overall effect if loosely drawn.

The course which would seem prudent would allow for the protection of health, but at the same time encourage the development of additives which might improve the nutritional values and organoleptic characteristics of preserved foods, which most certainly are in urgent need.

The creation and enforcement of laws requires the utmost skill and the most expert guidance from a multitude of disciplines. Food scientists and technologists have an important role to play in such activities.

REFERENCES

ACKER, L.W. 1969. Water activity and enzyme activity. Food Technol. 23, 1257.

ANGELINE, J.F., and LEONARDS, G.P. 1973. Food additives, some economic considerations. Food Technol. 27, No. 4, 40-50.

ANGELOTTI, R. 1973. FDA regulations promote quality assurance. Food Technol. 29, No. 11, 60-62.

ANON. 1956. Report of the Joint FAO/WHO Expert Committee on Food Additives, First Session, Rome, Dec. 3-10.

ANON. 1958. Hearings before a Subcommittee of the Committee on Interstate and Foreign Commerce, House of Representatives, 85th Congress, on Bills to Amend the Federal Food, Drug, and Cosmetic Act with Respect to Chemical Additives in Food. U.S. Gov. Print. Off., Washington, D.C.

ANON. 1961. Use of chemicals in food production, processing, storage and distribution. NAS/NRC Bull. 887.

ANON. 1962. Specifications for identity and purity of food additives. FAO, Rome.

ANON. 1965A. Chemicals used in food processing. National Academy of Science—National Research Council, Washington, D.C.

ANON. 1965B. Food Additive Control Series. FAO, Rome.

ANON. 1966. Food Chemical Codex. National Academy of Science—National Research Council, Washington, D.C.

ANON. 1968. Handbook of Food Additives. Chemical Rubber Co., Cleveland.

ANON. 1973. Code of Federal regulations. Fed. Regist. 38, No. 143, 20061.

BALDWIN, R.E., SIDES, K.G., and HEMPHILL, D.D. 1968. DDT and its derivatives in apples as affected by preparation procedures. Food Technol. 22, No. 11, 126-128.

BAUMAN, H.E. 1974. The HACCP concept and microbiological hazard categories. Food Technol. 28, No. 9, 30-34, 70.

BULLERMAN, L.B. 1974. Inhibition of aflatoxin production by cinnamon. J. Food Sci. 39, 1163-1165.

BULLERMAN, L.B., and OLIVIGNI, F.J. 1974. Mycotoxin producing potential of molds isolated from Cheddar cheese. J. Food Sci. 39, 1166-1168.

BYRAN, F.L. 1974. Microbiological food hazards today—based on epidemiological information. Food Technol. 28, No. 9, 52-66, 84.

CANTON, S.M. 1975. Patterns of use; Sweeteners—Issues and Uncertainties. National Academy of Sciences Forum, Washington, D.C.

CHRISTENSEN, C.M., NELSON, G.H., MICROCHA, C.J., and BATES, F. 1968. Toxicity to experimental animals of 943 isolates of fungi. Cancer Res. 28, 2293-2295.

CLAUSI, A.S. 1973. Improving the nutritional quality of food. Food Technol. 27, No. 6, 36-40.

DARBY, W.J. 1972. Fulfilling the scientific community's responsibilities for nutrition and food safety. Food Technol. 26, No. 8, 35-37.

DESROSIER, N.W. 1961. Attack on Starvation. AVI Publishing Co., Westport, Conn.

DUGGAN, R.R., and LIPECOMB, G.Q. 1969. Dietary intake of pesticide chemicals in the United States (II), June 1966-April 1968. Pestic. Monit. J. 2, 153.

FILER, L.J. 1976. Patterns of consumption of food additives. Food Technol. 30, No. 7, 62-70.

FOSTER, E.M. 1972. The need for science in food safety. Food Technol. 26, No. 8, 81-87.

FRANKFORT, H. 1951. The Birth of Civilization in the Near East. Indiana Univ. Press, Bloomington.

GOLDBLITH, S.A. 1971A. Pasteur and truth in labeling: "pro bono publico"— in the best of scientific tradition. Food Technol. 25, No. 3, 32-33.

GRAS, N.S.B. 1946. A History of Agriculture. F.S. Crofts, New York.

GUILD, L., DEETHARDT, D., and RUST, E., II. 1972. Data from meals selected by students. Nutrients in university food service meals. J. Am. Dietet. Assoc. 61, 38.

HALL, R.L. 1975. GRAS—concept and application. Food Technol. 29, No. 1, 48-53.

IFT EXPERT PANEL AND CPI. 1972A. Botulism. Food Technol. 26, No. 10, 63-66.

IFT EXPERT PANEL AND CPI. 1972B. Nitrites, nitrates, and nitrosamines in food—a dilemma. Food Technol. 26, No. 11, 121-124.

IFT EXPERT PANEL AND CPI. 1973. Organic Foods. Food Technol. 28, No. 1, 71-74.

IFT EXPERT PANEL AND CPI. 1974A. Nutrition labeling. Food Technol. 28, No. 7, 43-48.

IFT EXPERT PANEL AND CPI. 1974B. Shelf life of foods. Food Technol. 28, No. 8, 45-48.

IFT EXPERT PANEL AND CPI. 1975A. Naturally occurring toxicants in foods. Food Technol. 29, No. 3, 67-72.

IFT EXPERT PANEL AND CPI. 1975B. Sulfites as food additives. Food Technol. 29, No. 10, 117-120.

ITO, K.A. et al. 1973. Resistance of bacterial spores to hydrogen peroxide. Food Technol. 27, No. 11, 58-66.

JACOBS, M.B. 1951. The Chemistry and Technology of Food and Food Products. Interscience Publishers, New York.

JAHNS, F.D., HOWE, J.L., CODURI, R.J., and RAND, A.G., JR. 1976. A rapid visual enzyme test to assess fish freshness. Food Technol. 30, No. 7, 27-30.

JENSEN, L.B. 1954. Microbiology of Meats. Garrard Press, Champaign, Ill.

KAUFFMAN, F.L. 1974. How FDA uses HACCP. Food Technol. 28, No. 9, 51, 84.

KRAMER, A. 1973. Storage retention of nutrients. Food Technol. 28, No. 1, 50-60.

KRAMER, A., and FARQUHAR, J.W. 1976. Testing of time-temperature indicating and defrost devices. Food Technol. 30, No. 2, 50-53, 56.

LACHANCE, P.A., RANADIVE, A.S., and MATAS, J. 1973. Effects of reheating convenience foods. Food Technol. 27, No. 1, 36-38.

LABAW, G.D. 1954. The effect of naturally occurring constituents on the heat resistance of food spoilage organisms. Ph.D. Thesis, Purdue University, Lafayette, Ind.

LABREE, T.R., FIELDS, M.L., and DESROSIER, N.W. 1960. Effect of chlorine on spores of Bacillus coagulans. Food Technol. 14, 632-634.

LEUNG, H., MORRIS, H.A., SLOAN, E.A., and LABUZA, T.P. 1976. Development of an intermediate moisture processed cheese food product. Food Technol. 30, No. 7, 42-44.

LIBBY, W.F. 1951. Radiocarbon dates. Science 114, 291-296.

LIVINGSTON, G.E., ANG, C.Y.W., and CHANG, C.M. 1973. Effects of food service handling. Food Technol. 27, No. 1, 28-34.

MACGILLIVRAY, J.H. 1956. Factors affecting the world's food supplies. World Crops 8, 303-305.

MATCHES, J.R., and LISTON, J. 1968. Low temperature growth of Salmonella. J. Food Sci. 33, No. 6, 641-645.

MCARDLE, F.J., and DESROSIER, N.W. 1954. A rapid method for determining the pericarp content of sweet corn. Canner 118, 12-15.

MIDDLEKAUF, R.D. 1976. 200 Years of U.S. food laws: a gordian knot. Food Technol. 30, No. 6, 48-54.

MITCHELL, H.H., and BLOCK, R.J. 1946. Some relationships between the amino acid contents of proteins and their nutritive values for the rat. J. Biol. Chem. 163. 599-620.

MURPHY, E.W., PAGE, L., and WATT, B.K. 1970. Major mineral elements in Type A school lunches. J. Am. Dietet. Assoc. 57, 239.

NATL. ACAD. SCI. 1970. Evaluating the Safety of Food Chemicals. Appendix: Guidelines for estimating toxicologically insignificant levels of chemicals in food. National Academy of Sciences, Washington, D.C.

NAS/NRC. 1970. Evaluating the Safety of Food Chemicals. Food Protection Committee, National Academy of Sciences—National Research Council, Washington, D.C.

NAS/NRC. 1972. GRAS Survey Report. Food Protection Committee, National Academy of Sciencs—National Research Counci Washington, D.C.

NAS/NRC. 1973. Subcommittee on review of the GRAS list, Food Protection Committee. A comprehensive survey of industry on the use of food chemicals generally recognized as safe (GRAS). Natl. Tech. Inf. Serv. Rep. PB-221-925 and PB-221-939.

NAS/NRC. 1973. Toxicants Occurring Naturally in Foods, 2nd Edition. Food Protection Committee, National Academy of Sciences—National Research Council, Washington, D.C.

NATL. SCI. FOUND. 1973. President's science advisory committee panel on chemicals and health. Science and Technology Policy Office, Washington, D.C.

OUGH, C.S., ROESSLER, E.B., and AMERINE, M.A. 1960. Effects of sulfur dioxide, temperature, time, and closures on the quality of bottled dry white table wines. Food Technol. 14, 352-356.

PETERSON, M.S. 1963. Factors contributing to the development of today's food industry. In Food Technology the World Over, Vol. 1, M.S. Peterson, and D.K. Tressler (Editors). AVI Publishing Co., Westport, Conn.

PRESIDENT'S SCIENCE ADVISORY COMMITTEE. 1973. Report of the Panel on Chemicals and Health. U.S. Gov. Print. Off., Washington, D.C.

PUBLIC HEALTH SERV. 1965. Division of radiological health, Public Health Service: Radionuclides in institutional diet samples, April-June 1964. Radiol. Health Data 6, 31.

PURVIS, G.A. 1973. What do infants really eat? Nutrition Today 8, No. 5, 28.

SCOTT, P.M. 1973. Mycotoxins in stored grain, feed and other cereal products. In Grain Storage: Part of a System, R.N. Sinha, and W.E. Muir (Editors). AVI Publishing Co., Westport, Conn.

SMITH, C.A., JR., and SMITH, J.D. 1975. Quality assurance system meets FDA regulations. Food Technol. 29, No. 11, 64-68.

SMITH, E.S., BOWEN, J.F., and MACGREGOR, D.R. 1962. Yeast growth as affected by sodium benzoate, potassium sorbate and Vitamin K-5. Food Technol. 16, 93-95.

STEWART, R.A. 1971. Sensory evaluation and quality assurance. Food Technol. 25, No. 4, 103-106.

STOTT, W.T., and BULLERMAN, L.B. 1976. Instability of patulin in cheddar cheese. J. Food Sci. 41, 201-202.

STUNKARD, A.J. 1968. Environment and obesity: recent advances in our understanding of regulation of food intake by man. Fed. Proc. 27, No. 6, 1367-1374.

TANNER, F.W. 1953. Food Borne Infections and Intoxications. Twin City Printing Co., Champaign, Ill.

TRESSLER, D.K., and LEMON, J.M. 1951. Marine Products of Commerce. Reinhold Publishing Corp., New York.

ULRICH, W.F. 1969. Analytical instrumentation—its role in the food industry. Food Prod. Dev. 2, No. 6, 18-25.

VAUGHN, R.H., NAGEL, C.W., SAWYER, F.M., and STEWART, G.F. 1957. Antibiotics in poultry meat preservation. Food Technol. 11, 426-429.

VON SYDOW, E. 1971. Flavor—a chemical or psychophysical concept? Part I. Food Technol. 25, No. 1, 40-44.

WODICKA, V.O. 1971. The consumer protection team. Food Technol. 25, No. 10, 29-30.

WOODEN, R.P., and RICHESON, B.R. 1971. Technological forecasting: the Delphi technique. Food Technol. 25, No. 10, 59-62.

WHO. 1967. FAO/WHO Expert committee on food additives. Procedures for investigating intentional and unintentional food additives. WHO Tech. Rep. Ser. 348.

Chapter 13

Principles of Food Irradiation

DISCOVERY OF RADIOACTIVITY

While studying the phosphorescence of various materials in 1896, Henri Becquerel discovered radioactivity. He noticed that from several uranium salts an invisible radiation was emitted, which was capable of traversing thin layers of opaque materials and affecting a photographic plate. By storing the uranium salts in complete darkness for several months and noting no diminution of their ability to activate photographic emulsions, Becquerel concluded that the radiation was not phosphorescent, that is, was not dependent upon a primary exciting radiation. He also noted that the air close to the uranium salts was electrically excited and a charged electroscope could be discharged when in close proximity to the salts.

The discovery of artificially produced radiations called X-rays was reported in 1896 by Roentgen who gave a complete and accurate account of their properties. This served as a stimulus to workers observing the properties of radiation emitted by uranium compounds and other naturally occurring materials, which were not at first recognized.

In 1898 Schmidt and the Curies independently observed that similar radiations were emitted by compounds of thorium. At the same time, the Curies isolated from uranium salts a new element called *radium* (from the Latin word radius, meaning ray).

Alpha, Beta and Gamma Radiations

In 1899 it was shown independently by several researchers that the radiations of uranium compounds could be deflected and resolved in part when under the influence of strong magnetic fields. The portion not affected by the magnetic field was observed to be capable of traversing thick layers of matter. The entire undeflected part was at first termed *alpha radiation*. The deflected part of the radiation behaved like electrons and was termed *beta radiation*. In 1903 Rutherford demonstrated that if the applied magnetic field were strong enough, the alpha radiation itself could also be deflected and would behave as if positively charged. At almost the same time, however, it was found that a portion of the alpha radiation was highly penetrating and was undeflected in even the strongest magnetic field. This component was termed *gamma radiation* and was found to be similar to X-rays. The alpha ray was ultimately shown to be a particle of matter consisting of a helium atom stripped of the outer electrons and hence a positively charged nucleus; the beta particle a high energy electron and negatively charged; and the gamma radiation a non-corpuscular electromagnetic radiation of extremely short wavelength. Among other ionizing radiations of importance subsequently classified were: protons, single positively charged hydrogen nuclei; and neutrons, electrically neutral particles with mass of a hydgrogen nucleus.

Radioactive Decay.—Radioactive elements constantly decay, or lose radioactivity. The time required for a substance which is radioactive to lose 50% of its radioactivity is designated the half-life of the radioisotope. This decay is an example of a statistical process, i.e., the number of particles undergoing a reaction is proportional to the number of such particles present. The decay of radium is such that half the radium disappears in approximately 1600 years. Thus, starting with 1 g of radium, in 1600 years only 0.05 g would remain. In the following 1600 years, only 0.025 g would remain. The decay rate of a radioactive element is defined by its half-life, the time required to decompose such that one-half remains.

The interesting fact about radioactive decay is that the decay time of a radioactive substance is independent of the temperature, pressure, the presence of catalysts, or other factors commonly influencing chemical reactions. All evidence available indicates that this decay is absolutely constant for a particular radioisotope.

Units of Radiation.—Measurement of radiation involves the intensity of the source (characteristic solely of the source), the cumulative effect on the substrate, and the rate at which the effect is brought about. The source is characterized by the nature and energy distribution of the radiations, and how fast the radiation is being emitted (one curie equals 3.7×10^{10} disintegrations per sec). The original roentgen was defined in terms of ionization events but has conceptual difficulties. The rad is more useful because it is a unit based upon energy absorbed (100 ergs per gram) which is measurable. A roentgen of radiation (Glasstone 1950) is defined as:

(1) The quantity of radiation which produces one esu (electrostatic unit of positive or negative electricity per cm^3 of air at standard pressure and temperature, or,

(2) The quantity of radiation that will produce 2.083×10^9 ion pairs per cm^3 of dry air, or,

(3) The radiation received in 1 hr from a 1 g source of radium at a distance of one yard.

Example. A radiation source imparts a dose rate of 500,000 r per hr. In 10 hr, 5×10^6 r would be emitted.

A single charged ion carries 4.80×10^{-10} electrostatic units and one roentgen, abbreviated r, forms in 1 cm^3 of air 2.08×10^9 ions of either sign, or 2.08×10^9 ion pairs. At standard conditions, 1 cm^3 of air weighs 0.00129 g, so that in 1 g, air absorbing 1 r would result in the formation of 2.08×10^9 divided by 0.00129 or 1.61×10^{12} ion pairs.

The energy required to produce one ion pair in air is known to be about 32.5 electron volts. The energy required to form 1.61×10^{12} ion pairs is 5.23×10^{13} electron volts. Since one electron volt is equal to 1.60×10^{-12} ergs, the energy gained by absorption of 1 r in 1 g of air is about 83 ergs, for gamma and X-rays. The unit was designated for these radiations. Since the biological effects due to ionization from alpha, beta, proton, and neutron particles are similar in tissues, the term r has been extended to these radiations in addition to photons of electromagnetic radiations.

The term roentgen equivalent physical (rep) is defined as the quantity of ionizing radiation absorbed by soft tissues in gaining 83 ergs of energy per gram of tissue (Glasstone 1950). Actually the energy absorbed is greater in tissue than in air, being more than 90 ergs per gram. However, the term rep is roughly equivalent to r in soft tissue. In bone the energy absorbed is also greater than in air.

However, for most purposes it may be supposed that one rep of ionizing radiation produces the same energy change, the same physical effect, in soft tissues (Hannan 1956). This accounts for the name roentgen equivalent physical.

In order that a more convenient measuring unit be had, it has been agreed that the term rad should be used, meaning the absorption of 100 ergs per gram of substrate. Current literature finds the term rad used more frequently than the term rep.

All human beings and other living entities on earth are constantly living in an environment of radiations of low intensity without apparent harmful effects. All foodstuffs which man consumes are radioactive. There is a natural background of radioactivity on earth. Natural foodstuffs harvested over the face of the earth vary in radioactivity by as much as a factor of 10 and more. Man contains traces of two interesting radioactive elements, C^{14} and K^{40}. These decays yield 100,000 and 150,000 beta particles respectively per minute in an adult man, or 3×10^{-9} rads per min.

There is every reason to believe that man has a level of tolerance to radioactivity; man is constantly being exposed to low intensity radiations without any apparent harmful effects. In fact, man has been living in a radioactive environment throughout the ages. While it is impossible to establish completely the specific injury of radiation on a man, over a two week exposure the following indicated effects are anticipated:

less than 50 r—relatively little risk to man
    50-300 r—some injury accompanied with radiation sickness,
        with a few cases of delayed death
  300-500 r—serious injury and death, some in a day, some
        requiring weeks or a month; 50% of a population are
        expected to become fatalities
500-1,000 r—certain death, usually within a week
over 1,000 r—certain death, usually within a day

For personnel often exposed to radiation or who work in low intensity fluxes for long periods, a maximum exposure of 0.3 r per week has been suggested as the safe limit by the Atomic Energy Commission, although the figure is subject to constant revision as man obtains greater experience in evaluating the effects.

Ionization.—Radioactivity occurs in the breakdown of atomic structure resulting in nuclear emissions, etc., and does not include such radiations as heat, light, and radio waves. Alpha and beta particles and gamma photons are the radiations available for food preservation applications. Neutrons, while useful in reactors, are unwanted entities in food treatment due to their ability to induce or make elements radioactive (Fig. 13.1).

When an electron is knocked from an atom two charged particles or ions are formed. The electron is a negative ion. The remainder of the atom is therefore positively charged since one of its electrons has been removed. It is probably that two ions will be formed, so the formed ions are called ion pairs. This process must be distinguished from pair production.

The energy necessary to remove the most loosely bound electron from an atom or molecule has a definite value and is known as the

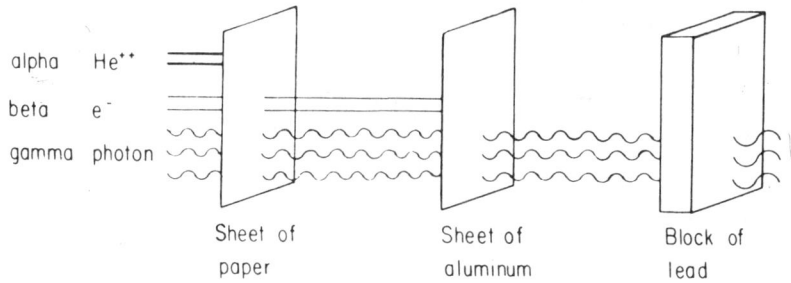

alpha	He⁺⁺			
beta	e⁻			
gamma	photon			

Sheet of paper   Sheet of aluminum   Block of lead

FIG. 13.1.   RELATIVE PENETRATING POWER OF ALPHA, BETA AND GAMMA
RADIATIONS

Usefulness of various radiations is dependent upon requirements. Where depth of
penetration is a factor, gamma radiations and high electrons must be used.

ionization energy. If the total energy of an ionizing radiation is
known, and the total number of ion pairs formed by it is established,
it is recognized that the energy lost per ion pair formed is greater
than the ionization energy requirement. This is due to the fact that
some collisions occur without ionization, but with energy losses to the
particle or photon. Also, in collisions where inner electrons of an
atom are removed, more energy than the ionization energy is required.
In air, the average energy loss per ion pair is about 32.5 electron volts
independent of particle or velocity. Hence a particle having an
energy of 1.5 Mev may produce in air 4.5×10^4 ion pairs. The
number of ions produced will be twice that figure.

Alpha particles cause much ionization along their paths. This is
due to their relative size and double positive charge carried. Alpha
particles slow rapidly in air and have little penetrating power
(Fig. 13.1), a sheet of paper being sufficient to stop them. In air
the particles have a range of a few centimeters. Lower energy
particles have correspondingly lower range. A 7 Mev particle is
capable of forming more than 200,000 ion pairs. Approximately 40,000
ion pairs are formed per centimeter of air traversed.

Beta particles have greater penetrating power, but lower
specific ionization ability. A 1.0 Mev particle can travel 200 cm in
air and cause about 150 ion pairs per cm. Gamma radiation has
lower specific ionization ability than even beta particles, but can
penetrate lead nearly equally as does a beta particle air.

This absorption of gamma radiation takes place in three ways,
depending upon the energy level of the incoming photons.

Photoelectric Effect.—With gamma photons of less than 0.1
Mev energy, irradiation of an atom finds the atom absorbing the
photon, by the ejection of an electron.

Compton Effect.—In the energy ranges between 0.1 and 10 Mev, gamma photons can be scattered. Energy is taken up from the incident photon in ejecting an electron, and the scattered photon is reradiated as a photon of lower energy, with a longer wavelength.

Pair Production.—When the energy of gamma radiation becomes greater than 1.02 Mev., pair production may occur. From 5 to 12 Mev and above, this is the primary method of energy absorption. A high energy photon converted into two electrically charged particles, an electron and a positron, occurs extremely close to the nucleus of the atom. The reaction is the conversion of energy into matter, following the Einstein equation $E = mc^2$. An electron has a mass of 9.107×10^{-28} g. The speed of light expressed by c is 3×10^{10} cm per sec. $E = 0.51$ Mev per particle.

An electron and a positron of equal weight and electrical charge are created at the same instant, so energy is converted into mass, in the form of two newly created particles. This effect cannot occur at less than 1.02 Mev. The positron formed has an extremely short life. By collision, and ionizing, atoms of matter in its path, the positron slows and undergoes a unique and final interaction with an orbital electron. Two electrically charged particles unite, annihilate themselves, and give off two 0.51 Mev photons. This annihilation process is the inverse of the action which produced the electron and positron to begin with.

Sources.—One of the many factors involved in the development of suitable radiation sterilization techniques is the selection of the type of radiation to be employed. Both beta (or cathode) and gamma radiations have been utilized. For cathode radiation treatment the resonant transformer, the Van de Graaff generator, or the linear accelerator have been used. In gamma radiation treatments, spent fuel elements (Fig. 13.2), radioisotopes such as cobalt[60], and mixed and/or separated fission products have been employed as sources. Beta or cathode rays with an energy of 5 Mev are capable of penetrating 1 in. of food product. Gamma rays have significantly greater penetrating power.

The source of ionizing radiations for nearly three decades following their initial discovery was the rare and laboriously obtained naturally occurring radioactive isotopes. The major advance in the developing of more available and applicable sources were machines such as electron generators, and more recently, artificially induced radioactive elements (i.e., Co[60] and products from the large-scale utilization of atomic energy). The large-scale artificial production of radiations awaited the development of equipment capable of producing voltages sufficiently high to deliver energies to electrons

FIG. 13.2. SPENT FUEL ELEMENTS MAY BE IN THE MILLION CURIE RANGE

An important source of radiation for processing, especially in separated elements. Arrow points to spent fuel rod. A—glow produced by intense gamma radiation in water.

or protons and carry them forward at accelerated speeds. Cockcroft and Walton, in 1930, using vacuum tubes and a voltage multiplying circuit, developed an apparatus which attained an output of 300,000 volts. In operation it delivered a beam of protons which on impact with lithium or beryllium targets gave an appreciable yield of alpha particles. This and similar machines were useful in studies of nuclear disintegration but were extremely bulky and expensive, and the samples irradiated were limited to extremely small sizes.

In 1931 Van de Graaff developed a high potential electrostatic generator capable of a steady output. Essentially, the Van de Graaff unit consisted of a large metal dome supported by an insulating column. Inside this column was an endless belt of conducting material which transported an electric charge from a voltage source at the base of the column to the dome. The dome then built up a large potential with respect to ground. Voltages up to 10 million volts have been obtained and utilized to accelerate electrons through a long cylindrical vacuum tube and out through a thin aluminum window. The source of electrons was a heated filament at the upper end of the tube. The high voltage was applied between the electron source and the window. By employing an oscillating magnetic field at the base of the vacuum tube it was possible to cause the beam to scan back and forth over the sample being irradiated. This feature would allow the processor to treat material uniformly under the beam on a conveyor in a continuous operation. Improved models of this machine are available.

Another source of high energy electrons, the linear accelerator, was proposed in principle by Wideroe in 1929. This consisted of a horizontal vacuumized cylinder with an electron source at one end and the target at the other end, either within the cylinder or just outside a thin aluminum window through which the high velocity electrons emerged. The electrons were accelerated by passage along a series of sections, each of which was one-half of the period of the electromagnetic wave which carried the electron along. The output energy of the first machine was 1.2 Mev. The penetration of such a beam into water or material of like density is approximately 0.05 cm per Mev of beam energy. Technical improvements have recently raised this beam energy to 25 Mev.

Of the particular types of ionizing radiations available, it has been rather universally agreed that the electron beam (synonyms: cathode ray and beta radiation) and gamma radiations (and X-rays) are the most applicable to processes such as food preservation.

The use of machine-produced ionizing radiations offers distinct advantages over those from fission products. The two main disadvantages of fission products are that they require extensive shielding and the dose supplied may be so low that prolonged exposure times are required. The machine-produced beams are unidirectional and can be turned off (Fig. 13.3).

The lower limit of energy range for application of machine-produced ionizing radiations will be determined by the economics of the value added to foodstuffs by irradiation, as compared to competitive methods of preserving food. This minimum has been estimated to be about 3 to 4 Mev, whereas the upper limitation will be imposed by the probability of induced radioactivities in food products. Fortunately, the probability of induced radioactivities of appreciable half-life and intensity are very low in the commercially interesting penetrating ranges of up to 10 to 25 Mev. The existing information on such effects is encouraging but further studies will be required before upper electron energy levels can be established useful for food preservation purposes.

Courtesy of High Voltage Engineering Corp.

FIG. 13.3. HIGH-ENERGY ELECTRON STERILIZATION OF PLASTIC CONTAINERS

A commercial application of high energy electron sterilization.

DOSIMETRY

The unit of radiation dose is the rad, previously defined as the absorption of 100 ergs per gram of product. Methods are available to measure the radiation absorbed by tissues.

Dosimeters of three types are required: (1) primary standard, (2) operating standards, and (3) production control devices.

A primary standard dosimeter is the calorimeter developed by the National Bureau of Standards. Calorimeter dosimetry does not lend itself to routine usage. Therefore, several operating dosimeter systems have been developed, which are standardized against calorimeter dosimeters. Modified ionization chambers can be used to monitor continuously the radiation source itself.

The operating standard dosimeters found useful in routine radiation processing control are cobalt glass, ferrous and cerous sulfate dosimeters. For example, cobalt glass attached to a container of food changes color in proportion to the radiation received. To establish the dose received, the changes in optical density are compared to standards.

One of the most important standards is the ferrous sulfate or Fricke dosimeter, which consists of an acid solution of a ferrous salt containing some chloride to inhibit side reactions and impurity effects. In this dosimeter the following reactions are believed to take place. Radiation ionizes water forming H_2O+. This dissociates to a hydroxyl radical OH and a proton H+. The hydroxyl radical oxidizes the ferrous ion producing hydroxyl ions and ferric ions.

$$OH + Fe^{++} \rightarrow OH^- + Fe^{+++}$$

The formation of ferric ions, Fe^{+++}, is the key effect which permits the determination of the radiation dose; the ferric ion is detected by its absorption band at wavelength of 305 mμ. The concentration dependence of the light absorption is exponential (Beer's law), so that the logarithm of the light attenuation is proportional to the concentration.

On the basis of the very careful experiments it has been determined that there are 15.45±0.11 ferric ions formed for every 100 electron volts of absorbed energy, so that a determination of the ferric ion concentration gives a direct measure of the energy absorbed. The yield of molecules or ions is called the G value, in this case $G = 15.45$.

Every unit or container of food treated with ionizing radiation can be monitored with go-no-go dosimeters. These production control devices may be as simple as colored markings on containers, or colored plastic tapes. Here again, color changes are related to dose received. Although the go-no-go dosimeters are not so precise as the others, such inexpensive controls are extremely useful in positively indicating that a radiation treatment was given to the food package. When this plus the information obtained with precise dosimeters, which are monitoring the radiation source itself, are coupled, there is little opportunity for error in establishing that a given food parcel received at least the radiation dose intended.

Dose Distribution

The administration of a dose of radiation to a food packet must be controlled. Dose distribution within a packet and between packets is a significant factor in quality control (Fig. 13.4). Nature places an absolute minimum variation which can be achieved, and economics dictates further compromise. Theoretical calculations indicate

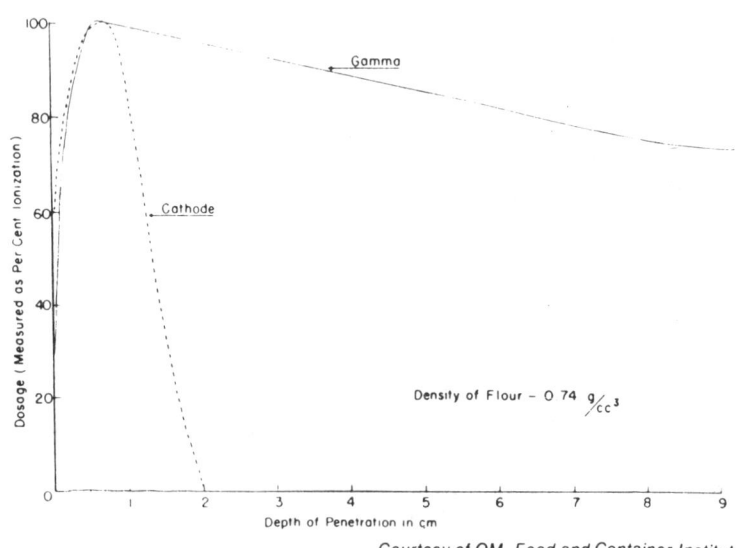

Courtesy of QM. Food and Container Institute

FIG. 13.4.　DOSAGE DISTRIBUTION VERSUS PENETRATION IN FLOUR

Important considerations in designing the radiation process and in choosing the radiation source.

that a 5% variation in a 6-in. diameter container will be encountered in a uniform gamma radiation flux. Experimental data indicate that variations of 10% are encountered. Such information can be obtained by judiciously placing cobalt glass strips in agar gels of desired size and shape, for irradiation.

Extensive studies indicate that variations in dose distribution, in many types of containers and with all types of sources, range between 100% and 125%, providing the container size is selected for the particular radiation source. If successful processing is to be obtained, no less than 100% of the desired dose must be administered to each food package.

Inasmuch as the radiation dose is directly correlated to changes in the food, variations must be maintained at a minimum. The range of 100 to 125% of dose required may be tolerable in foods, and appears to be compatible with radiation source performance and the economics of the process.

Induced Radioacitivity in Treated Foods

Inasmuch as all food is radioactive, and the background of radioactivity of foods varies depending upon where the food is grown, it is obvious that a treatise on preserving food with ionizing radiation must consider the induced activity in processed foods. Neutrons induce radioactivity in foods. Any radiation source to be used in food processing must have a low neutron flux. Sources which are able to treat foods in air are less prone to inducing activity than underwater treatments. The natural heavy water and tritium content of water are significant in this regard. Provisions must be taken to limit neutron fluxes in all radiation sources.

It is a fact that extremely short-lived induced activity is possible with electron energies over 2.3 Mev. Induced activities in foods at levels of 25 Mev and higher are found to be in the order of a thousand times less than commonly consumed and normally present in foodstuff. The gamma photon energies that have been employed in food processing are below 2.3 Mev, and negligible induction has been found.

On the basis of the above considerations, the statement is usually made that food preservation with these gamma photons and high energy electrons yield foodstuffs which have no radioactivity that can be discerned over background, providing the neutron flux in the former instance is controlled and is low.

In practice, the sterilized foods must be considered as having radioactivity, as does all foodstuff. There appears to be no danger in consuming radiation stabilized foods processed with gamma photons of less than 2.3 Mev.

The exposure of foods or other materials to high neutron fluxes results in radioactivity being induced. This is the method employed to make artificial radioisotopes, noted earlier.

The Herschman Equation.—An equation has been derived by Herschman to estimate the maximum radioactivity induced in irradiated food products:

$$Q = \frac{30\,DEn}{T}$$

$$\text{Activity in micro-micro curies} = \frac{30 \times \overset{\text{(Total dose in megarad)}}{} \times \overset{\text{(Energy of radiation Mev)}}{} \times \overset{\text{(Decay constant)}}{}}{\text{(Half-life of isotope formed)}}$$

This equation rigorously defines the maximum radioactivity which can be produced in a foodstuff, or other material, which has been

isotope formed, and the percentage composition of the food in regard to that isotope's parent atom. The magnitude of the activities calculated by the equation are in the realm of micro-micro-curies (10^{-12}). Such activity is far below any danger level; in fact it is nearly immeasurable in ordinary foodstuffs with today's most sensitive instruments. Even this theoretical maximum activity has not been found experimentally to date.

The radioisotope levels found acceptable in drinking water will no doubt be used eventually as a guide to acceptable levels of induced activity in irradiated foods. The French Government has recently taken such a position.

Mode of Action of Ionizing Radiations

When matter is traversed by any of the forms of ionizing radiations (beta, cathode, gamma, or X-rays), energy is absorbed and ion pairs are produced. Energy is absorbed by collision of the ionizing radiation with particles of the food, causing excitation of thousands of atoms in its path, as discussed previously, and occurring in time periods less than 0.001 sec. As a primary radiation distributes its energy throughout the volume of the absorber, many of the electrons which are stripped from atoms in the process of ionization may themselves possess sufficient energy to ionize other atoms. These excitation and ionization events produce a large portion of the biological effects of radiations.

Ion pair production by ionizing radiations is a very efficient process. Unlike the thermal process, very little of the energy of radiation is expended in the raising of the thermal energy of the absorber molecules. The energy required for sterilization by irradiation is about 1/50 that required for heat sterilization. This permits the sterilization of various substances with a temperature rise in the order of 2°C. Hence, the term cold sterilization was created.

Despite the negligible temperature rise, ionizing radiations do produce widespread chemical changes in irradiated materials. In fact, under ordinary conditions any oxidizable substance can be oxidized, and any reducible substance can be reduced. The biological effects of radiation occur as a result of discreet changes in the atomic and molecular structures of the irradiated material, perhaps less than 0.01% of the chemical bonds being affected.

Direct Effect.—Biologists have proposed the target theory of direct impact of the radiation with the substrate as being that principally responsible for irradiation effects. This was based on investigations of biological changes (cell multiplication, mutation, lethal effects, etc.). The theory proposed that if a swiftly moving charged particle struck a molecular complex of biological material, the biological function of the complex was altered and/or destroyed.

Within the past decade evidence has been presented to indicate that the direct "hits" may be responsible for some specific biological effects, but that many of the effects are caused in whole or part by radiation induced ionization of the solvent system of the biological material.

Indirect Effect.—The irradiation of a water-containing material causes the ionization of a portion of the water molecules, with the formation of highly reactive hydrogen and hydroxyl radicals. These radicals contribute substantially to the biological effects of ionizing irradiation. Hence, there is an indirect effect of irradiation of moist tissues, caused by these free radicals, due to "activated water."

The hydrogen and hydroxyl radicals are known to be chemically very reactive and can act as reducing and oxidizing agents as well as cleave carbon-to-carbon bonds. Secondary products of irradiation may be of equal importance, since in the presence of dissolved oxygen the hydrogen atom can combine with molecular oxygen to form the very reactive O_2H peroxide radical,

$$H + O_2 \rightarrow O_2H$$

which can form hydrogen peroxide,

$$2 O_2H \rightarrow H_2O_2 + O_2.$$

The hydroxyl radicals may also form hydrogen peroxide,

$$OH + OH \rightarrow H_2O_2.$$

The theories of direct and indirect action are capable of describing the mechanisms which apply simultaneously in the same system. The indirect theory, however, offers a wider basis for chemical change.

Effects on Food Protein.—Ionizing radiations produce changes in milk casein which result in an increase in its rennet coagulation time and a reduction of its stability to heat.

Similarities between the effects of radiation and of heat on milk are apparent in the influence of each on the rennet coagulation time. In both cases there is a decrease in rate of enzyme activity.

The influence on the rennet coagulation time of irradiated or heated milk due to the addition of calcium is significant. The effect indicates that in both treatments calcium is precipitated as tricalcium phosphate, a form in which it cannot function in the action of rennet. This precipitation of calcium is also of importance in the stability of casein to heat, acid, or other denaturing substances. The calcium balance is of particular importance in any condensing and canning process.

Cathode rays produce an effect on the calcium balance which is very similar to that which results from the heat treatment. The available calcium is reduced. It combines with activated protein molecules in such a way as to be inactive in the rennet reaction or even as a stabilizing element. Considering the extensive variety of excitation and ionization reactions which occur in irradiated substances, it is not difficult to visualize the calcium becoming combined into a complex of protein molecules. This seems more probable in the light of the results obtained in studying the effects of cathode rays on purified casein, for example, in aqueous solution. Here there is evidence, based upon changes in relative viscosity of the solution, of the association of casein molecules or fractions of molecules. The calcium is easily combined in such reactions. If the element is added before treatment (irradiation or heat) it does not affect the rennet reaction time after treatment. The surplus calcium is also made unavailable.

The changes in flavor resemble those which occur when milk is heated. Taste panel results indicate some similarity between the off-flavor of irradiated milk and the burnt flavor of overheated milk.

Example of Egg Protein.—The reduction in thickness of albumin by ionizing radiations seems to be in opposition to their effect on the relative viscosity of albumin in solution. Irradiation resulted in an increased relative viscosity of albumin in aqueous solution. The action on the protein in its natural environment is probably a result of the destruction or alteration of ovomucin, which is believed to be primarily responsible for the thickness of albumin in eggs. This thick layer is formed by albumin containing fibers of ovomucin. The structure of this thick albumin is very fragile and can be easily destroyed by agitation and other physical methods. Its destruction by ionizing radiations would be anticipated.

The obvious change in albumin thickness is accompanied by more subtle changes in the albumin molecules per se. This is confirmed by changes in the electrophoretic behavior of albumin when irradiated in natural whole egg white. The electrophoretic pattern and mobilities of albumin are altered due to molecular rearrangement in the protein. The occurrence of these two distinct types of protein alterations, one readily apparent usually

Control

Courtesy of McArdle and Desrosier

FIG. 13.5.   RADIATION EFFECTS OF 600,000 RADS
ON FRESH EGG

and the other detectable by electrophoretic apparatus, points out the obvious necessity of measuring protein changes objectively. Determinations by organoleptic methods are valuable, but they need to be supplemented by reliable objective methods (Fig. 13.5).

With the present method of grading eggs based upon the amount of thick albumin, its destruction by radiation presents the problem of decreased egg quality. The loss of thick albumin also impairs the quality of fried and boiled eggs. However, in some areas the sterilization or pasteurization of eggs by ionizing radiation shows considerable promise. Radiation sterilization might

TABLE 13.1

SEQUENCE OF PROTEIN BOND ATTACK BY IONIZING RADIATIONS

−S−CH₃
−SH
Imidazol
Indol
Alpha Amino
Peptide
Proline

be practical for eggs to be used in baking, since results indicate that the baking quality of eggs is not seriously impaired by sterilizing doses of ionizing radiations.

The results of experiments to determine the effects of cathode rays on aqueous solutions of casein and egg albumin indicate that in both proteins there is an unfolding of the molecules. The pattern followed differs in the two proteins, but the increase in free sulfhydryl groups demonstrates conclusively that molecular rearrangement does occur. Since this molecular change does not result in any marked increase in amino nitrogen, it is apparent that the peptide linkages are not attacked easily. The sulfur linkages are very definitely the site of a large share of the radiation effect (Table 13.1) but hydrogen bond linkages also are broken.

Radiation Effects on Enzyme Systems

Enzymes can be inactivated by either the direct effect or indirect effect of ionizing radiations. Both no doubt occur at the same time.

The destruction of systems of dry or frozen enzymes with ionizing radiations indicates that enzymes can be inactivated with direct or target theory effects. The decrease in inactivation dose in the presence of water is taken as a measure of the destruction by indirect effects.

Enzymes are more resistant to the effects of ionizing radiations in natural substrates than in pure solutions. This is taken as a measure of the natural protective substances (free radical acceptors) present in living tissues.

Enzyme reaction rates have been studied regarding the effect of irradiation upon the enzyme, the substrate, and their combination, on the resulting activity. A radiation dose which does not destroy the enzyme in a suitable substrate finds the reaction proceeding. If an untreated enzyme is allowed to react on an irradiated substrate, the reaction rate is increased, when compared with a non-irradiated substrate. The increase generally occurs, and perhaps is due to an activation of the substrate. With irradiated protein, ionization apparently causes an unfolding of the protein molecule, rendering the points of attack more accessible to the enzyme.

The low sensitivity of enzymes to radiation effects is of practical interest in sterilization because it is necessary to prevent enzymatic as well as microbial spoilage in preserving food. The inactivation of enzymes by moist heat usually coincides with the thermal destruction of vegetative forms of organisms. This is not the case with radiation preservation of foods. As a general rule, one can say that complete inactivation of enzymes requires approximately five times the dose required for the destruction of microorganisms.

Radiation Effect on Parasites and Insects

The ability of ionizing radiations to kill living entities has application in the field of food and drink disinfection. Amoebic cysts which are resistant to chlorination treatments are relatively nonresistant to gamma radiation. The parasites which infect man's food are within control with the application of radiation treatments. Included would be the parasites of animals transmitted to man, including trichina. The low radiation resistances of insects is also noteworthy.

Packaging of Radiation Stabilized Foods

One of the important considerations in radiation preservation of foods is packaging. If permanent preservation is to occur, food must be protected from recontamination. Therefore, hermetically sealed containers are required for sterile products. Pasteurized products require special packaging, related to each food product specifically.

Rigid Containers.—Metal, rigid containers such as tin and aluminum cans have been highly perfected. Tin coated containers have been used successfully for more than a century for sterile foods. The aluminum container has become more widely used in Europe than in the United States, and such containers are continuing to be perfected.

Effect on Base Metal.—At sterilizing dose levels, steel is stable. At doses ranging from 60,000,000 rads and higher, damage occurs in steel. The effect on aluminum is similar.

Effect on Can Coating.—Radiation has no influence in promoting tin rot or tin disease, the transition of rhombic to cubic crystalline structure of tin. Trace amounts of bismuth prevent this transition, in any event. Tin coatings over base steel are suitable for food irradiation.

Sealing Compounds.—End sealing compounds generally used in metal containers are actually improved slightly by irradiation. The exceptions are found in butyl rubber sealing compounds which are apparently depolymerized by irradiation.

Enamels.—Of the interior can enamels for tin plated cans, the oleoresinous enamels are unsatisfactory for high fat foods. Oleoresinous enamels appear satisfactory for enzyme inactivated foods in general.

Container Shape.—Container shape is important. Ideally cubic forms are desired for best radiation source utilization and dose distribution and control. Cylindrical rigid containers have found widespread use in the canning industry. Cylindrical cans are therefore the most commonly available, although rectangular cans are used for certain meat items such as luncheon meat loafs and sardines.

Flexible Containers.—Radiation processing of foods permits the storage of perishable commodities of high moisture content in plastic containers at usual room temperatures. Flexible containers have been developed for frozen and refrigerated foods, the latter of short duration in storage. Functional, economical plastic containers are being developed and should find application in radiation sterilization and pasteurization (Fig. 13.6).

Influence of Radiation on Plastic Packaging.—At doses near two million rads and less, radiation has no significant effects on the physical characteristics of flexible containers. At doses over two million rads, changes occur in physical properties of plastic films such as polyethylene, mylar, vinyl, and polystyrene. However, the changes are of minor consequence. Saran, pliofilm, and cellophanes are made brittle if the radiation exceeds three million rads.

Irradiation of most foods in plastic containers result in off-odors in the food. Nylon is very low in odor formation during sterilizing irradiation. Polyethylene irradiation at sterilizing doses evolves objectionable odorous compounds and short fragmentations of polymer, which are carried into the food itself.

Thin plastic films are objectionable, due to the ability of microorganisms to penetrate through microscopic flaws or openings caused by damage during sealing and rough handling. This may be overcome by increasing the thickness and improving sealing techniques.

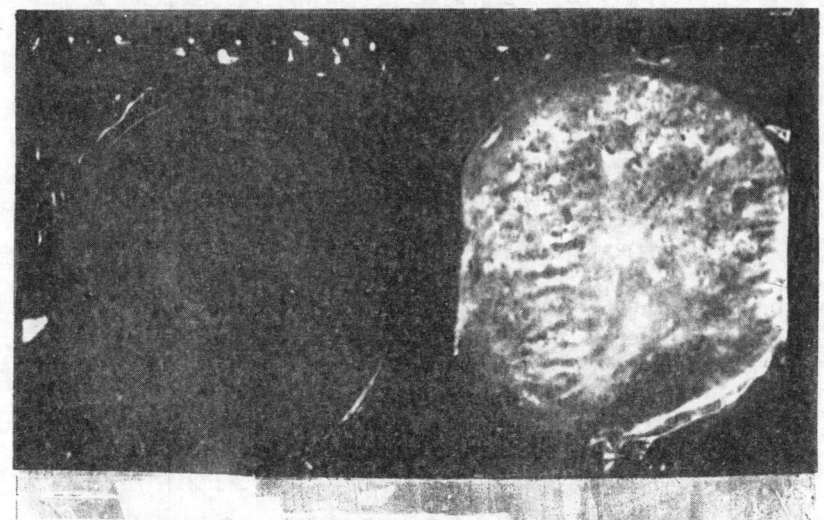

FIG. 13.6.   LOW-LEVEL IRRADIATION OF PACKAGED MEATS MAY IMPROVE ON THE
ECONOMICS OF FOOD DISTRIBUTION

Cold storage life may be increased by a factor of five to ten.

Laminated foils and plastics, and combination of the latter, are
functional. Some containers, such as Scotchpak, are as durable
in rough handling as metal cans, and perform satisfactorily up to
six megarads as a food package. The performance of standard
type flexible container materials is given in Table 13.2.

GENERAL METHOD FOR ESTABLISHING RADIATION STABILIZATION PROCESS FOR FOODS

Food may be stabilized by inactivating the microorganisms
and enzymes present, and by protecting the stabilized food from
recontamination and access to oxygen. The latter problem areas
are controlled by suitable packaging. The irradiation of a food
can destroy microorganisms and enzymes. It is more efficient
to employ ionizing radiations to kill microorganisms (Fig. 13.7)
than enzymes (Fig. 13.8). It may be desirable to inactivate
enzymes by other means, in complement to the irradiation action.

The sterilization of foods with ionizing radiations involves two
major considerations—the food product and a suitable radiation
source.

The characteristics of the food itself dictates the types of spoilage
organisms which are capable of spoiling the food. The acidity classification of foods is useful therefore not only in canning, but in radi-

TABLE 13.2

THE COMPATIBILITY OF VARIOUS CONTAINERS WITH THE RADIATION PROCESS AND PROCESSING EQUIPMENT

	Cans with Enamels		Semi-rigid Aluminum Dishes	Plastic Bag (Foil Lamination) in Carton					Wax Dip Coating
	Tinplate	Aluminum		Scotch-pak 4.5 Mil	Mylar 2 Mil	Poly-styrene 2 Mil	Nylon 2 Mil	Polyvinyl-chloride 2 Mil	
Radiation Process									
Change in physical properties (-10% max)	+	+	+	+	+	+	+	+	+
Development of off-odors and flavors	+	?	?	–	+	–	?	–	?
Extractive substances	?	?	?	?	?	?	?	?	?
Manufacturing Process									
Manufacturing equipment	+	+	+	+	+	+	+	+	+
Filling equipment	+	+	+	+	+	+	+	+	+
Sealability	+	+	+	+	–	–	–	+	+
Sealing equipment	+	+	–	+	–	–	–	+	+
Availability in desired shape and size	+	+	?	+	+	+	+	+	+

Note: + satisfactory; — unsatisfactory; ? unknown or borderline.

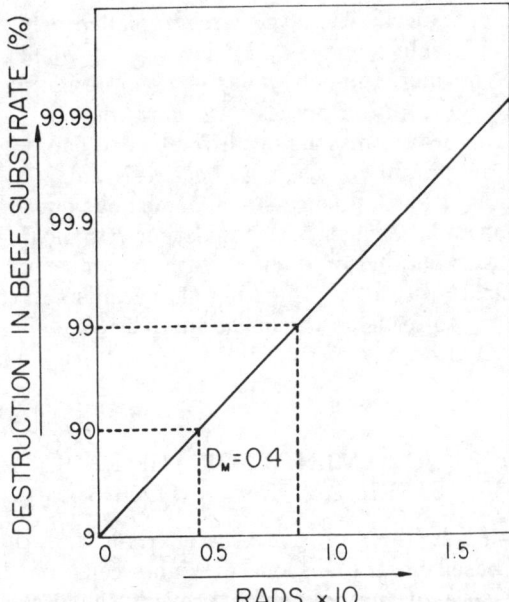

FIG. 13.7. THE D_M VALUE FOR *CL. BOTULINUM*
Unit of destruction of microorganisms

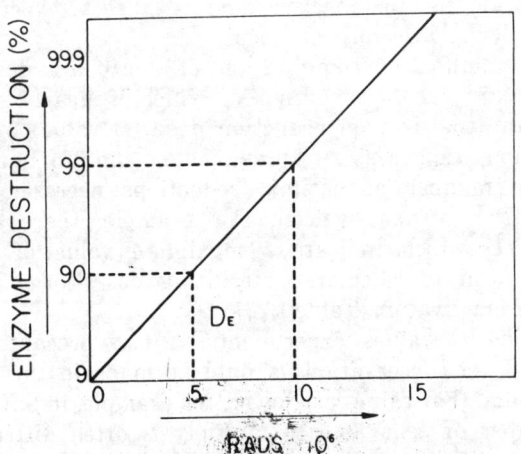

FIG. 13.8. THE D_E VALUE FOR ENZYMES
Unit of destruction of enzymes.

ation sterilization as well. The large category called food is subdivided into foods with pH values above and below 4.5. Above pH 4.5 *Cl. botulinum* must be considered, much the same as in designing the canning process. Unlike the canning process, *Cl. botulinum* is the most heat resistant spoilage organism causing difficulty. In canning, (see Chapter 6) certain other organisms are more heat resistant.

The irradiation of foods thus raises some obvious questions in public health. The Food and Agriculture Organization, Atomic Energy Branch, of the United Nations, in conjunction with a number of other international organizations, called a meeting of specialists in radiation and food microbiology to focus on the questions. The following is a general summary of the findings agreed upon by the participants of the meeting.

DOSE REQUIREMENTS FOR THE RADIATION
STERILIZATION OF FOODS

In non-acid foods (i.e., of pH greater than 4.5) the sterilizing dose must be based on the absence of viable cells of *Cl. botulinum* or on the presence of factors which prevent the occurrence of all predictable hazards.

When *Cl. botulinum* causes a hazard, a treatment or combination of treatments to obtain a reduction of the initial number of organisms by a factor of 10^{12} is required, in order to provide security over a sufficiently large number of unit packages. Since it is impracticable to explore such levels of inactivation by direct experiment, the indicated sterilizing dose must be calculated.

A suitable method of computation of sterilizing dose (suitable for *Cl. botulinum*, the destruction of which is nearly exponential) is to determine the decimal reduction dose (D value) over as wide an inactivation range as is practicable, and to multiply this value by the number of decimal reductions necessary to attain the desired level of inactivation. For example, there are various estimates of D which indicate a maximum value of about 0.37 Mrad, which lead to calculated sterilizing doses of 4.5 Mrad for a 10^{-12} level of inactivation (Table 13.3).

In estimating D values experimentally, two precautions should be observed. First, observations should be made in the food itself. It has been found that values are lower, for example, in buffer solution and the number of cells able to multiply is often different in the food and on artificial media. Second, it is desirable to experiment with the largest feasible initial number of cells and to follow their destruction to the lowest level which can be observed within

TABLE 13.3

APPROXIMATE MINIMAL DOSES OF GAMMA IRRADIATION FOR THE DESTRUCTION OF
SPECIFIC MICROORGANISMS AND TOXINS OF PUBLIC HEALTH SIGNIFICANCE

Microorganisms or Toxin	Medium	Inactivation Factor	Dose (Mrad)
C. botulinum, type A	canned meat	10^{12}	4.5
C. botulinum, type E (toxic strain)	broth, minced lean beef	10^6	1.5
C. botulinum, type E (non-toxic strain)	broth, minced lean beef	10^6	1.8
Toxin (C. bot., type A)	cheese	10^3	>7.0
Toxin (C. bot., type A and B)	broth	10^6 (based on mouse-units)	>3.0
Staphylococci (6 phage patterns)	broth, minced lean beef	10^6	0.35
Toxin (Staph. emetic-factor)	pork sausage	?	<1.0,<2.0
Toxin (Staph. emetic-factor)	water	?	0.72
Toxin (Staph. alpha lysin)	pork sausage	32	2.1
Salmonella (pullorum anatum, bareilly, manhattan, oranienburg, tennessee, thompson, typhimurium)	broth	10^6	0.32-0.35
Aerobacter	broth	10^6	0.16
E. coli	broth, minced lean beef	10^6	0.18
E. coli, adapted strain	broth, minced lean beef	10^6	0.35->1.2
M. tuberculosis	broth	10^6	0.14
Streptococcus faecalis	broth, minced lean beef	10^6	0.38
Streptococcus faecalis (adapted strain)	broth	10^6	0.6-1.3
Streptococcus faecalis		10^2	0.08-0.24
Streptococcus faecalis		10^2	0.17-0.65
Viruses: Herpes, mumps, influenza A and B, polio, type 2, vaccinia	tissue extracts	10^9 [1]	1.0
Polio, type 2	tissue-culture medium	10^9 [1]	2.0
Mumps	0.5% albumen	10^9 [1]	1.5
Influenza A	saline	50% reduction from 10^9 [1]	1.0
Influenza A	saline + 1% tryptophan	0	1.0
Polio, encephalitis	brain tissue	10^6 (L.D. 50)	3.5-4.0
Vaccinia	buffer	10^6 (L.D. 50)	1.5-3.0

[1] Virus particles in suspension at a concentration of 10^9 per ml.

the limits of statistical significance. For this latter purpose, a
procedure based on all-or-none survival in numerous containers
permits the exploration of the higher levels of inactivation. Consequently it is considered to be more realistic.

Almost all experimental work on this subject has been carried
out with gamma rays. It is desirable that similar experiments be
made with other types of radiation, particularly electrons, and
especially at high levels of inactivation.

Although this is not a question of microbiology, it should be noted that alterations of quality which cannot be accepted can be caused even by doses below 5 Mrad.

TECHNOLOGICAL ASPECTS OF THE RADIATION PASTEURIZATION OF FOODS

The aim in radiation pasteurization, as in heat pasteurization, is to diminish the total microbial flora or to eliminate pathogens. However, the organisms eliminated by radiation pasteurization are not necessarily the same as those eliminated by heat pasteurization.

In particular perishable foods, doses of the order of hundreds of kilorads can decrease the rate of microbial spoilage several-fold.

However, the extension of storage life of such foods is appreciable only if the irradiated foods are stored under conditions commonly used in commercial cold storage.

It should be noted that the temperature of storage should be sufficiently low to prevent the growth, on irradiated foods, of all pathogens and especially those with spores which survive irradiation. *Cl. botulinum* type E, for example, will grow at a temperature of 5°C and perhaps lower.

Though microbial spoilage may be delayed by irradiation, the storage life of the food may be determined by chemical and enzymatic changes (autolysis), some of which may be affected and even accelerated by irradiation (e.g., oxidative rancidity). Thus the storage life of the foods may be much less than that indicated by microblogical criteria.

Processes other than refrigeration (such as curing and the use of antibiotics) used in combination with irradiation may also help to delay microbial spoilage.

Doses of the order of hundreds of kilorads may also be used to eliminate particular pathogens in certain types of foods (e.g., semi-moist, frozen and dried foods). However, in such cases it is not the aim to use irradiation to prolong the storage life of the foods (Fig. 13.9).

More investigations are needed in order to determine which organisms are comparatively resistant to pasteurizing doses of ionizing radiations.

Radiation Resistant Organisms

Certain organisms possess unusual radiation resistance. For example, *Micrococcus radiodurans* survives doses which eliminate heat sensitive *Cl. botulinum*.

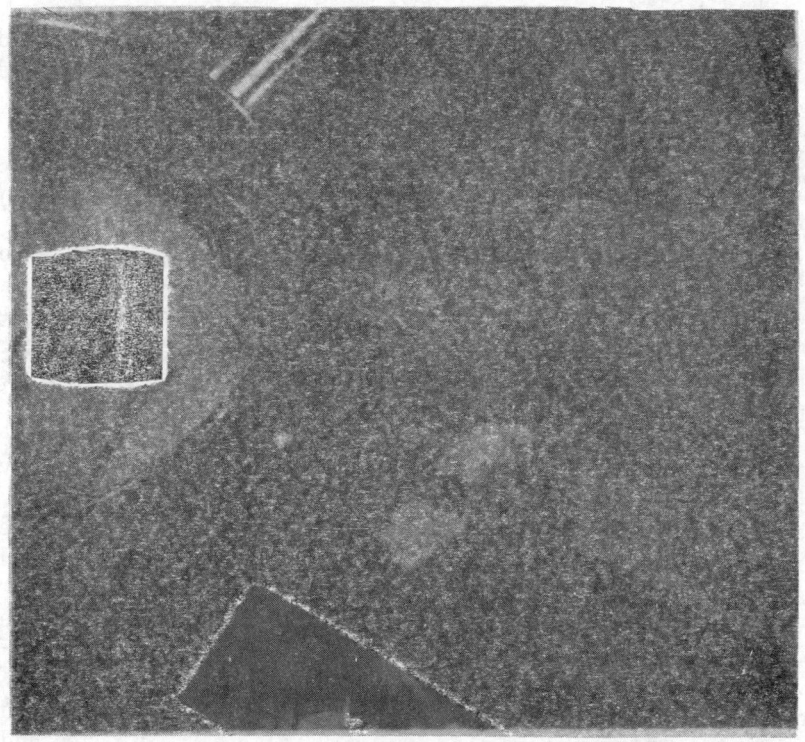

FIG. 13.9. SPROUT-INHIBITED POTATOES WITH 7500 RADS

Potatoes have been successfully stored for 2 years at ᵕ C and 85% RH, with approximately 10% losses in weight. Cured potatoes perform best. Wound healing occurs in treated tubers.

Similarly, certain other micrococci and faecal streptococci are so unusually resistant as to survive (with yeasts and molds) the doses usually given for radiation pasteurization.

Some pathogenic yeasts may survive doses considered as satisfactory for radiation pasteurization. Further research is needed to evaluate this potential health hazard, special emphasis being given to the radiation-resistant pathogenic yeasts which may be able to grow at 5°C or lower.

It has been established that in some instances the radiation resistance of some bacteria can be increased step-wise up to two-fold by a few repeated doses at a constant dose level, and several-fold by a progressive increase of applied dose.

FACTORS INFLUENCING THE SURVIVAL OF MICROORGANISMS FROM A RADIATION PROCESS

The Influence of the Type of Radiation on the Inactivation of Microorganisms

The heavy particle radiations (e.g., neutrons, alpha particles) have a substantially smaller effect on vegetative microorganisms than radiations from electron or gamma sources. In the case of spores, the position is less clear.

In any case, heavy particle radiation cannot be used for food irradiation because it is capable of inducing radioactivity in the target material at a significant though low level. This induced activity can persist for a significant time.

For the same reason it is undesirable to use electron or gamma sources of energy levels greater than about 10 Mev.

It is recommended that more microbiological investigations should be made with electron sources with energies of a few Mev.

The Influence of Dose Rate on the Inactivation of Microorganisms

At present there is little evidence that variations in dose rate within the usual ranges have any significant influence on the killing of microorganisms by radiation.

There is evidence that dose rate can influence the extent of certain chemical changes—high rates producing less effect.

It remains to be confirmed whether very high dose rates are equally effective in killing microorganisms, and even less damaging to the medium.

The effect of dose rate requires further investigation. A selection of sources of the same type which would provide a wide range of dose rates would be needed for accurate experimentation.

The Influence of Environmental Conditions on the Survival of Microorganisms from a Radiation Process

Oxygen.—The presence of oxygen increases two- or three-fold the sensitivity of vegetative bacteria to irradiation. During any investigation the oxygen tension should be kept constant, because variation in oxygen tension may greatly influence the results obtained.

Temperature.—The effects of irradiation are significantly different at temperatures above and below the freezing point of the system. Above the freezing point, the effect of temperatures up to about 20°C is small.

Freezing, especially to temperatures of −20°C or lower, reduces radiation effects in the food; however, it also reduces the effect on vegetative bacteria in a similar way. There is evidence, however, that the spores of *Cl. botulinum* are not similarly protected, so that there may be a theoretical advantage in using irradiation to sterilize meat at temperatures of −20°C or lower.

Of course foods can be preserved very well by storing them at these low temperatures (−20°C) without the use of irradiation.

Organic Compounds.—It is generally observed that microorganisms are more resistant to irradiation in organic media so that, for practical purposes, it is essential to make observations with food substrates. Irradiation in buffer or saline solutions, complex media, or foods may provide remarkably different survival data for a specific microorganism. Comparison of data, therefore, requires a statement of the nature of the suspending media used during irradiation.

Protective factors at work in organic media require more precise investigation.

Inorganic Salts.--Most salts, such as sodium chloride, generally do not have any important influence on the immediate effects of irradiation, but may have an influence on recovery.

A few salts such as nitrites and sulfites have been shown to be protective, but their influence on microorganisms requires more investigation.

Protective Compounds.—Many organic compounds, including SH compounds, protect microorganisms by influencing oxygen tension. Some particular SH compounds may also be protective in other ways which, as yet, are not understood. Moreover, there are indications that particular constituents of foods (e.g., catalase) may exert a protective action during or after irradiation.

Sensitizing Compounds.—An approach that would appear to be worthy of special attention would be to sensitize microorganisms to radiation. For this purpose, compounds which combine with SH groups are of particular interest. However, such compounds are often highly toxic to mammals (e.g., halogenated acetates). Thus, in carrying out any studies of the use in foods of compounds that combine with SH groups, it would be essential to investigate the possible toxicity of such compounds.

Another possibility is the use of chemical or physical processes to bring about the germination of all spores before applying the radiation process.

Combination Processes

When any irradiation process is used that does not result in complete sterilization, serious consideration must be given to decide which pathogenic organisms might develop in the food. Postirradiation conditions must be such as to inhibit them (e.g., by curing, drying, or storage at sufficiently low temperature).

When it is intended to irradiate meat without using in addition processes such as drying, curing, or storage at low temperatures, precooking may improve the stability of the final product by inactivating enzymes (hence reducing nonmicrobial deterioration) and also by decreasing the radiation dose necessary for a given amount of microbial inactivation. More investigation of this type of process is desirable.

With cured foods, the storage life of which may be prolonged by radiation, evidence is available to show that the curing process should not be reduced below that normally required, even when radiation doses exceeding one Mrad are used.

With fresh meat, tetracycline antibiotics used in combination with moderate doses of radiation may prolong storage life more than either agent separately, and the extension obtained may be approximately additive. Yeasts are important in the spoilage flora which eventually develops.

Combination processes employing irradiation and treatment with antibiotics should not be used commercially until more information is available concerning the incidence of drug resistance in various groups of microorganisms including radiation-induced mutants.

Conditions after Irradiation Affecting Survival and Recovery of Microorganisms

Usually (i.e., except in the case of infective pathogens), the survival of an individual microorganism is not significant unless it can multiply greatly. If the organism is unable to multiply, it will not harm the food, and it will not be possible to determine whether it has survived. Technically therefore it is necessary, by using plate counts or other appropriate techniques, to be able to detect every organism in an irradiated food which is capable of infinite multiplication.

It should be remembered that when the usual methods are employed in the enumeration of surviving organisms (plate counts, etc.) conditions may be very different from those obtained in the food itself. For example, it is possible that in the food there is a better opportunity for interactions of the kind which might influence survival (e.g., growth substances liberated from dead cells, or genetic recombinations).

The possibility of recovery of microorganisms in certain products in which growth conditions are suboptimal should be investigated.

The limited data which are available indicate that any changes which radiation might produce in the food itself have no significant effect on the subsequent growth of microorganisms. There is need for more work on this subject.

There is an increasing amount of evidence that the counting media used to determine post-irradiation survival is not large, and that the usual nutrient media may be used.

There is need for more work on the effect of different post-irradiation temperatures on survival.

With heat processing, the post-treatment effects of media and temperature are already well established and known. Although more investigation of these effects is needed in the case of radiation processing, the importance of these factors should not be exaggerated.

The Food Product—Enzyme Destruction

As previously noted, the enzymes found in foods are generally more resistant to ionizing radiations than are microorganisms, and by a factor of 5 to 10. The D_E value is the unit of destruction of an enzyme, taken as a value equal to 90% reduction in enzyme activity. The percentage destruction of enzymes with ionizing radiations is shown in a general manner in Fig. 13.8. It will be noted that the D value is much higher for enzymes than for microorganisms.

The degree of enzyme inactivation required in radiation stabilized foods is in the same order of magnitude as in other methods of preservation. It has been found that reductions in the order of four D units reduces activity to acceptable levels. If the D value is approximately 5 million rads, four times this value is 20 million rads. At this level great destruction in food quality can be anticipated. Supplementary methods of enzyme inactivation are therefore indicated.

The Process for Food Stabilization

Satisfactory stabilization of a food solely with ionizing radiations therefore will be obtained by applying either 12 D_M or 4 D_E levels of irradiation, whichever is greatest. In practice, the destruction of enzymes requires more ionizing radiation energy than the destruction of microorganisms.

Process—Heat Inactivation of Enzymes Plus Radiation Destruction of Microorganisms

Because the enzymes are generally radiation resistant, it is desirable to inactivate them by means other than radiation. The well established heat sensitivity of enzymes is an obvious solution. There may be many other methods of controlling enzymes. A combination treatment is indicated between ionizing radiation and heat inactivation of enzymes. If one D_E unit of destruction of enzymes can be obtained during the application of 12 D_M units of destruction of microorganisms, the application of sufficient heat to bring about 3 D_E units reduction in enzyme activity should result in stabilization. In practice this effect has been used with good results. Actually the application of less than 4 D_E units has been found to yield sufficient enzyme inactivation to permit several years of storage of irradiated foods without serious deterioration in food quality because of enzyme actions. Temperatures lower than 65°C for periods upwards to 30 min with beef has yielded acceptable levels of enzyme inhibition without sacrificing medium-rare appearance. The remaining enzyme destruction is accomplished when the beef is given 4.5 million rads.

The process and product specifications for various classes of radiation stabilized foods can now be formulated on the basis of available information (Table 13.4).

EXPERIMENTS WITH FRESH FISH

In recent years the extension of shelf-life of seafoods has received a great deal of consideration both in the United States and in many other countries of the world. Even greater stability is required than refrigeration alone can provide. Since many commercial food fish are taken from relatively cold waters, spoilage due to autolytic processes and microbial contamination continues even at cool temperatures. Thus the effectiveness of refrigeration for shelf-life extension is reduced. It is most noticeable in the flavor of fresh fish, its most perishable characteristic.

Connors and Steinberg (1966) have shown organoleptic acceptability of haddock fillets irradiated at 150 and 250 Krad and stored

TABLE 13.4

SUMMARY OF FOOD IRRADIATION PRODUCT AND PROCESS SPECIFICATIONS

Product	Process	Dose (rads)	Dosimeter	Package Requirements	Storage Temperature (°C)	Estimated Useful Storage Life
Potatoes	Sprout inhibition	7,500	Ferrous sulfate	Stored in open containers or in porous containers	5; 85% RH	2 years or more
					20	10 months or less
Flour	Insect destruction	50,000	Ferrous sulfate	Sealed paper or cloth bags, and overwrapped to prevent reinfestation	room temp.	2 years or more
					4	5 years or more
Berries	Pasteurization (mold destruction)	150,000	Cobalt glass	Sealed in oxygen, CO_2 permeable film	1	21 days or more
Sliced meat and fish	Pasteurization (bacteria, yeasts, mold, parasites, insects)	1,000,000	Ceric sulfate	Sealed in gas proof container	0	60 days or more
Animal, fish, and vegetable tissue	Heat inactivation of enzymes (75°C); radiation sterilization[1]	4,500,000	Ceric sulfate	Vacuum sealed in durable container (tin can) with odor scavenger	room temp.	2 years or more
					0	5 years
Fruits	Heat inactivation of enzymes (75°C); radiation sterilization	2,400,000	Ceric sulfate	Vacuum sealed in durable container (tin can) with odor scavenger	room temp.	2 years or more
					0	5 years

[1] Low temperature irradiation may improve eating qualities of certain products, including beef.

from 0 to 30 days at $-1°C$. They also found that irradiation flavors and odors were not detected in any of the taste tests. Angelini *et al.* (1975) reported on a study of irradiation and storage on haddock fillets by analyzing the volatile constituents. The concentration of volatiles in the haddock fillets irradiated at 2.8 Mrad was considerably lower than in the samples irradiated at 5.6 Mrad before storage. However, after 30 days in storage at $5°C$ the samples irradiated at 5.6 and 2.8 Mrad both contained low concentrations of volatile compounds. In view of these findings, samples analyzed in this study irradiated at 2.8 Mrad and stored at refrigerated temperature have good organoleptic quality. Mendelsohn *et al.* (1970) also found cod fillets to be organoleptically acceptable after irradiation at 4.5 Mrad at cryogenic temperatures. They concluded that:

(1) The microbial deterioration of haddock can be reduced by irradiation which decreases the spoilage microorganisms and thereby increases the shelf-life of haddock.

(2) The odor caused by volatiles formed in irradiated haddock decreases with storage time if the product is packed in a gas permeable container.

(3) The types and overall patterns of compounds produced by irradiation and microbial deterioration of fish are quite similar, but differ significantly with respect to hydrocarbons and amines.

(4) The expected shelf-life at a given storage temperature would depend on the irradiation dose.

SOME PUBLIC HEALTH ASPECTS OF THE MICROBIOLOGY OF IRRADIATED FOODS

The figures given in Table 13.3 illustrate the orders of magnitude of the doses necessary to destroy organisms significant from a public health point of view. Until more strains have been examined, however, these figures cannot be regarded as final.

It is evident that, because various Salmonellae are *more* resistant than the more common coliform organisms, the usual coliform tests will not be applicable to irradiated foods.

The faecal streptococci possess the highest resistance so far reported for enteric bacteria. However, it may be difficult to accept them as indicator organisms because their overall resistance may be unusually high.

The resistance of viruses to radiation is so high that it will probably be very difficult to eliminate them from foods by irradiation.

In this connection, much more investigation of food-borne viruses is needed. Meanwhile, if any irradiated food of animal origin is to be eaten without subsequent cooking, sufficient heat treatment will be necessary to guarantee the absence of viruses.

The mutagenic action of irradiation needs to be investigated in relation to the virulence, toxigenesis, antibiotic-resistance and environmental adaptation (including resistance to irradiation) of food-borne bacterial, mycotic and viral pathogens.

Experimentation is desirable to determine an effective method of identifying irradiated foods and some means of verifying that a specific food has received a treatment commensurate with refined antimicrobial objectives.

Studies should be encouraged to determine whether specific control methods could be recommended for the protection of consumers' interests as regards health and quality of food, with particular reference to the international movement of irradiated food.

CONCERN OVER MYCOTOXINS

Low dose gamma irradiation has potential to extend the shelf-life of certain foods. Hartung *et al.* (1973) applied a 50 Krad dose to flour and to bread made from the flour and obtained a reduction in the amount of visible and total mold that developed on the bread during storage up to 20 weeks. However, a small percentage of molds survived and were capable of outgrowth during storage. The predominate molds found were species of *Aspergillus* and *Penicillium*.

There have been reports that gamma irradiation doses below one Krad may induce or increase aflatoxin production by *A. flavus* in irradiated food. However, no stimulation of aflatoxin production has been found by this irradiation dose range of either *A. flavus* or *A. parasiticus* (Schindler *et al.* 1972; Bullerman and Hartung 1974).

If low level gamma irradiation is to be considered as a potential method of food processing, more information is needed concerning the effects of irradiation on microorganisms important to the public health, including mycotoxin-producing molds. Aside from the work with aflatoxins, too little is known of the effects of low level gamma irradiation on molds capable of producing other mycotoxins. Applegate and Chipley (1974) reported increased ochratoxin production by *Aspergillus ochraceus* after irradiation at 150 and 200 Krad. Bullerman and Hartung (1975) explored the effects of low level gamma irradiation on growth and patulin production by strains of *Penicillium patulum*, this organism having previously been isolated from irradiated bread.

Growth Variable

Bullerman and Hartung (1975) found that low level gamma irradiation of spores of *P. patulum* reduced subsequent growth in potato dextrose broth of both strains studied. After 7 days of incubation, the growth of strain NRRL 989 from spores irradiated at 100 Krad was 81% of the control. Growth from spores irradiated at 200 Krad was 40% of the control. Strain M 108 was somewhat more resistant to irradiation than strain NRRL 989. Growth of strain M 108 from spores irradiated at 100 and 200 Krad was 84% and 65% of the control, respectively. Irradiation of growing vegetative mycelia resulted in variable growth by subsequent cultures. The growth of strain NRRL 989 from irradiated mycelia was equal to or greater than the control at both the 100 and 200 Krad doses. The growth of strain M 108 was less than the control at the 100 Krad dose, but equal to the control at the 200 Krad dose. The final pH of all control cultures was lower than the final pH of cultures grown from irradiated inocula.

Patulin Production Reduced

Production of patulin by cultures grown from irradiated spores was substantially less than the control cultures (Table 13.5). Strain NRRL 989 produced 36% less patulin after irradiation at 100 Krad and 99+% less after irradiation at 200 Krad. Strain M 108 produced 68% less patulin at 100 Krad and 99+% less at the 200 Krad dose. Patulin production by cultures grown from irradiated vegetative mycelia produced even less patulin than cultures from irradiated spore cultures. Strain NRRL 989 did not produce detectable amounts of patulin after either irradiation treatment of mycelia while strain M 108 produced only trace amounts of patulin after irradiation of mycelia at 100 Krad (Table 13.6).

TABLE 13.5

PRODUCTION OF PATULIN BY PENICILLIUM PATULUM NRRL 989
AND M 108 ON POTATO DEXTROSE BROTH IN 7 DAYS OF INCUBATION
AT 25°C WHEN GROWN FROM IRRADIATED SPORES

Irradiation Level (Krad)	NRRL 989 Broth (μg/ml)	NRRL 989 Dry Mycelia (μg/mg)	M 108 Broth (μg/ml)	M 108 Dry Mycelia (μg/mg)
0 (Control)	137	24	376	67
100	88	19	122	26
200	0.2	0.09	2	0.6

Source: Hartung *et al.* (1973).

TABLE 13.6

PRODUCTION OF PATULIN BY PENICILLIUM PATULUM NRRL 989
AND M108 ON POTATO DEXTROSE BROTH IN 7 DAYS OF INCUBATION
AT 25°C WHEN GROWN FROM IRRADIATED VEGETATIVE MYCELIA

Irradiation Level (Krad)	NRRL 989		M 108	
	Broth (μg/ml)	Dry Mycelia (μg/mg)	Broth (μg/ml)	Dry Mycelia (μg/mg)
0 (Control)	32	9	13	3
100	ND[1]	—	0.7	0.2
200	ND[1]	—	ND[1]	—

Source: Hartung et al. (1973).
[1] ND = None detected.

There was no stimulation of patulin production observed in strain NRRL 989 or M 108 from spores or mycelia by either level of irradiation. The amount of patulin produced per milligram of dry mycelia declined steadily as the irradiation dosage increased. Complete inhibition of growth or patulin production was not achieved by irradiation of spores at levels up to 200 Krad. Irradiation of mycelia likewise did not eliminate growth, but did eliminate patulin production at the Krad dose level. These results suggest another area of potential benefit from irradiation of foods, which warrants further study.

Acceptability of Radiation Stabilized Foods

Providing the technology man has accumulated in food preservation is applied, many radiation stabilized foods can be made very acceptable. Demonstration of this fact can be seen in the tests conducted by volunteer troops in the U.S. Army. Pork loin, enzyme inactivated with mild heat, given 4.8 million rads of ionizing radiations, in suitable packages, and then stored for a year at room temperature was rated equal in acceptance to fresh pork loin by several hundred consumers. Thousands of consumers have eaten sterilized bacon, stored a year and more at room temperature; most consumers were unable to distinguish which item in a meal had been irradiated.

While there have been serious off-flavors in some radiation treated foods, this is not surprising. There is no reason to expect that the mere placing of a food in front of an irradiation source will yield acceptable products. However, precooked foods in meal assemblies have been produced with most satisfactory flavor and storage stability.

The addition of odor scavengers, such as a packet of carbon, in containers of food for irradiation results in greatly improved stabilized products after storage. The scavengers are particularly effective with precooked ready-to-eat foods.

The off-flavor so commonly found in irradiated foods may disappear on storage, such conditioning being most useful, and actually a part of the successful process.

The tenderness of meat given a sterilizing treatment with ionizing radiations is improved. In fact meat is prone to becoming too tender. Obvious metabolic methods are available in controlling tenderness of meat, in addition to the use of more chewy sections of carcasses and meat of lower grades. The process of rigor mortis can be used in controlling toughness of meats. The interference of normal rigor mortis in killed flesh by heat inactivation of enzymes is possible. With poultry the inactivation of enzymes immediately after killing results in flesh of too rubbery a texture. As a part of the process design for chickens, it is necessary to encourage the subsidence of rigor mortis prior to further treatment for enzyme inactivation and radiation sterilization. In beef the opposite appears to be of value. By allowing a degree of aging to occur, it is possible to control the tenderness of beef to some extent.

The texture of fruits and vegetables can be controlled with firming agents in much the same manner as is done in canning and other methods of food preservation.

The juiciness of meat can be controlled. The water binding capacity of the flesh may be improved by application of technologies developed by the meat industry in the curing methods of meat preservation (i.e., use of phosphates).

Quality Control with Radiation Stabilized Foods

The methods and technologies of quality control effective in other methods of food preservation are applicable generally in radiation stabilized foods. However, some further considerations are necessary because certain characteristics are different. For example, the color of radiation stabilized meats is unlike other preserved meats. Additional standards of quality are required for radiation treated foods.

Dosimeters are an important feature in quality control in radiation food processing. Sterility evaluation may be routinely performed as counterchecks to dosimeters and sanitary levels of products entering the plant.

New problems in quality control are encountered. An example

Courtesy of QM. Corps., U.S. Army

FIG. 13.10.   LINEAR ACCELERATOR COMPLEX
A schematic diagram of proposed food irradiation facility.

is shown in the fragility of chicken bones. The irradiation of very young chickens will result in fragile bones. Using older birds, this condition may be controlled.

Ionizing Radiations as a Unit Operation in the Food Industry

In addition to the ability of ionizing radiations to pasteurize and sterilize foods, this energy may be used in many food operations. Application areas are indicated in lower roasting requirements for coffee beans treated with low levels of ionizing radiations, probably with improved quality resulting from this treatment. The hydrolysis of starches and cellulose, the preparation of new flavoring compounds by treating selected substrates, the causing of yeasts to become giant sized when high yields of yeasts are desired, the aging of wines, the inhibition of sprouting of potatoes and numerous other crops, the increased yields of crops obtained by low level irradiation of seeds prior to planting, the synthesis of vitamin C, and numerous other applications indicate the great future ahead for ionizing radiations as a unit operation in the field of food and agriculture (Fig. 13.10 and 13.11).

SUMMARY

A new method of food preservation has been discovered in food irradiation with ionizing radiations. The potential benefits to

Courtesy of U.S. Department of Defense

FIG. 13.11.   ORIGINAL HIGH-INTENSITY GAMMA IRRADIATION FACILITY AT DUGWAY PROVING GROUNDS, DUGWAY, UTAH

A—Spent fuel elements are shown in cell, with containers being fed into it, indicated by arrows.

man are not yet realized. Only careful and continuing efforts by food scientists and food technologists can bring the needed developments into being. It is hoped that the new technology of food irradiation will take its due place alongside the other growing technologies of food preservation.

The health and welfare of mankind are intimately interwoven in the fabric called the economics of food distribution. The peaceful applications of atomic energy in the disinfection and stabilization of man's food directly involves the economics of food distribution. Radiation sterilization and radiation pasteurization of foods may have contributions to make to the health and welfare of all men. An acceptable solution to the successful applications of atomic energy in preserving foods still requires the best talents men can muster.

REFERENCES

ANGELINI, P., MERRITT, C., MENDELSOHN, J.M., and KING, F.J. 1975.
Volatile constituents of irradiated haddock. J. Food Sci. 40, 198-199.

ANON. 1965. Radiation Preservation of Food U.S. Dep. Commerce, Washington, D.C.

ANON. 1966. The technical basis for legislation on irradiated food. World Health Organ. Tech. Ser. 316.

ANON. 1968A. FDA rejection of irradiated ham. Food Chem. News, May 20, 3-5.

ANON. 1968B. Status of the food irradiation program. Joint Committee on Atomic Energy, Congress of the United States. U.S. Gov. Print. Off., Washington, D.C.

APPLEGATE, K.L., and CHIPLEY, J.R. 1973A. Increased aflatoxin production by Aspergillus flavus following ⁶⁰Co irradiation. Poultry Sci. 52, 1492.

APPLEGATE, K.L., and CHIPLEY, J.R. 1973B. Increased aflatoxin G₁ production by Aspergillus flavus via gamma irradiation. Mycologia 65, 1266.

APPLEGATE, K.L., and CHIPLEY, J.R. 1974. Effects of ⁶⁰Co irradiation on ochratoxin production by Aspergillus ochraceus. 74th Ann. Meeting Am. Soc. Microbiol., May 12-17, Chicago, Ill. (Abstr.)

BARRON, E.S.G., and DICKMAN, S. 1949. Studies on the mechanism of action of ionizing radiations. II. Inhibition of -SH enzymes by alpha, beta, and gamma rays. J. Gen. Physiol. 32, 395-405.

BARRON, E.S.G., and FINKLESTEIN, P. 1952. Studies on the mechanism of action of ionizing radiations. X. Effects of X-rays on some physiochemical properties of proteins. Arch. Biochem. Biophys. 41, 312-332.

BARRON, E.S.G., SEKI, L., and JOHNSON, P. 1952. The mechanism of action of ionizing radiations. VII. Effect of hydrogen peroxide on cell metabolism, enzymes, and proteins. Arch. Biochem. Biophys. 41, 188-202.

BATZER, O.F., and DOTY, D.M. 1955. Nature of undesirable odors formed by gamma irradiation of beef. J. Agric. Food Chem. 3, 64-68.

BELLAMY, W.D., and LAWTON, E.J. 1954. Problems in using high voltage electrons for sterilization. Nucleonics 12, 54-57.

BERAHA, L., and RAMSEY, G.B. 1956. Control of post harvest diseases of fruits and vegetables by radiation treatments. Quartermaster Food Container Inst. Prog. Rep. S-525, No. 2.

BONET-MAURY, P., and LEFORT, M. 1948. Formation of hydrogen peroxide in water irradiated with X and alpha rays. Nature 162, 381-382.

BRASCH, A., and HUBER, W. 1947. Ultra-short application time of penetrating electrons: a tool for sterilization and preservation of foods in the raw state. Science 105, 112-117.

BROWNELL, L.E. et al. 1955. Utilization of Gross Fission Products. Eng. Res. Inst., Univ. of Michigan, Ann Arbor.

BULLERMAN, L.B. and HARTUNG, T.E. 1974. Effect of low dose gamma irradiation on growth and aflatoxin production by Aspergillus parasiticus. J. Milk Food Technol. 37, 430.

BULLERMAN, L.B., and HARTUNG, T.E.   1975.   Effect of low level gamma irradiation on growth and patulin production by Penicillium patulum.   J. Food Sci. 40, 195-197.

BURNS, E.E., and DESROSIER, N.W.   1956.   Maturation changes in tomato fruits induced by ionizing radiations.   Food Technol. 11, 313-316.

CAIN, R.F., ANDERSON, A.W., and MALASPINA, A.S.   1958.   Effect of radiation on antibiotic treated meats.   Food Technol. 12, 582-584.

CAMPBELL, J.D., STOTHERS, S., VAISEY, M., and BERCK, B.   1968.   Gamma irradiation influences of the storage and nutritional quality of mushrooms   J. Food Sci. 33, No. 5, 540-543.

CARROLL, W.R., MITCHELL, E.R., and CALLAHAN, M.J.   1952.   Polymerization of serum albumen by X-rays.   Arch. Biochem. Biophys. 39, 232-233.

COCKCROFT, J.D., and WALTERS, E.   1930.   The production of alpha particles by high energy proton bombardment of lithium and beryllium.   Proc. Roy. Soc. 129, 477-485.

COLEBY, B.   1958.   Synthesis of vitamin C with ionizing radiation.   Personal communication. England.

CONNORS, T.J., and STEINBERG, M.A.   1966.   Preservation of fresh unfrozen fishery products by low-level radiation.   Food Technol. 20, 117.

CREAN, L.E., ISAACS, P.J., WEISS, G.J., and FAHROE, R.   1953.   Application of isotopic source to food and drug sterilization.   Nucelonics 11, 32-37.

CURTIS, H.J.   1951.   Biological effects of radiation.   Am. Rev. Physiol. 13, 41-56.

DAINTON, F.S.   1948.   On the existence of free atoms and radicals in water and aqueous solutions subjected to ionizing radiations.   J. Phys. Colloid Chem. 52, 490-517.

DALE, W.M.   1940.   The effects of X-rays on enzymes.   Biochem. J. 34, 1367-1373.

DALE, W.M.   1952.   The indirect action of ionizing radiations on aqueous solutions and its dependence on the chemical structure of the substrate.   J. Cell. Comp. Physiol. 39, 39-55.

DESROSIER, N.W.   1958.   Food irradiation research and its practical application in Europe and the U.S.A. Atoms for Peace Mission to Europe.   E.P.A.-O.E.E.C. Meeting on the Application of Atomic Science in Agriculture and Food. Paris, France, July.

DESROSIER, N.W.   1959.   A challenge to the food industry of the U.S.A. in the peaceful uses of atomic energy.   Food Process. 19, No. 2.

DESROSIER, N.W., MCARDLE, F.J., HOLLENDER, H.A., and MARION, W.W.   1955.   What cathode rays do to milk and eggs.   Food Eng. 26, 78-79.

DESROSIER, N.W., and ROSENSTOCK, H.M.   1960.   Radiation Technology in Food, Agriculture and Biology.   AVI Publishing Co., Westport, Conn. (Out of print.)

DILLON, L.G., BURRIS, L., and RODGERS, W.A.   1954.   Fission product kilocurie source: calculations of radiation intensity.   Nucleonics 12, 18-20.

DOTY, D.M., and WACHTER, J.P.   1955.   Radiation sterilization: influence of gamma radiation on proteolytic enzyme activity of beef muscle.   J. Agric. Food Chem. 3, 61-64.

DRAKE, M.   1958.   Disappearance of radiation induced flavor by high temperature conditioning. Personal communication.   Quartermaster Food and Container Inst., Chicago, Ill.

DUFFY, D.   1953.   Fission product potential of commercial reactors and their processes.   Nucleonics 11, 9-13.

DUGGAR, B.M. 1935. The Effects of Radiation on Bacteria. McGraw-Hill Book Co., New York.

EDWARDS, R.B., PETERSON, L.J., and CUMMINGS, D.G. 1954. The effect of cathode rays on bacteria. Food Technol. 8, 284-289.

ELLIS, N.K. et al. 1958. The feasibility of utilizing ionizing radiation to increase the storage life of potatoes. Final Rep. Proj. S-557. Quartermaster Food and Container Inst., Chicago, Ill.

ERDMAN, A.D., and WATTS, B.M. 1957. Radiation preservation of meats. Food Technol. 11, 349-353.

FANO, U. 1944. On the theory of ionization yield of radiations in different substrates. Phys. Rev. 70, 40-52.

FIELDS, M.L. 1959. Effect of cathode rays on food spoilage fungi. Ph.D. Dissertation, Purdue Univ., Lafayette, Ind.

FOSTER, F.L., DEWEY, D.R., and GALE, A.F. 1953. Van de Graaff accelerators for sterilization use. Nucleonics 11, 14-17.

FRICKE, H. 1952. Effect of ionizing radiations on protein denaturation. Nature 159, 965-966.

FRICKE, H., and HART, E.J. 1935. Oxidation of nitrite to the nitrate ion by X-rays. J. Chem. Phys. 3, 365-368.

GLASSTONE, S. 1950. Sourcebook on Atomic Energy. D.Van Nostrand Co., New York.

GOLDBLITH, S.A., KAREL, M., and OBERLE, E.M. 1953. Comparative bacterial effects of three types of high energy radiations. Food Technol. 7, 27-31.

GOMBERG, H.J., GOULD, S.E., and BETHELL, F.H. 1954. Control of trichinosis by gamma irradiation of pork. J. Am. Med. Assoc. 154, 654-658.

GOWEN, J.W. 1940. Inactivation of tobacco mosaic virus by X-rays. Proc. Natl. Acad. Sci. 26, 8-11.

HANNAN, R.S. 1954. Preservation of food with ionizing radiations. Food Sci. Abstr. 26, 121-126.

HANNAN, R.S. 1956. Science and Technology of Food Preservation by Ionizing Radiations. Chemical Publishing Co., New York.

HANNESSON, G. 1972. Objectives and present status of irradiation of fish and seafoods. Food Irradiat. Inf. 11, No. 1, 28.

HARTUNG, T.E., BULLERMAN, L.B., ARNOLD, R.G., and HEIDELBAUGH, N.D. 1973. Application of low dose irradiation to a fresh bread system for space flights. J. Food Sci. 38, 129.

HAUROWITZ, F., and TREMER, A. 1949. The proteolytic cleavage of irradiated proteins. Enzymologia 13, 229-231.

HAYNER, J.A., and PROCTOR, B.E. 1953. Investigations relating to the possible use of atomic fission products for food sterilization. Food Technol. 7, 6-10.

HINE, G.G., and BROWNELL, G.L. 1956. Radiation Dosimetry. Academic Press, New York.

HOLLAENDER, A. 1954. Radiation Biology, Vol. I. and II. McGraw-Hill Book Co., New York.

HOLMES, B. 1947. The inhibition of ribo- and thymonucleic acid synthesis in tumor tissue by irradiation with X-rays. Brit. J. Radiol. 20, 450-453.

HUBER, W. 1948. Electronic preservation of food by exposure to pulsed cathode rays from a capacitron: sterilization of drugs also possible. Electronics 21, 74-79.

HUBER, W., BRASCH, A., and WALY, A. 1953. The effect of processing conditions on organoleptic changes in foodstuffs sterilized with high intensity electrons. Food Technol. 7, 109-115.

INT. COMM. RADIOLOGY UNITS. 1951. Recommendations of International Commission. Radiology 56, 117-119.

JACKSON, J.M., and BENJAMIN, H.A. 1948. Sterilization of food. Ind. Eng. Chem. 40, 2241-2246.

JEMMALI, M., and GUILBOT, A. 1969. Influence de l'irradiation Υ des spores d' A. flavus sur la production d'aflatoxine B_1. C.R. Acad. Sci. 269, 2271.

JEMMALI, M., and GUILBOT, A. 1970A. Influence of gamma irradiation on the tendency of Aspergillus flavus spores to produce toxins during culture. Food Irradiat. 10, 15.

JEMMALI, M., and GUILBOT, A. 1970B. Influence de l'irradiation gamma de spores d'Aspergillus flavus sur la production d'aflatoxines. Congr. Int. Microbiol. Aug. 9-15, Mexico (Abstr.).

LATARJET, R., and CALDAS, L.R. 1952. Restoration induced by catalase in irradiated microorganisms. J. Gen. Physiol. 35, 455-470.

LEA, D.E. 1947. Actions of Radiations on Living Cells. Macmillan Co., New York.

LEA, D.E., SMITH, K.M., HOLMES, B., MARKHAM, R. 1944. Direct and indirect actions of radiations on viruses and enzymes. Parasitology 36, 110-115.

LEHMAN, A.J., and LANG, E.P. 1954. Evaluating the safety of radiation sterilized foods. Nucleonics 12, 52-54.

LEVIN, S. 1954. Radiological safety in cathode ray sterilization. Nucleonics 12, 54-57.

LICCIARDELLO, J.J., and NICKERSON, J.T.R. 1962. Effect of radiation environment on the thermal resistance of irradiated spores of Clostridium sporogenes type P.A. 3679. J. Food Sci. 27, 211-219.

LOVELL, R.T., and FLICK, G.J. 1966. Irradiation of Gulf Coast strawberries. Food Technol. 20, No. 7, 99-102.

MATCHES, J.R., and LISTON, J. 1968. Growth of Salmonellae on irradiated and non-irradiated seafoods. J. Food Sci. 33, No. 4, 406-410.

MCARDLE, F.J., and DESROSIER, N.W. 1955. Influence of ionizing radiations upon some protein components of selected foods. Food Technol. 9, 527-532.

MCARDLE, F.J., MARION, W.W., and DESROSIER, N.W. 1954. Cold sterilization of fresh shell eggs. Proc. Poultry Sci. Ann. Meeting, June.

MCARDLE, F.J., and NEHEMIAS, J.V. 1956. Effects of gamma radiation of the pectic constituents of fruits and vegetables. Food Technol. 10, 599-601.

MECKER, B.A., JR., and GROSS, R.E. 1951. Low temperature sterilization of organic tissue by high voltage cathode ray irradiation. Science 114, 283-285.

MEINKE, W.W. 1954. Does irradiation induce radioactivity in food? Nucleonics 12, 37-39.

MENDELSOHN, J.M., and BROOKE, R.O. 1968. Radiation processing and storage effects on the head space components of clam meats. Food Technol. 22, No. 9, 112-116.

MENDELSOHN, J.M., KING, F.J., and CUSHMAN, R.W. 1970. Preservation of uncooked fish muscle by cryogenic radiosterilization. Isotopes Radiat. Technol. 8, 81.

MORGAN, B.H., and REED, J.M. 1954. Resistance of bacterial spores to gamma radiations. Food Res. 19, 357-366.

MORGAN, G.W. 1949. Some practical considerations in radiation shielding. Circ. B-4, Isotopes Div., A.E.C., Oak Ridge, Tenn.

NELMS, A.T. 1953. Graphs of the Compton energy-angle relationship and Klein-Nishina formula from 10 Kev to 500 Mev. Natl. Bur. Stand. Circ. 542.

NICKERSON, J.T.R., LOCKHART, E.E., PROCTOR, B.E., and LICCIARDEL-LO, J.J. 1954. Ionizing radiations for control of fish spoilage. Food Technol. 8, 32-34.

NICKERSON, J.T.R., PROCTOR, B.E., and GOLDBLITH, S.A. 1953. Public health aspects of electronic food sterilizations. Am. J. Public Health 43, 554-560.

OLDENBERG, O. 1952. Resonance in collision processes. Phys. Rev. 87, 786-789.

O'MEARA, J.P. 1953. Radiation chemistry and sterilization of biological materials by ionizing radiations. Nucleonics 10, 19-23.

PEARSON, A.M., DAWSON, L.E., BRATZLER, L.J., and COSTILOW, J.G. 1958. The influence of short term high temperature storage with and without oxygen scavenger on the acceptability of precooked irradiated meat. Food Technol. 12, 616-619.

POLING, C.E. et al. 1955. Growth, reproduction, and histophathology of rats fed beef irradiated with electrons. Food Res. 20, 193-214.

POLLARD, E., and FARRA, F., JR. 1950. Cyclotron bombardment of enzymes and viruses. Phys. Rev. 78, 335-341.

POWELL, W.F., and POLLARD, E. 1953. Radiation sensitivity of enzymes in intact cells. Phys. Rev. 79, 494-496.

PROCTOR, B.E., and BHATIA, D.S. 1953. Mode of action of high voltage cathode rays on aqueous solutions of amino acids. Biochem. J. 53, 1-3.

PROCTOR, B.E., JOSLYN, R.P., NICKERSON, J.T.R., and LOCKHART, E.E. 1953. Elimination of Salmonella in whole egg powder by cathode ray irradiation of egg magma prior to drying. Food Technol. 7, 291-296.

PROG. REP. ATOMIC ENERGY RES. Hearings before the Committee on Research and Development of the Joint Committee on Atomic Energy, Congress of the United States, 4-8, June 1956. U.S. Gov. Print. Off. Washington, D.C.

RADOUCO-THOMAS, C. 1958. Meat irradiation in relation to other methods of preservation with special reference to enzymatic autolysis inhibition by adrenaline. European meeting on the use of ionizing radiations for food preservation, Atomic Energy Research and Establishment, Harwell, Eng., Nov.

RAHN, O. 1945. Physical methods of sterilization of microorganisms. II. Death of irradiation. Bacteriol. Rev. 9, 8-14.

ROBINSON, R.F. 1954. Some fundamentals of radiation sterilization. Food Technol. 8, 191-194.

RYER, R.R., III, and DRAKE, H. 1959. In package odor scavengers for irradiated foods. Personal communication. Quartermaster Food and Container Inst., Chicago, Ill.

SCHINDLER, A.F., ABADIE, A.N., and SIMPSON, R.E. 1972. Effect of low dose gamma radiation on the aflatoxin producing potential of Aspergillus flavus and A. parasiticus. 20th Ann. Meeting Radia. Res. Soc., May 14-18, Portland, Ore. (Abstr.).

SCHINDLER, A.F., and NOBLE, A. 1970. Enhanced aflatoxin production by Aspergillus flavus after gamma irradiation. Int. Congr. Microbiol., Aug. 9-15, Mexico (Abstr.).

SCHMIDT, C.F., LECKOWICH, R.V., and NANK, W.K. 1962. Radiation resistance of spores of Type E Clostridium botulinum as related to extension of the refrigerated storage of foods. J. Food Sci. 27, 85-90.

SCHMIDT, C.F., NANK, W.K., and LECKOWICH, R.V. 1962. Radiation sterilization of foods. II. Some aspects of the growth, sporulation and radiation resistance of spores of *Clostridium botulinum*, Type E. J. Food Sci. 27, 77-85.

SCHMITZ, J.V., and LOWTON, E.J. 1951. Initiation of vinyl polymerization by means of high energy electrons. Science 113, 718-719.

SCHULMAN, J.H., KLICK, C.C., and RABIN, H. 1955. Measuring high doses by absorption changes in glass. Nucleonics 13, 30-33.

SCHULTZ, H.W., CAIN, R.F., NORDAN, H.C., and MORGAN, B.H. 1956. Concomitant use of radiation with other processing methods for meat. Food Technol. 10, 233-235.

SCHWEIGERT, B.S., NEVIN, B.S., DOTY, C.F., and KRAYBILL, D.M. 1954. Cold sterilization of meat. Am. Meat. Inst. Found. Pamplet 10.

SHEFFNER, A.L., ADACHI, R., and SPECTOR, H. 1957. The effect of radiation processing upon the In Vitro digestibility and nutrient quality of proteins. Food Res. 22, 455-461.

SINNHUBER, R.O., LANDERS, M.K., and YU, T.C. 1968. Radiation sterilization of prefried cod and halibut patties. Food Technol. 22, No. 12, 74-77.

SPARROW, A.H., and CHRISTENSEN, E. 1954. Improved storage quality of potato tubers after exposure to cobalt-60 gammas. Nucleonics 12, 16-18.

SPENCER, L.V., and FANO, U. 1951. Penetration and diffusion of X-rays. Phys. Rev. 81, 464-465.

STEACIE, E.W.R. 1946. Atomic and Free Radical Reactions. Reinhold Publishing Corp., New York.

STOSIC, D.P. 1958. Rapid aging of wine with ionizing radiations. Personal communications. Belgrade.

TRIPP, G.E., and CROWLEY, J.P. 1957. Effect of radiation energy on flexible containers. Activities Report 9, No. 2, 112-122. Quartermaster Food and Container Inst., Chicago, Ill.

TRUMP, J.G., and VAN DE GRAAFF, R. 1948. Irradiations of biological materials by high energy X-rays and cathode rays. J. Appl. Phys. 19, 599-604.

TRUMP, J.G., WRIGHT, K.A., and CLARKE, A.M. 1950. Distribution of ionization in materials irradiated by two and three million volt cathode rays. J. Appl. Phys. 21, 345-349.

UNITED NATIONS. 1960. Report of the European Meeting on the Microbiology of Irradiated Foods held April 20-22, 1960, Paris. FAO, Rome.

U.S. ATOMIC ENERGY COMMISSION. 1955. Research Reactors. McGraw-Hill Book Co., New York.

URBAIN, W.M. 1953. Facts about cold sterilization. Food Eng. 25, 120-124.

VAN DE GRAAFF, R., TRUMP, J.G., and BUECHNER, W.W. 1948. Electrostatic generators for the acceleration of charged particles. Prog. Phys. Rep. 2.

WEISS, J. 1952. Chemical dosimetry using ferrous and ceric sulfates. Nucleonics 10, 28-31.

WORKMAN, M. 1959. On the inhibition of sprouting in potatoes. Personal communications. West Lafayette, Ind.

ZIRKLE, R.E. 1935. Biological effectiveness of alpha particles as a function of on concentration produced in their paths. Am. J. Cancer 23, 558-560.

Chapter 14

Principles of
Food Storage Stability

As we have seen, all food products are inherently unstable and quality retention depends upon a number of factors, including storage time and storage temperature. This is recognized in all new product work, in changing or improving existing products, and in modifying processes.

RELATIONSHIPS OF PRODUCT QUALITIES
AND STORAGE CONDITIONS

It becomes apparent that an enormous body of information is needed upon which the effective storage life of foods can be established. Data concerning changes in color, odor, flavor, texture, nutrients, moisture content, staling, rancidity, and overall changes in product acceptability are needed. Further data on the characteristics of rigid and flexible packages filled with various foods and their interaction under various storage conditions are required.

Fortunately voluminous literature has developed over the past few decades on these important subjects. Selected references on the various subjects are given at the end of this chapter.

OBJECTIVE TESTS OF QUALITY OF STORED FOOD

Objective tests of color are now well established for a range of products. In the following discussion note is made of the use of color and color difference measuring instruments. The Hunter instrument is shown in Fig. 14.1. Furthermore, we now see developments in objectively measuring odors and textures of foods, and their application in storage stability studies.

FIG. 14.1.   HUNTER COLOR DIFFERENCE METER IN USE TO
MEASURE COLOR CHANGES OBJECTIVELY

Objective Odor Measurements

A long-sought objective method for measuring volatile flavoring compounds in food products, together with the techniques for correlating the data obtained from expert and consumer sensory panels, is being perfected. Progress during recent years in the techniques of instrumental flavor analysis, sensory panel evaluation, the application of statistical analyses to flavor and taste data, and more recently, the use of computers to derive and analyze the data from these methods has provided Hoskins (1968) with a means of measuring and predicting storage age of food.

The importance of the volatile or odor compounds in product flavor has been defined by many researchers. Using a technique to recover, concentrate, and instrumentally measure these flavor volatiles, a major step in determining the total flavor can be taken.

Many techniques have been reported in the scientific literature, covering collection and instrumental measurement of flavor volatiles in food products. Highly satisfactory results have been obtained by Hoskins using modifications of the method of Jennings (1967). His procedure involves a "sweeping" of the sample product with charcoal-purified air, while holding the product under a vacuum and applying a minimum amount of heat. Flavor volatiles are swept from the flasks to a condenser where they are converted to a liquid phase. This liquid is then allowed to trickle over activated charcoal where the flavor volatiles are adsorbed and concentrated. After the distillation has been completed, the adsorbed volatiles are

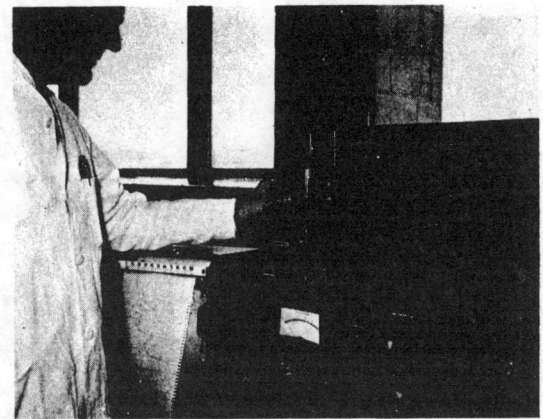

Courtesy of Nabisco

FIG. 14.2.   GAS CHROMATOGRAPH (GC) IN USE TO
STUDY AGING OF FOODS

washed from the charcoal by dissolving them in a minimum amount of solvent.

Several microliters of this concentrate are injected into a gas chromatograph (Fig. 14.2). The gas chromatograph (GC) uses a combination of specially prepared column, an inert carrier gas, and a programmed temperature increase to volatize and separate the individual components contained in the concentrated flavor sample. As each compound emerges from the column, a detector senses its presence and sends an electrical signal to the recorder which is displayed as a peak by the instrument recorder. The pattern of peaks becomes the flavor profile for the product under study. On the chromatograph each peak represents a specific isolated flavor volatile. The area under each peak is calculated to determine the amount of the flavor volatiles present in the product being investigated. Peak area information can also be fed directly from the GC unit into a computer, for calculation, storage or further correlation. Since some product samples produce hundreds of GC flavor peaks, the resulting ratios total thousands, so the use of a computer to store and analyze data becomes mandatory.

By running simultaneous taste panel studies on products of interest, a series of peak changes can be correlated to significant flavor changes detected by the panel. Subsequent investigation is then started to identify the compounds responsible for the peak/flavor changes. Identification of the peaks becomes important or useful to determine the mechanism causing the flavor change. Once definite sensory correlations have been established with peak profile chang-

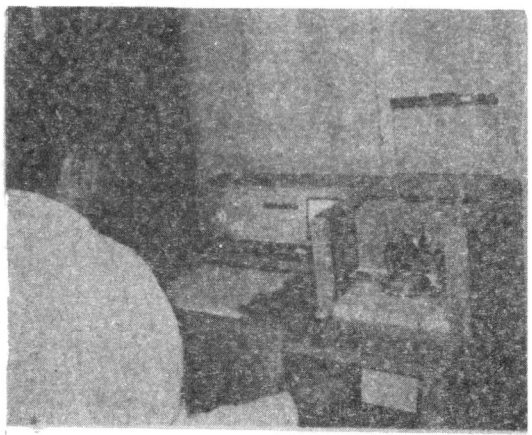

Courtesy of Nabisco

FIG. 14.3.   MECHANICAL MOUTH-TEXTUREMETER
IN USE TO CHARACTERIZE FOOD TEXTURES

es, these define limits and become indicators of flavor change.

By establishing the profile for a freshly made product, a basic foundation exists to measure changes due to aging. GC flavor profiles of aged product samples may be made while the keeping quality of these samples are investigated by the taste panel. Next, flavor changes can be correlated to instrumental data. As the correlations develop, and specific peaks or ratios of peaks are identified with flavor defects, it becomes possible to study the effects of aging or even predict the age of a product.

Mechanical Texturemeter

Texture is another food quality presently evaluated by organoleptic means. Here too, instruments are becoming available.

A mechanical "mouth," designed to simulate the mastication of a food material and express this as a curve on a recorder, is one of the first devices which measures the magnitude of effects as well as defining its meaning in terms of food texture.

The instrument consists of the masticator head, a sensing device (platform), a test plunger driven by an eccentric cam, and a two-speed strip chart recorder with suitable sensitivity to record very low magnitude force-time curves. These curves represent a texture profile and are characteristic of a food product (Fig. 14.3).

The sample to be tested is placed on a disc-like platform and contacted by an up-and-down motion of a plunger oriented so that a small, then progressively larger area of the plunger contacts the sample, essentially duplicating the mouth's motion during one phase of chewing, or biting. Force applied to the sample is at a maximum rate halfway between the up and down positions of the plunger, while at extreme positions the rate of application is at a minimum. Plunger speed can be adjusted to provide either 12 or 24 chews per minute, and 6 interchangeable plungers and 3 types of platform are available as accessories for different products. Deflection of the platform, which is a system of strain gauges balanced to a wheatstone bridge circuit, created the signal recorded on the strip chart. Voltage applied to the system is variable, and roughly controllable over 6 ranges from 0 to 20V, with fine adjustment using a potentiometer.

A "texture profile" can be generated by the instrument, defined in terms of 4 primary and 3 secondary texture parameters identified with a product. Since this facet of texture measurement and its relationship to panel data is becoming well documented, wide application is being found.

LONG-TERM STORAGE OF PRESERVED FOODS

Comprehensive studies of the storage stabilities of a wide range of military rations were undertaken by Woodroof at the Agricultural Experiment Station of the University of Georgia, and continued by Cecil and co-workers. These studies were summarized by Cecil and Woodroof (1962, 1968) and contain a wealth of information on long-term storage (upwards of seven years) of key bakery, dairy, cereal, beverage, pasta, confection, vegetable, fruit, meat and fish products. The products were well packaged and thought capable of storage at room temperature for at least five years. The application of this information to civilian food supplies is obvious.

An example of the amount of work involved becomes clear when they note that 93 types of products were studied over 9 temperature and humidity conditions for periods extending to 7 years. More than 144,000 units were tested, calling for more than 12,000 sensory evaluations, 76,000 chemical assays, 42,000 instrumental analyses, and examination of 110,000 individual flexible and rigid containers for defects in laminations, corrosion, and pin holes.

The results of these studies have broad meaning. They summarized the influence of the temperature of storage on food color, texture and nutrients. In addition, they estimated the cost to refrigerate and store foods for long periods. Some of their key conclusions follow.

Temperature of Storage

The most important result from these experiments was the knowledge concerning the extent by which the storage life of preserved foods can be increased by reducing storage temperatures. Each of the qualities—appearance, aroma, color, texture, flavor, acidity, drained weights (where applicable), and vitamins—of all products responded favorably to refrigeration.

Effect on Color.—The stability of color varied widely among the different products they studied with relation to temperature. The only significant change at $-17°$ or $-28°C$ was a slightly lighter appearance. There was no significant change in any product at $0°C$, and changes at $8°C$ result in significant changes, particularly in the less stable items.

At $21°$ and $37°C$ products tend to darken and brown and/or fade. Although there were exceptions, flavor changes generally tended to be associated with changes in color. Products which faded were frequently those which lost flavor or became stale, while off-flavors such as that of caramelization were often proportional to the extent of browning.

Effect on Texture.—Extremes of temperature had adverse effects on texture. Freezing damaged such items as tomatoes and beans enough to reduce them to substandard grades. Most of the sauces and gravies separated with freezing, and noodles, spaghetti or other starches tended to soften or slough.

Less serious freezing damage was noted in the softening of most vegetables and of cheese, and softening and slight sponging of frankfurter chunks. Beef and pork exhibited an increased tendency to break apart, and salmon to fragment with handling. Frozen and thawed ham and bacom were more easily separated into "strings" of fibers.

Their studies reveal softening of some meat products, "drying out" of certain bakery and confection products and the "syruping" of others, and the leaking of fats or oils from such items as cereal bars, meat bars, chocolates, and peanut butter at $21°$ and $37°C$, representing damage at the other temperature extremes. While softening or drying of texture were seldom as serious as color or flavor changes in the products involved, leaking of syrups or fats resulted in damage to wrappers, loosening of enamel linings of cans, and tended to increase development of stale or rancid flavors. Insect or rodent infestation is also encouraged when such products are packaged in nonrigid containers.

Natural Storage.—A wide range of natural temperature and humidity conditions exist in warehouses constructed of various mater-

FIG. 14.4.   UNDERGROUND STORAGE FACILITIES

ials, "dug outs" of many descriptions, shelters, and natural caves used for food storage. They can be highly satisfactory for food storage, if conditions are constant enough to allow predictions of storage life of the products (Fig. 14.4).

The temperature maintained in above-ground natural storage houses varied with site and geographical location. These may vary from below freezing to above 37°C at relative humidities from almost 100 to 20% or lower.

Temperatures in caves or other deeper excavations, such as certain mines or quarries, may have less than a 5°C annual fluctuation. Utilization of these or similar sources of natural cool storage can yield advantages over common above-ground storage in more uniform quality and increased storage life, or can greatly reduce the cost of holding food stocks under refrigerated storage (Fig. 14.5).

Nutrients

The results of vitamin determinations indicate that loss of nutritive values may be a more serious deterrent to very long storage than is loss of sensory appeal, particularly at temperatures in the 21° to 0°C range.

Cecil and Woodroof found that two vitamins—thiamin and ascorbic acid—decreased faster in storage than did either color or palata-

Courtesy of Casa Codornia

FIG. 14.5.   BOTTLE AGING IN A BODEGA IN NORTHERN
SPAIN

bility. Losses of 30 to 50% of these vitamins were determined in a few samples which were scored in taste panels almost as high as when storage tests began. Hence it is apparent that vitamin retention, rather than sensory quality, should be used as a basis for determining temperatures or time of storage of foods which are intended as primary or important secondary sources of thiamin or ascorbic acid.

During these tests other vitamins also were studied. There were no significant changes in B-carotene. Changes in niacin and riboflavin were relatively minor. In addition, none of these losses were as great as preceding or accompanying decreases in quality ratings; that is, these vitamins were at least as stable as the quality of products containing them.

Containers for Long-term Storage

Results from the study indicated that internal and external failure of containers frequently resulted in reducing quality or limiting storage life of the products. This was true with both rigid and flexible packages. Resistance to moisture, corrosion, leakage and package fatigue were found to need careful attention in order to withstand the effects of high temperatures or humidities or the corrosive action of certain products which were high in salt, fats, natural acids, or sulfur-containing compounds.

The results also indicated that damages to containers by excessive moisture, drying, heat or other unfavorable conditions were cumulative. Hence, selection and development of suitable containers can be as important as special formulation of foods for long-term storage.

VALUE OF EXCLUDING AIR

Table 14.1 shows the effect of nitrogen gas in the interstices of partially defatted peanuts on peroxide values of the oil in the peanuts during storage at 24°C (Pominski et al. 1975). Treatments for the three sets of storage samples were: (1) air cool, air hold; (2) air cool, nitrogen hold; and (3) nitrogen cool, nitrogen hold. After roasting, the initial peroxide value of the air cool, air hold peanuts was less than three. Use of either of the latter two methods considerably reduces the peroxide value. While peroxide value is not an exact measure for determining shelf-life, they believe it is generally accepted as an indication of deterioration.

The samples were evaluated for flavor and odor. Results are given in Tables 14.1 and 14.2. Flavor values are based on a hedonic scale of 9 to 1. At the end of 12 months the odors upon first opening the cans were respectively off, good, and good. Organoleptic ratings were higher for the two sets of defatted peanuts with nitrogen in the interstices.

Storage Costs

The storage of preserved foods for prolonged periods is receiving careful review. The broader use of refrigerated storage has also been shown to warrant attention. The feasibility of refrigerated storage depends upon the costs involved and the value received. The costs are divided into handling and storage. Since costs for handling are more or less fixed, warehousing rates decrease with the length of storage.

TABLE 14.1

EFFECTS OF NITROGEN GAS IN INTERSTICES OF PARTIALLY
DEFATTED PEANUTS ON FLAVOR AND ODOR DURING STORAGE AT 24°C

Conditions Prior to Vacuum Packaging	12 Months	
	Flavor[1]	Odor[2]
Air cool, air hold	5.8	Off
Air cool, nitrogen hold	6.8	Good
Nitrogen cool, nitrogen hold	7.3	Good

Source: Pominski et al. (1975).
[1] Based on hedonic scale of 9 to 1
[2] Odor when can first opened

As a rule of thumb, charges for *handling* food in refrigerated storage are about twice that for nonrefrigerated space. The *overall* cost of storage under refrigeration 0° to 2.2°C is about 30% higher than storage at room temperature (21°C), but 13% cheaper than freezer storage (−17.8° to −23°C). The average rate per hundred-weight for handling foods in nonrefrigerated space is about three times that for storage per month in the United States.

The cost of storage varies with the characteristics of the food item, particularly the density. Location and weather also influence storage costs, as do the temperatures maintained, type of equipment available, and the efficiency with which the equipment is used. An example is furnished by calculation made for operation of a 5-ton system, with electric rate at 2¢ per kwh. The cost of maintaining a temperature of 18°C for one year was increased only 8% if the temperature was lowered to 1°C and an additional 20% by decreasing the temperature to −17°C. In this instance, the use of equipment and space designed for −17°C storage was obviously a waste of money at 1°C and particularly

TABLE 14.2

EFFECTS OF NITROGEN GAS IN INTERSTICES OF PARTIALLY
DEFATTED PEANUTS ON PEROXIDE VALUES DURING STORAGE AT 24°C

Conditions Prior to Vacuum Packaging	Peroxide Values (meq/kg)		
	1 Month	4 Months	12 Months
Air cool, air hold[1]	28	29	19
Air cool, nitrogen hold	4	< 3	3
Nitrogen cool, nitrogen hold	< 3	< 3	< 3

Source: Pominski et al. (1975).
[1] Peroxide value immediately after roasting

at 18°C. Such examples illustrate the importance of selecting the most efficient facilities for the storage conditions proposed.

The cost of refrigerated storage depends upon the item. Cecil and Woodroof (1962) made a fascinating observation that canned meats could be held at 0° to 2°C for almost 40 years for the retail purchase price; cheese, fruit cakes and preserves may be held for about 25 years; candy 8 years; canned fruits and vegetables 4 to 8 years; but dried milk for less than 2 years. At −17°C canned meats may be held for 30 years for the purchase price; cheese may be held for 20 years; fruit cake 10 years; and candy 6 years.

Since most foods can be held in good condition for many years under certain conditions, it is important to establish the conditions and storage periods which afford the optimum balance between costs of storage and of changes in quality of stored products. Both the availability of food supplies, and the cost of frequent procurement and handling, and the availability and costs of available storage are factors which must eventually be balanced as food technology advances.

Some typical examples of storage changes in candy, cheese, meat and fish products are shown in Tables 14.3 through 14.7.

STORAGE STABILITY OF SELECTED FROZEN FOODS

Many food scientists and technologists have made careful studies of the stability of frozen foods. Results of earlier investigations on particular frozen commodities revealed to Van Arsdel of the Western Regional Research Laboratory, U.S. Dep. of Agriculture, that a new systematic study was needed to: (1) close several important gaps in knowledge, (2) broaden the coverage so as to take account of geographical and seasonal variations in products and of recently developed products and processing methods, and (3) obtain conclusive information about some questions on which the results of previous investigators had not been concordant.

The objectives of these studies were: (1) To determine how frozen food products behave under conditions of time and temperature, such as they may actually experience in the commercial distribution system, and in this way to provide information comprehensive enough to enable industry to (a) settle upon and specify allowable or tolerable deviations from "ideal" distribution conditions and (b) concentrate efforts on improvement of the more critical operations or areas in the system. (2) Upon the basis of the above results, to discover or devise improvements in raw material selection and handling, processing and packaging, such that the finished pro-

TABLE 14.3

THE EFFECT OF STORAGE TEMPERATURES ON CHANGES IN MOISTURE, TEXTURE
AND FLAVOR OF CANDY STORED TWO YEARS

Storage Temperature (°C)	Caramel Nougat		Coconut	Starch Jelly	
	Standard (Coated)	Special[1] (Coated)	Cream (Coated)	With Fruit[1]	Plain[1]
		Moisture			
	%	%	%	%	%
Initial	8.0	9.6	9.4	14.8	17.1
21	8.0	8.0	7.1	11.1	13.5
21 to 37	7.5	6.1	6.9	11.1	13.5
37	6.1	6.6	6.0	11.1	12.6
-23 to 37	6.1	6.7	7.0	10.9	13.0
-23	6.3	8.3	8.0	13.1	15.3
		Texture[2]			
	Score	Score	Score	Score	Score
Initial	7.3	7.6	7.6	6.9	7.9
21	4.9	6.2	5.8	5.7	6.2
21 to 37	3.7	4.9	4.4	4.6	6.0
37	4.2	5.0	4.6	4.6	6.4
-23 to 37	3.8	4.2	4.8	5.1	6.1
-23	6.7	6.8	6.8	6.9	7.2
		Flavor[2]			
	Score	Score	Score	Score	Score
Initial	7.4	7.8	7.6	7.0	7.6
21	4.8	5.8	4.2	6.0	6.3
21 to 37	3.0	3.0	3.0	4.3	4.9
37	2.8	3.2	2.0	4.1	6.1
-23 to 37	2.8	3.0	2.1	4.9	6.0
-23	6.5	6.6	6.3	7.0	7.2

Source: Cecil and Woodroof (1962).
[1] Average for two variations of this type
[2] On a 9 point scale

ducts will better withstand adverse subsequent experiences. (3) To
seek tests of product quality that can be applied to a frozen food at
any point in the distribution system to indicate how much it has
changed and, if possible, to estimate where it stands in the scale
of commercial acceptability.

Results

Details of these studies can be obtained from the U.S. Dep. of
Agriculture, Western Regional Research Laboratory, Albany, California.

DeEds (1959) summarized the findings as follows: (1) Seasonal or
regional variations have little effect on the sensitivity of a product
to damage due to temperature elevation or variation. Faulty processing or packaging has a major effect in increasing the sensitivity

TABLE 14.4

CHANGES IN VACUUM, HUNTER COLOR, AND ACIDITY VALUES OF CHEESE IN STORAGE

Storage		Cans Retaining Vacuum (%)[1]	Vacuum Retained Avg (cm²)	Hunter Color		Acidity	
Temperature (°C)	Time (Months)			a/b Ratio[2]	ΔE Units[3]	Titratable (%)[2]	Free Fatty Acids (%)[2]
37	6	0	0	0.29	2.0	1.69	2.51
21	6-12	97.2	14	0.27	1.3	1.71	1.82-0.71
	18-36	64.3	7	0.22	1.9	1.89	0.75
	60-72	31.0	6	0.28	2.3	1.77	1.03
8	6-12	100.0	13	0.26	1.0	1.54	1.27-0.47
	18-36	75.8	7	0.20	2.1	1.62	0.55
0	6-12	100.0	14	0.26	0.4	1.64	1.37-0.44
	18-36	83.9	10	0.22	0.9	1.72	0.49
	60-72	51.7	9	0.22	2.1	1.48	0.46
-17, -28	6-72	87.2	17-18	0.26-0.20	1.3	1.57-1.44	1.34-0.32

Source: Cecil and Woodroof (1962).
[1] Of cans examined during the period indicated.
[2] Mean for the period indicated.
[3] Values at 37°, 21°, 8°, 0°C are total color differences between samples at these temperatures and those at -17° and -28°C for the same periods. The value for -17° and -28°C is the mean color change between 6 and 72 months.

TABLE 14.5

TITRATABLE ACIDITY, FREE FATTY ACIDS, AND CONDITIONS OF INTERIOR SURFACE
OF CANS OF BEEF STEAKS IN STORAGE

Storage		Titratable Acidity As Lactic Acid (% in Lean)	Free Fatty Acids As Oleic Acid (% in Fat)	Condition of Cans Interior Surface Rating[1]
Temperature (°C)	Time (Months)			
37	6	0.92	2.0	8.0
	12	1.06	2.8	6.2
	18-24	1.33	4.6	5.3-4.7
21	6-30	0.98	1.5	8.0-6.3
	36-72	1.21	2.0	6.0-4.7
	84	1.00	2.3	4.4
8	6-12	0.94	1.6	8.5
	18-30	1.03	1.4	8.1
	36-48	1.11	1.5	8.0-7.0
0, -17, -28	6-12	0.97	1.5	8.6
	18-48	1.02	1.6	8.4-7.5
	60-84	1.00	0.4	7.1-6.0

Source: Cecil and Woodroof (1962).
[1] On scale of 9 (no damage) to 1 (complete corrosion) for inner can surfaces.

TABLE 14.6

COLOR CHANGES, RANCIDITY VALUES AND CONDITION OF INTERIOR SURFACES
OF CANS OF GROUND MEAT AND SPAGHETTI IN STORAGE

Storage		Hunter Color Changes		Rancidity Values		Can Surface Rating[1]
Temperature (°C)	Time (Months)	ΔE Units	a/b	Peroxides (M-mols/ kg)	Free Fatty Acids (% As Oleic)	
—	0	—	0.48	12.8	2.1	8.0
37	6-12	2.4	0.53	2.6	3.0[2]	7.1
	18-24	5.2	0.56	1.6	1.3	5.5
21	6-12	0.9	0.50	2.5	2.9[2]	7.5
	18-36	1.6	0.53	1.5	1.1	5.7
	60-72	-2.4	0.48	0.5	1.9	3.4
8	6-12	1.0	0.48	2.7	2.5[2]	7.9
	18-36	-0.2	0.50	1.4	1.0	6.6
0	6-12	0.4	0.48	2.8	2.3[2]	7.5
	18-36	-0.1	0.50	1.3	1.0	6.8
	60-72	-0.9	0.46	0.8	0.7	5.7
-17, -28	6-12	[3]	0.48	3.3	2.1[2]	8.0
	18-36		0.46	1.9	0.9	6.5
	60-72		0.47	0.7	0.6	6.5

Source: Cecil and Woodroof (1962).
[1] On a scale of 9 (no corrosion or discoloration) to 1 (surface completely corroded or discolored).
[2] Value at six months, after which free fatty acids decreased.
[3] Samples from these temperatures were used as references for the changes shown.

TABLE 14.7

HUNTER COLOR CHANGES, RANCIDITY VALUES, AND CONDITION OF CANS
OF SOCKEYE SALMON IN STORAGE

Storage		Hunter Color Changes		Rancidity Values		Fatty Condition of Cans Ratings[2]
Temperature (°C)	Time (Months)	ΔE Units[1]	a/b Ratio	Peroxides (M-mols/ kg)	Free Fatty Acids (% As Oleic)	
37	6	1.7	0.46	4.4	1.31	6.3
	12	2.8	0.36	1.5	1.65	5.6
	18	4.4	0.31	3.6	2.85	3.7
21	6-12	1.1	0.50	3.3	0.98	7.9
	18	3.2	0.38	1.6	1.10	7.8
	24-42	1.8	0.43	1.4	1.74	6.9
8	6-24	1.1	0.51	1.9	0.92	8.0
	30-42	1.1	0.52	1.3	0.94	7.7
0	6-42	0.7	0.51	2.1	0.87	7.8
	60-84	0.8	0.56	0.3	0.33	5.9
-17, -28	6-42	0.5[3]	0.51	1.7	0.87	7.8
	60-84	0.3[3]	0.54	0.3	0.26	7.0

Source: Cecil and Woodroof (1962).
[1] Calculated as vectorial sums of decreases in Rd and "a" and increases in "b" from values
 for the -28°C samples.
[2] On a scale of 9 (no discoloration) to 1 (completely discolored or corroded) for
 interior can surface.
[3] Includes -17°C only, as -28°C samples were used as reference.

of a product to damage. (2) Frozen products differ markedly in
their susceptibility to damage. For example, frozen orange juice
concentrate deteriorates rapidly under adverse conditions and so
do some precooked frozen items. Peas lose quality less rapidly, and
raspberries are even less sensitive to temperature change. (3) All
the changes characteristic of the deterioration of a given frozen
product are accelerated by any rise in temperature, and the magnitude of changes may be surprisingly great. A 5°C rise between −17° and
−1°C may make a big difference, often two- or three-fold and sometimes five-fold, depending on the product. In other words, a drop
of 5°C in temperature means it will take 2 to 5 times as long to produce
the same degree of deterioration. Moreover, each 5°C rise in temperature multiplies the effect. If one 5°C change produces a three-fold
change, 10°C will speed it up nine-fold, 15°C, twenty-seven-fold, and so
on. The effect of temperature rise in marketability of a frozen product
is obvious. (4) The effects of temperature abuse are additive, and the
sequence or spacing of the abuses makes no difference. However, a frozen product remembers its past history; damage once done cannot be
rectified by lowering the temperature, although further damage can
be prevented.

Guadagni (1961) reviewed the relationship between temperature

TABLE 14.8

TIME-TEMPERATURE COMBINATIONS WHICH CAUSE EQUIVALENT QUALITY
CHANGES IN SOME FROZEN FRUITS AND VEGETABLES

Temperature (°C)	Time
-17	1 year
-15	5 months
-12	2 months
-9	1 month
-6	2 weeks
-3	1 week
-1	3 days

Source: Guadagni (1961).

and the approximate average time required to cause a definite organoleptically measurable change in a number of frozen foods. He
reported that the time will vary for various products and even within a product there will be variations between lots. However, the
general exponential relation between temperature and time to produce this given degree of quality change holds for most frozen
products studied. For every 5°C that the temperature is increased,
the rate of speed of quality loss increases from about 2 to 2½ times.

He emphasized, however, that the quality change brought about
by each of these time-temperature combinations is essentially the
same. Therefore, if we start an experiment with the same quality
level, the sample held at −3°C for 1 week will be the same or indis-
tinguishable from the sample held at −17°C for one year, or from the
samples held at any of the other time-temperature combinations
listed in Table 14.8. From these data, it is clear that a week of −6°C
is certainly no worse than 6 months at −17°C. As a matter of fact,
on the average, neither of these time-temperature combinations
alone would cause a noticeable quality change, but the two experiences combined would have the same effect as 1 year at −17°C or two
weeks at −6°C.

Table 14.8 presents a simple straightforward relation between
steady temperature and time required to produce a definite quality
change. Guadagni stated that in handling frozen foods, however,
we are rarely, if ever, dealing with steady temperatures, but with
a series of variable temperature excursions.

The relative stabilities of selected types of frozen foods at various storage temperatures have been summarized from various
United States, Canadian, and German sources (Table 14.9). It is
to be emphasized that the products are still of good quality but detectable changes have occurred.

Pence and Heid (1960) reported on their studies of the stability

TABLE 14.9

RELATIVE STABILITIES OF SELECTED FROZEN FOODS[1]

Product	Storage Temperature (°C)							
	-23	-20	-17	-15	-12	-9	-6	-3
Lean fish	5.5[2]	4	3	2.5	2	1.5	1.2	1
Fat fish	3.5	2.5	2	1.5	1.2	<1	<1	<1
Chicken (well packed)	>30	>30	25	15	12	8	6	4.5
Chicken (fried)	4	2.5	2	1.5	1	<1	<1	<1
Pork	30	15	10	5	3	1.5	1	1
Beef	30	20	12	8	4	3	1.5	1
Turkey pies	>25	25	25	18	10	4	2	1
Frozen turkey dinners	>25	>25	25	18	10	4	2	1
Orange juice	>20	>20	20	15	9	5	3.5	2
Peas	25	15	9	5	2.5	1.5	1	<1
Green beans	25	15	9	5	2.5	1.5	1	<1
Cauliflower	>20	20	10	4	2	1	<1	<1
Spinach	15	8	5	.3	2	1	<1	<1
Raspberries	>30	>30	30	10	5	1.5	<1	<1
Strawberries	>20	20	10	4	2	1	<1	<1
Peaches	>20	>20	20	5	1.5	1	<1	<1

[1] Time for first meaningful changes, as detected by trained panels (see text).
[2] Time in months at various storage temperatures.

TABLE 14.10

STABILITY OF FROZEN CAKES[1]

Comparison	Flavor		Texture	
	Time	Preference	Time	Preference
Chocolate Layer				
-12°C vs Fresh	1	Neither	1	Neither
-17°C vs Fresh	3[2]	Neither	3[2]	Neither
-23°C vs Fresh	1	Frozen	1	Frozen
-34°C vs Fresh	1	Frozen	2	Neither
-12°C vs -23°C	4	-23°C	4	-23°C
Chiffon				
-12°C vs Fresh	1	Fresh	1	Fresh
-17°C vs Fresh	1	Fresh	3	Fresh
-23°C vs Fresh	5	Fresh	4	Fresh
-34°C vs Fresh	5	Fresh	5	Fresh
-12°C vs -23°C	1	-23°C	1	-23°C
Angel Food				
-12°C vs Fresh	1	Fresh	1	Fresh
-17°C vs Fresh	2	Fresh	2	Fresh
-23°C vs Fresh	2	Fresh	2	Fresh
-34°C vs Fresh	5	Neither	8	Neither
-12°C vs -23°C	2	-23°C	1	-23°C

Source: Pence and Heid (1960).

[1] Time in weeks required for first significant differences (5% point) between cakes exposed to different temperatures and freshly baked controls or between cakes held at -12°C and -23°C.

[2] Unavoidable circumstances prevented evaluation of cakes held at -17°C for two weeks.

of frozen cakes. Stability at several temperatures was determined for frozen yellow layer, chocolate layer, angel food, chiffon and pound cakes by means of taste panel comparisons with freshly baked and day-old unfrozen cakes. See Table 14.10 for typical results with selected products.

Texture changes in the frozen cakes were generally detected before changes in flavor. Principal texture changes were loss of original crumb resiliency and increases in crumbliness and harshness. Each type of cake exhibited texture changes of a somewhat characteristic nature. Flavor declined gradually to yield increasingly bland products. No characterizable stale or off-flavors were encountered.

Yellow and chocolate layer cakes exhibited relatively good stability at -12°C and greatly improved stability at -17°C. However, they reported little gain in stability was obtained at -23°C and -34°C. Layer cakes seemed to soften and become more moist at -17°C or below. This apparent softening eventually deteriorated into an unacceptable gumminess or pastiness. Layer cakes were by far the most stable of the cakes studied.

Pence and Heid (1960) found that pound cakes at temperatures of −17°C and below very rapidly developed differences in texture, but the initial loss of resiliency and increased firmness later appeared to become overshadowed by increased tenderness as the cake became more crumbly.

All types of cake showed highly significant differences from freshly baked cakes within 4 weeks at −17°C. Even after significant changes had occurred in the frozen cakes, however, their quality was still quite good. In all cases they were superior to day-old unfrozen cakes.

KINETICS OF NUTRIENT LOSS

Another significant development in storage technology has occurred. It deals with predicting storage changes from a nutritional point of view.

The loss of nutritional value during storage (Labuza 1973) has received inadequate study from a kinetic standpoint. He finds numerous papers report losses of vitamins or destruction of biological value of proteins during storage. However, very little information is available as to the rates of destruction, due to incomplete data or difficulty in analyzing the results.

Today, however, these data are needed desperately. Companies will eventually need to make a claim for the nutritional value of their product. Hence they will need to know, for example, the rate of loss of a particular vitamin.

To predict the extent of deterioration of important nutritive factors a knowledge of the reaction rate as a function of temperature and moisture content is needed. This is especially true during storage when water or oxygen, if involved in the reaction, slowly permeates through the packaging material.

A chemical reaction in a food can be quite complex, but for food deterioration in general, Labuza (1973) assumes a first-order reaction.

$$A \xrightarrow{\quad k \quad} B \qquad (14.1)$$

The loss of A can be described as:

$$-dA/d\theta = k[A] \qquad (14.2)$$

where $[A]$ = concentration of A; k = rate constant $= k_o e^{-Ea/RT}$; k_o = absolute rate constant; E_a = activation energy; R = gas constant = 1.986 cal/mole - °K; T = absolute temperature, °K; and θ = time. If $A = A_o$ at $\theta = 0$, then integration of Equation 14.2 yields:

$$2.3 \log A/A_o = -k\theta \qquad (14.3)$$

Thus a plot of the log (% of a nutrient remaining) versus time yields a straight line for a first-order reaction.

Many food reactions can be easily categorized in this manner, thus yielding useful information as to the shelf-life of the food with respect to a particular nutrient. The data can be easily collected by simply measuring the time for 50% loss of the nutrient, since:

$$k = 0.693/ O_{1/2} \qquad (14.4)$$

where $O_{1/2}$ is the half-life of the nutrient at a particular temperature (Labuza 1973).

In some cases, nutritional losses follow a zero-order reaction. This occurs if the loss of A is so small in the time period studied that the value of [A] does not change significantly. Thus, the right-hand side of Equation 14.2 is a constant.

Predicting Nutrient Losses

Labuza (1973) found that several problems exist in predicting nutrient losses. Each nutrient is destroyed at a different rate (k), and k is a function of temperature as controlled by the activation energy, E_a. The higher the activation energy, the more effect an increase in temperature has on the reaction.

This increase in rate with temperature is useful in that Labuza uses accelerated storage tests to predict what would happen at lower temperatures. He tests at three temperatures and plots the log k versus 1/T. However, in some cases he suggests using an accelerated test and a shelf-life prediction based on a reaction that becomes important only at high temperature.

More kinetics than illustrated can be handled. Overall, such methods are available, but the real problem is the dearth of data on nutrient destruction from a quantitative standpoint. Without that, he reports there is no capability to optimize from a nutrient standpoint.

MATHEMATICAL EXPRESSION OF STABILITY

In an excellent piece of work, Kwolek and Bookwalter (1971) studied the ability to predict storage stability in food products from time-temperature data, using changes in product quality. The following has been abstracted from their original work.

Product stability can be expressed mathematically in terms of three variables: Y, a measure of product quality, such as a taste score, a physical property, or the result of a chemical or microbiological assay or a feeding trial; t, the number of days stored; and T, the storage temperature, which may be expressed in degrees absolute, Centigrade or Fahrenheit.

The relation between Y and t for a constant T can often be described by a straight line, although sometimes it is necessary to use log Y or some other transformation. The slopes of several lines associated with different temperatures are some function of T. Thus we can relate the three measurements by the three equations:

$$Y = a + b_i t + u$$
$$b_i = f(T_i)$$
$$Y = a + tf(T_i) + u$$

where a is the response at zero time, b_i is the slope or rate of change in Y per unit change in t associated with the i temperature level, and u is a random error associated with the deviation of observed Y from the model.

Kwolek and Bookwalter (1971) further found that the relations $Y = a + b_i t$ can be determined by fitting least squares lines to Y versus t for each temperature level T_i. A relation between b_i and T_i can be determined by plotting b_i versus T_i (or C_i if temperature is in Centigrade). The functions $f(T_i)$ considered are

$$b_i = m + KT_i \qquad (14.5)$$
$$b_i = mT_i^K \qquad (14.6)$$
$$b_i = m/(K - C_i) \qquad (14.7)$$
$$b_i = me^{K/T}_i \qquad (14.8)$$
$$b_i = mK^{c_i} \qquad (14.9)$$

where T or C (subscripts will be omitted from here on) is the temperature, b is the slope relating Y and t, and m and K are constants. Generally, at least 3 temperature levels are required, with samples taken at least at 2 different times per temperature with a control sample at zero time. Figure 14.6 shows Equations 14.5-14.9 plotted to pass through points b = 0.00001 at C = 0 and b = 0.25 at C = 60".

Exponential Functions

Kwolek and Bookwalter (1971) found only minor differences

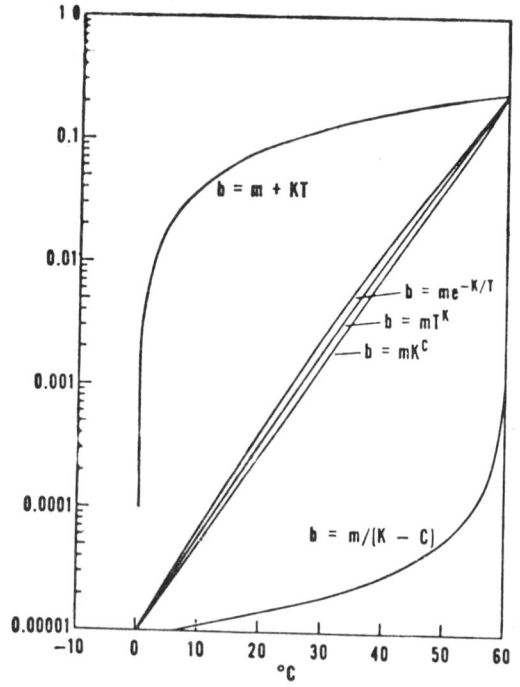

Courtesy of Kwolek and Bookwalter

FIG. 14.6.   GRAPHICAL REPRESENTATION OF b AS VARIOUS FUNCTIONS OF
STORAGE TEMPERATURE: b IS THE SLOPE OF THE PLOT
OF QUALITY RESPONSE VERSUS STORAGE TIME

among the 3 exponential functions (Equations 14.6, 14.8 and 14.9).
When log b is plotted against T or C, these exponential functions
yield practically straight lines, with Equation 14.8 showing the
most curvature. This function is associated with an Arrhenius plot,
with T being the absolute temperature. The usual linearizing plot
of the Arrhenius equation is log b vs $1/T$.

They indicate that the hyperbolic function (Equation 14.7) was
selected based on the argument that as temperature increased, the
slope would increase rapidly as the temperature approached a limiting temperature K. Over short suitable ranges of temperature, the
linear function (Equation 14.5) may prove satisfactory, depending
on the precision with which b is estimated.

Since the plots are essentially linear and C will seldom exceed 60°
(140°F) in stability trials, there is little empirical basis for choosing
among the three exponential models. Also, for values of $C/273$ less
than 0.22, the 3 exponential functions are satisfactorily approximat-
ed by a straight line.

Two Models Selected

Based on several fittings of these models Kwolek and Bookwalter (1971) selected two for further discussion:

$$Y = a + bt/(K - C) \qquad \text{(hyperbolic)} \qquad (14.10)$$
$$Y = a + bte^{-K/T} \qquad \text{(Arrhenius)} \qquad (14.11)$$

The model $Y = a + btK^c$ also provided as good a fit as the Arrhenius model. These models were selected based on slightly closer correlation between observed and predicted values, inclusion of the concept of a limiting temperature, and suitability over a wide range of temperature. The Arrhenius model is very likely the most satisfactory from a theoretical viewpoint. The linear model

$$Y = a + b_1 t + b_2 tT$$

based on Equation 14.10 can be handled by ordinary multiple regression techniques. However, they did not believe Equation 14.10 was a suitable relation, since one would expect b to increase rapidly at higher levels of C and slowly at lower levels of C, rather than at the same rate for all temperatures.

Kwolek and Bookwalter (1971) next set out to illustrate how stability data can be used to control product quality.

They chose a model to be fitted to the hyperbolic model $Y = a + bt/(K—C)$, where t is the number of days in storage, C is storage temperature, a is the initial or control value for Y at time zero, and b and K are constants. The term K is interpreted as a critical temperature at which the product tends to show rapid change depending on $b/(K—C)$. This model, as well as those previously mentioned, is difficult to fit by ordinary linear regression methods, since an exact least squares solution for estimating the constant K cannot be obtained. Rather a solution is based on a series of "guesses" that converge to a solution for K. However, they report, once K is determined, the problem is reduced to one of fitting a straight line to data. A solution for K is obtained when the following equality holds:

$$\frac{\Sigma yX}{n\Sigma X^2 - \Sigma X \Sigma X} = \frac{\Sigma yZ}{n\Sigma XZ - \Sigma X \Sigma Z} \qquad (14.12)$$

where
$$X = t/(K - C)$$
$$Z = X/(K - C)$$
$$y = Y - \Sigma Y/n$$

Equation 14.12 is the least squares solution for K after eliminating a and b from the normal equations. The solution for b is given by either side of Equation 14.12 multiplied by n; e.g.,

$$b = \frac{n \Sigma yX}{n \Sigma X^2 - \Sigma X \Sigma X}$$

and then

$$a = \frac{\Sigma Y}{n} - \frac{b}{n} \Sigma X$$

For the Arrhenius model $Y = a + bte^{-K/T}$

$$x = te^{-K/T}$$
$$Z = X/T$$
$$y = Y - \Sigma Y/n$$

and for the model $Y = a + btK^c$

$$X = tK^c$$
$$Z = CX/K$$
$$y = Y - \Sigma Y/n$$

The subscripts are omitted from these summations since all summations run from 1 to n, the total number of observations associated with a stored product. Successive values for K may be tried in Equation 14.12 until the equality is satisfied. For practical purposes, convergence is attained when both sides agree to 5 or 6 digits.

Adequacy of the Model

Kwolek and Bookwalter (1971) chose a model to be compared for stability based on the three parameters a, b, and K, and s^2, the mean square of deviations from the fitted equation. The adequacy of the model in fitting the data can be judged by comparing the variance,

$$s^2 = \sum^{n} (Y - \dot{Y})^2/(n - 3)$$

where $\dot{Y}$ is the predicted value for the observed Y, with an independent error variance s_1^2. The F-ratio s^2/s_1^2 would provide an approximate test.

They considered the model to be adequate if the variance s^2 is not statistically significantly different from an independent measure of variance s_1^2. When the deviations of observed values from predicted values lead to a significantly larger variance s^2 than expected, a further examination of deviations is worthwhile. Assuming three or more temperature levels are available, a one- or two-way analysis of variance of $(Y - Y)$ can be computed after grouping these deviations by temperature levels. The means of the deviations by temperature can then be tested for significant variation.

If this approximate test is not significant, the model is still acceptable, but variation is larger than usual. If there is significant variation between means of deviations grouped by temperature, the means can be plotted against T. If there is a consistent trend, the model is unacceptable and can perhaps be changed. If there is no consistency to the mean deviations, the model may still be acceptable. However, it is necessary to conclude that there is some unexplained "random" variation associated with temperature. Examination of the residuals will very likely be most useful. However, the validity and actual significance level of statistical tests may be somewhat questionable.

Specifying Quality Levels

Having determined a suitable equation, Kwolek and Bookwalter (1971) found another important application. When a specification of quality, Q, can be set on the product, the equation can be written

$$Q = a + \frac{bt}{K - C}$$

where Q is the specification value for Y. Solving for C or t the equation becomes

$$C = K - bt/(Q - a)$$

or

$$t = (Q - a)\frac{K}{b} - \frac{(Q - a)}{b}C$$

which are straight line relations in t and C. This result allows computation of contours of equal Q on the time-temperature plane. Points below the contour line given by this equation indicate storage conditions expected to meet the specification while points above the contour of equal Q would not meet the specification. Of course, some caution is necessary if an extrapolation is made to certain conditions outside the range of those tested. However, interpolations may be accepted with more confidence, they believe.

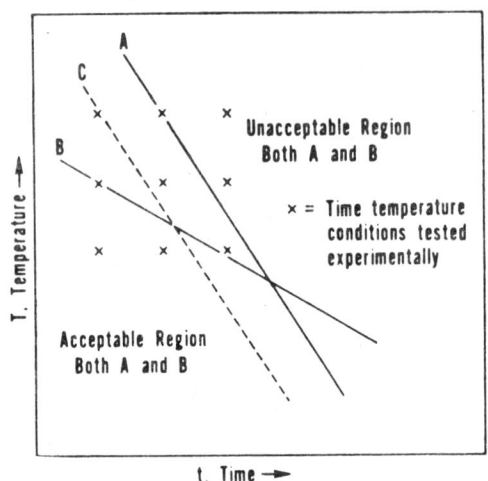

Courtesy of Kwolek and Bookwalter

FIG. 14.7. PRODUCT QUALITY CONTOURS
CAN BE EXPRESSED GRAPHICALLY ON A TIME
VERSUS TEMPERATURE PLOT

Lines A and B represent two quality level specifications;
line C represents quality level specification A minus
a safety factor; areas below contour lines represent acceptable quality.

Q may also be determined by selecting an acceptable level Y and then subtracting, say, d standard deviations; i.e., $Q = Y - ds$, where d depends upon the degrees of freedom (d.f.) for s, the standard deviation of the quality measurement, and the assurance desired. For example, a d of 2.4 (i.e., $Q = Y - 2.4s$) would provide 95% assurance that 95% of the population would find the product acceptable at a level Y, assuming s is based on 20 d.f. The term ds may be viewed as a safety factor for Y.

SPECIFYING QUALITY

Figure 14.7 shows the application of the fitted model to the problem of specifying product stability. Lines A and B can be considered two different response measurement contours for a product obtained by solving equations at specification levels Q_A and Q_B. The intersection of the lines defines four areas of product quality. Time-temperature combinations above both lines would yield unacceptable product quality, whereas combinations below both lines would yield acceptable products (Kwolek and Bookwalter 1971).

The figure also suggests that, at shorter time periods and higher temperatures, criterion B tends to be more important to product quality, while criterion A becomes limiting sooner than B at lower temperatures. The dashed line C demonstrates the position of the response contour A if a safety factor were added to the original definition of product response illustrated by line A. Thus, if a safety factor is included, acceptable storage conditions would be determined by lines C and B. The x's illustrate the combinations of temperature and time at which the product was actually stored.

Kwolek and Bookwalter (1971) were fairly confident of prediction for time-temperature conditions within the area of experimentation and slightly outside. However, extrapolation would involve increased risk.

Fitting Stability Data

The steps in fitting stability data reduce to the following:

(1) Fit the linear relations of Y or a function of Y versus t,

$$Y = a / b_i t$$

for each temperature level T_i.

(2) Check that there is significant variation in product response Y with storage time and temperature conditions. Compute the pooled mean square for deviations about separate lines. If variation with time and temperatures is not significant, stop.

(3) Plot b_i versus T_i and select a functional relation from Equations 14.6 to 14.9. One could fit all models and select that with the smallest s^2. However, they feel selecting the relation associated with the most linear plot will be satisfactory.

(4) For selected values of K, compute both sides of Equation 14.12. Plot the difference between the two sides of Equation 14.12 versus K and select K as the value for a difference of zero. The problem now becomes that of fitting a straight line with tf(T K) as the independent variable. They did not worry about the constant m, as it is incorporated into b.

(5) Calculate values for a, b, and s^2, where

$$s^2 = \Sigma [Y - a - btf(T)]^2 / (n - 3)$$

This method as outlined can be used to determine a and b or one can apply a linear regression computer program fitting

$$Y = a + bX$$

where

$$X = tf(T K)$$

and X varies over all possible times and temperatures in the trial, and K is as determined in step 4.

(6) Compare s^2 with the mean square computed in step 2 or with an independent measure of variation s_1^2; i.e., compute s^2/s_1^2 as an approximate F-test. If necessary, examine residuals before judging adequacy of the fit. Another measure of how well the data are fitted is to determine the multiple correlation coefficient R:

$$R = \left[1 - \frac{(n-3)s^2}{\Sigma\, y^2} \right]^{1/2}$$

R^2 is the proportion of variation in Y explained by the prediction equation. R^2 is statistically significant with these models at about the 0.05 level if it exceeds $6(n+1)/(n^2+15)$.

(7) Define levels of acceptability for Y. Apply the equation to determination of a figure showing time-temperature contours of product quality $Y = Q$, and areas of acceptable and unacceptable product quality.

Method Useful With Designed Experiments

Acceptable conditions for product storage and prediction of product stability can be determined by the procedures described with data taken in relatively small designed experiments. Several different criteria of quality can be used and plotted on the same figure to determine acceptable storage conditions.

SUMMARY

The combination of color measuring, odor identifying, and texture measuring instruments will find application in the objective

FIG. 14.8.   DESICCATORS—THE ORIGINAL STORAGE STA-
BILITY TESTING CHAMBERS

FIG. 14.9.   STORAGE STABILITY CABINETS WITH TEMPERA-
TURE AND HUMIDITY CONTROLS

Courtesy of Nabisco

FIG. 14.10.   WALK-IN STORAGE STABILITY ROOMS WITH
AUTOMATIC TEMPERATURE AND HUMIDITY PROGRAM-
MING SYSTEM

measurement of product characteristics and perhaps predict storage stability in the future.

Storage stability technologies have come a long way in a few
short years. Perhaps this can be seen visually in the evolution of
systems for holding products to study storage stability, first in
desiccators (Fig. 14.8), to programming storage cabinets (Fig. 14.9),
to storage rooms which can be programmed to duplicate the climate of any place on earth (Fig. 14.10).

REFERENCES

AD HOC COMM. NATL. MEAT CANNERS ASSOC. 1975. Field performance of metal food containers with easy-open ends. Food Technol. 29, No. 2, 48-49.

ADAMOVA, A.A., GINSBURG, S.K., and LEBEDEVA, M.A. 1950. Investigation of canned goods after several decades in the Arctic. Gigiena i Sanitaria 7, 28-29.

ADAMS, H.W., and OWENS, W. 1972. Internal pressure of cans agitated during heating and cooling. Food Technol. 26, No. 7, 28-30.

AGNEW, B. 1973. Clean air systems for aseptic packaging. Food Technol. 27, No. 9, 58-62.

ALGUIRE, D.E. 1973. Ethylene oxide gas sterilization of packaging materials. Food Technol. 27, No. 9, 64-67.

ANGELINE, J.F., and LEONARDS, G.P. 1973. Food additives, some economic considerations. Food Technol. 27, No. 4, 40-50.

ANGELOTTI, R. 1973. FDA regulations promote quality assurance. Food Technol. 29, No. 11, 60-62.

ANON. 1959. Storage of Quartermaster supplies. Dep. Army Tech. Manual TM 10-250. U.S. Gov. Print. Off., Washington, D.C.

ANON. 1971. Philosophy and guidelines for . . .nutritional standards for processed foods. Food Technol. 25, No. 1, 36-38.

BAKAN, J.A. 1973. Microencapsulation of foods and related products. Food Technol. 27, No. 11, 34-44.

BALLANTYNE, R.M., and ANGLIN, M.C. 1955. The effect of temperature on the stability of packaged ration items from the Canadian five-man Arctic ration pack RPX-1B. Def. Res. Med. Lab. Rep. 173-13, Toronto.

BAUMAN, H.E. 1974. The HACCP concept and microbiological hazard categories. Food Technol. 28, No. 9, 30-34, 70.

BEAZLEY, C.C., and INSALATA, N.F. 1971. The roles of fluorescent-antibody methodology and statistical techniques for salmonellae protection in the food industry. Food Technol. 25, No. 3, 24-26.

BONE, D. 1973. Water activity in intermediate moisture foods. Food Technol. 27, No. 4, 71-76.

BOOKWALTER, G.N., MOSER, H.A., PFEIFER, V.F., and GRIFFIN, B.L. 1968. Storage stability of blended food products. Food Technol. 22, No. 12, 85-89.

BOOKWALTER, G.N. et al. 1971. Full-fat soy flour extrusion cooked: properties and food uses. J. Food Sci. 36, 5.

BRODY, A.L. 1972. Aseptic packaging of foods. Food Technol. 26, No. 8, 70-74.

BUCK, P.A. 1971. Symposium participants express their opinions. Food Technol. 25, No. 10, 34-36.

BYRAN, F.L. 1974. Microbiological food hazards today—based on epidemiological information. Food Technol. 28, No. 9, 52-66, 84.

BYRNE, C.H. 1976. Temperature indicators—the state of the art. Food Technol. 30, No. 6, 66-68.

CECIL, S.R. 1968. Storage stability of Civil Defense shelter rations. Ga. Agric. Exp. Stn. Tech. Rep. 156 VI GES.

CECIL, S.R., and WOODROOF, J.G. 1962. Long term storage of military rations. Ga. Agric. Exp. Stn. Tech. Bull. 25.

CLAUSI, A.S. 1973. Improving the nutritional quality of food. Food Technol. 27, No. 6, 36-40.

COBB, T.P. 1955. Sixty year old corn. Food Packer 36, No. 2, 28.

COSLER, H.B. 1957. Storage of candies. Manuf. Confect. 37, No. 6, 37-44.

COSLER, H.B. 1958. Prevention of staleness, rancidity, in nut meats and peanuts. Peanut J. Nut World 37, No. 11, 10-11, 15.

COSLER, H.B., ALIKONIS, J.J., and MCCORMICK, R.D. 1959. Edible coating prevents staleness, rancidity in nutmeats, peanuts. Food Proc. 20, No. 2, 46-48.

COSLER, H.B., WOODROOF, J.G., and GRANT, B. 1953. The stability of confections in military rations under varying temperatures. Manuf. Confect. 33, No. 10, 19-20, 22, 24-26.

DANIELS, R.W. 1974. Handbook No. 8 and nutritional labeling. Food Technol. 28, No. 1, 46-47, 60.

DARBY, W.J. 1972. Fulfilling the scientific community's responsibilities for nutrition and food safety. Food Technol. 26, No. 8, 35-37.

DAVIS, R.B., LONG, F.E., and ROBERTSON, W.F. 1972. Engineering considerations in retort processing of flexible packages. Food Technol. 26, No. 8, 65-68.

DEEDS, F. 1959. Time-temperature tolerance in frozen foods. J. Am. Dietet. Assoc. 35, No. 2, 128-130.

DEFIGUEIREDO, M.P. 1971. Quality assurance of liquid eggs. Food Technol. 25, No. 7, 70-76.

FENNEMA, O. 1976. The U.S. frozen food industry: 1776-1976. Food Technol. 30, No. 6, 56-61, 68.

FILER, L.J., JR. 1976. Patterns of consumption of food additives. Food Technol. 30, No. 6, 62-70.

FOSTER, E.M. 1972. The need for science in food safety. Food Technol. 26, No. 8, 81-87.

FRIES, J.A., and GRAHAM, D.M. 1972. Reconstituting preplated frozen meals with integral heat. Food Technol. 26, No. 11, 76-82.

GRASCOIGNE, C.E. 1971. Frozen food standards and regulations—an industrial view. Food Technol. 25, No. 5, 68-72.

GUADAGNI, D.G. 1961. Integrated time-temperature experience as it relates to frozen food quality. ASHRAE. Frozen Food Handling Symp. Feb. 13, 16, Chicago.

HALL, H.E. 1971. The significance of Escherichia coli associated with nut meats. Food Technol. 25, No. 3, 34-36.

HALL, R.L. 1971. GRAS review and food additive legislation. Food Technol. 25, No. 5, 12-16.

HALL, R.L. 1975. GRAS—concept and application. Food Technol. 29, No. 1, 48-53.

HARPER, J.M., and JANSEN, G.R. 1973. Nutrient standard menus. Food Technol. 27, No. 6, 48-52.

HARPER, W.J. 1974. In-plant control of dairy wastes. Food Technol. 28, No. 6, 50-52.

HARRIS, R.S., and KARMAS, E. 1975. Nutritional Evaluation of Food Processing, 2nd edition. AVI Publishing Co., Westport, Conn.

HEDRICK, T.I. 1973. Aseptic packaging in paperboard containers. Food Technol. 27, No. 9, 46-48.

HENIZ, Y., and MANHEIM, C.H. 1971. Drum drying of tomato concentrate. Food Technol. 25, No. 2, 59-62.

HOTCHNER, S.J. 1974. Easy-open feature prevents bimetallic can corrosion. Food Technol. 28, No. 11, 54-64.

HOWARD, H.W. 1971. Easier said than done. . .ingredient labeling. Food Technol. 25, No. 5, 18-20.

HU, K.H. 1972. Time-temperature indicating system "writes" status of product shelf life. Food Technol. 26, No. 8, 56-62.

IFT EXPERT PANEL AND CPI. 1972A. Botulism. Food Technol. 26, No. 10, 63-66.

IFT EXPERT PANEL AND CPI. 1972B. Nitrites, nitrates, and nitrosamines in food—a dilemma. Food Technol. 26, No. 11, 121-124.

IFT EXPERT PANEL AND CPI. 1974A. Nutrition labeling. Food Technol. 28,No. 7, 43-48.

IFT EXPERT PANEL AND CPI. 1974B. Shelf life of foods. Food Technol. 28, No. 8, 45-48.

ITO, K. 1974. Microbiological critical control points in canned foods. Food Technol. 28, No. 9, 46-50.

JAHNS, F.D., HOWE, J.L., CODURI, R.J., and RAND, A.G., JR. 1976. A rapid visual enzyme test to assess fish freshness. Food Technol. 30, No. 7, 27-30.

JAMES, D.E. 1971. Managing raw materials and good manufacturing. Food Technol. 25, No. 10, 30-31.

JENNINGS, W.G. 1967. Gas chromatograph analysis of dilute aqueous systems. Anal. Chem. 39, 521-523.

KAREL, M. 1974. Packaging protection for oxygen-sensitive products. Food Technol. 28, No. 8, 50-60, 65.

KAREL, M., MIZRAHI, S., and LABUZA, T.P. 1971. Computer prediction of food storage. Mod. Packaging, Aug., p. 54.

KAUFFMAN, F.L. 1974. How FDA uses HACCP. Food Technol. 28, No. 9, 51, 84.

KOHL, W.F. 1971. A new process for pasteurizing egg whites. Food Technol. 25, No. 11, 102-110.

KRAMER, A. 1971. A systems approach to quality assurance. Food Technol. 25, No. 10, 28-29.

KRAMER, A. 1973. Storage retention of nutrients. Food Technol. 28, No. 1, 50-60.

KRAMER, A., and FARQUHAR, J.W. 1976. Testing of time-temperature indicating and defrost devices. Food Technol. 30, No. 2, 50-53, 56.

KROCHTA, J.M., TILLIN, S.J., and WHITEHAND, L.C. 1975. Ascorbic acid content of tomatoes damaged by mechanical harvesting. Food Technol. 29, No. 7, 28-30, 38.

KWOLEK, W.F., and BOOKWALTER, G.N. 1971. Predicting storage stability from time-temperature data. Food Technol. 25, No. 10, 51-57, 63.

LABUZA, T.P. 1973. Effects of dehydration and storage. Food Technol. 27, No. 1, 20-26.

LABUZA, T.P. 1976. Drying food: technology improves on the sun. Food Technol. 30, No. 6, 37-46.

LACHANCE, P.A., RANADIVE, A.S., and MATAS, J. 1973. Effects of reheating convenience foods. Food Technol. 27, No. 1, 36-38.

LAMPI, R.A., SCHULZ, G.L., CIAVARINI, T., and BURKE, P.T. 1976. Performance and integrity of retort pouch seals. Food Technol. 30, No. 2, 38-48.

LANGE, L. 1973. Aseptic bag-in-box packaging. Food Technol. 27, No. 10, 80-82.

LAZAR, M.E., LUND, D.B., and DIETRICH, W.C. 1971. A new concept in balancing IQB reduces pollution while improving nutritive value and texture of processed foods. Food Technol. 25, No. 7, 24-26.

LEININGER, H.V., SHETON, L.R., and LEWIS, K.H. 1971. Microbiology of frozen cream-type pies, frozen cooked-peeled shrimp, and dry food-grade gelatin. Food Technol. 25, No. 3, 28-30, 33.

LEUNG, H., MORRIS, H.A., SLOAN, A.E., and LABUZA, T.P. 1976. Development of an intermediate-moisture processed cheese food product. Food Technol. 30, No. 7, 42-44.

LIVINGSTON, C. 1971. Assuring quality of processed foods from field to table. Food Technol. 25, No. 10, 31-34.

LIVINGSTON, G.E., ANG, C.Y.W., and CHANG, C.M. 1973. Effects of food service handling. Food Technol. 27, No. 1, 28-34.

LORENZ, K.L., CHARMAN, E., and DILSAVER, W. 1973. Baking with microwave energy. Food Technol. 27, No. 12, 28-36.

LUND, D.B. 1973. Effects of heat processing. Food Technol. 27, No. 1, 16-18.

MAJORACK, F.C. 1971. FDA's quality assurance programs: tools for compliance. Food Technol. 30, No. 10, 38-42.

MCFARREN, E.F. 1971. Assay and control of marine biotoxins. Food Technol. 25, No. 3, 38-48.

MEADE, R.E. 1973. Combination process dries crystallizable materials. Food Technol. 27, No. 12, 18-26.

MEER, G., MEER, W.A., and TANKER, J. 1975. Water-soluble gums, their past, present, and future. Food Technol. 29, No. 11, 22-30.

MERMELSTEIN, N.H. 1973. Nutrient labeling and the independent laboratory. Food Technol. 27, No. 6, 42-46.

MERMELSTEIN, N.H. 1976. The retort pouch in the U.S. Food Technol. 30, No. 2, 28-37.

MIDDLEKAUF, R.D. 1976. 200 Years of U.S. food laws: a gordian knot. Food Technol. 30, No. 6, 48-54.

MITSUDA, H., KAWAI, F., and YAMAMOTO, A. 1972. Underwater and underground storage of cereal grains. Food Technol. 26, No. 3, 50-56.

NAGEL, A.H. et al. 1972. Forum: voluntary food standards. Food Technol. 26, No. 11, 57-64.

OSMAN, E.M. 1975. Interaction of starch with other components of food systems. Food Technol. 29, No. 4, 30-35, 44.

PARSONS, P.C. 1971. Canners adopt new attitude. Food Technol. 25, No. 10, 31.

PENCE, J.W., and HEID, M. 1960. Effect of temperature on stability of frozen cakes. Food Technol. 14, No. 2, 80-83.

PERYAM, D.R. 1964. Consumer preference evaluation of the storage stability of foods. Food Technol. 18, 214.

PETERSON, A.C., and GUNNERSON, R.E. 1974. Microbiological critical control points in frozen foods. Food Technol. 28, No. 9, 37-44.

PETROWSKI, G.F. 1975. Food-grade emulsifiers. Food Technol. 27, No. 7, 52-62.

PLOUGH, I.C., HARDING, R.S., GERHARD, J.I., and FRIEDMANN, T.E. 1958. The effect of high temperature storage on the acceptability, digestibility, and composition of the U.S. Army ration, individual, combat. U.S Army Med. Res. Dev. Command. Rep. 228.

POMINSKI, J., PEARCE, H.M., VIX, H.L.E., and SPADARO, J.J. 1975. Improvement of shelf life of partially defatted peanuts by intromission of nitrogen into the interstices of the peanuts. J. Food Sci. 40, 192-194.

RASMUSSEN, C.L., and OLSON, R.L. 1972. Freezing methods as related to cost and quality. Food Technol. 26, No. 12, 32-47.

REY, L.R. 1971. The role of industry in meeting the challenge of future food needs. Food Technol. 25, No. 11, 26-32.

RIESTER, D.W. 1973. FDA's view on chemical sterization of aseptic packaging containers. Food Technol. 27, No. 9, 56, 62.

RINSCHLER, R.A. 1954. Underground storage test. Phase I. Summary Rep. Quartermaster Corps. Natick, Mass.

ROSS, K.D. 1975. Estimation of water activity in intermediate moisture foods. Food Technol. 29, No. 3, 26-34.

SCHOEN, H.M., and BYRNE, C.H. 1972. Defrost indicators. Food Technol. 26, No. 10, 46-50.

SHEPHERD, A.D. 1960. Report on Frozen Food Quality. Western Regional Research Laboratory, U.S. Dep. Agric. Albany, Calif. Summary reprinted in Food Process. No. 1, 1961.

SIMON, S. 1972. Continuous processing systems for skinless frankfurters. Food Technol. 26, No. 12, 50-59.

SINNAMON, H.L., ACETO, N.C., and SCHOPPET, E.F. 1971. The development of vacuum foam-dried whole milk. Food Technol. 25, No. 12, 52-64.

SMITH, C.A., Jr., and SMITH, J.D. 1975. Quality assurance system meets FDA regulations. Food Technol. 29, No. 11, 64-68.

SNEDECOR, G.W., and COCHRAN, W.G. 1968. Statistical Methods, 6th Edition. Iowa State Univ. Press, Ames.

SPLITTSTOESSER, D.F. 1973. The microbiology of frozen vegetables. Food Technol. 27, No. 1, 54-56, 60.

STEWART, R.A. 1971. Sensory evaluation and quality assurance. Food Technol. 25, No. 4, 103-106.

TOMPKIN, R.B. 1973. Refrigeration temperature. Food Technol. 27, No. 12, 54-58.

TRACY, P.H., HETRICK, J.H., and KRIENKE, W.S. 1950. Relative storage qualities of frozen and dried milk. J. Dairy Sci. 33, 832-841.

TRESSLER, D.K., VAN ARSDEL, W.B., and COPLEY, M.J. 1968. The Freezing Preservation of Foods, 4th Edition, Vols. 1, 2, 3, and 4. AVI Publishing Co., Westport, Conn.

VAN ARSDEL, W.B. 1957. The time-temperature tolerance of frozen foods. 1. Introduction—The problem and the attack. Food Technol. 11, No. 1, 28-33.

WIERZCHOWSKI, J., and SEVERIN, M. 1958. Influence of storage time on the tin and iron content in canned fish products. Roczniki Panstwowego Zakladu Hig. 8, 481-494, 1957 (Chem. Abstr. 52, 7563c).

WILLICH, R.K., MORRIS, N.J., and FREEMAN, A.F. 1954. Peanut butter. V. The effect of processing and storage of peanut butters on the stability of their oils. Food Technol. 8, 101-104.

WOODROOF, J.G. 1955. Ration storage requirements. Establishing Optimum Conditions for Storage and Handling of Semiperishable Subsistence Items. Series IV. No. 1, 21-34. Dep. Army, Off. Quartermaster Gen., Washington, D.C.

WOODROOF, J.G.   1961.   Long-term storage of operational rations.   QMFCI/ Univ. of Ga., File R-301: Proj. 7-84-06-030-7-91-03-012; Prog. Rep. 13-17, 19-23. June 1953-January 1955. Proj. 7-84-13-002C; Prog. Rep. 24-28. February-November 1955. Proj. 7-84-13-002C; Prog. Rep. 16-18, 20-23, 25-29; Interim Rep. 19, 24; Term. Rep. 30. December 1955-August 1958. Proj. 7-84-13-002B; Prog. Rep. 1-8, Term. Rep. 9.

WOODROOF, J.G., and LEBEDEFF, O.K.   1960.   Foods for Shelter Storage, a Literature Review for the Office of Civil and Defense mobilization.   CDM-SR-59-31, Rep. 6, Ga. Exp. Stn.

WOODROOF, J.G., and MALCOM, H.R.   1958.   Moisture migration in frozen canned bread.   Food Technol. 12, No. 6, 268-269.

ZIMMERMAN, P.L., ERNST, L.J., and OSSIAN, W.F.   1974.   Scavenger pouch protects oxygen-sensitive foods.   Food Technol. 28, No. 8, 63-65.

Chapter 15

Principles of
Food Quality Assurance

Food manufacturers usually have two stated levels of quality for products marketed. One deals with a product's quality established as company policy to meet consumer needs. The other deals with product quality in terms of meeting governmental regulations and laws.

Branded products marketed by a company are matters requiring the most careful attention by company management. It is a general rule that company policy relating to branded product quality is more rigid than that required to meet governmental regulations. Company policy statements generally include a statement demanding that all products marketed meet the laws and regulations of all federal, state and local governments.

THE NEED

The epidemiology of food-borne hazards has been summarized by the U.S. Center for Disease Control. Their analysis of where foods are mishandled is given in Table 15.1. While the percentage of cases traced to food processing plants is low (6%), one factory can create widespread difficulty compared to a home or a restaurant.

The outbreaks traced to foods from food processing plants are summarized in Table 15.2. Inadequate refrigeration was the chief operational procedure that contributed to these outbreaks. Other procedures involved were: preparing foods too far in advance; foods reinfected after the final heat processing; inadequate heat processing; and holding foods at temperatures that favor bacterial growth. Food-

TABLE 15.1

PLACES WHERE FOODS WERE MISHANDLED IN SUCH A WAY
THAT FOOD-BORNE DISEASE OUTBREAKS RESULTED

Place	Number	Percentage
Food service establishments	589	37
Homes	230	14
Food processing plants	104	6
Unknown or unspecified	692	43
Total	1615	100

Source: Byran (1974).

borne outbreaks in which processed foods were incriminated and the relationship between reported diseases and processes by various food industries are shown in Table 15.3.

Processed foods that received no heat treatment were often made up of contaminated raw ingredients. The source of contamination with salmonellae was raw ingredients. In the few cases of contamination with trichinellae (increasingly rare in the United States), the incoming pork was infested (Byran 1974).

Heat process failures were common. Processes, such as smoking, often failed to kill salmonellae or trichinellae that were on or in the product. *Clostridium botulinum* cells multiplied and produced neurotoxin in canned or vacuum-packed foods after their spores survived improper heat processing.

In most instances *S. aureus* was introduced by workers after foods were heat processed. Hepatitis viruses and shigellae came from workers. Post-processing contamination with salmonellae and trichinellae, by cross-contamination from raw products to heat processed food by equipment or workers during subsequent handling, was another significant source.

The frequency with which certain food processing plants have produced foods that have been incriminated in food-borne disease outbreaks and the factors that led to contamination, survival or multiplication of pathogens are reviewed in Table 15.4. Incoming raw materials (usually foods of animal origin) are revealed as hazards in the processing of meat, poultry, eggs, baked goods containing eggs, milk, fish salads and confections.

Interpretation of data accumulated by the Center for Disease Control discloses that certain operations in food processing plants may permit contamination, may allow pathogens to survive processing, and may even promote bacterial growth. The cause-and-effect relationship between these inadequate operational procedures and out-

TABLE 15.2

OUTBREAKS INVOLVING FOODS MANUFACTURED IN PROCESSING PLANTS CITED
IN SCIENTIFIC LITERATURE OR SURVEILLANCE REPORTS AND CONTAINING
INFORMATION ABOUT FAULTY OPERATIONS

Disease	Food Incriminated	Type of Processing
Trichinosis	Spiced bacon roll	Meat processing, curing
Botulism	Smoked fish	Smoking, vacuum packaging
Staphylococcal intoxication	Ham	Baking (catering)
Staphylococcal intoxication	Meat loaf	Meat processing
Salmonellosis	Dietary supplement containing dried eggs	Egg processing, blending
Salmonellosis	Chicken salad	Salad processing
Salmonellosis	Eclairs	Baking
Salmonellosis	Eclairs	Baking
Salmonellosis	Eclairs	Baking
Calcium chloride poisoning	Popsicles	Freezing
Salmonellosis	French-vanilla ice cream	Freezing
Salmonellosis	French-vanilla ice cream	Freezing
Salmonellosis	Eclairs	Baking
Trichinosis	Smoked sausage	Meat processing, smoking
Staphylococcal intoxication	German chocolate cake	Baking
Salmonellosis	Cream-filled doughnuts	Baking
Trichinosis	Summer sausage	Meat processing, smoking
Staphylococcal intoxication	Ham	Baking
Botulism	Tunafish	Fish processing, canning
Botulism	Smoked fish	Fish processing, smoking, vacuum packing
Botulism	Liver paste	Meat processing, canning
Salmonellosis	Dried whole eggs	Drying
Salmonellosis	BBQ pork	Meat processing
Salmonellosis	Dried whole eggs	Egg processing
Trichinosis	Smoked sausage	Meat processing
Staphylococcal intoxication	Eclairs	Baking
Salmonellosis	Dietary food supplement containing cotton seed protein and brewer's yeast	Drying, blending
Salmonellosis	Cream-filled bakery products	Baking, egg processing
Salmonellosis	Headcheese	Meat processing
Salmonellosis	Cream pies	Baking, egg processing
Salmonellosis	Cream pies	Baking, egg processing
Myocardosis and cardiac failure (cobalt poisoning)	Beer	Brewing, foam stabilization
Salmonellosis	Carmine dye	Extraction, food color and preservatives
Salmonellosis	Beef jerky	Meat processing, preserving, drying
Salmonellosis	Bakery products	Baking
Staphylococcal intoxication	German chocolate cake	Baking, icing
Salmonellosis	Dried milk	Spray drying, instantizing
Salmonellosis	Smoked whitefish	Smoking, fish processing
Botulism	Chicken liver sauce	Pasteurizing, (canning)
Shigellosis	Apple cider	Pressing, bottling
Viral hepatitis	Glazed doughnuts	Baking, glazing
Salmonellosis	Smoked turkey	Poultry processing, smoking
Salmonellosis	Cooked beef	Meat processing
Trichinosis	Smoked summer sausage	Meat processing, smoking

TABLE 15.2 (Continued)

Disease	Food Incriminated	Type of Processing
Trichinosis	Smoked sausage	Meat processing, smoking
Trichinosis	Smoked sausage	Meat processing, smoking
Trichinosis	Smoked bacon	Meat processing, smoking
Trichinosis	Smoked summer sausage	Meat processing, smoking
Salmonellosis	Imitation ice cream made of eggs	Freezing, egg processing
Salmonellosis	Beef jerky	Meat processing, preservation, drying
Botulism	Tomato meatball sauce	Canning
Salmonellosis	Custard-filled pastry (maple bars)	Baking, egg processing
Trichinosis	Smoked sausage	Meat processing, smoking
Trichinosis	Smoked bacon	Meat processing, smoking
Staphylococcal intoxication	Genoa salami	Meat processing, fermentation
Staphylococcal intoxication	Genoa salami	Meat processing, fermentation
Botulism	Vichyssoise	Canning
Vibrio parahaemolyticus gastroenteritis	Steamed crabs	Steaming, delivery
Cadmium poisoning	Candy love beads	Candy manufacturing (imported)
Salmonellosis	Dietary supplement	Drying, blending
Trichinosis	Smoked pork butts	Meat processing, smoking
Trichinosis	Cold smoked sausage (Kolbasy)	Meat processing
Staphylococcal intoxication	Cured bacon	Meat processing, curing, frying
Shigellosis	Poi	Peeling, grinding, fermentation, packaging
Tin poisoning	Tomato juice	Canning, excessive nitrification of fields
Salmonellosis	Headcheese	Meat processing, cooking
Botulism	Oil-packed hot peppers	Canning
Brucellosis	Goat cheese	Fermentation, cheese making (imported)
Scombroid poisoning	Tuna fish	Canning
Salmonellosis	Chocolate candy	Candy manufacturing

Source: Byran (1974).

breaks of food-borne disease is apparent. It provides a basis for judging the need for suitable quality assurance programs in the food processing industry in spite of costs incurred.

A ROLE FOR GOVERNMENT

Responsibility for the safety, wholesomeness and nutritional quality of food rests with the food industry, not with the Food and Drug Administration. The task of the FDA (Angelotti 1975) is to monitor industry to determine whether it is meeting its responsibilities. The FDA has the role of motivating compliance, but does not act as a company quality assurance division. The FDA takes appropriate corrective actions when industry fails to meet its responsibilities.

A number of techniques are used by the FDA in determining the

TABLE 15.3

FACTORS CONTRIBUTING TO OUTBREAKS IN WHICH FOODS PREPARED AT PROCESSING PLANTS SERVED AS VEHICLES

Disease	Number of Outbreaks	Contamination (If No Terminal Heat Process)			Heat Process Failure	Post-process Contamination					Factors Permitting Growth in Plant			Evidence of Growth After Product Left Plant
		Raw Product	Additive	During Process		Raw Product	Equipment	Cross-Contamination	Workers	Cooling Water	Inadequate Refrigeration	Fermentation	Anaerobic Packaging	
Salmonellosis	30	20			13(3)		3	5	2		10			5
Trichinosis	13	9(4)		1	9(3)		1(1)	1(1)						
Staphylococcal intoxication	9	1			1				4		3	2		2
Botulism	8	1(1)			5(3)					(1)			8	1
Shigellosis	2			2							1	1		1
Vibrio para-haemolyticus gastroenteritis	1					1		1						
Brucellosis	1	1										1		1
Scombroid poisoning	1	1												
Viral hepatitis	1								1					
Chemical poisoning	4		3	1										
Total	70	33(5)	3	4	28(9)	1	4(1)	7(1)	7	(1)	14	4	8	10

Source: Byran (1974).

Note: Figures in parenthesis refer to situations in which contributory factors were impaired.

TABLE 15.4

FACTORS CONTRIBUTING TO OUTBREAKS IN WHICH FOODS PREPARED AT PROCESSING PLANTS SERVED AS VEHICLES

Industry	Number of Outbreaks	Contamination (If No Terminal Heat Process) Raw Product	Additive	During Process	Heat Process Failure	Post-process Contamination Raw Product	Equipment	Cross-Contamination	Workers	Cooling Water	Factors Permitting Growth in Plant Inadequate Refrigeration	Fermentation	Anaerobic Packaging	Evidence of Growth After Product Left Plant
Meat processing	24	11(4)			13(3)		2(1)	2(1)	1		2	2	1	4
Baking	16	8			4(1)		1	1	4		7			
Canning	8	1(1)	1		3(3)					(1)			6	3
Fish processing	6	3(1)			2(2)	1		1		(1)	1		3	
Egg processing	6	3(2)		1	5(1)						1			
Milk product processing	4	3			4		1		1		1	1		
Dietary supplement manufacturing	3	2			(1)									1
Candy manufacturing	2	1	1					1						
Poultry processing	2	1						2	1					
Food color manufacturing	1	1					1				1			1
Poi processing	1			1							1	1		1
Cider manufacturing	1			1										
Salad processing	1	1						1	1					
Brewing	1			1										
Popsicle manufacturing	1		1											
Total	77	35(8)	3	4	31(11)	1	5(1)	8(1)	8	(2)	14	4	10	10

Source: Byran (1974).

manner in which industry accepts its responsibilities. These include: (1) Establishment Inspection, which may vary in comprehensiveness and intensity from that of the Hazard Analysis and Critical Control Point (HACCP) inspection to that of a less-comprehensive key indicator inspection. (2) Sample Collection and Analysis of a product in process and of finished product in distribution channels. (3) Surveillance intended to identify new problems as well as to quantify the extent and significance of known problems that may be associated with processing, the environment, and other factors.

Though these techniques are useful, the best hope for safety and quality in food lies in the development and maintenance of adequate in-plant quality assurance programs. Promoting quality assurance at the plant level is thus a primary goal in FDA regulation.

What Are Good Manufacturing Practices?

The FDA has recently embarked upon a course of public rule-making as a means of obtaining industry-wide compliance with the responsibilities of industry. Public rule-making developed with the input of all interested and affected parties. It provided the means by which to inform all persons of what is considered appropriate.

Angelotti (1975) describes one regulation which attempted to do this. The initial Good Manufacturing Practice Regulations for foods was published in 1969, and is often referred to as the Umbrella GMP. It made use of such words as "adequate," "proper," and "sufficient." Despite the fact that many provisions in the GMP were written as mandates, compliance was often difficult because of the vagueness of the subjective terms. In spite of these shortcomings, the most important objective of the Umbrella GMP was to provide guidance relative to long-range improvement programs directed to plant facilities and practices. This general regulation served to encourage the adoption of quality assurance systems, and to indicate the need for such systems where they were absent.

The FDA has now embarked on a program of expanding the GMPs to include regulations that will ultimately apply specifically to all the major segments of the food industry. Detailed and specific GMP regulations which have been publically developed are now in effect for many areas of concern.

A typical GMP outline has similarities to a quality assurance system. A preamble introduces the subject, discussing the background of the industry and why a GMP is considered necessary, and reports on information gathered in plant inspections. It also stresses reasons why certain features or requirements are needed.

Each regulation has a definition section. Each definition states that the word "shall" means that the requirement is mandatory and that the word "should" refers to an item that is desirable but not absolutely essential for the carrying-out of an operation. The next section refers to the Umbrella GMP regulation and is followed by specific sections pertaining to the industry. If there is concern for contamination from outside the factory or the possibility of cross-contamination between different operations, note is made. The next section deals with equipment and utensils and stresses general overall design criteria for equipment that is unique for that industry, or that requires special controls for safety and sanitation. A section on personnel sanitation facilities is frequently included. Sections on the cleaning and sanitizing of equipment follow. The FDA believes the most important part of the GMP involves processes and controls that are vital for the particular type of food manufacturing operation. A section usually is devoted to records and record-keeping that cover at least the average life of the product in distribution.

FDA believes that the Good Manufacturing Practice regulations materially reduce the probability of release of food products that are not in compliance (i.e., distribution of an unsafe product, exactly the aim of any effective quality assurance system).

An example of a GMP that delineates specific quality assurance requirements is the low-acid canned food GMP—"Thermally Processed Low-Acid Foods Packaged in Hermetically Sealed Containers." This GMP details the operations that must be conducted to assure the production of a safe low-acid canned food, according to the FDA. It contains the same elements found in quality assurance systems throughout industry. Quality assurance systems in industry dovetail the low-acid canned food GMP into their operations.

MICROBIOLOGICAL STANDARDS

Another FDA regulatory area having an impact upon quality assurance systems is the establishment of microbiological quality standards. Acceptable microbiological levels for food products at the retail or consumer level, taking into account post-production abuses, are established. This is done by conducting a nationwide statistical survey of the food under evaluation. The data are reviewed, and a proposed standard is developed that considers both the producer's and consumer's risks in realistic terms. Once a proposed standard is developed it is published in the *Federal Register* as a proposal with provision for comment. Each comment received is reviewed and is responded to through subsequent publication in the *Federal Register*, and changes indicated by comments are made. When the proposed standard is finalized, it is a mandatory standard that must be met under the provisions of Section 401 of the Food, Drug and Cosmetic Act.

The projected impact of microbiological quality standards upon the food processing industry's quality assurance system greatly affects the distribution and retail outlet industries. The establishment of certain microbiological quality standards requires a reevaluation of some quality assurance systems and causes the introduction of such systems in areas that now operate without such controls.

Good manufacturing practices and microbiological quality standards are examples of FDA regulatory initiatives that affect quality assurance systems. The FDA indicates it supports efforts to require all food processors to establish and use quality assurance systems in their production operations, with emphasis upon monitoring and disclosure of hazards associated with the product and the means employed to control the hazards, in the public interest.

A Role for Industry

There are some ten elements of production safety if strict compliance with FDA safety standards is maintained. The minimum a food manufacturer must do is to assure the quality and ensure the safety of marketed products.

Product Safety Analysis.—Food safety analyses must be conducted on all products to assess their microbiological, physical and chemical safety. All aspects of a product's makeup are evaluated (e.g., product formulation, processing, distribution and final use) to ensure safety in use.

New products should not be manufactured, distributed or sold until a rigorous review of all aspects of a product's composition has been completed. The purpose of such a review is to assure that the proper formulation, processing and distribution of every product results in the offering for sale of unadulterated foods for consumption that meet or exceed federal regulations and requirements.

Product Specifications.—Each product must have certain specifications for manufacture. These specifications must cover all safety, quality and regulatory requirements. They must specify the uses of processes, ingredients, and acceptance tests, packaging materials and labels, and include descriptions of utilized processes and finished products. Such specifications serve as the vehicle through which a company can communicate federal standards and regulations to factory production.

Depending on the size of the manufacturing operation within a given company, the amount of documentation required for each product, packaging material, ingredient specification and testing procedure is compatible with simple specification files. However, as the size of a company's operations expand, the magnitude of the files is such that an alternative means of recording specifications may be necessary. It is not difficult to computerize this process, once the specification files reach a certain size. Computerization creates an effective and efficient storage and retrieval system of documentations of product specifications.

Physical Systems Hazard Control.—Industry is required to make and maintain inventories of food processing systems and their environments and the possible hazards that may occur within the context of each system. A hazard control system is usually generated by the particular facility in which food processing or production takes place. It is the responsibility of the manufacturer to prepare and maintain these documents so that clear identification of all physical systems hazards are known, and a complete and lucid understanding of each step in the processing and packaging of manufactured products is available. Flow diagrams are particularly suited for this purpose, because they can show where possible hazards may occur in a particular process or part thereof. A schematic representation of the manufacturing procedures and processes allows for their exposure on a level where hazards may otherwise pass unnoticed.

The following procedure is useful in trouble-shooting for physical systems hazards:

(1) Develop and maintain flow diagrams to cover all food processing and physical systems and environments.

(2) Identify all physical systems hazards to the safety and integrity of products.

(3) Establish and document systems of control for all hazards, whether actual or possible.

(4) Maintain records of control procedures for all physical systems hazards that are critical to product safety.

Whenever possible, correctible hazards must be eliminated. Physical plant and systems hazards must be reviewed before a food product is manufactured in a facility.

Purchasing Requirements.—Purchase of raw materials should be limited to approved suppliers who can offer an acceptable continuous guarantee of a material's quality. Further, it is in the company's interest to require the supplier to submit proof of ability to supply the appropriate quantities of materials of certain predetermined quality and safety standards. Raw materials must pass a plant inspection. Routine inspections assure that raw material quality is kept at the appropriate level. A supplier who frequently delivers products below the stated requirement levels should be covered by contracts describing specifications and safety analyses. The facilities used should be submitted to regular inspections by those who must approve any and all such products.

GMP Compliance.—High sanitation standards must be maintained, documented and rigorously observed in all production, storage and distribution facilities. Sanitation procedures instruct personnel on how to comply with sanitation standards. They must know the essential technical aspects of these procedures and why they must be followed. The knowledge behind these practices is given to personnel during their in-plant training.

Product Recall System.—A tracing system is needed so that all products sold can be accounted for and located if it is necessary to recall products in retail distribution. Either manual or computerized systems suit these purposes, depending upon the size of the overall operation. Tests can be conducted periodically to ensure the workability of the system. Results should be recorded and documented for future reference or for possible improvements in the system employed. Some tests that can be conducted to measure the capabilities of a recall system are:

(1) What is the least traceable unit of distribution, i.e., date, shift, batch, low, etc.?

(2) How effective is the method used in determining the amount of product to be traced?

(3) How long does it take to make a complete trace of a product?

Customer Service.—Means of recording and responding to consumer and customer complaints are needed. This is not for public relations but is a means of detecting safety and/or deficiencies in products. Prompt attention is essential.

Inspections and Safety Incidents.—Means of recording and responding to all safety or regulatory incidents are needed. A regulatory incident is a visit by federal, state or local inspectors or any regulatory agency to any facility. A company facility is any of the following: plant, mill, warehouse, restaurant, research and development center, factory, etc. Some of the regulatory agencies that might make such a visit are: FDA, USDA, EPA, OSHA, FEA, military veterinary corps, or state and local health inspectors.

A means of instant communication on all regulatory incidents is needed so that those involved may be able to intelligently respond if a letter or notice is received from an inspection agency. Safety incidents should also be communicated, such as: products presenting a threat to the consumer; personnel safety incidents where serious infectious disease, injury or death have occurred; environmental incidents where hazardous material(s) is released into the environment, etc. Keep a comprehensive written record of all such incidents, containing the date, time, location, description of incident, and action taken to correct the condition if possible.

Auditing.—All processing plants, warehouses, and other storage facilities should be routinely and periodically audited by company personnel to gauge the level of compliance with company standards regarding specifications, product safety and regulatory requirements. Regular audits for the purpose of evaluating the adequacy of safety, documentation and the degree of compliance with established company procedures and standards are needed.

Product Integrity.—To maintain a high standard of product excellence, a company must have an organized and centralized means of surveying and controlling the contributing processes and procedures. A company must be on guard against violations. Therefore, it is imperative that a company have an effective and efficient program to protect itself and the public it serves.

DESIGN OF COMPANY QA PROGRAM

Quality assurance (QA) is a major management function, usually organized at the senior level, reporting directly to the president of a company.

For illustrative purposes, assume a company composed of divisions, each operating factories, each factory having a Quality Control (QC) Laboratory. The company has a central Purchasing Department.

Kramer and Twigg (1973) applied the systems approach to quality assurance. They found such an approach was indispensable to a successful quality assurance program. The approach is to conceive of a program in the form of two cycles (Fig. 15.1 and 15.2). In the control cycle customer specifications for each quality factor are established, then means of measuring them are developed. In the production cycle control is maintained over incoming raw materials and finished product. They believe such an approach prevents products which do not meet customer specifications from entering the channels of trade.

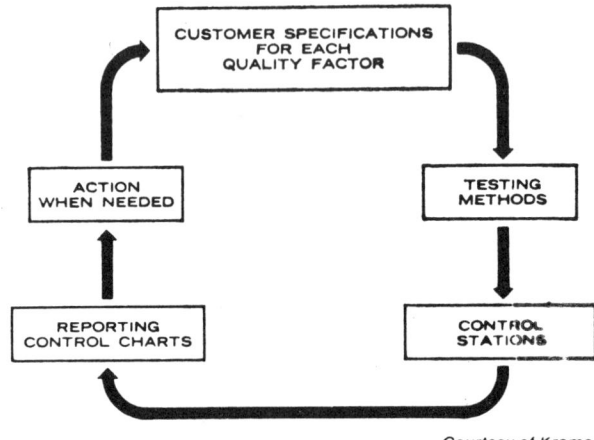

Courtesy of Kramer

FIG. 15.1.   QUALITY CONTROL CYCLE

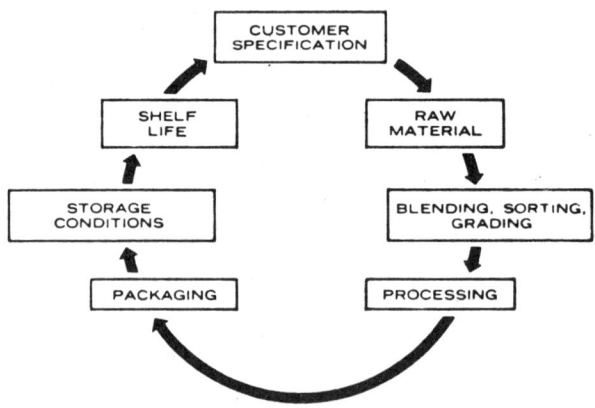

Courtesy of Kramer

FIG. 15.2.   PRODUCTION CYCLE

The QA activity generally operates as follows.

Objectives

To establish, issue and maintain standards and specifications for all raw materials used and for all finished products sold by the company.

To assure that all ingredients and all finished products adhere to the company's quality standards and to recommend corrective action as required. This includes microbiological purity and nutritional integrity of all products, and conformance to state and federal regulations.

To be of service to the company in all areas related to product quality. These areas include: trouble-shooting when quality problems exist; visiting production facilities for review and updating of QC Programs; designing laboratory installations and expansions; and training QC personnel in routine and new testing procedures. Further service is rendered by answering consumer nutritional inquiries and in reviewing consumer complaints concerning product quality.

Raw Material Quality Assurance

Specifications.—All approved raw materials are covered by tentative or permanent specifications. As new raw materials are required for new products, tentative specifications are developed to cover their purchase. All specifications are constantly being updated and, as conditions permit, the limits of acceptance are made more strict. Bacteriological conditions are constantly stressed and limits are reviewed to be kept at least as strict as governmental regulations. New and/or revised specifications are generally prepared each month for raw materials.

Survey Program.—All company units sample and inspect raw materials as received. Most of the tests are simple and/or organoleptic in nature. More thorough and sophisticated tests are performed at the Company Central Analytical Laboratory. Raw material surveys are conducted periodically. Greatest attention is given to so-called "critical" raw materials. Physical, chemical and bacteriological examinations are made not only to be sure that all public health regulations are met, but also to establish conformance with established company specifications.

The average number of raw material samples handled monthly might be:

Current raw materials	100
Alternate sources of supply	50
New raw materials	10

Service.—In the area of raw materials, QA is in daily communication with production units, the Purchasing Department, and current or potential suppliers. With the number of ingredients used and the multitude of suppliers and production units involved, it is normal for many minor, yet important, problems to develop. These problems must be handled quickly and effectively. A considerable portion of the effort by QA for the company is related to these "little" day to day jobs which usually require numerous telephone calls, confirming letters, sample reviews, and discussions to arrive at the proper recommendations for resolving these problems.

Vendor's Appraisal.—An important aspect of raw material procurement is reliability of the supplier. In order to implement the program of Certified Analyses (which is particularly important for critical ingredients) visits are made by QA personnel to vendors' production plants to observe sanitation conditions and technical facilities. While such activities are, generally, more directly related to the Purchasing Department, the Operating unit shares in their values, since the continued and efficient production of quality merchandise is possible only with sound and reliable raw materials.

Complaint Handling.—Shipments of raw materials received at the factories in an unsatisfactory condition are rejected by management. Copies of the complaint form plus a sample of the material is sent to the laboratory for concurrence and for filing (so that Vendor Histories can be kept current). A report and/or letter is sent to the Purchasing Department with copies to Division Management and Factory Production with observations and/or recommendations as needed.

Shipments which are questionable are temporarily isolated and the factory laboratory acts as the monitor. Complaint forms and samples are expedited for examination. Based on the findings QA will recommend rejection or, if the violation is not serious, discuss rejection or acceptance with the Production and Purchasing Departments, based upon the specifics involved.

In-process Quality Assurance

Obviously the use of high quality raw materials does not automatically result in high quality finished products. To achieve the latter, well balanced formulas must be followed accurately and proper mixing and processing procedures must be maintained with the minimum variation at all times.

In order to control these processing procedures, programs for "In-process" QC are developed by QA for implementation at the various factories. As new products and new processing techniques (such as extrusion) become a part of the production scene in a number of the factories, the role of In-process controls takes on added importance.

Plant Quality Coordination (a department of QA) receives weekly QC reports from all factories. All reports are reviewed and specified production periods are discussed with Factory Production. Monthly status reports giving statistical analyses of specialized process conditions are prepared and issued by QA. Such reviews make it possible to follow long-range trends and to alert Factory Production so that corrective actions may be taken.

Finished Product Quality Assurance

Finished Product Monitoring.—With many of the newer type products, QA has developed definite Finished Product Quality Assurance Programs which are followed in the company plants. Daily QC reports are received by QA and monthly statistical status reports are issued to show the degree of conformance with established standards and the range of variation from the average.

Special Finished Product Survey.—QA develops a program of Finished Product Surveys, which includes all products produced by the company. Since some of these products may be enriched or will carry label claims concerning vitamin and mineral content, a frequent check for these factors must be made. All products in this program are submitted periodically to taste panels for organoleptic evaluation and may be subjected to objective tests for color and texture. Compositional analyses are made on a scheduled basis so that up-to-date information will be available for company publications covering this subject. Bacteriological analyses are conducted on all products so that the sanitary and public health aspects are known and recorded. A detailed report of the finished products tested and all analyses performed is prepared and distributed. The modern approach to total QA embraces both the product and the package containing the product. Therefore, the quality condition of the package is reviewed and all defects are noted and reported. Periodic reports are issued to transmit this information to all sectors of the company.

Special Studies.—In addition to routine examination of finished products, QA is involved in frequent "crash" programs to assist in the resolution of various aspects of finished product quality; including, for example, off-taste, a lack of color development, too high a pH, excessive moisture content, out of specification free fatty acid content, etc. These problems may require special plant visits, or the assistance may be limited to frequent telephone calls and follow-up letters. All of this effort requires many man hours of discussion, sample review and communication.

Factory Visits.—In order to standardize the functions and activities of all company Quality Control Laboratories, factory visits are made to assure that they are fully informed and to ascertain that they are equipped and up-to-date in methodology so that they can carry out their programs efficiently and accurately, QA personnel visit all plant laboratories on a well established schedule.

Other Services.—Also related to operating division activity is the effort expended by QA in reviewing and answering consumer complaints. Most replies are made directly to the Treasurer's Department, but copies are given to Division Production so that follow-up can be made to possibly eliminate the sources of such complaints.

Another consumer service is rendered by QA for the Operating Division in the area of answering nutritional inquiries, and other related consumer problems—allergies, special diets, etc. Again this is not always direct service to a division, but an informed and contented consumer is an asset to any production unit.

Central Analytical Services.—In order to implement the examinations and surveys programmed for Operating Division in the areas of raw materials and finished products and to answer urgent daily requests, a very large number of individual physical, chemical and bacteriological analyses are required. The Central Analytical Services (usually another department of QA) carries out these tests. In addition to this activity, collaborative studies are carried out periodically with all company Division Laboratories. Analytical personnel are also involved in field trips to review new methods and to train plant personnel in their application.

A typical summary of the number of tests carried out for a Division might be:

Analyses	Avg per Month
Chemical	2000
Nutritional	1000
Bacteriological	500

Communications.—An excellent rapport is needed among Division, Factory and QA personnel. There is usually a constant and continuing exchange of telephone conversations .which might concern raw material and/or finished product evaluation and/or testing, questions on specifications or plant trials.

These exchanges are daily and often take up a major part of the working time of specialized personnel in the critical areas of the Operating Division. All decisions and related matters should be recorded in Memorandum-For-The-Record, copies being distributed to all personnel involved for future reference.

REFERENCES

AMMERMAN, G.R. 1957. Effect of equal lethal heat treatments at various times and temperatures upon selected food constituents. Ph.D. Dissertation, Purdue Univ., Lafayette, Ind.

ANGELINE, J.F., and LEONARDS, G.P. 1973. Food additives, some economic considerations. Food Technol. 27, No. 4, 40-50.

ANGELOTTI, R. 1975. FDA regulations promote quality assurance. Food Technol. 29, No. 11, 60-62.

ANON. 1973. Code of Federal regulations. Fed. Regist. 38, No. 143, 20061.

BAUMAN, H.E. 1974. The HACCP concept and microbiological hazard categories. Food Technol. 28, No. 9, 30-34, 70.

BULLERMAN, L.B. 1974. Inhibition of aflatoxin production by cinnamon. J. Food Sci. 39, 1163-1165.

BULLERMAN, .L.B. and OLIVIGNI, F.J. 1974. Mycotoxin producing potential of molds isolated from Cheddar cheese. J. Food Sci. 39, 1166-1168.

BYRAN, F.L. 1974. Microbiological food hazards today—based on epidemiological information. Food Technol. 28, No. 9, 52-66, 84.

CHRISTENSEN, C.M., NELSON, G.H., MICROCHA, C.J., and BATES, F. 1968. Toxicity to experimental animals of 943 isolates of fungi. Cancer Res. 28, 2293-2295.

CLAUSI, A.S. 1973. Improving the nutritional quality of food. Food Technol. 27, No. 6, 36-40.

DARBY, W.J. 1972. Fulfilling the scientific community's responsibilities for nutrition and food safety. Food Technol. 26, No. 8, 35-37.

DESROSIER, N.W. 1961. Attack on Starvation. AVI Publishing Co., Westport, Conn. (Out of print).

DUGGAN, R.R., and LIPECOMB, G.Q. 1969. Dietary intake of pesticide chemicals in the United States (II), June 1966-April 1968. Pestic. Monit. J. 2, 153.

FILER, L.J. 1976. Patterns of consumption of food additives. Food Technol. 30, No. 7, 62-70.

FOSTER, E.M. 1972. The need for science in food safety. Food Technol. 26, No. 8, 81-87.

FRANKFORT, H. 1951. The Birth of Civilization in the Near East. Indiana Univ. Press, Bloomington.

GOLDBLITH, S.A. 1971. Pasteur and truth in labeling: "pro bono publico"— in the best of scientific tradition. Food Technol. 25, No. 3, 32-33.

GRAS, N.S.B. 1946. A History of Agriculture. F.S. Crofts, New York.

GUILD, L., DEETHARDT, D., and RUST, E. II. 1972. Data from meals selected by students. Nutrients in university food service meals. J. Am. Dietet. Assoc. *61*, 38.

HALL, R.L. 1975. GRAS—concept and application. Food Technol. *29*, No. 1, 48-53.

IFT EXPERT PANEL AND CPI. 1972A. Botulism. Food Technol. *26*, No. 10, 63-66.

IFT EXPERT PANEL AND CPI. 1972B. Nitrites, nitrates, and nitrosamines in food—a dilemma. Food Technol. *26*, No. 11, 121-124.

IFT EXPERT PANEL AND CPI. 1973. Organic Foods. Food Technol. *28*, No. 1, 71-74.

IFT EXPERT PANEL AND CPI. 1974A. Nutrition labeling. Food Technol. *28*, No. 7, 43-48.

IFT EXPERT PANEL AND CPI. 1974B. Shelf life of foods. Food Technol. *28*, No. 8, 45-48.

IFT EXPERT PANEL AND CPI. 1975A. Naturally occurring toxicants in foods. Food Technol. *29*, No. 3, 67-72.

IFT EXPERT PANEL AND CPI. 1975B. Sulfites as food additives. Food Technol. *29*, No. 10, 117-120.

ITO, K.A., et al. 1973. Resistance of bacterial spores to hydrogen peroxide. Food Technol. *27*, No. 11, 58-66.

JAHNS, F.D., HOWE, J.L., CODURI, R.J., and RAND, A.G., JR. 1976. A rapid visual enzyme test to assess fish freshness. Food Technol. *30*, No. 7, 27-30.

JENSEN, L.B. 1954. Microbiology of Meats. Garrard Press, Champaign, Ill.

KAUFFMAN, F.L. 1974. How FDA uses HACCP. Food Technol. *28*, No. 9, 51, 84.

KRAMER, A. 1973. Storage retention of nutrients. Food Technol. *28*, No. 1, 50-60.

KRAMER, A., and FARQUHAR, J.W. 1976. Testing of time-temperature indicating and defrost devices. Food Technol. *30*, No. 2, 50-53, 56.

KRAMER, A., and TWIGG, B.A. 1973. Quality Control for the Food Industry, 3rd Edition, Vols. 1 and 2. AVI Publishing Co., Westport, Conn.

KWOLEK, W.F., and BOOKWALTER, G.N. 1971. Predicting storage stability from time-temperature data. Food Technol. *25*, No. 10, 51-63.

LACHANCE, P.A., RANADIVE, A.S., and MATAS, J. 1973. Effects of reheating convenience foods. Food Technol. *27*, No. 1, 36-38.

LEUNG, H., MORRIS, H.A., SLOAN, E.A., and LABUZA, T.P. 1976. Development of an intermediate moisture processed cheese food product. Food Technol. *30*, No. 7, 42-44.

LIBBY, W.F. 1951. Radiocarbon dates. Science *114*, 291-296.

LIVINGSTON, G.E., ANG, C.Y.W., and CHANG, C.M. 1973. Effects of food service handling. Food Technol *27*, No. 1, 28-34.

MACGILLIVRAY, J.H. 1956. Factors affecting the world's food supplies. World Crops *8*, 303-305.

MATCHES, J.R., and LISTON, J. 1968. Low temperature growth of Salmonella. J. Food Sci. *33*, No. 6, 641-645.

MCARDLE, F.J., and DESROSIER, N.W. 1954. A rapid method for determining the pericarp content of sweet corn. Canner *118*, 12-15.

MIDDLEKAUF, R.D. 1976. 200 Years of U.S. food laws: a gordian knot. Food Technol. *30*, No. 6, 48-54.

MITCHELL, H.H., and BLOCK, R.J. 1946. Some relationships between the amino acid contents of proteins and their nutritive values for the rat. J. Biol. Chem. 163, 599-620.

MURPHY, E.W., PAGE, L., and WATT, B.K. 1970. Major mineral elements in Type A school lunches. J. Am. Dietet. Assoc. 57, 239.

NATL. ACAD. SCI. 1970. Evaluating the Safety of Food Chemicals. Appendix: Guidelines for estimating toxicologically insignificant levels of chemicals in food. National Academy of Sciences, Washington, D.C.

NAS/NRC. 1970. Evaluating the Safety of Food Chemicals. Food Protection Committee, National Academy of Sciences—National Research Council, Washington, D.C.

NAS/NRC. 1972. GRAS Survey Report. Food Protection Committee, National Academy of Sciences—National Research Council, Washington, D.C.

NAS/NRC. 1973. Subcommittee on review of the GRAS list, Food Protection Committee. A comprehensive survey of industry on the use of food chemicals generally recognized as safe (GRAS). Nat. Tech. Inf. Serv. Rep. PB-221-925 and PB-221-939.

NAS/NRC. 1973. Toxicants Occurring Naturally in Foods, 2nd Edition. Food Protection Committee, National Academy of Sciences—National Research Council, Washington, D.C.

NATL. SCI. FOUND. 1973. President's science advisory committee panel on chemicals and health. Science and Technology Policy Office, Washington, D.C.

PETERSON, M.S. 1963. Factors contributing to the development of today's food industry. In Food 'Technology the World Over, Vol. 1, M.S. Peterson, and D.K. Tressler (Editors). AVI Publishing Co., Westport, Conn.

PRESIDENT'S SCIENCE ADVISORY COMMITTEE. 1973. Report of the Panel on Chemicals and Health. U.S. Gov. Print. Off., Washington, D.C.

PUBLIC HEALTH SERV. 1965. Division of radiological health, Public Health Service: Radionuclides in institutional diet samples, April-June 1964. Radiol. Health Data 6, 31.

PURVIS, G.A. 1973. What do infants really eat? Nutrition Today 8, No. 5, 28.

SCOTT, P.M. 1973. Mycotoxins in stored grain, feed and other cereal products. In Grain Storage: Part of a System, R.N. Sinha, and W.E. Muir (Editors). AVI Publishing Co., Westport, Conn.

SMITH, C.A., JR., and SMITH, J.D. 1975. Quality assurance system meets FDA regulations. Food Technol. 29, No. 11, 64-68.

STEWART, R.A. 1971. Sensory evaluation and quality assurance. Food Technol. 25, No. 4, 103-106.

STOTT, W.T., and BULLERMAN, L.B. 1976. Instability of patulin in cheddar cheese. J. Food Sci. 41, 201-202.

STUNKARD, A.J. 1968. Environment and obesity: recent advances in our understanding of regulation of food intake by man. Fed. Proc. 27, No. 6, 1367-1374.

TANNER, F.W. 1953. Food Borne Infections and Intoxications. Twin City Printing Co., Champaign, Ill.

TRESSLER, D.K., and LEMON, J.M. 1951. Marine Products of Commerce. Reinhold Publishing Corp., New York.

ULRICH, W.F. 1969. Analytical instrumentation—its role in the food industry. Food Prod. Dev. 2, No. 6, 18-25.

VON SYDOW, E. 1971. Flavor—a chemical or psychophysical concept? Part I. Food Technol. 25, No. 1, 40-44.

WODICKA, V.O. 1971. The consumer protection team. Food Technol. 25, No. 10, 29-30.

WOODEN, R.P., and RICHESON, B.R. 1971. Technological forecasting: the Delphi technique. Food Technol. 25, No. 10, 59-62.

WHO. 1967. Joint FAO/WHO Expert committee on food additives. Procedures for investigating intentional and unintentional food additives. WHO Tech. Rep. Ser. 348.

Chapter 16

Application of Technology

The food industries of today have found and hopefully will continue to find increasing opportunities in successfully producing and marketing wholesome new products, which are made increasingly more stable in storage, to growing numbers of consumers in expanding domestic and global markets. It has not been the purpose of this book to cover all the technological foundations supporting the whole food industry, since the building blocks themselves are each a specialization of their own. It is the purpose of this book, however, to unfold the technology of food preservation as it is known and applied. In this chapter the application of the technology of food preservation in new product development will be explored.

Obviously there is no one system or method of product development followed by all the food industry. Nevertheless, there is a distinct phenomenon called new product development; it is recurrent and it can be understood. The system to be described herein is employed by many large and small companies, in one variation or another, and found to be effective. The growing technology of food preservation continues to offer manifold possibilities upon which to base new products. Their success will be proven in the marketplace.

INTRODUCTION TO INDUSTRIAL PRODUCT DEVELOPMENT

The successful and expeditious development of new products which meet marketing and production requirements is a major function of industrial food research laboratories. This function, however, involves most other elements of a company and, to be effective, must truly be a company "team" effort.

Since many research, marketing, production, purchasing, legal and financial personnel become involved, product development must be a coordinated and controlled company or corporate function. Its effectiveness will be contingent on close working relationships among all members of the team and the smooth integration of their efforts toward desired ends.

For the purpose of this discussion, the words "Marketing Unit" will be used broadly, and will include Marketing and Market Research units. From a research laboratory view the Marketing Unit is the seat of corporate responsibility for introducing a proposed new product, the "intelligence" unit for identifying company interests in a proposed new product area, and, eventually, the unit that recommends that such interests be pursued.

In the ordinary course of events the recommendation eventually will take the form of a laboratory research project.

FOOD PRODUCT DEVELOPMENT TOOL— RESEARCH GUIDANCE PANELS

In general, once a new product project has been approved and assigned to a development food technologist, it becomes of utmost importance that he focus his attention on the product concept under study and that the decisions he makes concerning his progress in developing the product become as objective as possible. It is at this point that product development research guidance in the form of organoleptic panels of various types, sizes, and compositions become available. Typical taste panel forms are shown in Fig. 16.1. For new product development purposes another type of panel is needed industrially, beyond laboratory taste panels, called generally the Research Guiding Panel (RGP).

In order to put the RGP system into operation, many laboratories require three types of organoleptic testing panels, preliminary to the RGP. A typical taste panel arrangement is shown in Fig. 16.2.

Expert Profile Panels—An Analytical Tool.—Several highly trained experts are organized to identify and establish intensity values for sensory characteristics of products, i.e., what is the major flavor note in sample X? (The sample might have either a mint-chocolate or a chocolate-mint flavor.)

Primary Sensory Panels—A Product Development Tool.—Groups of product development personnel organized into panels evaluate their products at each step in development. The purpose is to yield the best product possible.

Secondary Sensory Panels—A Decision Tool for Measuring Differences.—Samples are tested for significance from control samples (Fig. 16.2). If the product meets design criteria it is ready to move into

LIKE
EXTREMELY   9

LIKE
VERY MUCH   8

LIKE
MODERATELY   7

LIKE
SLIGHTLY   6

NEITHER LIKE
NOR DISLIKE   5

DISLIKE
SLIGHTLY   4

DISLIKE
MODERATELY   3

DISLIKE
VERY MUCH   2

DISLIKE
EXTREMELY   1

AFTER EATING, PLEASE CHECK WHICHEVER BOX(ES) MOST NEARLY DESCRIBE THE COMPONENT AND
CIRCLE ON THE SCALE THE POINT AT WHICH YOU RATE THE ITEM QUALITY.
(USE REVERSE SIDE FOR ADDITIONAL REMARKS, IDENTIFYING EACH ITEM BY NUMBER)

MEAT COMPONENT

1   The proportion of gravy is: ☐ Too much   ☐ Too little   ☐ Just right

2   The meat is: ☐ Tender   ☐ Moist   ☐ Dry   ☐ Fatty   ☐ Tough   ☐ Stringy
     ☐ Other (Specify)

3.   The meat and gravy are: ☐ Well seasoned   ☐ Overseasoned   ☐ Too bland
                           ☐ One seasoning too predominant   ☐ Other (Specify)

4   Please describe off flavor, if any

Extremely Poor	Very Poor	Poor	Below Fair Above Poor	Fair	Below Good Above Fair	Good	Very Good	Excellent
1	2	3	4	5	6	7	8	9

Courtesy of QM. Food and Container Institute

FIG. 16.1.   SCORE SHEETS USED IN EVALUATION OF FOOD SAMPLES

Food acceptance score sheet (left) and food quality evaluation sheet (right) are examples
of those used in food research.

FIG. 16.2.   TYPICAL TASTE PANEL ARRANGEMENT

Taste panels are decision-making bodies.

the pilot plant. At this point a new type of testing panel becomes needed.

Research Guidance Panels

Purpose.—Research Guidance Panels represent a relatively inexpensive mechanism for measuring consumer acceptance of products under

development by a research laboratory. Information of this type is necessary at a fairly early stage in product development, and is useful in the continuing modifications of properties designed to secure the maximum consumer acceptability within the predetermined cost. Products so developed will generally meet all the criteria demanded in a broader sense.

Panel Organization.—RGPs can readily be set up, especially in a metropolitan area where a broad spectrum of people of various racial origins and foreign and native derivation exists. Voluntary panels of suitable size are organized. Test samples are prepared and distributed, along with a questionnaire (Fig. 16.3).

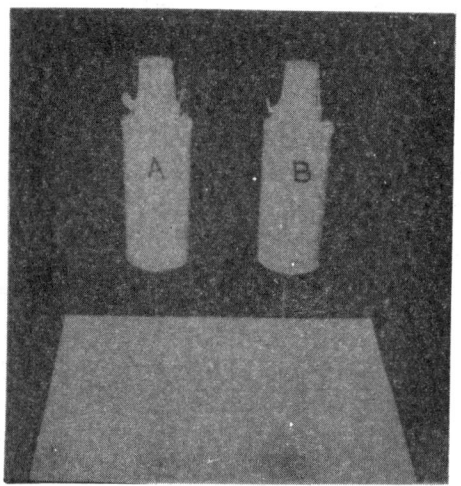

FIG. 16.3.   RESEARCH GUIDANCE
PANEL (RGP) TEST SAMPLES

Utility of Results.—There is an excellent agreement between RGP testing and broad scale consumer surveys with wide geographical distribution, using the same paired samples but with much more elaborate questionnaires.

One important point is the early availability of results—weeks as compared to months for the average consumer survey. Another is the cost. An example of the type of result received from a product is shown in Table 16.1.

TYPES OF NEW PRODUCTS FOR A FOOD COMPANY

The development of new products for a company usually falls technologically into one of four categories, in decreasing order of difficulty:

TABLE 16.1

EXAMPLE OF RGP RESULTS. MONADIC TEST ON FROZEN STRAWBERRY PIE,
SUCH AS MIGHT BE OBTAINED

	Adults	Children	Combined
	Flavor of Filling (%)		
Too strong	47.6	33.0	38.6
Just right	39.7	52.8	47.7
Too low	12.7	14.2	13.7
	Sweetness of Filling (%)		
Too sweet	73.0	47.8	57.5
Just right	27.0	49.8	41.0
Not sweet enough		2.4	1.5

	Flavor Remarks (No.)		
Good/like crust	46	Dislike filling	31
Good filling	17	Too sweet	16
Good/tasty/excellent	12	Filling too sweet	15
Liked	10	Dislike	8
Real strawberry flavor	9	Artificial strawberry	7
Just sweet enough	3	filling	
		Flavor too strong	5
	Texture Remarks (No.)		
Crisp	6	Dry Crust	4
Flaky crust	5	Crumbly	3

(1) New factory—new technology and equipment (i.e., new raw materials, new processing equipment, new packaging, warehousing, and distribution needs, etc.).

(2) Existing factory—new types of technology and equipment (i.e., new process and/or new packaging equipment, etc.).

(3) Existing factory—some minor new technology and equipment (i.e., new step in process, or new packaging equipment, etc.).

(4) Existing factory—new entry using existing technology and equipment (i.e., new canned food, new frozen food, new dry mix, new cookie, etc.).

Category I—Completely New Product

Comprehensive market analysis is often undertaken as the initial phase of a program. Next, Research Laboratory personnel, in conjunction with Market Research Specialists, develop new product ideas and concepts embracing potential consumer interests and needs revealed by the market analysis. The most meaningful and fruitful are technological innovations which would in time lead to a whole family of additional products.

Concept Testing

These concepts are then screened for technical feasibility, developed to prototype stages, and consumer researched to establish prototype acceptability. The decision more often than not revolves around the compatibility of the new product and long-range company plans, and the choices available for commitment of available resources.

The following outline gives the general sequence of events in more detail for a formal development program. Close coordination is required at each stage with all operating elements involved.

Prototype Product

This type of product development deals with an entirely new product concept where no known product on the market meets the anticipated need. A "prototype" product must be developed, and the optimum method of preservation identified. (A typical product development laboratory is shown in **Fig. 16.4.**)

Courtesy of Nabisco

FIG. 16.4.   TYPICAL PRODUCT DEVELOPMENT AREA;
NOTE ABSENCE OF LARGE EQUIPMENT

Process Development

When a new product reaches the development stage where it appears feasible, the Laboratory begins to plan an economical process for its industrial manufacture.

In process development many factors must receive consideration, such as:

(1) Can it be made in an existing factory?

(2) Should it be a continuous (Fig. 16.5), semi-continuous, or batch process?

(3) What physical properties of the ingredients, mixes prepared during formulation, and the product itself are needed for process equipment design? Such matters as temperature-viscosity relations, heat transfer coefficients, emulsion stabilities, density range, flow rates, and specific heat data are among the factors frequently requiring study.

FIG. 16.5. AN EXAMPLE OF CONTINUOUS FLOW EQUIPMENT

Product can be preheated, sterilized at high temperatures and pressures, then cooled for discharge at atmospheric pressure.

A more complete checklist of the factors requiring specific attention are presented later in this chapter. The specialized knowledge of production personnel may also be called upon at this stage, as required. However, for our present purpose, the following sequence generally occurs: (1) establish tentative process specifications; (2) organize necessary pilot plant facilities; (3) produce pilot plant products and submit to taste panels and run RGP; (4) coordinate with Marketing Unit; (5) produce sufficient pilot plant quantities of products for consumer research; (6) coordinate with Production, Purchasing and Legal Units; and (7) formalize product and process specifications by preparing tentative: (a) raw material and packaging specifications, (b) process specifications, (c) quality control manuals, (d) product and process operating manuals, and (e) patent disclosures covering inventions.

Public Health Clearance

Complete studies to establish (1) absence of public health hazards in all raw and finished products; and (2) obtain laboratory microbiological clearance.

Packaged Product Storage Studies

Establish storage life of packaged products by (1) actual tests under several temperature and humidity conditions, taking careful note of the information presented in Chapter 14; (2) ensure package integrity by abuse tests; and (3) perform needed organoleptic tests.

Finalize Specifications

Correct any defects in packaged products (1) revealed by the above studies; (2) complete all specifications; and (3) submit specifications for establishment of costs.

Develop Advertising Claims

In conjunction with Marketing Unit (1) develop statements concerning desirable characteristics and uses of developed products for potential uses in advertising; and (2) provide data to support advertising claims.

Preproduction Runs

In many companies the production organization has an engineering group for design and maintenance of plant.

As an example, Research will generally accumulate all the data on physical properties required for process equipment selection or design. Production Engineering can then lay out the designed process with due regard to available space, available equipment, and other production considerations.

Research will generally have worked with small-scale pilot equipment to obtain the necessary physical information. Presumably also the pilot plant will have used either specially built or available commercial equipment in producing pilot lots of the proposed product. It is well to keep in mind that small pilot operations frequently give differing results from plant operations, where there may be from ten- to a hundredfold increase in size.

Market Research

Provide Products for Field Testing.—All necessary information concerning product acceptance based on actual home-use tests must be completed successfully. Here is where the idea of how many times a week the product will be used (which gives some indication of the production volumes to anticipate), what the consumer feels the price should be, and consumer reaction to the product overall are unfolded. This process of finalization is undertaken with potential consumers in test market areas. Typical results of Market Research are shown in Table 16.2.

TABLE 16.2
TYPICAL CONCEPT TEST SUMMARY

| | Prototype—Child Food Product | | | |
| | A | B | C | D |
	N = 188 (%)	N = 188 (%)	N = 186 (%)	N = 186 (%)
Definite purchase intent	34	44	32	31
Total purchase intent	72	77	72	65
Inclined to switch	33	32	38	33
Preference				
Housewives	38	52	46	45
Children	28	55	29	53
Product Ratings[1]				
Overall appeal	7.3	7.7	7.1	7.0
Natural rather than artificial taste	7.9	7.7	7.9	7.4
Retaining freshness	8.9	9.0	9.0	9.1
Having an attractive color	8.1	8.5	8.1	8.4
Having a pleasant texture	8.1	8.1	7.9	7.8
Having an attractive appearance	7.0	9.0	7.2	8.9
Best overall taste	7.3	7.5	7.6	6.9
Having an appealing shape	7.0	9.0	7.1	8.7

[1] 1—10 rating, with 10 being highest possible score.

Timing

When existing production and packaging facilities are available, the production of test market quantities may be anticipated in less than 50 weeks from the time of initiation of a study.

New proposals requiring new facilities and equipment may take from 2 to 4 years to reach the marketplace regionally. It will take approximately a year from this point to complete regional test market studies prior to launching nationally.

Overall it generally requires from 3 to 5 years to bring a completely new product into national distribution. It may take longer.

CATEGORY II—NEW PRODUCT FOR COMPANY-EXISTING COMPETITOR PRODUCT

The Idea Sources

First of all the idea for the improved products must be uncovered. Experience indicates that ideas frequently come from unexpected sources. Some may come from various areas in a company. Still others may be brought to the attention of a company from outside. Advertising people may make surveys leading to suggestions for new products. All should receive attention with due regard to the legal aspects of company liability, particularly in the case of submissions from outside a company.

"Must Have"—"Would Like" Specifications

Once a novel product idea has passed marketing and research scrutiny, the next step is for the Marketing Unit to set up specifications for the product they would like to market. This also takes the form of a description of properties it should have from their point of view. The preliminary specifications should embrace the "must have" and "would likes." The immediate goal is to develop a "prototype" product and to learn by means of suitable testing what the consumer reaction may be. (See section on Research Guidance Panels.) Here the simplest problems concern a product which is designed to fulfill a given need better than any product currently on the market. RGP "blind" testing of the "prototype" versus the leading product will serve to establish consumer likes, dislikes and preference patterns.

CATEGORIES III AND IV

New products for a company falling into these categories are undertaken with less difficulty. While all the steps and testings are performed, these products are usually based on minor technological changes. They can be brought about with more dispatch and less involvement of personnel and resources, since the area has been successfully explored before. Therefore, the risks are reduced in most stages of development.

INDUSTRIAL APPLICATION OF NEW TECHNOLOGY

While science and technology are thought of as the relation of things and processes to each other, when attempts are made to relate them to man another component is added—common sense. The application of science and technology eventually involves people and their harmonious interrelationships. Some examples follow:

Information Required Prior to Launching a New Product

Some feeling for the myriad of details and decisions required prior to reaching the marketplace with a new product can be obtained from the following generalized listings:

The Product.—Its specifications; formulation tolerance; analytical techniques for specifications; sampling techniques, for specification control; patents on the product itself; toxicity clearance; quality control manual; uniformity of production, from panel evaluations.

Product Storage Specifications.—At normal conditions; at accelerated conditions; correlation between normal and accelerated storage.

Packaging.—Specifications for carton development—type, size and materials; label development; label declarations; recipe preparation for label and design of label; case specifications; shipping tests; field storage and handling of packaged product.

Marketing.—Consumer acceptance of the product through home-use and market tests; consumer acceptance of product versus nearest competition; determination of quality/price specification—consumer desires; determination of consumer habits in use of product, i.e., full package, part of package, half package, etc.; what is best size package; time to start filling distribution channels; advertising—when to initiate; trade markets; identification of potential competing products; study of market demand; volume expected, location of market demand; legal requirements for sale, use of patents, of materials involved, and other requirements; seasonability of market.

Raw Materials Specifications.—For each ingredient—size, purity, composition, contaminants, variety, color, others, and the variations permitted in each specification; analytical procedures for identifying specifications; sampling procedures; toxicity clearance; availability of material meeting specifications, locationwise—geographically and suppliers; timewise—seasonal or periodically variable supply within seasons or periods, variable supply from season to season or period to period, others; characteristics of material available, technical or chemical, physical, varietal, others; list of sources; adequacy of supply; market influence on supply and resulting fluctuations; costs—fluctuations, compositionwise, qualitywise; transportation, government, market speculation influences; selection of material for processing—techniques needed to establish whether material from an aggregate contains some out-of-specification material, and the possible uses of out-of-specification material; handling—equipment needed, techniques required, requirements during transportation, personnel training required, hazards involved; aid required to suppliers—either industrial or agricultural, legal problems, tax, tariff, and patent problems; purchasing—contractual agreements, special training or knowledge required, and the amount of effort required.

Raw Material Storage.—Containers and conditions required, normal, special, effect of condition on quality and quantity; facilities investigation.

Process Development.—Process conditions—research, pilot plant, production, optimum conditions of processing, tolerance of processing conditions, in-process specifications, peculiarities of the process, balance between quality and possible cost reduction, raw material effects on process and product, uniformity of in-process control, process formulations, basic formula development, tolerance limits of formula, substitutions or variations of formula possible; process description, adequate presentation of information, subsequent modifications; yield—sources, causes, importance, methods of reducing losses; personnel requirements—number, training, job descriptions, points of high useage, methods of reducing the number of people in time; utility requirements—quantity, both normal and maximum demand, quality, and possibility of reducing them; process sanitation; process safety and problems; patents—patent infringement of other processes, patents on new process; licenses required; waste disposal—amounts of waste, composition of wastes, possible combinations of waste for treatment, possible treatments; time-motion studies—yield versus labor, quality versus labor, quality versus yield, and others; equipment performance—capacity, process problems; ease of operating, stability, and maintenance problems; development of quality control techniques, quality control manual; capital costs; process costs; process requirements for special building facilities.

Facilities.—Definition—including flowsheet and specifications, equipment, utilities, labor, area requirements, building, land, costs, design; construction; delivery times for equipment; installation schedule.

Plant Location.—General area of plant location, specific area, location within a general geographical locality, location within a given plant; land characteristics—virgin or filled land, cavernous, rocky, sandy, or other; utilities available in the desired area—power, water, telephone, gas, and others; suitability and adequacy of water supply—for processing, for power house, others, and treatment needed; availability of required fuel; labor survey—available, qualified, labor laws, and other data; living conditions; effect of processed plant on nearby industries, effect of climate on process; waste disposal—legal requirements, fitting process requirements to legal requirements; air pollution requirements of community; transportation availability; taxes and other fees; legal requirements; relation between raw material supply and the location of markets; problems peculiar to the plant construction at that locality; acceptability by the local people.

Costs.—Material, process raw materials, packaging materials, auxiliary materials, and cleaning; processing—direct labor, indirect costs, supervision, marketing; other costs; overhead.

Plant Operation.—Job descriptions, personnel acquiring, personnel training, plant startup; operating manual preparation—initial, revisions; sanitation check; check on safety and health conditions; arrangements for materials for production; raw materials, packaging materials, and miscellaneous materials.

Normal Production.—Periodic review of market performance—are objectives being met; and maintenance—periodic updating of product/process to yield "new improved" product.

It is obvious that these areas need not all become the subject of new study for a company, and there also could be other items added to these lists. However, it also should be obvious that an organization involving a host of professional personnel is needed.

INTERLOCKING ACTIVITIES OF PEOPLE AND ORGANIZATIONS

An insight into the division of responsibility and efforts, and the communications needed between organizations, in bringing out a new product may be found in just one aspect of it—establishing the package for a new product. In large food manufacturing companies the following pattern is often found.

Marketing Unit.—Decision on the type of package, cost, design, labeling, and other information for casing of shelf units.

Purchasing Unit.—Work with suppliers to obtain submissions of new or modified types of packaging for Marketing Unit inspection and for preliminary consideration by the Research Laboratory and Production Unit.

Production Unit.—Preliminary advice on the feasibility of proposed packaging changes as related to operation of packaging equipment and final testing of operating characteristics on available equipment in the event of acceptance by the Marketing Unit.

Laboratory.—The preliminary and final determination of the adequacy of the proposed package to protect the product. It should also assist Marketing and Purchasing Units in the obtaining of preliminary samples of packaging materials. This assistance should be on request of these two groups and generally will take the form of discussions with suppliers' technical people on needs and problems as they arise.

Legal Unit.—Generally the Legal Unit has responsibility to ensure that labeling meets the requirements of FDA, FTC, and Meat Inspection Service or other government agencies that may be involved in the United States. In other countries similar governmental agencies frequently become involved.

The effectiveness of the team will be revealed in the dispatch with which they bring out new products which meet established goals.

Industrial research laboratories generally devote a major portion of their resources and personnel to new and improved product and process projects.

The flow of knowledge to sustain these activities originates from the scientific and technical laboratories both inside and outside a company.

COMPETITIVE ACTIONS

Another factor to receive serious consideration is that inventions, being what they are, often can occur to more than one inventing team or person in more than one company at nearly the same time. It must be assumed that this will occur and plans drawn accordingly.

Given success in launching a new product, various means are available to measure the progress of the product as competitors and markets grow. One of the widely used services to keep abreast of market activity and product improvement is the Nielsen Food Index. In a regularly updated review, product and competitor shares of markets, regionally and nationally, are estimated. Based on these and internally produced information in a company, future plans and needs unfold. The general approach used in such reports is shown in Table 16.3.

SUMMARY

The technology of food preservation is one of the key building blocks of the food industry. In one way or another all food products draw upon this technology, which rigidly defines the parameters within which successful new products can exist, and existing products can be improved, if they are to be stable.

The technology of food preservation itself continues to grow and expand, one discovery leading to another, unfolding new possibilities. New product development has been seen as a recurring phenomenon and is therefore subject to analysis. These studies have yielded a methodology which is becoming widely recognized as capable of expediting new food product developments and reducing the risks involved.

TABLE 16.3

TYPE OF INFORMATION GENERATED BY MARKET ANALYSIS, CONCERNING
PRODUCT MOVEMENT THROUGH RETAIL CHANNELS, ONCE PRODUCT[1]
REACHES NATIONAL DISTRIBUTION

| | % Total Instant Beverage Sales | | | |
| | Product A | | Product B | |
	May-June	July-August	May-June	July-August
U.S. Sales[2]				
Total	8.3	8.2	8.8	8.9
U.S. Sales[3] Regional Shares				
New England	18	17	20	21
Mid-Atlantic	25	24	30	30
East Central	20	20	24	24
West Central	15	15	10	10
Southeast	5	5	7	7
Southwest	4	4	1	1
Pacific	10	10	7	6
Other	3	4	1	1
U.S. Sales[4] Selected Cities				
New York	11.5	10.8	18.9	18.9
Chicago	16.2	16.9	10.1	9.5
Los Angeles	7.3	6.6	12.5	13.2

[1] For purpose of example, assume products are Instant Beverages, and Product A is being challenged by Product B.
[2] Share of total market.
[3] Distribution of share regionally.
[4] Percentage of total sold (in selected cities) of new products compared to the total instant beverages sold in those cities.
Note: Other minor competitive products also exist. Analysis focused on two main entries.

An outline of the present methodology in food product development has been presented in this chapter. Further interest in the subject might be pursued by following the leads presented in references which follow.

REFERENCES

ANGELUS, T.L. 1970. Improving the success ratio in new products. Food Technol. 24, No. 4, 29-36.

ANON. 1970. Finance. Analysis of corporate earnings, stockholder relations, small business financing and new techniques. Harvard Bus. Rev. Reprint Serv., Soldier's Field, Boston.

ANON. 1976. ASTM Committee E-18 on Sensory Evaluation of Materials and Products, Am. Society for Testing and Materials, Manual on Sensory Testing Methods, STP 434. Am. Soc. for Testing and Materials, Philadelphia.

BASS, W.L. 1967. The planning of innovation. Chem. Ind. 39, 1671-1675.

BATES, B. 1967. Venture group approach to new products. Assoc. Natl. Advertisers Workshop, New York.

BRIGHT, W.M. 1968. Science and mergers and acquisitions in corporate growth. Food Prod. Dev. 2, No. 4, 43-50.

CALVIN, L.D., and SATHER, L.A. 1959. A comparison of student preference panels with a household consumer panel. Food Technol. 13, 469-472.

CARSON, G.D. 1970. Product development decisions using Bayesian decision theory. Food Prod. Dev. 2, No. 6, 46-50.

CAUL, J.F., and TAYMOND, S.A. 1967. Comparison of local and national consumer panels in a paired preference test. I. Statistical data. J. Soc. Cosm. Chem. 18, 123-133.

CAVALIER, P.A. 1967. The financial function in product development. Food Prod. Dev. 1, No. 6, 28-30.

CAVALIER, P.A. 1968. The Balance Sheet. Food Prod. Dev. 2, No. 1, 32-34.

DESROSIER, N.W., and DESROSIER, J.N. 1971. Economics of New Food Product Development. AVI Publishing Company, Westport, Conn.

DETHMERS, A.E. 1968. Experienced vs. inexperienced judges for preference testing. Food Prod. Dev. 2, No. 5, 23-25.

FREDRICKSON, E.B., and LAWSON, E.W. 1970. The role of technology transfer in product development and marketing. Res. Manage. 13, No. 4, 265-272.

GOODMAN, S.R. 1970. Using decision guides for research and development. Food Technol. 24, No. 4, 42-47.

GREYSER, S.A. 1964. The case of unproductive products. Harvard Bus. Rev. 42, No. 4, 38-47.

LAVIDGE, R.J. 1968. Equating market opportunities with product development potential. Food Prod. Dev. 2, No. 5, 28-31.

LISTON, M.I. 1970. Some basic requisites for research productivity. J. Home Econ. 62, No. 4, 234-245.

MILLER, P.G., NAIR, J.H., and HARRIMAN, A.J. 1955. A household and a laboratory type of panel for testing consumer preference. Food Technol. 9, 445-449.

ORMAN, A.D. 1967. A systematic approach to screening new product ideas. Assoc. Natl. Advertisers Workshop on New Products and Services, New York.

SAMPSON, R.T. 1969. Sense and sensitivity in pricing. Harvard Bus. Rev. 42, No. 6, 45-49.

SCHEUBLE, P.A. 1964. ROI for new product planning. Harvard Bus. Rev. 42, No. 6, 118-128.

Appendix

Useful Tables

TABLE A.1

RECOMMENDED DAILY DIETARY ALLOWANCES[1]
Designed for the maintenance of good nutrition of practically all healthy people in the U.S.A.

	Age (Yr)	Weight (Kg)	Weight (Lb)	Height (Cm)	Height (In.)	Energy (Kcai)[2]	Protein (Gm)	Vitamin A Activity (RE)[3]	Vitamin A (IU)	Vitamin D (IU)	Vitamin E Activity[5] (IU)
Infants	0.0-0.5	6	14	60	24	kg × 117	kg × 2.2	420[4]	1,400	400	4
	0.5-1.0	9	20	71	28	kg × 108	kg × 2.0	400	2,000	400	5
Children	1-3	13	28	86	34	1,300	23	400	2,000	400	7
	4-6	20	44	110	44	1,800	30	500	2,500	400	9
	7-10	30	66	135	54	2,400	36	700	3,300	400	10
Males	11-14	44	97	158	63	2,800	44	1,000	5,000	400	12
	15-18	61	134	172	69	3,000	54	1,000	5,000	400	15
	19-22	67	147	172	69	3,000	54	1,000	5,000	400	15
	23-50	70	154	172	69	2,700	56	1,000	5,000		15
	51+	70	154	172	69	2,400	56	1,000	5,000		15
Females	11-14	44	97	155	62	2,400	44	800	4,000	400	12
	15-18	54	119	162	65	2,100	48	800	4,000	400	12
	19-22	58	128	162	65	2,100	46	800	4,000	400	12
	23-50	58	128	162	65	2,000	46	800	4,000		12
	51+	58	128	162	65	1,800	46	800	4,000		12
Pregnant						+300	+30	1,000	5,000	400	15
Lactating						+500	+20	1,200	6,000	400	15

Source: Natl. Acad. Sci.–Natl. Res. Council. (1974).

[1] The allowances are intended to provide for individual variations among most normal persons as they live in the United States under usual environmental stresses. Diets should be based on a variety of common foods in order to provide other nutrients for which human requirements have been less well defined. See text for more detailed discussion of allowances and of nutrients not tabulated.
[2] Kilojoules (kJ) = 4.2 × kcal.
[3] Retinol equivalents.
[4] Assumed to be all as retinol in milk during the first six months of life. All subsequent intakes are assumed to be half as retinol and half as β-carotene when calculated from international units. As retinol equivalents, $\frac{3}{4}$ are as retinol and $\frac{1}{4}$ as β-carotene.

SI Units and Conversion Factors[1]

Since 1960, most of the countries of the world have made formal commitments to convert to the International System of Units (Système International d'Unités), abbreviated as SI.

This has been written to assist in the presentation of quantities in SI metric. The existence of other metric units, which do not agree with recently adopted SI units, is a source of confusion to many. As some have been accustomed to using the non-SI and/or British Imperial units, factors have been included for converting both to SI.

The International System of Units (SI)

The SI system consists of seven base units, two supplementary

---

[1] The following pages are adapted from ASHRAE's 1976 Systems Handbook with permission from the American Society of Heating, Refrigerating and Air-Conditioning Engineers, Inc., New York, N.Y.

TABLE A.1 *(Continued)*

Ascorbic Acid (Mg)	Folacin[6] (µg)	Niacin[7] (Mg)	Riboflavin (Mg)	Thiamin (Mg)	Vitamin B-6 (Mg)	Vitamin B-12 (µg)	Calcium (Mg)	Phosphorus (Mg)	Iodine (µg)	Iron (Mg)	Magnesium (Mg)	Zinc (Mg)
			Water-soluble Vitamins						Minerals			
35	50	5	0.4	0.3	0.3	0.3	360	240	35	10	60	3
35	50	8	0.6	0.5	0.4	0.3	540	400	45	15	70	5
40	100	9	0.8	0.7	0.6	1.0	800	800	60	15	150	10
40	200	12	1.1	0.9	0.9	1.5	800	800	80	10	200	10
40	300	16	1.2	1.2	1.2	2.0	800	800	110	10	250	10
45	400	18	1.5	1.4	1.6	3.0	1,200	1,200	130	18	350	15
45	400	20	1.8	1.5	2.0	3.0	1,200	1,200	150	18	400	15
45	400	20	1.8	1.5	2.0	3.0	800	800	140	10	350	15
45	400	18	1.6	1.4	2.0	3.0	800	800	130	10	350	15
45	400	16	1.5	1.2	2.0	3.0	800	800	110	10	350	15
45	400	16	1.3	1.2	1.6	3.0	1,200	1,200	115	18	300	15
45	400	14	1.4	1.1	2.0	3.0	1,200	1,200	115	18	300	15
45	400	14	1.4	1.1	2.0	3.0	800	800	100	18	300	15
45	400	13	1.2	1.0	2.0	3.0	800	800	100	18	300	15
45	400	12	1.1	1.0	2.0	3.0	800	800	80	10	300	15
60	800	+2	+0.3	+0.3	2.5	4.0	1,200	1,200	125	18+[8]	450	20
80	600	+4	+0.5	+0.3	2.5	4.0	1,200	1,200	150	18	450	25

Total vitamin E activity, estimated to be 80% as α-tocopherol and 20% other tocopherols.
The folacin allowances refer to dietary sources as determined by *Lactobacillus casei* assay. Pure forms of folacin may be effective in doses less than $1/4$ of the recommended dietary allowance.
Although allowances are expressed as niacin, it is recognized that on the average 1 mg of niacin is derived from each 60 mg of dietary tryptophan.
This increased requirement cannot be met by ordinary diets; therefore, the use of supplemental iron is recommended.

units, and a number of derived units. The term, unit of measurement, SI symbol, and formula are given in Table A. 2.

Base units are defined as follows:

1. *Length:* The *metre* is the length equal to 1 650 763.73 wavelengths in vacuum of the radiation corresponding to the transition between the levels $2p_{10}$ and $5d_5$ of the krypton-86 atom.

2. *Mass:* The *kilogram* is the unit of mass; it is equal to the mass of the international prototype of the kilogram.

3. *Time:* The *second* is the duration of 9 192 631 770 periods of the radiation corresponding to the transition between the two hyperfine levels of the ground state of the cesium-133 atom.

4. *Amount of Substance.* The *mole* is the amount of substance of a system that contains as many elementary entities as the number of atoms in 0.012 kilogram of carbon-12.

Supplementary units are as follows:

1. *Plane Angle:* The *radian* is the unit of measure of a plane angle with its vertex at the center of a circle and subtended by an arc equal in length to the radius.

TABLE A.2

SI UNITS

Term	Unit	Symbol	Formula
Base Units			
length	metre	m	
mass	kilogram	kg	
time	second	s	
electric current	ampere	A	
thermodynamic temperature	kelvin	K	
amount of substance	mole	mol	
luminous intensity	candela	cd	
Supplementry Units			
plane angle	radian	rad	
solid angle	seradian	sr	
Derived Units with Special Names			
electric capacitance	farad	F	C/V
quantity of electricity	coulomb	C	A·s
electric potential difference	volt	V	W/A
electric resistance	ohm	Ω	V/A
electrical conductance	siemens	S	A/V
energy	joule	J	N·m
force	newton	N	kg·m/s²
frequency (cycles per second)	hertz	Hz	cycle/s
illuminance	lux	lx	lm/m²
inductance	henry	H	Wb/A
luminous flux	lumen	lm	cd·sr
magnetic flux	weber	Wb	V·s
magnetic flux density	tesla	T	Wb/m²
power	watt	W	J/s
pressure	pascal	Pa	N/m²
stress	pascal	Pa	N/m²
Drived Units without Special Names			
acceleration— angular	radian per second squared		rads²
linear	metre per second squared		m/s²
area	square metre		m²
density	kilogram per cubic metre		kg/m³
luminance	candela per square metre		cd/m²
magnetic field strength	ampere per metre		A/m
moment of a force	newton-metre		N·m
permeability	henry per metre		H/m
permittivity	farad per metre		F/m
specific heat capacity	joule per kilogram-kelvin		J/(kg·K)
thermal capacity (entropy)	joule per kelvin		J/K
thermal conductivity	watt per metre-kelvin		W/m·K

2. *Solid Angle:* The *steradian* is the unit of measure of a solid angle with its vertex at the center of a sphere and enclosing an area of the spherical surface equal to that of a square with sides equal in length to the radius.

Derived units are as follows:

1. *Force:* The *newton* is that force which, when applied to a body having a mass of one kilogram, gives it an acceleration of one metre per second per second.

2. *Energy:* The *joule* is the work done when the point of application of a force of one newton is displaced a distance of one metre in the direction of the force.

3. *Power:* The *watt* is the power which gives rise to the production of energy at the rate of one joule per second.

BASIC RULES

In using the SI system, the understanding of application of prefixes, basic rules of expression, and methods of conversion and rounding will be helpful.

Prefixes

Prefixes for multiple and submultiple units are listed in Table A.3. Only one multiple or submultiple prefix is applied at one time to a given unit and is printed immediately preceding the unit symbol.

However, to maintain the coherence of the system, multiples or submultiples of SI units should not be used in calculations.

When expressing a quantity by a numerical value and a unit, prefixes should be chosen so that the numerical value lies between 0.1 and 1000, except where certain multiples or submultiples have been agreed to for particular use.

Multiple and submultiple prefixes representing steps of 1000 are recommended. Show force in mN, N, kN, and length in mm, m, km, etc. Use of centimetres should be avoided unless a strong reason exists.

Prefixes should not be used in the denominator of compound units, except for the kilogram (kg). Since the kilogram is a base unit of SI, this is not a violation.

With SI units of higher order such as m^2 and m^3, the prefix is also raised to the same order; for example mm^3 is 10^{-9} m^3 not 10^{-3} m^3. In such cases, the use of cm^2, cm^3, dm^2, dm^3, and similar nonpreferred prefixes is permissible.

TABLE A.3

MULTIPLE AND SUBMULTIPLE UNITS

Multiplication Factors		Prefix	SI Symbol
1 000 000 000 000	= 10^{12}	tera	T
1 000 000 000	= 10^6	giga	G
1 000 000	= 10^6	mega	M
1 000	= 10^3	kilo	k
100	= 10^2	hecto[1]	h
10	= 10^1	deka[1]	da
0.1	= 10^{-1}	deci[1]	d
0.01	= 10^{-2}	centi[1]	c
0.001	= 10^{-3}	mili	m
0.000001	= 10^{-6}	micro	μ
0.000 000 001	= 10^{-9}	nano	n
0.000 000 000 001	= 10^{-12}	pico	p
0.000 000 000 000 001	= 10^{-15}	femto	f
0.000 000 000 000 000 001	= 10^{-18}	atto	a

[1] To be avoided if possible.

Mass, Force, and Weight

The principal departure of SI from the gravimetric form of metric engineering units is the separate and distinct units for mass and force.

The kilogram is restricted to the unit of *mass*. The mass of a body never varies; it is independent of gravitational force.

The newton is the unit of *force* and should be used in place of kilogram-force. The newton represents the force required to support a mass against a gravitational attraction, or the force required to produce a change in motion. It is the force required to accelerate a one kilogram mass one metre per second every second.

The newton, instead of kilogram-force, should be used in combination units which include force. For example, pressure or stress ($N/m^2 = Pa$); energy ($N \cdot m = J$); power ($N \cdot m/s = W$).

In scientific use the term *weight* of a body usually means the force which if applied to the body would give it acceleration equal to the local acceleration of free fall. In commercial and everyday use, the term weight nearly always means mass. Because of this dual use, the term weight should be avoided except under circumstances in which its meaning is completely clear. When the term is used, it is important to know whether mass or force is intended, and to use SI units properly.

The relationship between newtons, kilograms, metres, and seconds given by the first law of motion ($F = ma$) should be used to simplify units (that is, $N = kg \cdot m/s^2$). Thus, the combination unit of foot-pound-force per pound-mass used to measure efficiency may be converted to newton-metre per kilogram ($N \cdot m/kg$) or joule per kilogram (J/kg), but it must not be converted to metre since pound-force and pound-mass are not equivalent units.

When non-SI units are used, a distinction should be made between (1) force and (2) mass, for example, lbf to denote force in gravimetric engineering units and lb for mass.

Common use has been made of the metric ton, also called *tonne* (exactly 1 Mg) in previously metric countries. This use is strongly discouraged, and such large masses should be measured in megagrams.

CONVERSION FACTORS

Table A.4 can be used to convert either English (U.S. customary) units or non SI metric units to SI.

Conversion and Rounding

Multiply the specified quantity by the conversion factor exactly as given in Table A.4 and then round to the appropriate number of significant digits. For example, to convert 11.4 ft to metres: 11.4 × 0.3048 = 3.474 72, which rounds to 3.47 m. Do not round either the conversion factor or the quantity before performing the multiplication, as accuracy would be reduced.

The product will usually imply an accuracy not intended by the original value. Proper conversion technique includes rounding this converted quantity to the proper number of significant digits commensurate with its intended precision.

Table A. 5 is convenient for temperature interconversion for values between −40 and +298. For values below or above this range, use conversion factors given in Table A.4.

TABLE A.4
CONVERSION FACTORS TO SI UNITS

To convert from an English (customary U.S.) unit to its SI equivalent (listed in the center heading of that section) multiply by the SI factor to the left of the center line. To convert from an non-SI metric unit to SI, use the factor to the right of the line.

Example: 1 ft/sec² X 3.048 000 E−01 = 0.3048 000 E−01 = 0.3048 m/s²; 5 gal X 1.000 E−02 = 0.05 m/s².

English Unit	Symbol	Multiplier		Multiplier	Symbol	Non-SI Metric Unit
		Acceleration (Linear), *metre per second² (m/s²)				
foot/second²	ft/sec²	3.048 000*E−01		1.000*E−02	gal	galileo
inch/second²	in./sec²	2.540 000*E−02				
free fall, standard		9.806 650*E+00				
		Area, metre² (m²)				
acre (U.S. survey)B		4.046 873 E+03		1.000 000*E+02	a	acre
circular mil		5.067 075 E−10		1.000 000*E−28	b	barn
foot²	ft²	9.290 304*E−02		1.000 000*E+04	ha	hectare
inch²	in.²	6.451 600*E−04				
mile² (International)		2.589 988 E+06				
mile² (U.S. survey)B		2.589 998 E+06				
yard²	yd²	8.361 274 E−01				
		Bending Moment or Torque, newton-metre (N · m)				
pound force-inch	lbf-in.	1.129 848 E−01		1.000 000*E−07	dyn-cm	dyne-centimetre
pound force-foot	lbf-ft	1.355 818 E+00		9.806 650*E+00	kgf-m	kilogram force-metre
ounce force-inch	ozf-in.	7.061 552 E−03				
		Bending Moment or Torque per Length, newton-metre per metre (N · m/m)				
pound force-foot/inch	lbf-ft/in.	5.337 866 E+01				
pound force-inch/inch	lbf-in./in.	4.448 222 E+00				

TABLE A.5

Temperature Conversion

The numbers in boldface type in the center column refer to the temperature, either in degree Celsius or Fahrenheit, which is to be converted to the other scale. If converting Fahrenheit to degree Celsius, the equivalent temperature will be found in the left column. If converting degree Celsius to Fahrenheit, the equivalent temperature will be found in the column on the right.

Temperature			Temperature			Temperature			Temperature		
Celsius	°C or F	Fahr	Celsius	°C or F	Fahr	Celsius	°C or F	Fahr	Celsius	°C or F	Fahr
−40.0	−40	−40.0	−31.7	−25	−13.0	−23.3	−10	+14.0	−15.0	+5	+41.0
−39.4	−39	−38.2	−31.1	−24	−11.2	−22.8	−9	+15.8	−14.4	+6	+42.8
−38.9	−38	−36.4	−30.6	−23	−9.4	−22.2	−8	+17.6	−13.9	+7	+44.6
−38.3	−37	−34.6	−30.0	−22	−7.6	−21.7	−7	+19.4	−13.3	+8	+46.4
−37.8	−36	−32.8	−29.4	−21	−5.8	−21.1	−6	+21.2	−12.8	+9	+48.2
−37.2	−35	−31.0	−28.9	−20	−4.0	−20.6	−5	+23.0	−12.2	+10	+50.0
−36.7	−34	−29.2	−28.3	−19	−2.2	−20.0	−4	+24.8	−11.7	+11	+51.8
−36.1	−33	−27.4	−27.8	−18	−0.4	−19.4	−3	+26.6	−11.1	+12	+53.6
−35.6	−32	−25.6	−27.2	−17	+1.4	−18.9	−2	+28.4	−10.6	+13	+55.4
−35.0	−31	−23.8	−26.7	−16	+3.2	−18.3	−1	+30.2	−10.0	+14	+57.2
−34.4	−30	−22.0	−26.1	−15	+5.0	−17.8	0	+32.0	−9.4	+15	+59.0
−33.9	−29	−20.2	−25.6	−14	+6.8	−17.2	+1	+33.8	−8.9	+16	+60.8
−33.3	−28	−18.4	−25.0	−13	+8.6	−16.7	+2	+35.6	−8.3	+17	+62.6
−32.8	−27	−16.6	−24.4	−12	+10.4	−16.1	+3	+37.4	−7.8	+18	+64.4
−32.2	−26	−14.8	−23.9	−11	+12.2	−15.6	+4	+39.2	−7.2	+19	+66.2

TABLE A.5 (Continued)

The numbers in boldface type in the center column refer to the temperature, either in degree Celsius or Fahrenheit, which is to be converted to the other scale. If converting Fahrenheit to degree Celsius, the equivalent temperature will be found in the left column. If converting degree Celsius to Fahrenheit, the equivalent temperature will be found in the column on the right.

Celsius	°C or F	Fahr		Celsius	°C or F	Fahr		Celsius	°C or F	Fahr		Celsius	°C or F	Fahr
−6.7	+20	+68.0		+26.7	+80	+176.0		+60.0	+140	+284.0		+93.3	+200	+392.0
−6.1	+21	+69.8		+27.2	+81	+177.8		+60.6	+141	+285.8		+93.9	+201	+393.8
−5.5	+22	+71.6		+27.8	+82	+179.6		+61.1	+142	+287.6		+94.4	+202	+395.6
−5.0	+23	+73.4		+28.3	+83	+181.4		+61.7	+143	+289.4		+95.0	+203	+397.4
−4.4	+24	+75.2		+28.9	+84	+183.2		+62.2	+144	+291.2		+95.6	+204	+399.2
−3.9	+25	+77.0		+29.4	+85	+185.0		+62.8	+145	+293.0		+96.1	+205	+401.0
−3.3	+26	+78.8		+30.0	+86	+186.8		+63.3	+146	+294.8		+96.7	+206	+402.8
−2.8	+27	+80.6		+30.6	+87	+188.6		+63.9	+147	+296.6		+97.2	+207	+404.6
−2.2	+28	+82.4		+31.1	+88	+190.4		+64.4	+148	+298.4		+97.8	+208	+406.4
−1.7	+29	+84.2		+31.7	+89	+192.2		+65.0	+149	+300.2		+98.3	+209	+408.2
−1.1	+30	+86.0		+32.2	+90	+194.0		+65.6	+150	+302.0		+98.9	+210	+410.0
−0.6	+31	+87.8		+32.8	+91	+195.8		+66.1	+151	+303.8		+99.4	+211	+411.8
0	+32	+89.6		+33.3	+92	+197.6		+66.7	+152	+305.6		+100.0	+212	+413.6
+0.6	+33	+91.4		+33.9	+93	+199.4		+67.2	+153	+307.4		+100.6	+213	+415.4
+1.1	+34	+93.2		+34.4	+94	+201.2		+67.8	+154	+309.2		+101.1	+214	+417.2
+1.7	+35	+95.0		+35.0	+95	+203.0		+68.3	+155	+311.0		+101.7	+215	+419.0
+2.2	+36	+96.8		+35.6	+96	+204.8		+68.9	+156	+312.8		+102.2	+216	+420.8
+2.8	+37	+98.6		+36.1	+97	+206.6		+69.4	+157	+314.6		+102.8	+217	+422.6
+3.3	+38	+100.4		+36.7	+98	+208.4		+70.0	+158	+316.4		+103.3	+218	+424.4
+3.9	+39	+102.2		+37.2	+99	+210.2		+70.6	+159	+318.2		+103.9	+219	+426.2
+4.4	+40	+104.0		+37.8	+100	+212.0		+71.1	+160	+320.0		+104.4	+220	+428.0
+5.0	+41	+105.8		+38.3	+101	+213.8		+71.7	+161	+321.8		+105.6	+222	+431.6
+5.5	+42	+107.6		+38.9	+102	+215.6		+72.2	+162	+323.6		+106.7	+224	+435.2
+6.1	+43	+109.4		+39.4	+103	+217.4		+72.8	+163	+325.4		+107.8	+226	+438.8
+6.7	+44	+111.2		+40.0	+104	+219.2		+73.3	+164	+327.2		+108.9	+228	+442.4
+7.2	+45	+113.0		+40.6	+105	+221.0		+73.9	+165	+329.0		+110.0	+230	+446.0
+7.8	+46	+114.8		+41.1	+106	+222.8		+74.4	+166	+330.8		+111.1	+232	+449.6
+8.3	+47	+116.6		+41.7	+107	+224.6		+75.0	+167	+332.6		+112.2	+234	+453.2
+8.9	+48	+118.4		+42.2	+108	+226.4		+75.6	+168	+334.4		+113.3	+236	+456.8
+9.4	+49	+120.2		+42.8	+109	+228.2		+76.1	+169	+336.2		+114.4	+238	+460.4

+464.0	+240	+115.6	+338.0	+170	+76.7	+230.0	+110	+43.3	+122.0	+50	+10.0
+467.6	+242	+116.7	+339.8	+171	+77.2	+231.8	+111	+43.9	+123.8	+51	+10.6
+471.2	+244	+117.8	+341.6	+172	+77.8	+233.6	+112	+44.4	+125.6	+52	+11.1
+474.8	+246	+118.9	+343.4	+173	+78.3	+235.4	+113	+45.0	+127.4	+53	+11.7
+478.4	+248	+120.0	+345.2	+174	+78.9	+237.2	+114	+45.6	+129.2	+54	+12.2
+482.0	+250	+121.1	+347.0	+175	+79.4	+239.0	+115	+46.1	+131.0	+55	+12.8
+485.6	+252	+122.4	+348.8	+176	+80.0	+240.8	+116	+46.7	+132.8	+56	+13.3
+489.2	+254	+123.3	+350.6	+177	+80.6	+242.6	+117	+47.2	+134.6	+57	+13.9
+492.8	+256	+124.4	+352.4	+178	+81.1	+244.4	+118	+47.8	+136.4	+58	+14.4
+496.4	+258	+125.5	+354.2	+179	+81.7	+246.2	+119	+48.3	+138.2	+59	+15.0
+500.0	+260	+126.7	+356.0	+180	+82.2	+248.0	+120	+48.9	+140.0	+60	+15.6
+503.6	+262	+127.8	+357.8	+181	+82.8	+249.8	+121	+49.4	+141.8	+61	+16.1
+507.2	+264	+128.9	+359.6	+182	+83.3	+251.6	+122	+50.0	+143.6	+62	+16.7
+510.8	+266	+130.0	+361.0	+183	+83.9	+253.4	+123	+50.6	+145.4	+63	+17.2
+514.4	+268	+131.3	+363.2	+184	+84.4	+255.2	+124	+51.1	+147.2	+64	+17.8
+518.0	+270	+132.2	+365.0	+185	+85.0	+257.0	+125	+51.7	+149.0	+65	+18.3
+521.6	+272	+133.3	+366.8	+186	+85.6	+258.8	+126	+52.2	+150.8	+66	+18.9
+525.2	+274	+134.4	+368.6	+187	+86.1	+260.6	+127	+52.8	+152.6	+67	+19.4
+528.8	+276	+135.6	+370.4	+188	+86.7	+262.4	+128	+53.3	+154.4	+68	+20.0
+532.4	+278	+136.7	+372.2	+189	+87.2	+264.2	+129	+53.9	+156.2	+69	+20.6
+536.0	+280	+137.8	+374.0	+190	+87.8	+266.0	+130	+54.4	+158.0	+70	+21.1
+539.6	+282	+138.9	+375.8	+191	+88.3	+267.8	+131	+55.0	+159.8	+71	+21.7
+543.2	+284	+140.0	+377.6	+192	+88.9	+269.6	+132	+55.6	+161.6	+72	+22.2
+546.8	+286	+141.1	+379.4	+193	+89.4	+271.4	+133	+56.1	+163.4	+73	+22.8
+550.4	+288	+142.2	+381.2	+194	+90.0	+273.2	+134	+56.7	+165.2	+74	+23.3
+554.0	+290	+143.3	+383.0	+195	+90.6	+275.0	+135	+57.2	+167.0	+75	+23.9
+557.6	+292	+144.4	+384.8	+196	+91.1	+276.8	+136	+57.8	+168.8	+76	+24.4
+561.2	+294	+145.6	+386.6	+197	+91.7	+278.6	+137	+58.3	+170.6	+77	+25.0
+564.8	+296	+146.7	+388.4	+198	+92.2	+280.4	+138	+58.9	+172.4	+78	+25.6
+568.4	+298	+147.8	+390.2	+199	+92.8	+282.2	+139	+59.4	+174.2	+79	+26.1

TABLE A.6

PECTIN EQUIVALENTS BASED ON JELLY GRADE[1]

	Amount of Pectin Equivalent						
1	2	3	4	5	6	7	8
1 pound 150 Grade	1 pound 160 Grade	1 pound 120 Grade	1 pound 100 Grade	1 pound 80 Grade	1 pound 50 Grade	1 pound 40 Grade	1 pound 5 Grade
15 ounces 160 Grade	1 lb. 1 oz. 150 Grade	12 ounces 160 Grade	10 ounces 160 Grade	8 ounces 160 Grade	5 ounces 160 Grade	4 ounces 160 Grade	1/2 ounce 160 Grade
1 lb. 4 oz. 120 Grade	1 lb. 5 1/3 oz. 120 Grade	12 3/4 ounces 150 Grade	10 2/3 ounces 150 Grade	8 1/2 ounces 150 Grade	5 1/3 ounces 150 Grade	4 1/4 ounces 150 Grade	1/2 ounce 150 Grade
1 lb. 8 oz. 100 Grade	1 lb. 9 2/3 oz. 100 Grade	1 lb. 3 1/4 oz. 100 Grade	13 1/3 ounces 120 Grade	10 3/4 ounces 120 Grade	6 2/3 ounces 120 Grade	5 1/3 ounces 120 Grade	2/3 ounce 120 Grade
1 lb. 14 oz. 80 Grade	2 pounds 80 Grade	1 lb. 8 oz. 80 Grade	1 lb. 4 oz. 80 Grade	12 3/4 ounces 100 Grade	8 ounces 100 Grade	6 1/2 ounces 100 Grade	4/5 ounce 100 Grade
3 pounds 50 Grade	3 lbs. 3 1/4 oz. 50 Grade	2 lbs. 6 1/2 oz. 50 Grade	2 pounds 50 Grade	1 lb. 9 2/3 oz. 50 Grade	10 ounces 80 Grade	8 ounces 80 Grade	1 ounce 80 Grade
3 lbs. 12 oz. 40 Grade	4 pounds 40 Grade	3 pounds 40 Grade	2 lbs. 8 oz. 40 Grade	2 pounds 40 Grade	1 lb. 4 oz. 40 Grade	12 3/4 ounces 50 Grade	1 2/3 ounces 50 Grade
30 pounds 5 Grade	32 pounds 5 Grade	24 pounds 5 Grade	20 pounds 5 Grade	16 pounds 5 Grade	10 pounds 5 Grade	8 pounds 5 Grade	2 ounces 40 Grade

Source: Courtesy of Sunkist.

[1] The mathematical expression for determining the amount of a certain grade of pectin equivalent to a given amount of pectin of another grade is as follows:

$$(\text{Amount of opectin to be used}) = \frac{(\text{Wt. of pectin being used}) \times (\text{Grade of pectin being used})}{(\text{Grade of pectin to be used})}$$

For example, if you wish to know how much 60 grade pectin is equivalent to 12 ounces of 100 grade pectin, the correct amount is obtained as follows:

$$\text{Amount of 60 grade pectin in ounces} = \frac{(12)(100)}{(60)} = 20 \text{ ounces.}$$

TABLE A.7

INDEX OF REFRACTION OF SOLUTIONS OF SUCROSE SUGAR

Sugar (Sucrose) (%)	Index of Refraction at 20°C	Sugar (Sucrose) (%)	Index of Refraction at 20°C	Sugar (Sucrose) (%)	Index of Refraction at 20°C
5	1.3403	32	1.3847	59	1.4396
6	1.3418	33	1.3865	60	1.4418
7	1.3433	34	1.3883	61	1.4441
8	1.3448	35	1.3902	62	1.4464
9	1.3463	36	1.3920	63	1.4486
10	1.3478	37	1.3939	64	1.4509
11	1.3494	38	1.3958	65	1.4532
12	1.3509	39	1.3978	66	1.4555
13	1.3525	40	1.3997	67	1.4579
14	1.3541	41	1.4016	68	1.4603
15	1.3557	42	1.4036	69	1.4627
16	1.3573	43	1.4056	70	1.4651
17	1.3589	44	1.4076	71	1.4676
18	1.3605	45	1.4096	72	1.4700
19	1.3622	46	1.4117	73	1.4725
20	1.3638	47	1.4137	74	1.4749
21	1.3655	48	1.4158	75	1.4774
22	1.3672	49	1.4179	76	1.4799
23	1.3689	50	1.4200	77	1.4825
24	1.3706	51	1.4221	78	1.4850
25	1.3723	52	1.4242	79	1.4876
26	1.3740	52	1.4264	80	1.4901
27	1.3758	54	1.4285	81	1.4927
28	1.3775	55	1.4307	82	1.4954
29	1.3793	56	1.4329	83	1.4980
30	1.3811	57	1.4351	84	1.5007
31	1.3829	58	1.4373	85	1.5033

Source: Courtesy of Sunkist.

TABLE A.8

BOILING POINTS OF TYPICAL FRUIT JUICE-SUGAR MIXTURES AT VARIOUS ALTITUDES

Soluble Solids (%)	Sea Level	500	1000	1500	2000	2500	3000	3500	4000	4500	5000
					Boiling Point in °F at Altitude (ft) Indicated						
50	216.0	215.2	214.2	213.2	121.3	211.4	210.4	209.4	208.5	207.6	206.6
52	216.6	215.8	214.8	213.8	212.9	212.0	211.0	210.0	209.1	208.2	207.2
54	217.0	216.2	215.2	214.2	213.3	212.4	211.4	210.4	209.5	208.6	207.6
56	217.5	216.7	215.7	214.7	213.8	212.9	211.9	210.9	210.0	209.1	208.1
58	218.1	217.3	216.3	215.3	214.4	213.5	212.5	211.5	210.6	209.7	208.7
60	218.7	217.9	216.9	215.9	215.0	214.1	213.1	212.1	211.2	210.3	209.3
62	219.4	218.6	217.6	216.6	215.7	214.8	213.8	212.8	211.9	211.0	210.0
64	220.2	219.4	218.4	217.4	216.5	215.0	214.6	213.6	212.7	211.8	210.8
66	221.1	220.3	219.3	218.3	217.4	216.5	215.5	214.5	213.6	212.7	211.7
68	222.2	221.4	220.4	219.4	218.5	217.6	216.6	215.6	214.7	213.8	212.8
70	223.5	222.7	221.7	220.7	219.8	218.9	217.9	216.9	216.0	215.1	214.1
72	225.0	224.2	223.2	222.2	221.3	220.4	219.4	218.4	217.5	216.6	215.6
74	226.8	226.0	225.0	224.0	223.1	222.2	221.2	220.2	219.3	218.4	217.4
76	229.1	228.3	227.3	226.3	225.4	224.5	223.5	222.5	221.6	220.7	219.7

Source: Courtesy of Sunkist.

Index

543